Introductory Structured COBOL Programming

Introductory Structured COBOL Programming

Gary S. Popkin
New York City Technical College

Kent Publishing Company / Boston

A DIVISION OF WADSWORTH, INC.

Cover photo courtesy of Melvin L. Prueitt, Los Alamos Scientific Laboratory, University of California, Los Alamos, New Mexico. Work performed under the auspices of the Department of Energy.

Library of Congress Catalog Card Number: 80-51061
ISBN: 0-442-23166-0

Printed in the United States of America

4 5 6 7 — 85 83 82

To Louise

For more support and encouragement
than anyone should have needed

Preface

Introductory Structured COBOL Programming is suitable for students with little or no knowledge of computers as well as for those students who know one or more programming languages and wish to learn COBOL. The book is designed for students learning COBOL in Computer Science and Data Processing departments and in Schools of Business. The book begins with an elementary program in Chapter 1 and introduces additional COBOL features of increasing logical complexity in successive chapters. Thirty-six complete, working COBOL programs and part of a thirty-seventh are shown in facsimile, in addition to the inputs and outputs. All together, the figures in *Introductory Structured COBOL Programming* present a wide variety of examples of what works in COBOL.

At each step, only those COBOL features needed for a particular program are discussed, so that complications and unnecessary detail are avoided. The programs are written to comply with the highest level of the 1974 ANSI standard and with IBM Full American National Standard COBOL, Version 4. They should therefore work on most COBOL compilers in use today. Exceptions to this approach are taken only where the ANSI standard imposes requirements that would present unnecessary complications in the first program. Those complications are the requirement of a CONFIGURATION SECTION and the requirement that the Procedure Division begin with a section name or a paragraph name. Where the ANSI standard and the IBM compiler are incompatible, as in the use of the NOMINAL KEY clause, incompatibilities were resolved in favor of the compiler so that the programs shown in this book would run.

Differences among COBOL compilers are pointed out in this text. A reference to "some older COBOL systems" means that the feature being discussed has been dropped from the 1968 standard and should be avoided. A reference to "some newer COBOL systems" means that the feature is new in the 1974 standard and may not work on old compilers. The phrase "some of the newest COBOL systems" refers to features expected to be included in the 1981 standard, and "some COBOL systems" refers only to current differences among compilers which do not fit into any of the other already listed categories.

Program design techniques are used when suitable to particular programs. Flowcharts, control charts, pseudocode, and hierarchy diagrams are all used in appropriate circumstances. The largest and most complicated programs use hierarchy diagrams to show the big picture in the most concise and practical way. Dogmatic regard for any particular design technique has been avoided.

Chapters 1 through 8 and parts of Chapter 9 constitute a one-semester course for continuing students. For students not likely to continue beyond the introductory COBOL course, instructors may select topics from Chapters 9 through 12 in place of some topics in earlier chapters.

For continuing students of COBOL, Chapters 9 through 12, supplemented by a COBOL manual, provide a thorough coverage of some of the most commonly used advanced COBOL techniques. An Instructor's Manual contains answers to the chapter exercises, additional exercises for each chapter, a lesson plan for each chapter, and numerous transparency masters that can be reproduced for classroom use. The publisher will provide on request a free Test Data Deck, containing test data for the major programming assignments, to each school adopting *Introductory Structured COBOL Programming.* One hundred and fifty standard 80-column punched cards will be provided and may be reproduced for distribution to each instructor and student, or they may be catalogued.

I wish to thank a number of people for their help in putting this book together: Ellen M. Longo, then a student at New York City Technical College, who worked long hours to produce many of the program examples; Phil Rubenstein, also then a student at NYCTC, who worked with me from early in the project and produced many programs; Louise Moran of New York Life Insurance Company and Michael L. Trombetta of Queensboro Community College for supplying research materials and programming ideas and suggestions; Frank M. Rand, Chairman of the Data Processing Department at NYCTC, who provided me with the necessary time and equipment; my daughter, Deborah Popkin, who helped with some of the artwork, and to Deborah and to Charles Dreiss, another NYCTC student, for help in copying and manuscript preparation; Mike Marlow, Parkland College, Robert D. Chenoweth, County College of Morris, Howard Granger, City College of San Francisco, and Alzebeta Hardgrave, University of Maryland, for reading the manuscript and making many useful suggestions; Margaret Glassman, Olga Saunders, and my mother, Mona Popkin, who helped with the typing.

The acknowledgement required by the American National Standards Institute for using copyrighted material follows:

by Sperry Rand Corporation; IBM Commerical Translator Form No. F 28-8013, copyrighted 1959 by IBM; FACT, DSI 27A5260-2760, copyrighted 1960 by Minneapolis-Honeywell

have specifically authorized the use of this material in whole or in part, in the COBOL specifications. Such authorization extends to the reproduction and use of COBOL specifications in programming manuals or similar publications.

Gary S. Popkin

Contents

Introductory Structured COBOL Programming

Computer Programming with COBOL

1

KEY POINTS

Here are the key points you should learn from this chapter:

1. the steps a person would have to follow in order to obtain useful results from a computer;
2. what a computer program is, and what a programmer is;
3. how to write an elementary program using COBOL.

KEY WORDS

Key words to recognize and learn:

design	ENVIRONMENT DIVISION	LABEL RECORDS ARE OMITTED
instruction	DATA DIVISION	record name
coding	PROCEDURE DIVISION	PICTURE
COBOL	PROGRAM–ID	reserved word
program	record	programmer-supplied name
programmer	file	user-defined word
memory	FILE SECTION	OPEN
storage	FD	statement
output	file description	MOVE
input	file name	WRITE
division header	label records	CLOSE
IDENTIFICATION DIVISION		STOP RUN

A computer, by itself, cannot solve a problem or do anything useful. People can solve problems, using computers as tools. A person must do the following in order to obtain useful results from a computer:

1. Understand the problem to be solved. Only certain kinds of problems lend themselves to solution by computer. As we go through this book, you will see what some of those problems are.

A person must thoroughly understand a problem if a computer is to be used in its solution.

2. Plan how the computer can help to solve the problem. Since computers can do only certain things and not others, it is up to the human problem-solver to figure out exactly what things the computer must do to contribute to the solution of the problem. This is called **designing** a solution. In this book you will study several modern design techniques now being used by computer professionals in industry.

3. Write **instructions** on a piece of paper, in a language the computer understands, that tell the computer what to do. This step is called **coding.**

   There are many languages in existence today that can be used to tell computers what to do. The language that you will study in this book is called **COBOL** (rhymes with snowball). COBOL stands for **CO**mmon **B**usiness-**O**riented **L**anguage. It is called common because the original designers of the COBOL language in 1959 envisioned a COBOL that all computers could understand. Now, after many years of growth of the language, different computers understand slightly different versions of the language. Any troubles that might arise from differences among different COBOL systems are usually easy to avoid and will be pointed out as we go along.

   A complete set of instructions that a computer uses in solving a particular problem is called a **program.** The person who prepares the program is called a **programmer.** In this book there will be many opportunities for you to create programs. When you have completed all of the exercises in this book, you will be a COBOL programmer.

4. Get the program into the computer and have the computer carry out your instructions. The program must be placed into the computer's **memory,** or **storage,** in order to be executed. Then you wait while the computer does exactly what you told it to do.

In real life, errors can creep into this process in any of the four steps. Errors in any of the steps will cause the programmer to have to redo some of the work. Depending on the severity of the error and the step where it occurs, more or less reworking may be required.

Some errors are so easy to find that even the computer can do it. In step 4, if you make some purely mechanical error in physically getting the program into the computer, the computer will usually object in one way or another, and the error can be fixed in minutes. Or, in step 3, if you write an instruction which does not follow COBOL's rules of grammar, the computer will point out the error to be corrected.

Scope of the Book

In most of the programs in this book, the computer presents the results of its processing by printing its **output** onto paper. You have probably seen computer-printed output. Most computers can print their output results onto continuous-form, fan-fold paper, the kind with sprocket holes along both sides.

Computers can present their output in ways other than by printing. You have probably seen results come out of computers onto screens that look a lot like television screens. Computers can present their output in still other ways, and in Chapter 11 we will study several COBOL programs that deal with some of those ways.

We also have to have some way of getting programs (and also data) into computer storage. We already know that a program, once written, must be put into the computer to be executed. One common way to get a program into a computer is to punch the program onto cards and have the computer read the cards. An example of a COBOL instruction punched onto a card is shown in Figure 1.1.

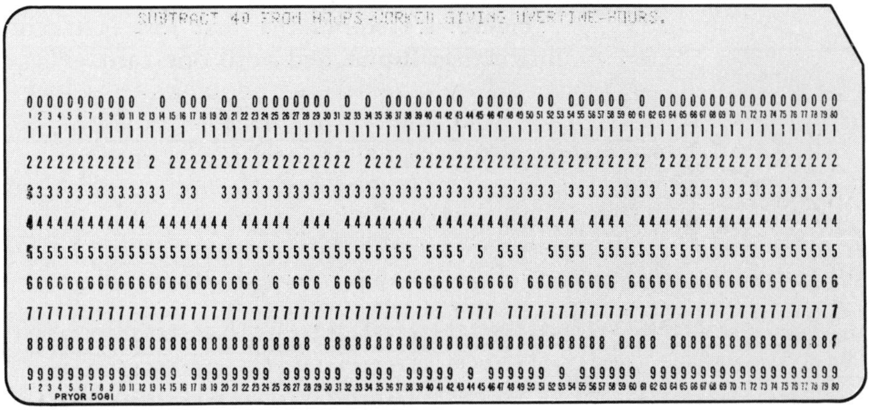

SUBTRACT 40 FROM HOURS-WORKED GIVING OVERTIME-HOURS.

Figure 1.1
A COBOL instruction punched in a card.

Other methods of getting programs into storage involve keying the program on some kind of key device which is connected directly to the computer.

You will study some other **input** methods in Chapter 11.

A First COBOL Program

To lend concreteness to the ideas so far discussed, we will now do a complete working COBOL program. The techniques needed for doing this program will be presented without unnecessary details. In this program, and with all programs in this book, there will be presented only as much detail as is needed at that stage to write the particular program being discussed. So if some of the material seems incomplete at this time, just wait and it will all be filled out in later chapters.

Let us now set to work on Program 1, which is a program that does nothing but print out the author's name and address. Figure 1.2 shows the output as it was produced by the program.

```
G. S. POPKIN
1921 PRESIDENT ST.
BROOKLYN, NY   11221
```

Figure 1.2
Output from Program 1.

The design for this program was done by the author and will not be shown here. We will save discussion of program design for more complicated programs.

Figure 1.3 shows the first few instructions for Program 1. Each instruction is punched onto one card.

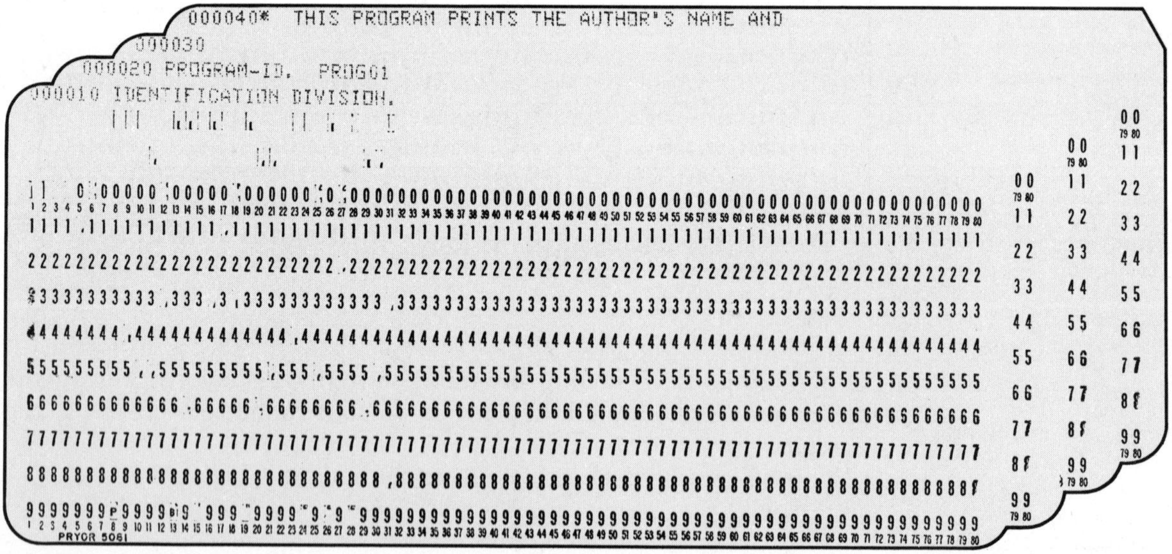

```
000040*  THIS PROGRAM PRINTS THE AUTHOR'S NAME AND
000030
000020 PROGRAM-ID.  PROG01
000010 IDENTIFICATION DIVISION.
```

Figure 1.3 The first few instructions of Program 1 punched in cards.

You can see that each card contains a sequence number in columns 1 through 6. The sequence numbers are in intervals of 10 to allow for later insertions if necessary.

For ease in reading, we will show all programs in the form of a list rather than as the actual punched cards. Program 1 complete is shown in Figure 1.4.

```
4-CB2 V4 RELEASE 1.5 10NOV77          IBM OS AMERICAN NATIONAL STANDARD COBOL


  000010 IDENTIFICATION DIVISION.
  000020 PROGRAM-ID.  PROG01.
  000030
  000040*  THIS PROGRAM PRINTS THE AUTHOR'S NAME AND
  000050*     (FICTITIOUS) ADDRESS, ON THREE LINES.  ZIP CODE IS
  000060*     INCLUDED IN THE ADDRESS.
  000070*
  000080******************************************************************
  000090
  000100 ENVIRONMENT DIVISION.
  000110 INPUT-OUTPUT SECTION.
  000120 FILE-CONTROL.
  000130     SELECT COMPLETE-ADDRESS ASSIGN TO UT-S-SYSPRINT.
  000140
  000150******************************************************************
  000160
  000170 DATA DIVISION.
  000180 FILE SECTION.
  000190 FD  COMPLETE-ADDRESS
  000200     LABEL RECORDS ARE OMITTED.
  000210
  000220 01  ADDRESS-LINE           PICTURE X(120).
  000230
  000240******************************************************************
  000250
  000260 PROCEDURE DIVISION.
  000270     OPEN OUTPUT COMPLETE-ADDRESS.
  000280     MOVE 'G. S. POPKIN'          TO ADDRESS-LINE.
  000290     WRITE ADDRESS-LINE.
  000300     MOVE '1921 PRESIDENT ST.'  TO ADDRESS-LINE.
  000310     WRITE ADDRESS-LINE.
  000320     MOVE 'BROOKLYN, NY  11221' TO ADDRESS-LINE.
  000330     WRITE ADDRESS-LINE.
  000340     CLOSE COMPLETE-ADDRESS.
  000350     STOP RUN.
```

Figure 1.4
Program 1.

Program 1 consists of four divisions, as do all COBOL programs. Each division is identified by a **division header.** In the program shown, each division header starts in column 8 but could have begun anywhere between columns 8 and 11. The four division headers are **IDENTIFICATION DIVISION, ENVIRONMENT DIVISION, DATA DIVISION,** and **PROCEDURE DIVISION.** The four division headers must be punched with the exact spellings given, and each must be followed by a period. Whenever a period is used in a COBOL program, it must be followed by a space but not preceded by a space.[1]

THE IDENTIFICATION DIVISION

The Identification Division is used to give the program a name and to provide other descriptive information about the program. The **PROGRAM–ID** entry is required and must include a name for the program made up by the programmer. In this program the made-up program name is PROG01. The rules for making up program names will be given later in this chapter. Following the PROGRAM–ID is a brief description of the program. Each line of description has an asterisk in column 7, identifying the line as descriptive comment.

THE ENVIRONMENT DIVISION

The Environment Division serves a number of purposes in COBOL. In this program, the entries shown are used only to indicate that the output being produced by the program is to be printed on the high-speed printer. Your instructor will give you the Environment Division entries necessary for your particular computer.

THE DATA DIVISION

The Data Division can also be used for a number of purposes. In this program, it is used to describe some details about the output to be printed. As you may recall, the output is to be three lines of name and address. Each line of printed output is considered to be one **record,** and the complete printed output is a **file.** In the **FILE SECTION** of the Data Division, we describe the output record and the output file. The words FILE SECTION must appear exactly as shown, followed by a period.

The letters **FD** stand for **file description.** There must be one FD entry for every file being used by the program. In this program, there is only one file, the printer file, so we need only one FD. The required letters FD must begin between columns 8 and 11 and be followed by a **file name** made up by the programmer. In this program, the made-up file name is COMPLETE–ADDRESS. The rules for making up file names are given later in this chapter. The file name and all other portions of the FD entry must begin between columns 12 and 72. In this program, they begin in column 12. Since this is a printer file, it contains no **label records.** Label records are special records discussed in Chapter 11; they are never present on punched card or printer files. The file name must be followed by some indication that label records are omitted. One correct way of indicating that label records are omitted is to include the clause **LABEL RECORDS ARE OMITTED.** Other ways will be shown later. There must be only one period at the end of the entire FD entry.

1. Some newer COBOL systems permit a period to be preceded by a space.

The 01 on the next line of the program indicates that we are now describing some characteristics of a single record of the output file. In this case, each record is a line of print. The 01 in column 8 must be followed by a **record name** made up by the programmer. Here the record name is ADDRESS–LINE. Rules for making up record names will be given later in this chapter. The **PICTURE** clause shown tells the computer that the line of print may contain up to 120 characters. If your computer has a larger printer, such as a 132-character printer, you may use a larger number in the parentheses. Your instructor will give you the information you need for this.

Before going on to the next division, we can draw some generalizations from what we have seen so far. First, in COBOL there are certain words that must be used exactly as given, and there are some words that must be made up by the programmer. Those that must be used as given are called **reserved words,** and those that are made up by the programmer are called **programmer-supplied names** or **user-defined words.** A reserved word must never be used where a programmer-supplied name is required. Also, some entries must begin between columns 8 and 11, and some must begin between columns 12 and 72. The area from columns 8 through 11 is called area A, and that from columns 12 through 72 is called area B. Finally, lines which are entirely blank (except for their sequence numbers) and lines which contain an asterisk in column 7 may be included freely anywhere in the program to improve its appearance and readability.

THE PROCEDURE DIVISION

We come now finally to the Procedure Division, which consists of the step-by-step procedure that the computer must follow to produce the desired output. The required words PROCEDURE DIVISION are reserved words and must begin in area A. We then have an instruction to **OPEN** the output file. All files must be opened before being used for an input or output operation, and the OPEN **statement** must say whether the file is going to be used for input or for output or for both. In our case, the file will be used only for output to be printed on. There then follow pairs of **MOVE** and **WRITE** statements. Each pair of statements causes a single line of output to be printed. Following the instructions to print all three lines of output, we **CLOSE** the file. All files must be closed when we are through using them. The **STOP RUN** statement is the last executable statement in a program. Execution of the STOP RUN statement signals the computer to go on to the next job.

There are two important things to notice about this Procedure Division. First, in the OPEN and CLOSE statements, we use the name of the file as we made it up in the Data Division. The name of the file in the OPEN and CLOSE statements must be spelled exactly as it was made up. The computer does not understand English, and, if even one letter is wrong in spelling the file name, the computer will not understand what we mean. So when you make up programmer-supplied names, make them whatever you like (within the rules), and then use them exactly the way you made them up.

Second, the MOVE and WRITE statements use the record name that was made up in the Data Division. These usages are required by the rules of COBOL; that is, OPEN and CLOSE statements use file names, and MOVE and WRITE statements use record names. All of these rules may be a bit confusing now, but as you begin to write programs they will become automatic and even reasonable.

Rules for Program Names, File Names, and Record Names

A program name may be up to 30 characters long, and the characters may be any of the digits 0–9, the letters A–Z, and/or the hyphen. A program name may consist entirely of numbers, and it must not begin or end with a hyphen. Many COBOL systems ignore any characters in the program name beyond the sixth or eighth character.

File names and record names may each be up to 30 characters long; they may consist of any of the numbers, letters, and/or the hyphen, but must not begin or end with a hyphen, and must contain at least one letter. All COBOL systems recognize all of the characters in file names and record names.

Exercise 1. Which of the following are valid program names, which are valid as file names and record names as well, and which are just invalid?

a. PROG–01
b. PROGRAM–ONE
c. LIST ONE
d. 001–050
e. 001–A50
f. –001A50

Exercise 2. Write and execute a COBOL program to print your own name and address. Your address may be three or more lines as needed. Make up new programmer-supplied names for the program name, the file name, and the record name.

Summary

Before a programmer can write a computer program, the problem to be solved must be completely understood. Then the programmer must plan how the computer can be used in the solution. The instructions that make up the program can first be written onto a piece of paper. Then the instructions may be transferred to computer storage by means of punched cards or by being keyed directly into the computer.

COBOL stands for COmmon Business-Oriented Language. COBOL

programs always consist of four divisions. The IDENTIFICATION DIVISION can be used to give the program a name. The ENVIRONMENT DIVISION helps to relate programmer-supplied names to hardware devices. The DATA DIVISION describes all the details of the files and records used by the program. The PROCEDURE DIVISION contains the steps the computer must execute in order to solve the given problem.

Files must be OPENed before being used and CLOSEd after the last operation on them. A WRITE verb is used to place data on an output file.

Fill-in Exercises

1. The word COBOL stands for ________ ________ ________ ________.
2. The set of instructions that a computer follows is called a ________.
3. The person who makes up the computer's instructions is called a ________.
4. The printed results of executing a computer program are called the program's ________.
5. The four divisions of every COBOL program are the ________ DIVISION, the ________ DIVISION, the ________ DIVISION, and the ________ DIVISION.
6. The PROGRAM–ID entry is part of the ________ DIVISION.
7. The FD entry is part of the ________ DIVISION.
8. An OPEN statement would be found in the ________ DIVISION.
9. The ________ and ________ statements have file names as their objects.
10. The number used for defining a record name in the FILE SECTION is ________.

Programs Using Input, Output, and Reformatting

2

KEY POINTS

Here are the key points you should learn from this chapter:

1. how to use the COBOL Coding Form;
2. the rules of COBOL for reading card input files;
3. how to program for a variable number of card input records.

KEY WORDS

Key words to recognize and learn:

COBOL Coding Form
section header
paragraph name
field
data name
FILLER
level numbers
READ
AT END
source field
literal
nonnumeric literal
numeric literal
assign
carriage control character
WORKING-STORAGE SECTION
initialize
VALUE
PIC
priming READ
PERFORM
paragraph
AFTER ADVANCING
CONFIGURATION SECTION
SPECIAL-NAMES

In this chapter we will be writing several programs which show more of the basic techniques of data handling. But first we will look at the **COBOL Coding Form.**

COBOL Coding Form

The COBOL Coding Form is helpful to a programmer writing COBOL programs. The forms are available in pads of about 50 at most college bookstores and wherever computer programming supplies are sold. Figure 2.1 shows a COBOL Coding Form.

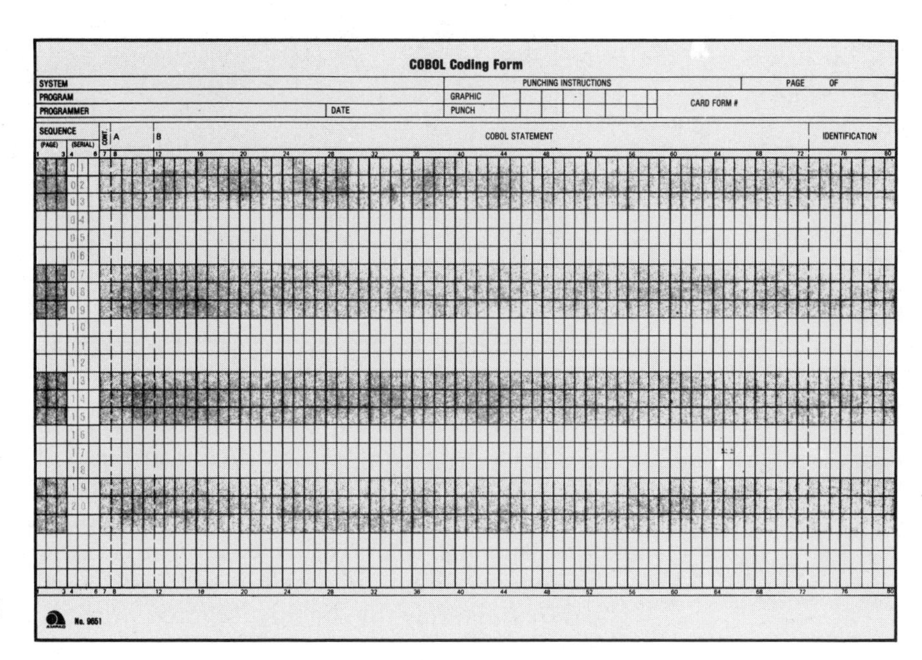

COBOL Coding Form

SYSTEM
PROGRAM
PROGRAMMER
DATE
PUNCHING INSTRUCTIONS
PAGE OF
GRAPHIC
PUNCH
CARD FORM #
SEQUENCE (PAGE) (SERIAL)
CONT.
A
B
COBOL STATEMENT
IDENTIFICATION
No. 9651

Figure 2.1
A COBOL Coding Form.

The form permits a programmer to write coding line for line exactly as it will be keyed. The form has room at the top for identifying information, and space below for 24 lines of COBOL code. You can see that separate areas are marked off for the sequence number for each line, column 7 for the asterisk and the A and B areas. Columns 73–80, marked Identification, can be used by the programmer for anything and are often used for the program name. Figure 2.2 shows a COBOL Coding Form partially filled in with some of the coding of Program 1.

Remember that the A area is the place to begin division headers, 01-level entries, and level indicators like FD. **Section headers** and **paragraph names,** discussed later in this chapter, also begin in the A area. All other statements must begin in the B area.

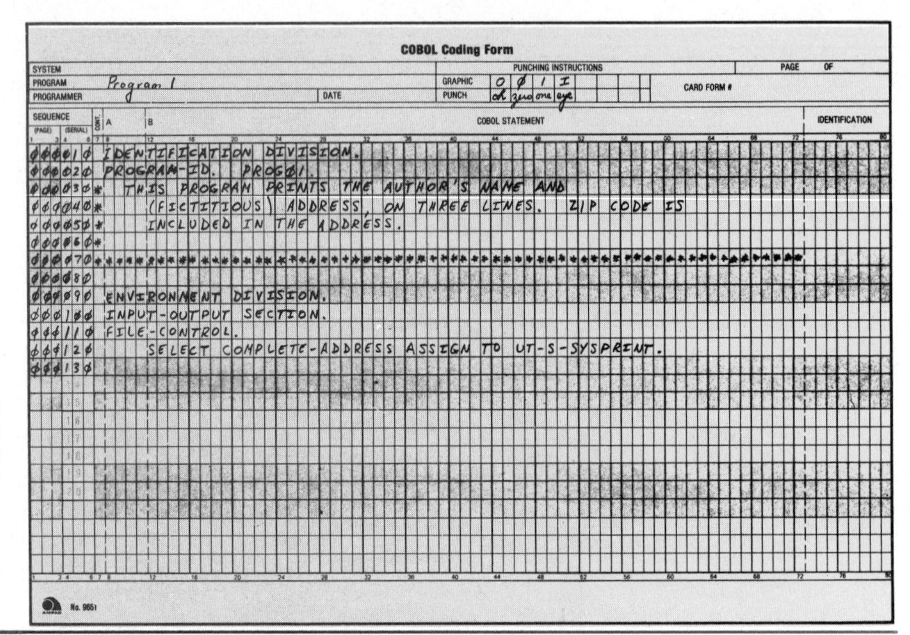

COBOL Coding Form

SYSTEM		PUNCHING INSTRUCTIONS					PAGE	OF
PROGRAM	Program 1	GRAPHIC	O	Ø	1	I	CARD FORM #	
PROGRAMMER	DATE	PUNCH	oh	zero	one	eye		

```
SEQUENCE  A   B                    COBOL STATEMENT                        IDENTIFICATION
ØØØØ1Ø IDENTIFICATION DIVISION.
ØØØØ2Ø PROGRAM-ID.   PROGØ1.
ØØØØ3Ø*   THIS PROGRAM PRINTS THE AUTHOR'S NAME AND
ØØØØ4Ø*   (FICTITIOUS) ADDRESS ON THREE LINES.  ZIP CODE IS
ØØØØ5Ø*   INCLUDED IN THE ADDRESS.
ØØØØ6Ø*
ØØØØ7Ø*************************************************************
ØØØØ8Ø
ØØØØ9Ø ENVIRONMENT DIVISION.
ØØØ1ØØ INPUT-OUTPUT SECTION.
ØØØ11Ø FILE-CONTROL.
ØØØ12Ø     SELECT COMPLETE-ADDRESS ASSIGN TO UT-S-SYSPRINT.
ØØØ13Ø
```

Figure 2.2
A COBOL Coding Form partially filled in with some of the coding from Program 1.

A Program to Process One Input Data Card

We now try a program slightly more complicated than Program 1. This program will print a three-line name and address, but this time the program will obtain the name and address to be printed by reading in a data card containing them. The previous program could print only the name and address that were written into the program in the MOVE statements, but this program will now be able to print any name and address merely by being provided with a data card containing the information to be printed.

A data card can be considered a single record by a COBOL program, just as a line of print is. Let's assume the data card to be read by the program contains a name and address in the format shown in Figure 2.3.

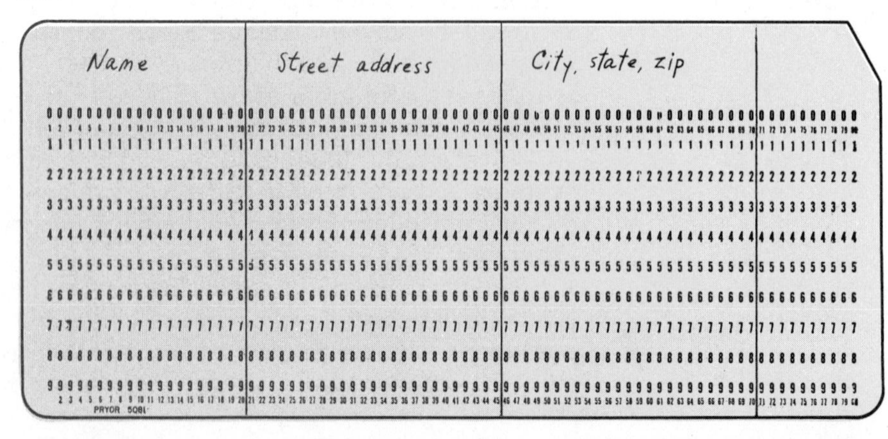

Figure 2.3
Data card format for Program 2.

A typical input data card punched in that format is shown in Figure 2.4.

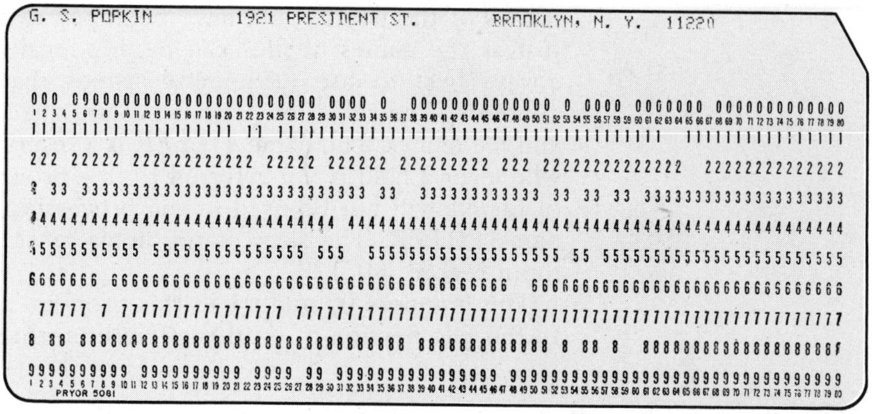

G. S. POPKIN 1921 PRESIDENT ST. BROOKLYN, N. Y. 11220

Figure 2.4
Typical input data card for Program 2.

The program to read one card and print its contents is shown in Figure 2.5. Program 2 differs from Program 1 in several ways. In this program the output file is called THREE–LINE–ADDRESS, and the input file (which consists of only the one card record) has been named INPUT–FILE. Since we now have two files we must have two SELECT sentences (in the Environment Division) and two FD entries. The 01 entry for the output file is the same as in Program 1, but the 01 entry for the input card introduces some new ideas.

Since the input record is formatted in **fields,** we must describe those fields to the computer. The fields are described in this program by the

Figure 2.5 Program 2, Part 1 of 2.

```
4-CB2 V4 RELEASE 1.5 10NOV77          IBM OS AMERICAN NATIONAL STANDARD COBOL



  000010 IDENTIFICATION DIVISION.
  000020 PROGRAM-ID.  PROG02.
  000030
  000040*  THIS PROGRAM READS ONE DATA CARD AND PRINTS
  000050*     ITS CONTENTS ON THREE LINES.
  000060*
  000070*****************************************************************
  000080
  000090 ENVIRONMENT DIVISION.
  000100 INPUT-OUTPUT SECTION.
  000110 FILE-CONTROL.
  000120     SELECT THREE-LINE-ADDRESS ASSIGN TO UT-S-SYSPRINT.
  000130     SELECT INPUT-FILE         ASSIGN TO UT-S-SYSIN.
  000140
  000150*****************************************************************
```

05 entries. Each 05 entry allows the programmer to make up a name for the field and indicate its length. Here the names of the three fields of input are NAME, STREET–ADDRESS, and CITY–STATE–ZIP. The names of the fields could have been any legal **data names,** of course, just as the names of files can be any legal made-up names, but it is always best to use meaningful names that well describe the field contents. The rules for data names are the same as for record names and file names. The name **FILLER** is a reserved word and may be used whenever a field is not referred to in a program. In this case, the field of blanks will not be used in the processing, and so can be given the name FILLER. The sizes of the fields must add up to the size of the input record, 80.

The numbers 05 and 01 in this program are called **level numbers.** In the File Section of the Data Division, the 01 level always describes

Figure 2.5 Program 2, Part 2 of 2.

```
000160
000170 DATA DIVISION.
000180 FILE SECTION.
000190 FD  THREE-LINE-ADDRESS
000200     LABEL RECORDS ARE OMITTED.
000210
000220 01  ADDRESS-LINE                 PICTURE X(120).
000230
000240
000250 FD  INPUT-FILE
000260     LABEL RECORDS ARE OMITTED.
000270
000280 01  INPUT-RECORD.
000290     05 NAME                      PICTURE X(20).
000300     05 STREET-ADDRESS            PICTURE X(25).
000310     05 CITY-STATE-ZIP            PICTURE X(25).
000320     05 FILLER                    PICTURE X(10).
000330
000340*****************************************************************
000350
000360 PROCEDURE DIVISION.
000370     OPEN INPUT  INPUT-FILE,
000380          OUTPUT THREE-LINE-ADDRESS.
000390     READ INPUT-FILE
000400         AT END
000410             CLOSE INPUT-FILE,
000420                   THREE-LINE-ADDRESS
000430             STOP RUN.
000440     MOVE NAME              TO ADDRESS-LINE.
000450     WRITE ADDRESS-LINE.
000460     MOVE STREET-ADDRESS TO ADDRESS-LINE.
000470     WRITE ADDRESS-LINE.
000480     MOVE CITY-STATE-ZIP TO ADDRESS-LINE.
000490     WRITE ADDRESS-LINE.
000500     CLOSE INPUT-FILE,
000510           THREE-LINE-ADDRESS.
000520     STOP RUN.
```

a record. Fields within the record may use any level number from 02 through 49. Level numbers 02 through 49 must be written in area B.

The Procedure Division in this program differs slightly from the Procedure Division of Program 1 but in significant ways. In this program we have both an input file and an output file. Both must be OPENed before they can be used and CLOSEd after we are through using them. The OPEN statement in this program opens both files and indicates that INPUT–FILE is to be OPENed as an INPUT file and that THREE–LINE–ADDRESS is to be OPENed as an OUTPUT file. The **READ** statement causes the computer to read one record from the file named in the READ and to make that record available to the program under the record description entries associated with the file name. In this program, the file being read by the READ statement is INPUT–FILE, and the record read from the file is made available to the program under the names INPUT–RECORD, NAME, STREET–ADDRESS, and CITY–STATE–ZIP.

Every READ statement must have an **AT END** clause, to tell the computer what to do if the file contains no more input data. In this case, since we expect this file to contain only one record and this is the one and only READ statement to be executed on this file, the AT END clause tells the computer what to do if the file accidentally contains no input record. If our input file doesn't contain data, then there is very little we can do, and the AT END clause in this program simply closes both files and executes a STOP RUN. Notice that there is only one period at the end of the entire READ statement.

Of course, if all is well, our file will contain the expected input record and the computer can go on to the MOVE and WRITE statements which follow. Notice how the MOVE statements in this program differ from those in the previous program. In the earlier program, the sending field of the MOVE was some actual piece of data enclosed in quotes. Here, the sending field, or **source field,** of the MOVE is a data name, so what will be moved is the NAME, STREET–ADDRESS, and CITY–STATE–ZIP fields that were read from the input data card, and not the words 'NAME', 'STREET–ADDRESS', or 'CITY–STATE–ZIP'. This very important difference between having a data name in a statement as distinguished from writing the data itself literally into the statement will be further discussed shortly.

After the three lines of output are moved and written, we close the input and output files with a CLOSE statement and execute a STOP RUN in the usual way.

The output produced by Program 2 is shown in Figure 2.6.

Figure 2.6
Output from Program 2.

```
G. S. POPKIN
1921 PRESIDENT ST.
BROOKLYN, N. Y.  11220
```

Literals and Assignment

In the first program, where the sending fields of the MOVE statements were pieces of data enclosed in quotes, the pieces of data are called **literals.** There are two categories of literals in COBOL. The ones we have used in the first program are called **nonnumeric literals.** A nonnumeric literal is any string of characters enclosed in quotes, except that in some COBOL systems a quote sign cannot itself be part of a nonnumeric literal. In those systems, 'G. S. POPKIN' is a valid nonnumeric literal, but 'B. O'MALLEY' and 'DON'T WALK' are not. In other COBOL systems, the quote sign in a nonnumeric literal would be designated by two consecutive quote signs, so that 'B. O"MALLEY' and 'DON"T WALK' would be valid. When these latter literals print as output, only one quote sign would print. Nonnumeric literals may be up to 120 characters long. A further complication is that some COBOL systems use the single quote also known as the apostrophe (') to enclose the literal, as we have already done in Program 1, and some use the regular double quotes ("). For safety we will not use quotes as part of our nonnumeric literals.

The other category of literals is the **numeric literal.** We have not yet used any numeric literals. The complete rules for forming numeric literals will be given later when we need them.

In Program 2, Figure 2.5, the sending fields of the MOVE statements were not literals but data names. In such a case, the data that is moved is whatever data happens to be **assigned** to the data name at the time. In our program, data were assigned to the data names by the READ statement; that is, the READ statement brought in an input data card and assigned the first field on the card to the name NAME, the second field to the name STREET–ADDRESS, and the third field to the name CITY–STATE–ZIP. There are many ways that data can be assigned to data names and they will be mentioned as needed. For now, the very important difference between

```
MOVE 'DOG' TO OUTPUT-LINE
```

and

```
MOVE DOG TO OUTPUT-LINE
```

is that in the first case the word DOG will be MOVEd to the output line, but in the second, what will be MOVEd to the output line is whatever data value happens to have been assigned to the data name DOG by some earlier part of the program. The MOVE does not change the value assigned to tne sending field; that is, the value assigned to DOG after the MOVE is the same as the value it had before the MOVE.

Exercise 1. Write a program to process a single input data card to the following specifications:

Input
One data card containing a name and address in the following format:

Cols.	Field
1–20	Name
21–35	Title
36–50	Company name
51–65	Street address
66–80	City and state

Output
A five-line address containing each on its own line a name, title, company name, street address, and city and state.

A Program to Process Multiple Input Data Cards

The usual situation in data processing is that a program will read many input records and process each one, but the number of input data records to be processed is not known at the time the program is being written. In fact, it is usual for any program to be run many times, and in each running it can be expected to have to process different numbers of input records. So we must have some way to accommodate this variability in the number of input records to be read. We will do this by using the AT END clause of the READ statement in a new way.

This program will use data cards in the same format as did Program 2, the program shown in Figure 2.5. The format of that data card is shown in Figure 2.3. This new program, Program 3, will READ any number of data cards, whereas Program 2 read only one. The contents of each card will be printed in the usual three-line format, single spaced as before, but now the program will double-space between addresses. Whenever we have a situation where we want the program to control the spacing of the printed output, we must provide room at the beginning of the output record for one character to be used for control of the paper carriage, called the **carriage control character.** The program doesn't have to do anything with this extra character; it only has to provide space for it. And so you can see, in the program in Figure 2.7, the output record, which has again been called ADDRESS–LINE, is made up of the single space for carriage control and an ADDRESS–ITEM of 120 characters. Some COBOL systems require the extra character even to give ordinary single spacing. From now on we will always provide the extra character at the beginning of the output record for any special carriage control we might want.

Figure 2.7 Program 3.

```
4-CB2 V4 RELEASE 1.5 10NOV77        IBM OS AMERICAN NATIONAL STANDARD COBOL


000010 IDENTIFICATION DIVISION.
000020 PROGRAM-ID. PROG03.
000030
000040*     INPUT - CARDS CONTAINING NAMES AND ADDRESS IN THE FOLLOWING
000050*     FORMAT:
000060*
000070*     COLS. 1-20                  NAME
000080*     COLS. 21-45                 STREET ADDRESS
000090*     COLS. 46-65                 CITY, STATE, ZIP
000100*
000110*     OUTPUT - THREE-LINE ADDRESSES, SINGLE SPACED, WITH A DOUBLE
000120*     SPACE BETWEEN ADDRESSES.
000130*
000140*****************************************************************
000150
000160 ENVIRONMENT DIVISION.
000170 INPUT-OUTPUT SECTION.
000180 FILE-CONTROL.
000190     SELECT THREE-LINE-ADDRESSES ASSIGN TO UT-S-SYSPRINT.
000200     SELECT ADDRESSES-IN         ASSIGN TO UT-S-SYSIN.
000210
000220*****************************************************************
000230
000240 DATA DIVISION.
000250 FILE SECTION.
000260 FD  THREE-LINE-ADDRESSES
000270     LABEL RECORDS ARE OMITTED.
000280
000290 01  ADDRESS-LINE.
000300     05 FILLER                  PIC X.
000310     05 ADDRESS-ITEM            PIC X(120).
000320
000330 FD  ADDRESSES-IN
000340     LABEL RECORDS ARE OMITTED.
000350
000360 01  INPUT-RECORD.
000370     05 NAME                    PIC X(20).
000380     05 STREET-ADDRESS          PIC X(25).
000390     05 CITY-STATE-ZIP          PIC X(20).
000400     05 FILLER                  PIC X(15).
000410
000420 WORKING-STORAGE SECTION.
000430 01  MORE-INPUT                 PIC X(3) VALUE 'YES'.
000440
000450*****************************************************************
```

Figure 2.7 Program 3 (continued).

```
000460
000470 PROCEDURE DIVISION.
000480      OPEN INPUT  ADDRESSES-IN,
000490           OUTPUT THREE-LINE-ADDRESSES.
000500      READ ADDRESSES-IN
000510          AT END
000520              MOVE 'NO' TO MORE-INPUT.
000530      PERFORM MAIN-LOOP UNTIL MORE-INPUT IS EQUAL TO 'NO'.
000540      CLOSE ADDRESSES-IN,
000550            THREE-LINE-ADDRESSES.
000560      STOP RUN.
000570
000580 MAIN-LOOP.
000590      MOVE NAME TO ADDRESS-ITEM.
000600      WRITE ADDRESS-LINE AFTER ADVANCING 2 LINES.
000610      MOVE STREET-ADDRESS TO ADDRESS-ITEM.
000620      WRITE ADDRESS-LINE AFTER ADVANCING 1.
000630      MOVE CITY-STATE-ZIP TO ADDRESS-ITEM.
000640      WRITE ADDRESS-LINE AFTER ADVANCING 1.
000650      READ ADDRESSES-IN
000660          AT END
000670              MOVE 'NO' TO MORE-INPUT.
```

PROCESSING A VARIABLE NUMBER OF INPUT CARDS

The really new things in this program are those needed to handle the unknown variable number of input cards. First, we need a new section in the Data Division, the **WORKING–STORAGE SECTION.** The Working Storage Section can be used for holding and naming any data that is neither read in nor written out; that is, any data that is not part of the File Section of the Data Division. In this program we will use a single field, called MORE–INPUT, to indicate to the program when there are no more cards to be processed. MORE–INPUT is a programmer-supplied data name made up by the author and conforms to the rules for making up data names. The field has been made three characters long and is **initialized,** or started out, with a value of 'YES', by the use of the **VALUE** clause. This indicates that as the program begins execution, we expect to find more input, since we have so far processed none at all. This program also shows our first use of the abbreviation **PIC** for PICTURE. PIC is a reserved word specifically provided as an abbreviation for PICTURE. Only a few reserved words in COBOL have authorized abbreviations, and only those words for which abbreviations are given may be abbreviated. You cannot make up your own abbreviations for reserved words.

The Procedure Division in this program shows a conventional form that we will use from now on to process an unknown variable number of input records. The OPEN and READ statements at the beginning of the Procedure Division are similar to those we have seen in an earlier program. In this program, however, the READ statement that comes right after the OPEN statement will be used to read only the first record from the file. This is called the **priming READ** statement. In this program all the other records will be read from the file by another READ statement elsewhere in the Procedure Division.

The AT END clauses in the READ statements in this program are different from the AT END clause used in the previous program. In that program, if the first (and only) READ statement discovered that there was accidentally no input data the AT END clause directed the computer to close the files and STOP RUN. We use here a more general procedure in both of the READ statements in this program. Each AT END clause, if or when it discovers that there is no input data or no more input data, moves the word 'NO' to the working storage field MORE–INPUT.

The next statement in the program after the priming READ directs the computer to **PERFORM** the **paragraph** called MAIN–LOOP until all of the input has been processed; that is, until the field MORE–INPUT has been made equal to 'NO'. MAIN–LOOP is a programmer-supplied name. Paragraph names are made up in accordance with the rules for making up program names, except that COBOL does not ignore any of the characters in a paragraph name.

The PERFORM statement in this program causes the paragraph MAIN–LOOP to be executed as many times as necessary, until the specified condition has been met. We will look at the paragraph MAIN–LOOP shortly. When the PERFORM is done and there is no more input to be processed, the statements following the PERFORM are executed. They just CLOSE the files and STOP RUN.

THE MAIN LOOP

The main loop is executed after the priming READ has been executed. The loop will process the first record (already read) and READ the second record; then the main loop will execute again to process the second record and READ the third. The main loop executes over and over, processing each record and reading the next, until finally there is no more data to be read, and the main loop sets MORE–INPUT to 'NO'.

The actions carried out by the main loop are ones we have already seen, except for the double-spacing of the first line of each address. The pairs of MOVE and WRITE statements print the input data, and the READ statement (to read the second and subsequent records) is identical to the priming READ. Of course, while we expect the priming READ never to execute its AT END clause, the READ statement in MAIN–LOOP will eventually find no more input data and will carry out the instruction in its AT END clause.

The WRITE statement that prints the first line of each address contains the clause **AFTER ADVANCING** 2 LINES. This is the most straightforward way of obtaining double spacing. But the rules of COBOL state that if any WRITE statement in a program uses the

AFTER ADVANCING clause, then every WRITE statement to that same file must have an AFTER ADVANCING clause. So to obtain single spacing for the remaining lines of the address other than the first, we would have to say something like AFTER ADVANCING 1 LINE in the WRITE statement for those address lines. In some COBOL systems, though, the word LINE is not allowed in an AFTER ADVANCING clause (LINES is allowed but not LINE). So AFTER ADVANCING 1 LINE is not legal. AFTER ADVANCING 1 LINES is legal, but the poor grammar involved is distracting. LINES is an optional word, however, and can be omitted altogether, so we use AFTER ADVANCING 1 to obtain single spacing.

A listing of the input cards used for this program is shown in Figure 2.8, and the program output is shown in Figure 2.9. (Notice that the quote sign is always legal in input data.)

Figure 2.8
Input data for Program 3.

```
G. S. POPKIN        1921 PRESIDENT ST.        BROOKLYN, NY  11221
L. MORGAN           11 W. 42 ST.              NEW YORK, NY  10010
C. W. TROTSKY       44 W. 10TH ST.            NEW YORK, NY  10036
S. O'MALLEY         121 5TH AVE.              BROOKLYN, NY  11217
```

Figure 2.9
Output from Program 3.

```
G. S. POPKIN
1921 PRESIDENT ST.
BROOKLYN, NY  11221

L. MORGAN
11 W. 42 ST.
NEW YORK, NY  10010

C. W. TROTSKY
44 W. 10TH ST.
NEW YORK, NY  10036

S. O'MALLEY
121 5TH AVE.
BROOKLYN, NY  11217
```

Exercise 2. Write a program to read and process a deck of data cards to the following specifications:

Input
A deck of data cards in the following format:

Cols.	Field
1–20	Name
21–35	Title
36–50	Company name
51–65	Street address
66–80	City and state

Output
For each input card, a five-line address containing each on its own line a name, title, company name, street address, and city and state. The first line of each address is to be double-spaced, and the address itself single-spaced.

A Program with Output Line Formatting

In the programs we have done so far, only one field appeared on each output line of print. It is more usual to have several fields printed on a single line, and those fields must be properly spaced across the page to make the output easily readable and useful. We will now consider a program to read a deck of cards and print the entire contents of each card on one line. The format of the input cards is:

Cols.	Field
1–9	Social security number
10–14	Employee number
15–21	Annual salary (dollars and cents)
22–46	Employee name

as shown in Figure 2.10.

There are no blanks between fields. There may be blanks within the Employee name field, however, as part of the name, as shown in the input data listing in Figure 2.11. We will now try to write a program that will read a data card and print all four fields from the card on one output line, then read the next card and print the four fields from that card on a line, and so on. Since there are no blanks between fields in the input card, if we print the card just as it is the output will be difficult to read. Fortunately, in COBOL, it is easy not only to insert

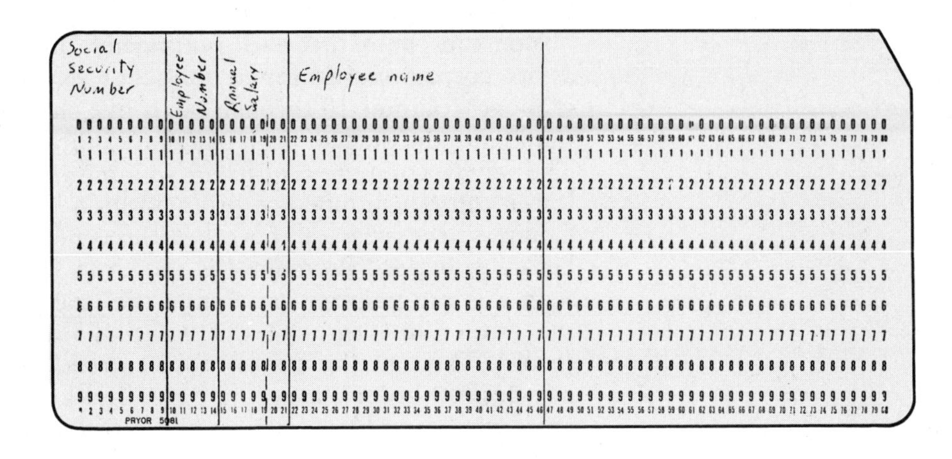

Figure 2.10
Data card format for Program 4.

Figure 2.11
Input data for Program 4.

```
1000400021050350000OMORALES, LUIS
1018500051089046510000JACOBSON, MRS. NELLIE
2011100081127743020000GREENWOOD, JAMES
2095600111166439530000COSTELLO, JOSEPH S.
3018100141205136040000REITER, D.
304870017124383255000MARRA, DITTA E.
4017100201282529060000LIPKE, VINCENT R.
4073900231321225570000KUGLER, CHARLES
5020700261359922080000JAVIER, CARLOS
5056800291398618590000GOODMAN, ISAAC
6049100321437315100000FELDSOTT, MS. SALLY
6082500351476011610000BUXBAUM, ROBERT
7031000381514708120000DUMAY, MRS. MARY
7080200411553404630000SMITH, R.
8032200441592101140000VINCENTE, MATTHEW J.
9010500471630842350000THOMAS, THOMAS T.
```

spaces between the fields for printing but also to rearrange the fields on the output line in any desired way. Let us say that we would like to print each line with the fields in the following print positions:

Print Pos.	Field
5–9	Employee number
12–20	Social security number
25–49	Employee name
52–58	Annual salary

Now, not only have we specified spaces between the fields when printed, but also that the fields be in a different order on the output line from what they are on the input card. Formatting output lines in this way is very easy in COBOL.

You can check that each output field has been made the same size as its corresponding input field. For example, the input Employee number, columns 10 through 14, is five characters long. The Employee number in the output, print positions 5 through 9, is also five characters. The way to compute the size of an input or output field is: subtract the lower column number or print position from the higher, and add 1. Compute the sizes of the other three fields in the input and output before going on. Program 4 is shown in Figure 2.12.

In this program we can define the exact placement of the fields in the output line through the use of 05-level entries for the output record. Through the use of those entries, we can arrange the fields in any desired order across the page, and by using FILLERs we can get any desired number of blank print positions between fields. If you now study the 05-level entries for the EMPLOYEE–LINE–OUT, you will see that the organization of the fields agrees with the requirements for the output line, as given in the statement of the problem. Of course, the very first 05-level FILLER is for carriage control as usual and does not occupy a print position.

Figure 2.12 Program 4.

```
4-CB2 V4 RELEASE 1.5 10NOV77          IBM OS AMERICAN NATIONAL STANDARD COBOL



 000010 IDENTIFICATION DIVISION.
 000020 PROGRAM-ID. PROG04.
 000030
 000040*     THIS PROGRAM READS DATA CARDS IN THE FOLLOWING FORMAT:
 000050*
 000060*     COLS. 1-9             SOCIAL SECURITY NUMBER
 000070*     COLS. 10-14           EMPLOYEE NUMBER
 000080*     COLS. 15-21           ANNUAL SALARY
 000090*     COLS. 22-46           EMPLOYEE NAME
 000100*
 000110*     FOR EACH CARD, ONE LINE IS PRINTED IN THE FOLLOWING FORMAT:
 000120*
 000130*     PP 5-9                EMPLOYEE NUMBER
 000140*     PP 12-20              SOCIAL SECURITY NUMBER
 000150*     PP 25-49              EMPLOYEE NAME
 000160*     PP 52-58              ANNUAL SALARY
 000170*
 000180*****************************************************************
 000190
 000200 ENVIRONMENT DIVISION.
 000210 INPUT-OUTPUT SECTION.
 000220 FILE-CONTROL.
 000230     SELECT EMPLOYEE-DATA-OUT ASSIGN TO UT-S-SYSPRINT.
 000240     SELECT EMPLOYEE-DATA-IN  ASSIGN TO UT-S-SYSIN.
 000250
 000260*****************************************************************
```

Figure 2.12 Program 4 (continued).

```
000270
000280 DATA DIVISION.
000290 FILE SECTION.
000300 FD  EMPLOYEE-DATA-IN
000310     LABEL RECORDS ARE OMITTED.
000320
000330 01  EMPLOYEE-RECORD-IN.
000340     05 SS-NO-IN                PIC X(9).
000350     05 IDENT-NO-IN             PIC X(5).
000360     05 ANNUAL-SALARY-IN        PIC X(7).
000370     05 NAME-IN                 PIC X(25).
000375     05 FILLER                  PIC X(34).
000380
000390 FD  EMPLOYEE-DATA-OUT
000400     LABEL RECORDS ARE OMITTED.
000410
000420 01  EMPLOYEE-LINE-OUT.
000430     05 FILLER                  PIC X.
000440     05 FILLER                  PIC X(4).
000450     05 IDENT-NO-OUT            PIC X(5).
000460     05 FILLER                  PIC X(2).
000470     05 SS-NO-OUT               PIC X(9).
000480     05 FILLER                  PIC X(4).
000490     05 NAME-OUT                PIC X(25).
000500     05 FILLER                  PIC X(2).
000510     05 ANNUAL-SALARY-OUT       PIC X(7).
000520
000530 WORKING-STORAGE SECTION.
000540 01  MORE-INPUT                 PIC X(3) VALUE 'YES'.
000550
000560******************************************************************
000570
000580 PROCEDURE DIVISION.
000590     OPEN INPUT  EMPLOYEE-DATA-IN,
000595          OUTPUT EMPLOYEE-DATA-OUT.
000600     READ EMPLOYEE-DATA-IN
000602         AT END
000604             MOVE 'NO' TO MORE-INPUT.
000610     PERFORM MAIN-LOOP UNTIL MORE-INPUT IS EQUAL TO 'NO'.
000620     CLOSE EMPLOYEE-DATA-IN,
000625           EMPLOYEE-DATA-OUT.
000630     STOP RUN.
000640
000650 MAIN-LOOP.
000660     MOVE SPACES            TO EMPLOYEE-LINE-OUT.
000670     MOVE SS-NO-IN          TO SS-NO-OUT.
000680     MOVE IDENT-NO-IN       TO IDENT-NQ-OUT.
000690     MOVE ANNUAL-SALARY-IN  TO ANNUAL-SALARY-OUT.
000700     MOVE NAME-IN           TO NAME-OUT.
000710     WRITE EMPLOYEE-LINE-OUT AFTER ADVANCING 1.
000720     READ EMPLOYEE-DATA-IN
000722         AT END
000724             MOVE 'NO' TO MORE-INPUT.
```

Let us now examine the processing paragraph, which in this program is called MAIN–LOOP. It begins with a MOVE SPACES statement that assigns blanks to all the fields in EMPLOYEE–LINE–OUT; that is, it blanks out the entire output line area. This statement is needed mainly to blank out the FILLER areas in EMPLOYEE–LINE–OUT, for there is no other way to do it. The MOVE SPACES statement is followed by four MOVE statements, each moving a single field from the input area to the output area. Notice that there are no intervening WRITE statements as there were in the previous program. That is because each WRITE statement causes a line of output to print, and in this program we don't want an output line until we have MOVEd all four fields of input to the output area. In this program the four fields could be MOVEd to the output area in any order; none of the MOVEs affects any of the others, and we don't WRITE a line until all four are MOVEd. In previous programs, the fields had to be MOVEd to the output area in the order in which we wanted them to print, because we wrote a line after every MOVE.

The output from Program 4 is shown in Figure 2.13.

Figure 2.13
Output from Program 4.

```
10503  100040002   MORALES, LUIS                 5000000
10890  101850005   JACOBSON, MRS. NELLIE         4651000
11277  201110008   GREENWOOD, JAMES              4302000
11664  209560011   COSTELLO, JOSEPH S.           3953000
12051  301810014   REITER, D.                    3604000
12438  304870017   MARRA, DITTA E.               3255000
12825  401710020   LIPKE, VINCENT R.             2906000
13212  407390023   KUGLER, CHARLES               2557000
13599  502070026   JAVIER, CARLOS                2208000
13986  505680029   GOODMAN, ISAAC                1859000
14373  604910032   FELDSOTT, MS. SALLY           1510000
14760  608250035   BUXBAUM, ROBERT               1161000
15147  703100038   DUMAY, MRS. MARY              0812000
15534  708020041   SMITH, R.                     0463000
15921  803220044   VINCENTE, MATTHEW J.          0114000
16308  901050047   THOMAS, THOMAS T.             4235000
```

***Exercise 3.** Write a program to read and process a deck of data cards to the following specifications:

Input
A deck of data cards in the following format:

Cols.	Field
1–20	Name
21–35	Title
36–50	Company name
51–65	Street address
66–80	City and state

*Solutions to starred exercises are needed in later exercises. Save them for later use.

Output
For each input card, print the contents of the card on one line in the following format:

Print Pos.	Field
6–20	Company name
23–37	Street address
40–54	City and state
60–74	Title
76–95	Name

Page Skipping

If you have done the programming exercises in this chapter, you will have seen that the printed output produced by your program comes out on the same page with other output produced by the COBOL system. We will now try to modify Program 4 so that the program will skip to a new page before printing the first line of name and address output. With some COBOL systems it is also necessary to skip to a new page after printing the last line of output to prevent the system from putting unwanted printing at the bottom of the program output. In this program, Program 5, we will see how to skip to a new page both before printing the first line of output and after printing the last.

SPECIAL NAMES

To enable a program to skip the output listing to a new page, we must first make up a name for the top of a page and tell the COBOL system what the name is. The name can be any legal data name, such as NEW–PAGE, TOP–OF–PAGE, or FIRST–LINE–OF–PAGE. For Program 5 we will use TO–NEW–PAGE. To tell the COBOL system that we have chosen the name TO–NEW–PAGE to stand for the top of a page, we need a new section in the Environment Division, and that is the **CONFIGURATION SECTION.**

The Configuration Section is optional in some COBOL systems, and we have gotten along without it so far. When the Configuration Section is used, it must appear before the Input–Output Section in the Environment Division. The Configuration Section and the Input–Output Section are the only two sections of the Environment Division. The words CONFIGURATION SECTION, as all section headers, must begin in the A area and be followed by a period, as shown in Figure 2.14.

Figure 2.14 Program 5.

```
4-CB2 V4 RELEASE 1.5 10NOV77          IBM OS AMERICAN NATIONAL STANDARD COBOL



000010 IDENTIFICATION DIVISION.
000020 PROGRAM-ID. PROG05.
000030
000040*     THIS PROGRAM READS DATA CARDS IN THE FOLLOWING FORMAT:
000050*
000060*     COLS. 1-9             SOCIAL SECURITY NUMBER
000070*     COLS. 10-14           EMPLOYEE NUMBER
000080*     COLS. 15-21           ANNUAL SALARY
000090*     COLS. 22-46           EMPLOYEE NAME
000100*
000110*     FOR EACH CARD, ONE LINE IS PRINTED IN THE FOLLOWING FORMAT:
000120*
000130*     PP 5-9                EMPLOYEE NUMBER
000140*     PP 12-20              SOCIAL SECURITY NUMBER
000150*     PP 25-49              EMPLOYEE NAME
000160*     PP 52-58              ANNUAL SALARY
000170*
000180*     SKIP TO THE TOP OF A NEW PAGE BEFORE PRINTING THE FIRST
000190*     LINE OF OUTPUT AND AFTER PRINTING THE LAST.
000200*
000210*****************************************************************
000220
000230 ENVIRONMENT DIVISION.
000240 CONFIGURATION SECTION.
000250 SPECIAL-NAMES.
000260     C01 IS TO-NEW-PAGE.
000270 INPUT-OUTPUT SECTION.
000280 FILE-CONTROL.
000290     SELECT EMPLOYEE-DATA-OUT ASSIGN TO UT-S-SYSPRINT.
000300     SELECT EMPLOYEE-DATA-IN  ASSIGN TO UT-S-SYSIN.
000310
000320*****************************************************************
000330
000340 DATA DIVISION.
000350 FILE SECTION.
000360 FD  EMPLOYEE-DATA-IN
000370     LABEL RECORDS ARE OMITTED.
000380
000390 01  EMPLOYEE-RECORD-IN.
000400     05 SS-NO-IN                 PIC X(9).
000410     05 IDENT-NO-IN              PIC X(5).
000420     05 ANNUAL-SALARY-IN         PIC X(7).
000430     05 NAME-IN                  PIC X(25).
000435     05 FILLER                   PIC X(34).
000440
000450 FD  EMPLOYEE-DATA-OUT
000460     LABEL RECORDS ARE OMITTED.
000470
000480 01  EMPLOYEE-LINE-OUT.
```

Figure 2.14 Program 5 (continued).

```
000490        05 FILLER                     PIC X.
000500        05 FILLER                     PIC X(4).
000510        05 IDENT-NO-OUT               PIC X(5).
000520        05 FILLER                     PIC X(2).
000530        05 SS-NO-OUT                  PIC X(9).
000540        05 FILLER                     PIC X(4).
000550        05 NAME-OUT                   PIC X(25).
000560        05 FILLER                     PIC X(2).
000570        05 ANNUAL-SALARY-OUT          PIC X(7).
000580
000590 WORKING-STORAGE SECTION.
000600 01   MORE-INPUT                      PIC X(3) VALUE 'YES'.
000610
000620*****************************************************************
000630
000640 PROCEDURE DIVISION.
000650      OPEN INPUT  EMPLOYEE-DATA-IN,
000655           OUTPUT EMPLOYEE-DATA-OUT.
000660      MOVE SPACES TO EMPLOYEE-LINE-OUT.
000670      WRITE EMPLOYEE-LINE-OUT AFTER ADVANCING TO-NEW-PAGE.
000680      READ EMPLOYEE-DATA-IN
000682          AT END
000684              MOVE 'NO' TO MORE-INPUT.
000690      PERFORM MAIN-LOOP UNTIL MORE-INPUT IS EQUAL TO 'NO'.
000700      MOVE SPACES TO EMPLOYEE-LINE-OUT.
000710      WRITE EMPLOYEE-LINE-OUT AFTER ADVANCING TO-NEW-PAGE.
000720      CLOSE EMPLOYEE-DATA-IN,
000725            EMPLOYEE-DATA-OUT.
000730      STOP RUN.
000740
000750 MAIN-LOOP.
000760      MOVE SPACES             TO EMPLOYEE-LINE-OUT.
000770      MOVE SS-NO-IN           TO SS-NO-OUT.
000780      MOVE IDENT-NO-IN        TO IDENT-NO-OUT.
000790      MOVE ANNUAL-SALARY-IN TO ANNUAL-SALARY-OUT.
000800      MOVE NAME-IN            TO NAME-OUT.
000810      WRITE EMPLOYEE-LINE-OUT AFTER ADVANCING 1.
000820      READ EMPLOYEE-DATA-IN
000822          AT END
000824              MOVE 'NO' TO MORE-INPUT.
```

Within the Configuration Section, it is the **SPECIAL–NAMES** entry that assigns the name TO–NEW–PAGE to the top of a page. In the particular COBOL system being used to produce the programs shown in this book, the conventional hardware designation for the top of a page is C01 (C-zero-one). In the SPECIAL–NAMES entry in Figure 2.14, you can see that the hardware name C01 has been made equivalent to our name TO–NEW–PAGE. You will shortly see how the name TO–NEW–PAGE is used in the Procedure Division of the program. The SPECIAL–NAMES entry has other uses, but we will not discuss any of them in this book.

SKIPPING TO A NEW PAGE

We come now to the entries in the Procedure Division that will cause the program to skip to a new page at the desired times. There is no statement in COBOL which will simply skip to a new page, but there is one which will print a line of data after skipping. So if we make the line of data all blanks, we get the desired result of just skipping. This can be accomplished with the two statements:

```
MOVE SPACES TO EMPLOYEE–LINE–OUT.
WRITE EMPLOYEE–LINE–OUT AFTER ADVANCING TO–NEW–PAGE.
```

You can now see the use of the name that was made up for the top of the page. If any different name had been made up and written in the SPECIAL–NAMES entry, the AFTER ADVANCING clause would have had to use it exactly as it was made up.

Now where should these two statements be inserted into the program to produce page skipping once and only once before the first data line is printed? Then, once that is solved, where should the same pair of statements be inserted to produce page skipping once and only once after the last data line is printed? The answer suggests itself if we remember that everything in the MAIN–LOOP paragraph is executed once for each data line that is printed. Since we certainly don't want to skip to a new page with the printing of each output line, the statements that cause skipping must be in the main control paragraph, not in MAIN–LOOP. Since the writing of all output data lines is under control of the PERFORM statement, statements placed before the PERFORM will be executed before any output lines are written, and statements placed after the PERFORM will be executed after the last output line is written. You can see in Figure 2.14 where the additional statements have been inserted to cause skipping to a new page before and after printing the data lines. The pair of statements placed before the PERFORM also happens to be before the priming READ, but it could have been after the READ, so long as it remained before the PERFORM.

Some important ideas should be emphasized here before going on. Each statement in the main control paragraph is executed only once during the entire execution of the program. More importantly, every statement in the MAIN–LOOP paragraph is executed once for each output line written. Later on we will see situations where one or more processes must be executed only once during each execution of the entire program, and we will know that they can't be placed in the repetitive loop. Another idea that will prove useful later is that all statements placed after the PERFORM are executed only after the end of the file has been detected on the input data. There will be situations where it is necessary to wait for end-of-file before doing certain steps, and statements for them will have to be placed after the PERFORM in the main control paragraph.

Exercise 4. Modify your solution to Exercise 3 so that your program skips the output listing to a new page before printing the first output line and after printing the last. Make up your own name for the top of a page.

Summary

The COBOL Coding Form permits a programmer to write coding line for line as it will be keyed or punched.

Records may be described as being composed of fields, whether in the File Section or the Working Storage Section of the Data Division. Level numbers are used to indicate the relationships between record names and field names.

The READ verb is used to obtain data from input files. The MOVE verb is used to transfer data from one area in the program to another.

Reformatting of input data is obtained easily by describing the output record in the desired format. The AFTER ADVANCING clause can be used to further improve the readability of output by controlling the movement of the printed page.

Fill-in Exercises

1. Columns 12 through 72 are marked off on the COBOL Coding Form for the ________ area.
2. An input data card can be treated as a single ________ in the Data Division.
3. Fields may be described in the Data Division with level numbers in the range ________ through ________.
4. The reserved word ________ may be used to name fields not referred to in the processing.
5. Two kinds of literals in COBOL are ________ literals and ________ literals.
6. The ________ Section is used to describe data which is not related to any input or output file.
7. The ________ clause is used to initialize data in working storage.
8. The statement that obtains the first input record from an input file is called a ________ READ.
9. The ________ clause can be used to control the vertical spacing of printed program output.
10. The ________ entry is used to give a name to the top of the page.

Review Exercises

1. Name the two SECTIONs of the ENVIRONMENT DIVISION in the order in which they must appear.
2. In which division of a COBOL program would each of the following be found?
   a. SPECIAL–NAMES
   b. a READ statement
   c. the WORKING–STORAGE SECTION
   d. a PERFORM statement
   e. an AFTER ADVANCING clause
3. Which of the following statement(s) must refer to a file name, and which must refer to a record name?
   a. OPEN b. CLOSE c. READ d. WRITE
4. Name the five kinds of entries that must begin in the A area.
5. Find and correct the error in the following statement:

   WRITE A–LINE AFTER ADVANCING 1 LINE.

*6. Write a COBOL program to the following specifications:

Input
A deck of data cards in the following format:

Cols.	Field
1–8	Part number
9–28	Part description
30–32	Quantity on hand

Output
A listing of the contents of the cards, one line per card, in the following format:

Print Pos.	Field
9–16	Part number
20–22	Quantity on hand
41–60	Part description

Skip to a new page before printing the first line and after the last.

The IF Statement

3

KEY POINTS

Here are the key points you should learn from this chapter:

1. how to use several forms of the IF statement;
2. how to use the printer spacing chart;
3. flowcharts and control charts for the IF statement.

KEY WORDS

Key words to recognize and learn:

printer spacing chart	program flowchart	assumed decimal point
alphanumeric	control chart	ELSE
numeric	FROM	NEXT SENTENCE
conditional	edited numeric	sentence
condition	zero suppression	
relational operator	insertion characters	

In this chapter we will study COBOL statements that permit the computer to examine the data that it is working on and to decide which one out of two or more processes should be executed. For example, the computer might examine an employee's number of hours worked and decide whether or not to compute overtime pay. But first we will look at a second kind of form that is helpful when writing COBOL programs, the **printer spacing chart.**

Printer Spacing Chart

Figure 3.1 shows a printer spacing chart. Printer spacing charts can usually be bought wherever you buy your coding forms. The chart can be used by the programmer to plan the format of an output report or listing before starting to code a program. You can see in Figure 3.1 the print positions numbered across the top of the chart. The print position numbers are in groups of 10 for ease of locating any particular position.

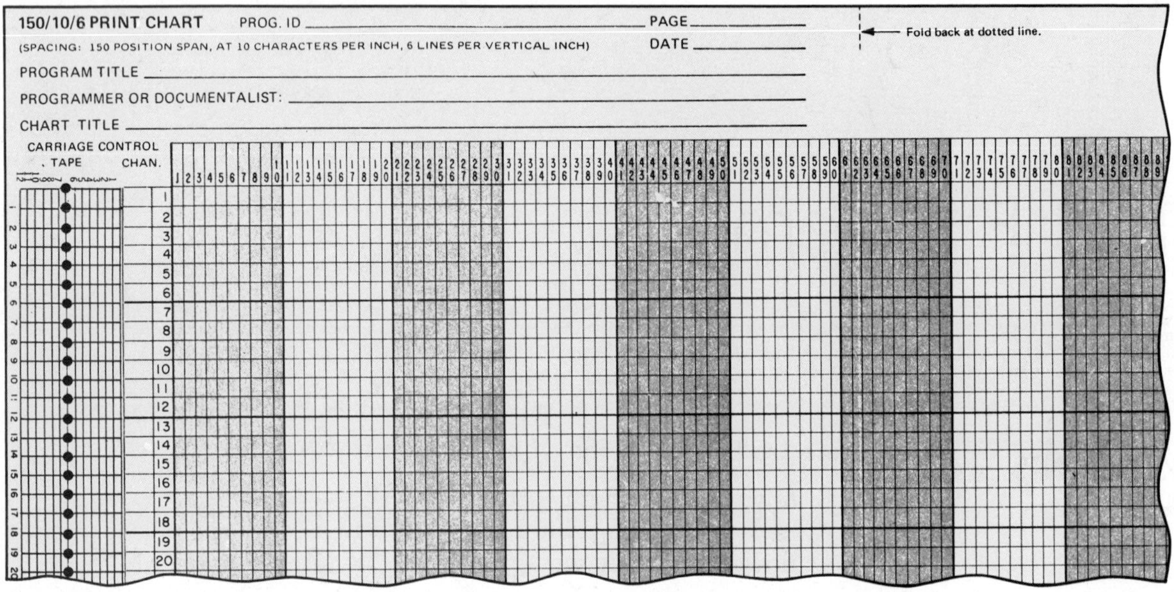

150/10/6 PRINT CHART PROG. ID PAGE DATE

(SPACING: 150 POSITION SPAN, AT 10 CHARACTERS PER INCH, 6 LINES PER VERTICAL INCH)

Fold back at dotted line.

PROGRAM TITLE

PROGRAMMER OR DOCUMENTALIST:

CHART TITLE

CARRIAGE CONTROL TAPE CHAN.

Figure 3.1 Printer spacing chart.

In planning an output report, the programmer determines the print positions in which the various fields of output are to appear and indicates those fields on the spacing chart. The fields should be indicated on the chart using the same character that would be used in a PICTURE for that field in a COBOL program. To give an example of the use of a printer spacing chart, we will use the output specifications of the program in Exercise 3, Chapter 2, which are repeated here for convenience:

Print Pos.	Field
6–20	Company name
23–37	Street address
40–54	City and State
60–74	Title
76–95	Name

Since all of these fields have PICTURE characters of X, we will use Xs on the printer spacing chart. In Figure 3.2 you can see that several lines have been filled in in the print positions where the fields Company name, Street address, City and State, Title, and Name would appear.

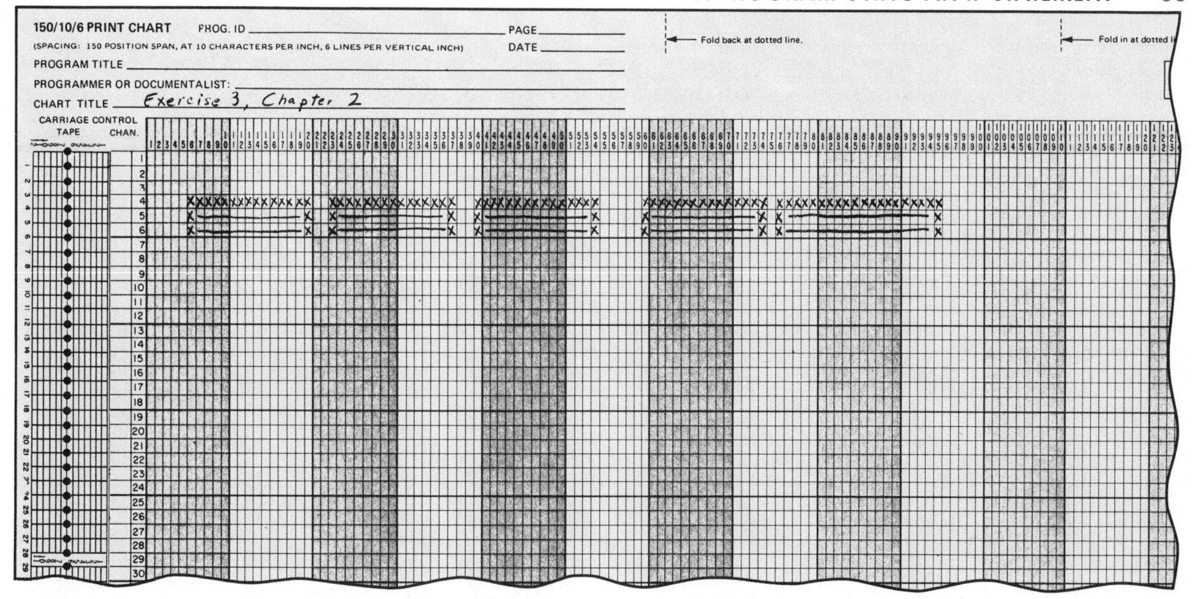

Figure 3.2 A printer spacing chart showing the output specifications for Exercise 3, Chapter 2.

Ordinarily, a printer spacing chart would not be used merely to visualize an output specification that has already been developed. Instead, it is usually used to develop the output specification; that is, to plan what the output will look like. So for the next program, you will be given the output requirements not in the form used in Chapter 2, but in the form of a printer spacing chart. Later on, you will be asked to develop your own output specifications using a printer spacing chart.

Exercise 1. Using a printer spacing chart, show the output specification for Program 4 (Chapter 2, page 23).

Exercise 2. Given the printer spacing chart in Figure 3.E2, write output specifications in the form used in Chapter 2.

A Program Using an IF Statement

The next program, Program 6, reads a deck of input cards in the format shown in Figure 3.3 and lists the contents of each card on one line. Each card is a record for one part number showing, for each part number, an English description of the part and the quantity on hand.

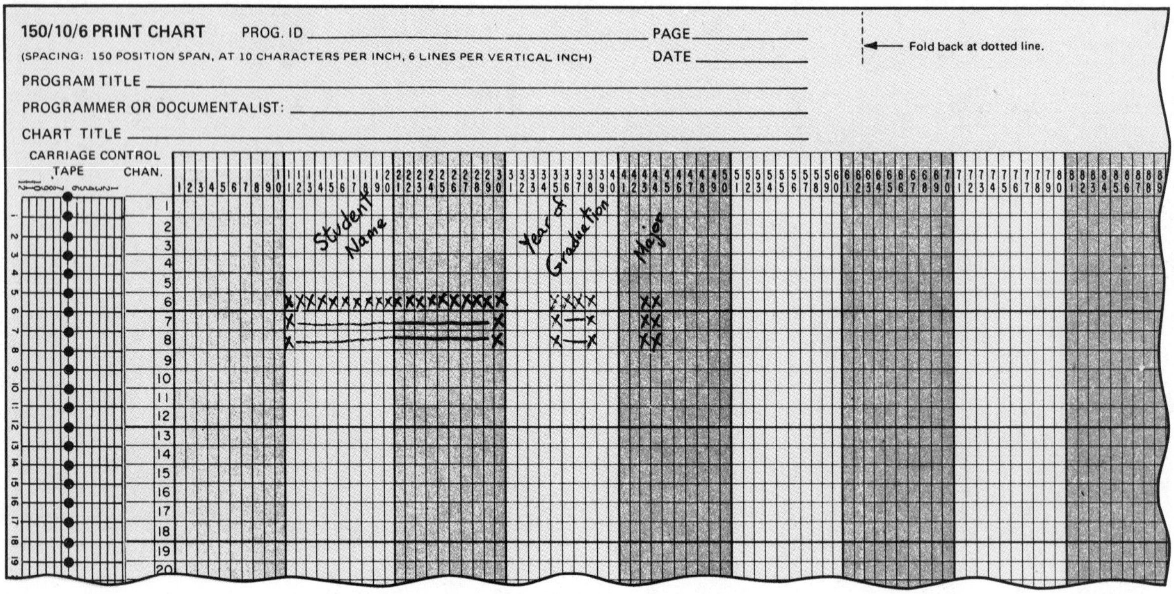

150/10/6 PRINT CHART PROG. ID ________ PAGE ________

(SPACING: 150 POSITION SPAN, AT 10 CHARACTERS PER INCH, 6 LINES PER VERTICAL INCH) DATE ________

Fold back at dotted line.

PROGRAM TITLE ________

PROGRAMMER OR DOCUMENTALIST: ________

CHART TITLE ________

CARRIAGE CONTROL TAPE CHAN.

Student Name

Year of Graduation

Major

Figure 3.E2 Printer spacing chart for Exercise 2.

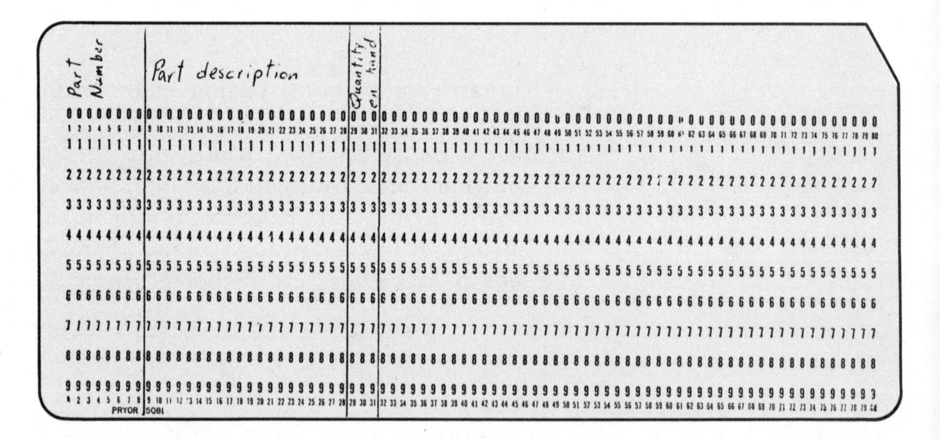

Figure 3.3 Data card format for Program 6.

We are now for the first time dealing with a numeric quantity, the quantity on hand. Up to now, the only **class of data** that we have used is **alphanumeric.** Alphanumeric data may consist of any of the computer's characters. **Numeric** input data, on the other hand, may consist only of the digits 0 through 9 and may contain a plus or a minus sign. The sign of the number may be punched either immediately to the left or right of the number or may be punched over the leftmost or rightmost digit of the number. Unsigned numeric data is assumed to be positive. Numeric input data never contains an explicitly punched decimal point. If decimal fractions or mixed numbers are to be part of the input data, the location of the decimal point is provided through programming in the Data Division. We will see how later in this chapter.

A numeric input data item may be up to 18 digits long, unless the sign of the number appears as a separate character; then the sign is counted as one of the 18.[1] The particular pieces of numeric data to be used in this program are all three-digit, unsigned integers, the quantity on hand. The PICTURE character for integer data is 9, just as the PICTURE character for alphanumeric data is X.

Program 6, in addition to listing the contents of each card, will examine the quantity on hand of each part and print the word REORDER next to any part with fewer than 400 on hand. The output format is shown in the printer spacing chart in Figure 3.4, with Xs showing where alphanumeric data will print and 9s showing where numeric data will print. Notice that some of the lines have the word REORDER and some do not, indicating that the printing of the word is **conditional** on something in the data. Since the printer spacing chart shows only a sample of what the output might look like, we would not expect the actual output to correspond line for line with the chart. At the time the program is being written, we may not know what actual data values will be used, and the program must work correctly for any combination of inputs having more or fewer than 400 on hand, or even exactly 400 on hand. (Should the word REORDER print when there are exactly 400 on hand?)

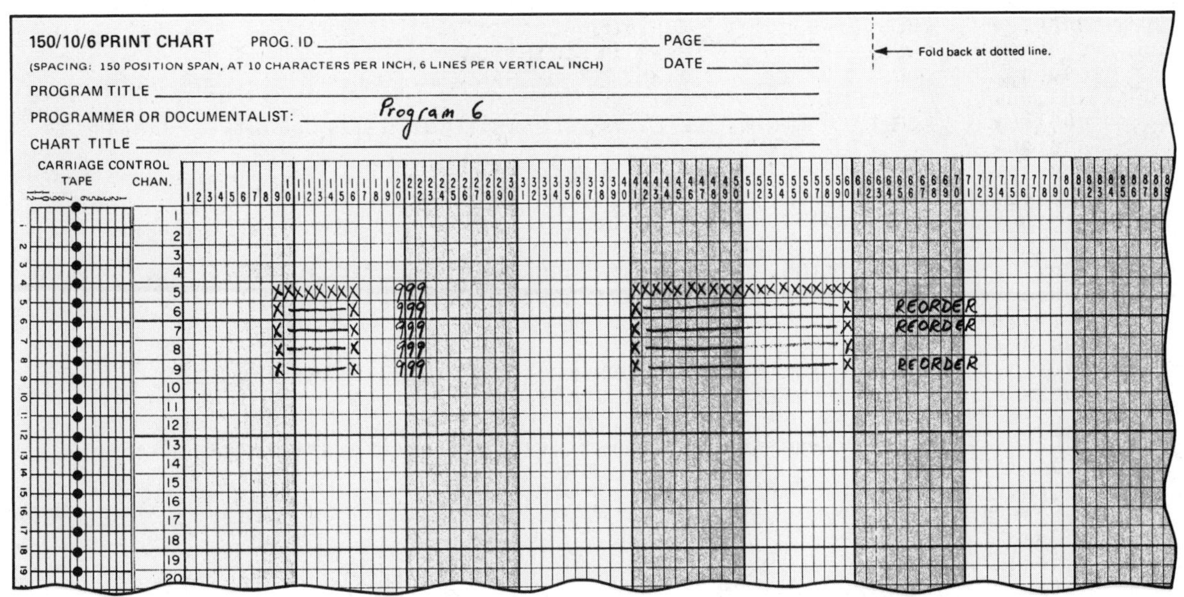
150/10/6 PRINT CHART PROG. ID PAGE
(SPACING: 150 POSITION SPAN, AT 10 CHARACTERS PER INCH, 6 LINES PER VERTICAL INCH) DATE
Fold back at dotted line.
PROGRAM TITLE
PROGRAMMER OR DOCUMENTALIST: Program 6
CHART TITLE
CARRIAGE CONTROL TAPE CHAN.

Line	Positions 13–20	Positions 28–30	Positions 59–86	Positions 107–113
5	XXXXXXXX	999	XXXXXXXXXXXXXXXXXXXXXXXXXXXX	
6	X——X	999	X——X	REORDER
7	X——X	999	X——X	REORDER
8	X——X	999	X——X	
9	X——X	999	X——X	REORDER

Figure 3.4 Printer spacing chart showing output format for Program 6.

1. Some COBOL systems also use a form of numeric data called floating-point numbers, with different rules from the ones just given. Floating-point numbers are not covered in this book.

A program to carry out the required processing is shown in Figure 3.5. There is nothing new in the Identification or Environment Divisions. In the Data Division the two quantity on hand fields, called QUANTITY–IN and QUANTITY–OUT, have been defined with 9 instead of X, and a place called MESSAGE–SPACE has been allocated in print positions 65–71 of the output line. When a line is printed with fewer than 400 parts on hand, MESSAGE–SPACE will be filled with the word REORDER, but when the number of parts is 400 or greater and the message is not wanted, the MESSAGE–SPACE will be filled with blanks. The coding in the Procedure Division will see to it that MESSAGE–SPACE is properly filled.

Figure 3.5 Program 6.

```
4-CB2 V4 RELEASE 1.5 10NOV77          IBM OS AMERICAN NATIONAL STANDARD COBOL

000010 IDENTIFICATION DIVISION.
000020 PROGRAM-ID. PROG06.
000030*AUTHOR. PHIL RUBENSTEIN.
000040*
000050*     THIS PROGRAM READS DATA CARDS IN THE FOLLOWING FORMAT:
000060*
000070*     COLS. 1-8           PART NUMBER
000080*     COLS. 9-28          PART DESCRIPTION
000090*     COLS. 29-31         QUANTITY ON HAND
000100*
000110*     FOR EACH CARD, ONE LINE IS PRINTED IN THE FOLLOWING FORMAT:
000120*
000130*     PP 9-16             PART NUMBER
000140*     PP 20-22            QUANTITY ON HAND
000150*     PP 41-60            PART DESCRIPTION
000160*
000170*     IN ADDITION, THE WORD 'REORDER' IS TO BE PRINTED
000180*     IN PRINT POSITIONS 65-71 NEXT TO ANY PART WHOSE
000190*     QUANTITY ON HAND IS LESS THAN 400.
000200*
000210*     SKIP TO THE TOP OF A NEW PAGE BEFORE PRINTING THE FIRST
000220*     LINE OF OUTPUT AND AFTER PRINTING THE LAST.
000230*
000240*****************************************************************
000250
000260 ENVIRONMENT DIVISION.
000270 CONFIGURATION SECTION.
000280 SPECIAL-NAMES.
000290     C01 IS TO-NEW-PAGE.
000300 INPUT-OUTPUT SECTION.
000310 FILE-CONTROL.
000320     SELECT INVENTORY-LIST     ASSIGN TO UT-S-SYSPRINT.
000330     SELECT INVENTORY-FILE-IN  ASSIGN TO UT-S-SYSIN.
000340
000350*****************************************************************
```

Figure 3.5 Program 6 (continued).

```
000360
000370 DATA DIVISION.
000380 FILE SECTION.
000390 FD  INVENTORY-FILE-IN
000400      LABEL RECORDS ARE OMITTED.
000410
000420 01  INVENTORY-RECORD-IN.
000430      05 PART-NUMBER-IN          PIC X(8).
000440      05 PART-DESCRIPTION-IN     PIC X(20).
000450      05 QUANTITY-IN             PIC 9(3).
000460      05 FILLER                  PIC X(49).
000470
000480 FD  INVENTORY-LIST
000490      LABEL RECORDS ARE OMITTED.
000500
000510 01  INVENTORY-LINE.
000520      05 FILLER                  PIC X.
000530      05 FILLER                  PIC X(8).
000540      05 PART-NUMBER-OUT         PIC X(8).
000550      05 FILLER                  PIC X(3).
000560      05 QUANTITY-OUT            PIC 9(3).
000570      05 FILLER                  PIC X(18).
000580      05 PART-DESCRIPTION-OUT    PIC X(20).
000590      05 FILLER                  PIC X(4).
000600      05 MESSAGE-SPACE           PIC X(7).
000610
000620 WORKING-STORAGE SECTION.
000630 01  MORE-INPUT                  PIC X(3) VALUE 'YES'.
000640
000650*****************************************************************
000660
000670 PROCEDURE DIVISION.
000680      OPEN INPUT  INVENTORY-FILE-IN,
000690           OUTPUT INVENTORY-LIST.
000700      MOVE SPACES TO INVENTORY-LINE.
000710      WRITE INVENTORY-LINE AFTER ADVANCING TO-NEW-PAGE.
000720      READ INVENTORY-FILE-IN
000730          AT END
000740              MOVE 'NO' TO MORE-INPUT.
000750      PERFORM MAIN-PROCESS UNTIL MORE-INPUT IS EQUAL TO 'NO'.
000760      MOVE SPACES TO INVENTORY-LINE.
000770      WRITE INVENTORY-LINE AFTER ADVANCING TO-NEW-PAGE.
000780      CLOSE INVENTORY-FILE-IN,
000790            INVENTORY-LIST.
000800      STOP RUN.
000810
000820 MAIN-PROCESS.
000830      MOVE SPACES                TO INVENTORY-LINE.
000840      MOVE PART-NUMBER-IN        TO PART-NUMBER-OUT.
000850      MOVE PART-DESCRIPTION-IN   TO PART-DESCRIPTION-OUT.
000860      MOVE QUANTITY-IN           TO QUANTITY-OUT.
000870      IF QUANTITY-IN IS LESS THAN 400
000880          MOVE 'REORDER' TO MESSAGE-SPACE.
000890      WRITE INVENTORY-LINE AFTER ADVANCING 1.
000900      READ INVENTORY-FILE-IN
000910          AT END
000920              MOVE 'NO' TO MORE-INPUT.
```

The main control paragraph of the Procedure Division contains nothing new. The main loop, in this program called MAIN–PROCESS, begins in the usual way by moving blanks to the output line. In that MOVE blanks are moved to MESSAGE–SPACE as well. Now the IF statement will move the word REORDER to MESSAGE–SPACE only when the quantity on hand is less than 400. The expression

QUANTITY–IN IS LESS THAN 400

is referred to as the **condition** in the IF statement, and the phrase

IS LESS THAN

is called a **relational operator.** Other relational operators that you may use are:

IS NOT LESS THAN
IS GREATER THAN
IS NOT GREATER THAN
IS EQUAL TO
IS NOT EQUAL TO

Figure 3.6 shows the input, and Figure 3.7 the output of Program 6.

Figure 3.6
Input data for Program 6.

```
10010334HARD RUBBER WASHER   489
20033547TWELVE INCH WRENCH   301
30048929FORTY FOOT LADDER    400
34554311SEARCH LIGHT         259
55599432HAMMER               999
00012349PLIERS               557
00123435AXE                  207
01256749LEVEL                399
32109886PAINT BRUSH          019
56677216PAINT                456
02023406PAINT REMOVER        367
```

Figure 3.7
Output from Program 6.

```
10010334    489            HARD RUBBER WASHER
20033547    301            TWELVE INCH WRENCH      REORDER
30048929    400            FORTY FOOT LADDER
34554311    259            SEARCH LIGHT            REORDER
55599432    999            HAMMER
00012349    557            PLIERS
00123431    207            AXE                     REORDER
01256742    399            LEVEL                   REORDER
32109885    019            PAINT BRUSH             REORDER
56677212    456            PAINT
02023400    367            PAINT REMOVER           REORDER
```

The number 400 in the IF statement is a numeric literal. Numeric literals were mentioned briefly in Chapter 2. Numeric literals may contain up to 18 digits, may contain a plus or a minus sign at the left end of the number, and may contain a decimal point anywhere in the number, except at the right end. If the literal contains no sign, it is assumed to be positive. If the literal contains no decimal point, as our 400 doesn't, it is an integer.[2] Numeric literals are never enclosed in quotes. If a numeric quantity is enclosed in quotes, it is treated as a nonnumeric literal.

The rules for numeric literals are quite different from those for numeric input data. The rules for numeric data were given earlier in the chapter. Table 3.1 compares the two sets of rules.

	Numeric input data	Numeric literal
Plus or minus sign	May appear immediately before or after the number, or punched over the rightmost or leftmost digit of the number.	May appear at the left end of the number only.
Decimal point	Never punched as part of the number.	May appear anywhere except at the right end of the number.
Comma	Never punched as part of the number.	Never punched as part of the number.
Size	Up to 17 or 18 digits, depending on the location of the sign.	Up to 18 digits, not counting the sign.

Table 3.1 Summary of the rules for numeric input data and for numeric literals.

We use a numeric literal in Program 6, because the field that it is being compared to, QUANTITY–IN, is defined as numeric. Although not always absolutely necessary, it is usually safest to operate on like classes of data.

2. Floating-point literals may contain in addition the character E and a second plus or minus sign.

Program Design with Conditions

Now that our programs involve the computer in making logical decisions, we can begin to look at program design techniques. Each of the several program design tools that we will use in this book is suited to a different kind of design task. **Program flowcharts** and **control charts** are useful for designing a small program like Program 6. In later chapters the programs become big enough so that flowcharts and control charts are too cumbersome, so in Chapter 4 we will look at other design methods.

The program flowchart is one of the oldest tools for program design. It has drawbacks which have caused it to fall into disrepute in recent years, but it is still the most useful design tool for small programs and for small pieces of larger ones. A flowchart may show in diagram form each IF statement in a program (or portion of a program), the processes that the program carries out if the condition is True, and the processes that it carries out if the condition is False. For example, the IF statement in Program 6 can be flowcharted as shown in Figure 3.8. An IF decision is always flowcharted as a diamond with two paths coming out of it. The two paths can be labelled Yes and No, True and False, or with any other complementary pair of outcomes. Any processing steps that are carried out in the True and False paths are shown in rectangles. In this flowchart only the Yes path contains processing.

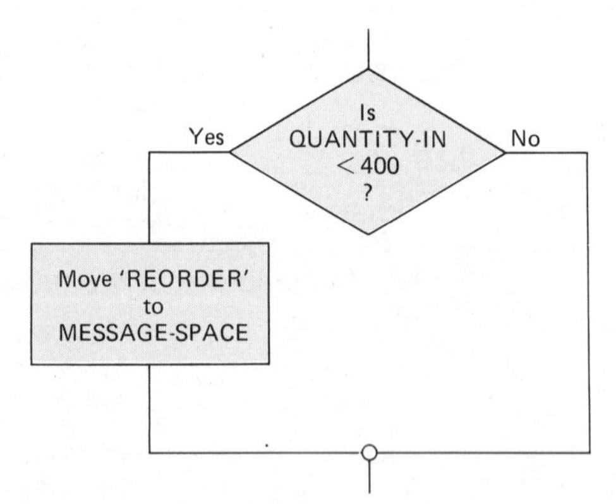

Figure 3.8
Flowchart of IF statement in Program 6.

In a flowchart of an IF statement, either the True path or the False path, or both, may contain any amount of processing. In Figure 3.8 the Yes path contains one processing step and the No path contains no processing. The two paths of the decision must rejoin before the flowchart can continue.

A decision may also be shown in control chart form. A control chart for the decision shown in Figure 3.8 is given in Figure 3.9. All rules and options that apply to decisions in flowcharts apply to decisions in control charts.

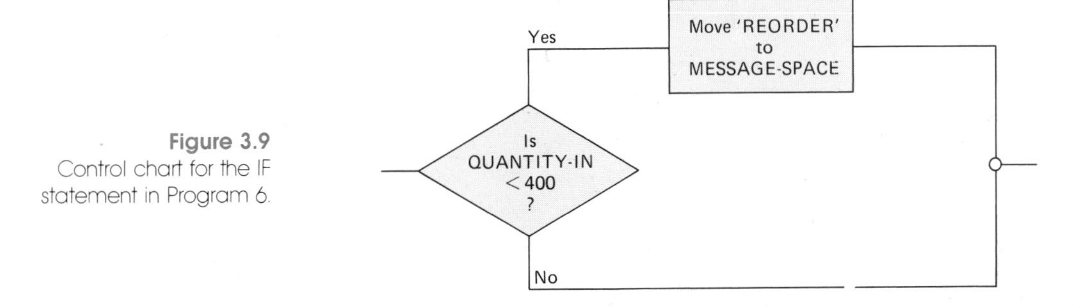

Figure 3.9 Control chart for the IF statement in Program 6.

Let us now consider a problem in which there is more than one processing step in the Yes path of a decision. Program 7: Write a program to read the same input data as Program 6. For each part with a quantity on hand of less than 400, print the contents of the card on one line. For cards with a quantity of 400 or more, print nothing. (Notice that we don't print the word REORDER this time, because the entire output is a list of parts to be reordered.)

A flowchart of the MAIN–PROCESS loop for Program 7 is shown in Figure 3.10. This flowchart shows the use of a special symbol for input and output operations—the parallelogram.

The flowchart shows how we manage to print nothing when the quantity is 400 or more. The coding of the flowchart is shown in Figure 3.11.

Figure 3.11 shows how an IF statement can be written when there is more than one processing step in the Yes path. The several processing steps are simply written one after the other, indented as shown, with a period at the end of the last step only. There can be, by this method, as many processing steps as needed, but there must be only one period at the end of the last step only. The period is what tells the COBOL system that the IF is done[3] and that the Yes and No paths rejoin at that point. The COBOL system pays no attention to indenting. Indenting is used only to make it easy for the programmer to see where the paths rejoin.

3. In the very newest COBOL systems, END–IF or a period may be used to terminate an IF.

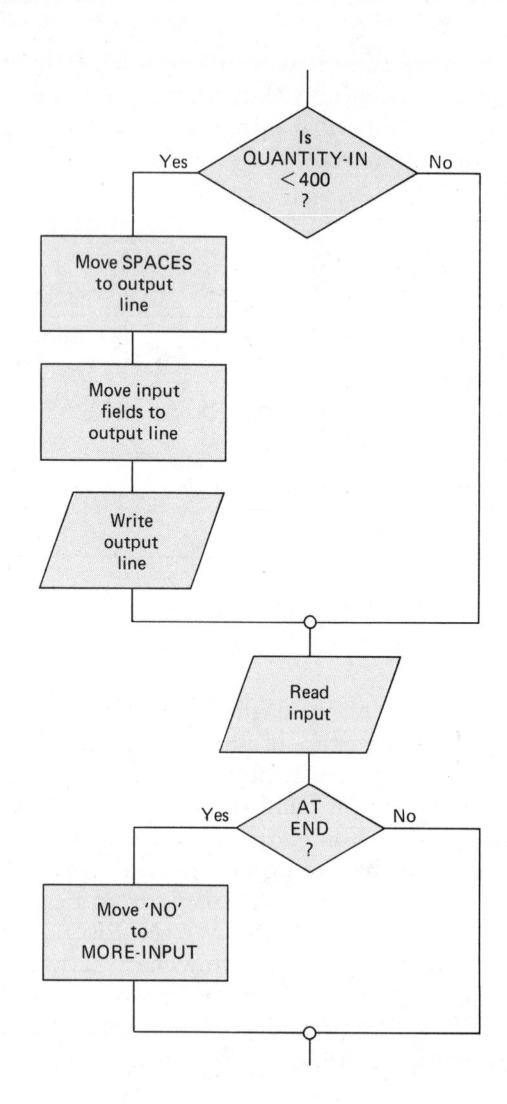

Figure 3.10
Flowchart of the MAIN–PROCESS loop in Program 7.

Figure 3.11 MAIN–PROCESS loop in Program 7.

```
000730 MAIN-PROCESS.
000740     IF QUANTITY-IN IS LESS THAN 400
000750         MOVE SPACES                TO INVENTORY-LINE
000760         MOVE PART-NUMBER-IN        TO PART-NUMBER-OUT
000770         MOVE PART-DESCRIPTION-IN   TO PART-DESCRIPTION-OUT
000780         MOVE QUANTITY-IN           TO QUANTITY-OUT
000790         WRITE INVENTORY-LINE AFTER ADVANCING 1.
000800     READ INVENTORY-FILE-IN
000802         AT END
000804             MOVE 'NO' TO MORE-INPUT.
```

The output from Program 7 is shown in Figure 3.12.

```
20033547    301                 TWELVE INCH WRENCH
34554311    259                 SEARCH LIGHT
00123435    207                 AXE
01256749    399                 LEVEL
32109886    019                 PAINT BRUSH
02023406    367                 PAINT REMOVER
```

Figure 3.12
Output from Program 7.

Exercise 3. Write a program to read cards in the format shown in Figure 3.E3.1 and print output in the format shown in Figure 3.E3.2. Print one line for each card read, and in addition, print the word OVERTIME for those employees who have more than 40 hours worked.

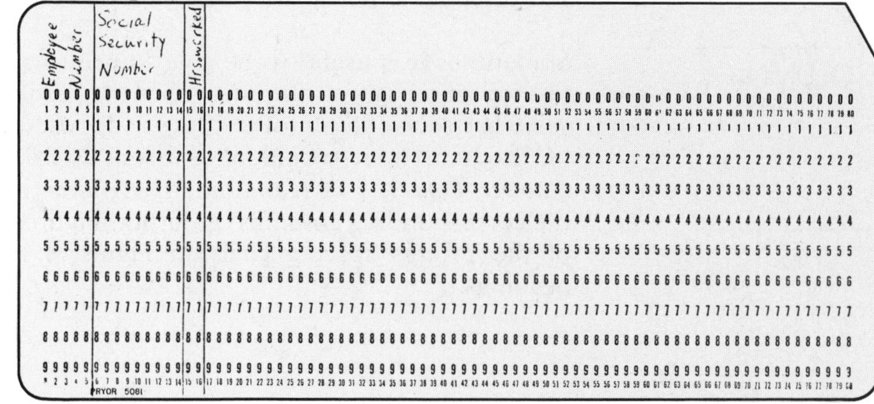

Figure 3.E3.1
Data card format for Exercise 3.

150/10/6 PRINT CHART PROG. ID ________ PAGE ________
(SPACING: 150 POSITION SPAN, AT 10 CHARACTERS PER INCH, 6 LINES PER VERTICAL INCH) DATE ________
Fold back at dotted line.
PROGRAM TITLE ________
PROGRAMMER OR DOCUMENTALIST: ________
CHART TITLE *Exercise 3, Chapter 3*
CARRIAGE CONTROL TAPE CHAN.

Employee Number — Hours Worked

Line	Employee Number	Hours Worked	
6	XXXXX	99	OVERTIME
7	X———X	99	
8	X———X	99	OVERTIME
9	X———X	99	OVERTIME
10	X———X	99	

Figure 3.E3.2 Output format for Exercise 3.

*Exercise 4. Using the same input and output formats as in Exercise 3, write a program that reads all the cards and prints only those employees who have worked fewer than 25 hours. Do not print the word OVERTIME for any employees.

Output Formatting

As we go on to more difficult problems with more elaborate output, it will be helpful to be able to use some COBOL features that make output more easily readable. Let us look at two of these features now and then use both in the next program.

COLUMN HEADINGS

Sometimes it is useful to be able to print fixed heading information at the top of a page of output so that the reader of the output can easily see what each column of output is supposed to contain. In the output of Program 6, it would have been useful to have the words *Part Number, Quantity on Hand,* and *Description* appear approximately over the corresponding columns of output. Figure 3.13 shows a modification of the printer spacing chart in Figure 3.4, which includes column headings.

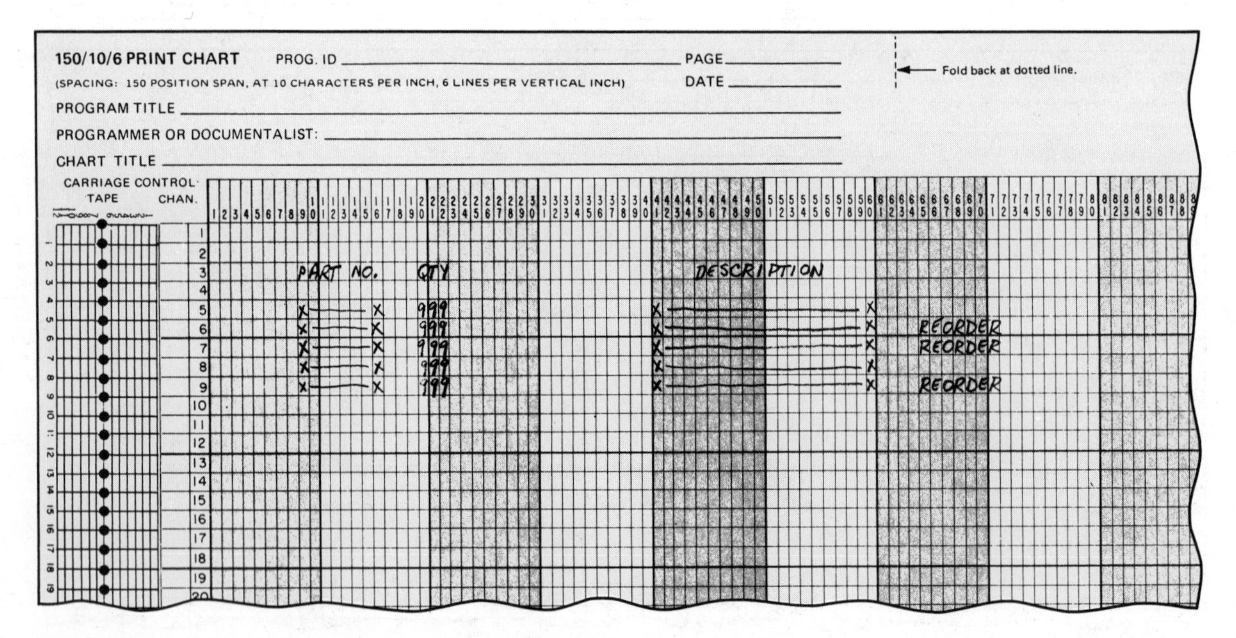

Figure 3.13 Printer spacing chart showing column headings.

The printed line containing the column headings is all constant information, and the usual way to set up such a constant is in the Working Storage Section. The heading line consists of the space for the carriage control character, some blank spaces at the left end of the print line, then some characters (the words Part No.), then some more blanks, then Qty., then more blanks, and the word Description. Figure 3.14 shows how such a line of constants can be described in working storage. It is legal to combine the FILLER for the carriage control character with the FILLER for the initial SPACES on the line, but we leave them separate as a reminder.

```
000010 01   COL-HD.
000020      05 FILLER          PIC X.
000030      05 FILLER          PIC X(8)  VALUE SPACES.
000040      05 FILLER          PIC X(8)  VALUE 'PART NO.'.
000050      05 FILLER          PIC X(3)  VALUE SPACES.
000060      05 FILLER          PIC X(3)  VALUE 'QTY'.
000070      05 FILLER          PIC X(22) VALUE SPACES.
000080      05 FILLER          PIC X(11) VALUE 'DESCRIPTION'.
```

Figure 3.14 Working storage description of the column headings shown in Figure 3.13.

To print a line that has been defined in working storage, we need a WRITE statement. But WRITE statements always have as their object some record name in the File Section, so a WRITE statement with the **FROM** option can be used, as:

```
WRITE INVENTORY-LINE FROM COL-HD.
```

INVENTORY–LINE was used as the object of the WRITE statement, because it is the record name associated with the output file (see Figure 3.5). The 05-level definitions in INVENTORY–LINE have no effect on the column headings, and the only restriction on INVENTORY–LINE is to define it as having at least as many character positions as does COL–HD. In this case, INVENTORY–LINE is 71 characters long and COL–HD is only 55 characters, so INVENTORY–LINE is long enough. If INVENTORY–LINE had been shorter than COL–HD, only the leftmost characters of COL–HD would print. If it had been necessary, INVENTORY–LINE could have been expanded at the end with FILLER to bring it up to the size of COL–HD.

Since we want the column headings to print at the top of a page, we can use:

```
WRITE INVENTORY-LINE FROM COL-HD AFTER ADVANCING TO-NEW-PAGE.
```

and there would be no need for skipping to a new page by the method used in the earlier programs. You will see an example of the printing of column headings in context later, in Program 8.

EDITING OF NUMERIC OUTPUT

COBOL provides facilities for rendering numeric output more readable by making it easy for the programmer to insert decimal points and commas into numbers that are to be printed and also to eliminate unwanted leading zeros in numeric printed output. To produce **edited numeric** output the programmer need only specify the desired editing in the PICTURE clause of the output field to be edited. Then, whenever a number is assigned to that output field, COBOL will perform the desired editing with no further effort on the part of the programmer. The PICTURE characters , (comma) and . (decimal point) can be used to obtain insertion of commas and decimal points, respectively, and the PICTURE character Z can be used to obtain suppression of insignificant leading zeros. For example, in Program 7 we could have printed the QUANTITY–OUT field with **zero suppression** to eliminate unwanted zeros when the quantity amount was small by using PICTURE Z(3) instead of PICTURE 9(3). The use of Z(3) instead of 9(3) would in no way affect the printing of three-digit quantities. Table 3.2 shows how different values of QUANTITY–OUT would print using the two PICTUREs 9(3) and Z(3).

9(3)	Z(3)
800	800
395	395
035	35
006	6
000	

Table 3.2. **How a three-digit integer would print using the two PICTUREs 9(3) and Z(3).**

You can see a slight difficulty with the PICTURE Z(3): When the quantity is zero, the entire field is zero-suppressed, and nothing at all prints. Usually in a listing where we are showing the quantity of each part in the list, if the quantity on hand is zero, we would probably want a zero to print. We can obtain this result by telling COBOL to zero-suppress only the first two positions of the field and print the third, regardless of what it is. We would use PICTURE ZZ9 to indicate only two positions of zero suppression in a three-digit field.

The **insertion characters** comma and decimal point can be used in a PICTURE to show where commas and decimal points should appear in printed output. Commas and decimal points are, as we know, never part of numeric input data, nor are they part of the data as COBOL is processing them. Since COBOL cannot perform arithmetic nor do numeric comparisons on numbers that contain commas or decimal points, these two characters can be inserted into numbers only right before the numbers are printed. Thus it is the PICTURE of the output field as it is to be printed that tells COBOL where to insert commas and decimal points.

To use the insertion characters, the programmer simply places them in the PICTURE as they are to appear in the printed output. For example, to print a four-digit number and insert a comma after the first digit, we can use PICTURE 9,999. To obtain zero suppression of the first three digits in case the number is small and also to allow for the insertion of a comma in case the number has four digits, we can use PICTURE Z,ZZ9. Table 3.3 shows how several numbers would print under the edit PICTURE Z,ZZ9.

Value of data	Z,ZZ9
8000	8,000
3657	3,657
0657	657
0022	22
0000	0

Table 3.3 **How several four-digit numbers would print with an output edit PICTURE Z,ZZ9.**

Decimal point insertion may be obtained in a similar way: The programmer writes the PICTURE with a decimal point showing where it should appear in the printed output. To print a five-digit number as a dollars-and-cents figure, we could use PICTURE 999.99; to obtain zero suppression up to the decimal point, we would use PICTURE ZZZ.99. Zero suppression can be stopped before it reaches the decimal point with the use of PICTURE ZZ9.99 or Z99.99. Table 3.4 shows how different money amounts would print under four different PICTUREs.

Amount	999.99	ZZZ.99	ZZ9.99	Z99.99
657‸00	657.00	657.00	657.00	657.00
075‸25	075.25	75.25	75.25	75.25
006‸50	006.50	6.50	6.50	06.50
000‸29	000.29	.29	0.29	00.29
000‸00	000.00	.00	0.00	00.00

Table 3.4 How different money amounts would print under four different output edit PICTUREs.

Zero suppression, commas, and decimal points may be used in any desired combination in a PICTURE except that there may be not more than one decimal point in a PICTURE, it may not be at the right end, and, if Zs appear to the right of a decimal point, they must occupy all the positions to the right of the point. In such a case, if the data is zero, then the printed field will contain all blanks. If the data is not zero, then a PICTURE with Zs to the right of the point is the same as if the Zs stopped immediately to the left of the point.

There are many more editing features for both numeric and nonnumeric fields that will be discussed as we need them. Still, we will not cover all of the many detailed editing rules. The reader should refer to any COBOL programming manual for complete editing specifications.

Exercise 5. Figure 3.E5 shows column headings. Write the working storage entries that describe them.

Exercise 6. Figure 3.E6 shows column headings that occupy two print lines. Write two separate Working Storage entries, each beginning at the 01 level, that describe the two-line headings. Call the first 01-level COLUMN–HEADS–1 and the second COLUMN–HEADS–2.

Exercise 7. Write edit PICTUREs for each of the following output fields:

a. a six-digit integer, comma after the third digit, no zero suppression;
b. a seven-digit, dollars-and-cents money amount, comma after the second digit, zero suppression up to the decimal point;
c. a seven-digit integer, comma after the first and fourth digits, entire field zero-suppressed (blank when zero);
d. a nine-digit, dollars-and-cents amount, comma after the first and fourth digits, zero suppression up to one place before the decimal point;
e. a six-digit number with three decimal places, zero suppression up to one place before the decimal point.

150/10/6 PRINT CHART PROG. ID ________ PAGE ________
(SPACING: 150 POSITION SPAN, AT 10 CHARACTERS PER INCH, 6 LINES PER VERTICAL INCH) DATE ________
Fold back at dotted line.
PROGRAM TITLE ________
PROGRAMMER OR DOCUMENTALIST: ________
CHART TITLE ________
CARRIAGE CONTROL TAPE CHAN.
1 2 3 4 5 6 7 8 9 10 11 12

1 1 1 1 1 1 1 1 1 1 2 2 2 2 2 2 2 2 2 2 3 3 3 3 3 3 3 3 3 3 4 4 4 4 4 4 4 4 4 4 5 5 5 5 5 5 5 5 5 5 6 6 6 6 6 6 6 6 6 6 7 7 7 7 7 7 7 7 7 7 8 8 8 8 8 8 8 8 8
1 2 3 4 5 6 7 8 9 0 1 2 3 4 5 6 7 8 9 0 1 2 3 4 5 6 7 8 9 0 1 2 3 4 5 6 7 8 9 0 1 2 3 4 5 6 7 8 9 0 1 2 3 4 5 6 7 8 9 0 1 2 3 4 5 6 7 8 9 0 1 2 3 4 5 6 7 8 9 0 1 2 3 4 5 6 7 8 9

1
2
3 STUDENT NAME YEAR MAJOR
4
5
6
7
8
9
10
11
12
13
14
15
16
17
18
19
20

1 2 3 4 5 6 7 8 9 10 11 12 13 14 15 16 17 18 19

Figure 3.E5 Printer spacing chart for Exercise 5.

150/10/6 PRINT CHART PROG. ID ________ PAGE ________
(SPACING: 150 POSITION SPAN, AT 10 CHARACTERS PER INCH, 6 LINES PER VERTICAL INCH) DATE ________
Fold back at dotted line.
PROGRAM TITLE ________
PROGRAMMER OR DOCUMENTALIST: ________
CHART TITLE ________
CARRIAGE CONTROL TAPE CHAN.
1 2 3 4 5 6 7 8 9 10 11 12

1 1 1 1 1 1 1 1 1 1 2 2 2 2 2 2 2 2 2 2 3 3 3 3 3 3 3 3 3 3 4 4 4 4 4 4 4 4 4 4 5 5 5 5 5 5 5 5 5 5 6 6 6 6 6 6 6 6 6 6 7 7 7 7 7 7 7 7 7 7 8 8 8 8 8 8 8 8 8
1 2 3 4 5 6 7 8 9 0 1 2 3 4 5 6 7 8 9 0 1 2 3 4 5 6 7 8 9 0 1 2 3 4 5 6 7 8 9 0 1 2 3 4 5 6 7 8 9 0 1 2 3 4 5 6 7 8 9 0 1 2 3 4 5 6 7 8 9 0 1 2 3 4 5 6 7 8 9 0 1 2 3 4 5 6 7 8 9

1
2
3 STUDENT YEAR OF MAJOR
4 NAME GRADUATION CODE
5
6
7
8
9
10
11
12
13
14
15
16
17
18
19
20

1 2 3 4 5 6 7 8 9 10 11 12 13 14 15 16 17 18 19

Figure 3.E6 Printer spacing chart for Exercise 6.

Processing in both the Yes and No Paths

The next program, Program 8, shows how to code in COBOL a decision that requires one kind of processing if the outcome is Yes and another if it is No. The format of the input data for this program is shown in Figure 3.15. Each card is a record for one account, and the account number is shown in columns 2–6 of the card. The amount shown in the card can be either a debit or a credit, and the indicator column, column 1, tells which it is. If there is a – (dash) in column 1, it means that the amount is a debit; a blank in column 1 means the amount is

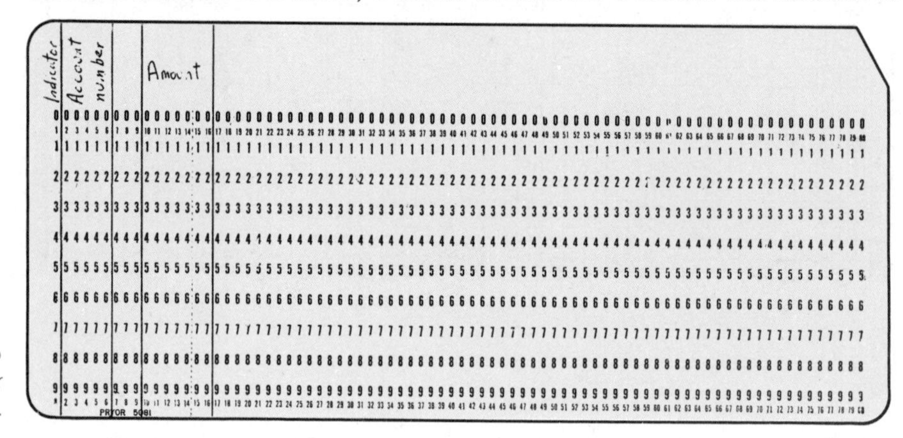

Figure 3.15 Data card format for Program 8.

150/10/6 PRINT CHART PROG. ID. PAGE

(SPACING: 150 POSITION SPAN, AT 10 CHARACTERS PER INCH, 6 LINES PER VERTICAL INCH) DATE

Fold back at dotted line.

PROGRAM TITLE PROGRAM 8

PROGRAMMER OR DOCUMENTALIST:

CHART TITLE

CARRIAGE CONTROL TAPE CHAN.

Line	Content
2	ACCT. NO. DEBITS CREDITS
4	XXXXX ZZ,ZZZ.99
5	XXXXX ZZ,ZZZ.99
6	XXXXX ZZ,ZZZ.99
7	XXXXX ZZ,ZZZ.99

Figure 3.16 Output format for Program 8.

a credit. Program 8 READs each card and prints the contents of each on one line, according to the format shown in Figure 3.16. All debit amounts read in will print in the column headed "Debits," and all credits will print in the "Credits" column. A flowchart of the MAIN–PROCESS to carry out the required processing is shown in Figure 3.17.

To define the output line for this program, we will need two places for the amount—one in the debits' column and one in the credits'. The

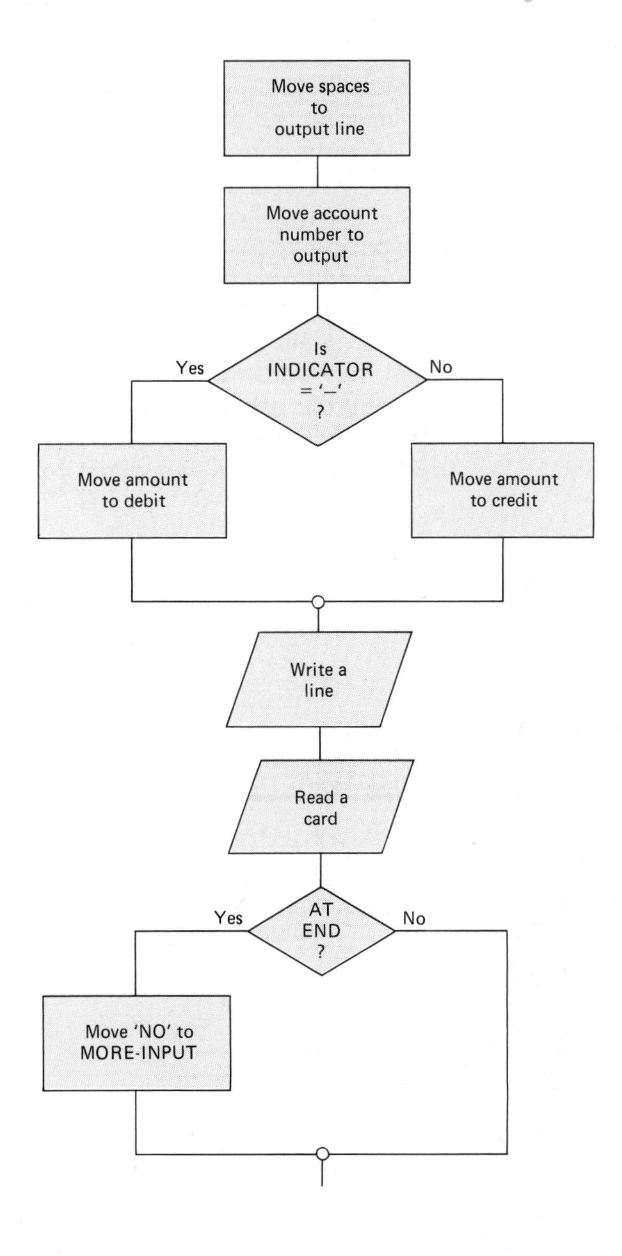

Figure 3.17
Flowchart for Program 8.

Procedure Division coding will fill the appropriate column for each card read.

The money amount fields are shown on the printer spacing chart with zero suppression and a decimal point. The input data of course contains no decimal point in the amount field, so the PICTURE for the output field must show where the decimal point is to be inserted. But COBOL has to have some way of knowing where the decimal point is to be assumed in the input field. The decimal point is not punched in the input data, and without some indication of where the point is supposed to be, COBOL can't tell whether the seven-digit amount field is dollars and cents or four integer places and three decimal places or some other combination of integer and decimal places adding up to seven. The PICTURE character V is used to tell COBOL where the **assumed decimal point** is located. V is used whenever the data represents a decimal amount, and there is no actual decimal point in the data; the PICTURE character . is used only for output editing when the printed result is to contain a printed decimal point. Figure 3.18 shows Program 8.

In Program 8 there is nothing new in the Identification or Environment Divisions. In the Data Division, the input record has a field showing the use of the PICTURE character V. The PICTURE for the amount field shows that five digits precede the assumed location of the decimal point and two digits follow. The PICTURE could also, and

Figure 3.18 Program 8, Part 1 of 2.

```
4-CB2 V4 RELEASE 1.5 10NOV77          IBM OS AMERICAN NATIONAL STANDARD COBOL



  000010 IDENTIFICATION DIVISION.
  000020 PROGRAM-ID. PROG08.
  000030*AUTHOR. PHIL RUBENSTEIN.
  000040*
  000050*    THIS PROGRAM READS DATA CARDS IN THE FOLLOWING FORMAT:
  000060*
  000070*    COL.  1            INDICATOR
  000080*    COLS. 2-6          ACCOUNT NO.
  000090*    COLS. 10-16        AMOUNT
  000100*
  000110*    FOR EACH CARD, ONE LINE IS PRINTED IN THE FOLLOWING FORMAT:
  000120*
  000130*    PP 13-17           ACCOUNT NO.
  000140*    PP 24-32           DEBITS
  000150*    PP 35-43           CREDITS
  000160*
  000170*    PRINT COLUMN HEADINGS AS SHOWN ON
  000180*    THE PRINTER SPACING CHART.
  000190*
  000200*****************************************************************
```

Figure 3.18 Program 8, Part 1 (continued).

```
000210
000220 ENVIRONMENT DIVISION.
000230 CONFIGURATION SECTION.
000240 SPECIAL-NAMES.
000250     C01 IS TO-NEW-PAGE.
000260 INPUT-OUTPUT SECTION.
000270 FILE-CONTROL.
000280     SELECT ACCOUNT-LIST     ASSIGN TO UT-S-SYSPRINT.
000290     SELECT ACCOUNT-FILE-IN ASSIGN TO UT-S-SYSIN.
000300
000310*****************************************************************
000320
000330 DATA DIVISION.
000340 FILE SECTION.
000350 FD  ACCOUNT-FILE-IN
000360     LABEL RECORDS ARE OMITTED.
000370
000380 01  ACCOUNT-RECORD-IN.
000390     05 INDICATOR              PIC X.
000400     05 ACCOUNT-NUMBER-IN      PIC X(5).
000410     05 FILLER                 PIC X(3).
000420     05 AMOUNT                 PIC 99999V99.
000430     05 FILLER                 PIC X(64).
000440
000450 FD  ACCOUNT-LIST
000460     LABEL RECORDS ARE OMITTED.
000470
000480 01  ACCOUNT-LINE.
000490     05 FILLER                 PIC X.
000500     05 FILLER                 PIC X(12).
000510     05 ACCOUNT-NUMBER-OUT     PIC X(5).
000520     05 FILLER                 PIC X(6).
000530     05 DEBITS                 PIC ZZ,ZZZ.99.
000540     05 FILLER                 PIC X(2).
000550     05 CREDITS                PIC ZZ,ZZZ.99.
000560
000570 WORKING-STORAGE SECTION.
000580 01  MORE-INPUT                PIC X(3) VALUE 'YES'.
000590
000600 01  COLUMN-HEADS.
000610     05 FILLER                 PIC X.
000620     05 FILLER                 PIC X(10) VALUE SPACES.
000630     05 FILLER                 PIC X(9)  VALUE 'ACCT. NO.'.
000640     05 FILLER                 PIC X(6)  VALUE SPACES.
000650     05 FILLER                 PIC X(6)  VALUE 'DEBITS'.
000660     05 FILLER                 PIC X(4)  VALUE SPACES.
000670     05 FILLER                 PIC X(7)  VALUE 'CREDITS'.
000680
000690*****************************************************************
```

more probably, have been written 9(5)V99. The PICTUREs for the output fields do not use the V for the decimal point but a dot, because we want an actual decimal point to print on the output. Remember: The dot is used as a decimal point only for printing final output on paper; V is used for everything else.

Figure 3.18 Program 8, Part 2 of 2.

```
000700
000710 PROCEDURE DIVISION.
000720      OPEN INPUT  ACCOUNT-FILE-IN,
000730           OUTPUT ACCOUNT-LIST.
000740      WRITE ACCOUNT-LINE FROM COLUMN-HEADS
000750                                    AFTER ADVANCING TO-NEW-PAGE.
000760      MOVE SPACES TO ACCOUNT-LINE.
000770      WRITE ACCOUNT-LINE AFTER ADVANCING 1.
000780      READ ACCOUNT-FILE-IN
000790          AT END
000800              MOVE 'NO' TO MORE-INPUT.
000810      PERFORM MAIN-PROCESS UNTIL MORE-INPUT IS EQUAL TO 'NO'.
000820      CLOSE ACCOUNT-FILE-IN,
000830            ACCOUNT-LIST.
000840      STOP RUN.
000850
000860 MAIN-PROCESS.
000870      MOVE SPACES TO ACCOUNT-LINE.
000880      MOVE ACCOUNT-NUMBER-IN TO ACCOUNT-NUMBER-OUT.
000890      IF INDICATOR IS EQUAL TO '-'
000900          MOVE AMOUNT TO DEBITS
000910      ELSE
000920          MOVE AMOUNT TO CREDITS.
000930      WRITE ACCOUNT-LINE AFTER ADVANCING 1.
000940      READ ACCOUNT-FILE-IN
000950          AT END
000960              MOVE 'NO' TO MORE-INPUT.
```

The MAIN–PROCESS in this program shows our first use of an IF statement with processing in both paths. The word **ELSE** separates the Yes path from the No, and only one period at the end of the whole thing terminates the IF and joins the Yes and No paths. In this program there is only one processing step in the Yes path and one in the No path, but there can be as many steps as required in either or both paths. In each case the several steps would be written, aligned as in Program 7, before or after the ELSE as appropriate, with only one final period.

Sample input data for Program 8 is shown in Figure 3.19 and the output is shown in Figure 3.20.

```
 20202   6085720
-45672   0030000
-01590   0400002
 35470   0000579
-98735   0000062
 67024   0000005
-39674   6900204
 76839   0682100
 57573   0011101
-32323   0001000
```

Figure 3.19 Input data for Program 8.

ACCT. NO.	DEBITS	CREDITS
20202		60,857.20
45672	300.00	
01590	4,000.02	
35470		5.79
98735	.62	
67024		.05
39674	69,002.04	
76839		6,821.00
57573		111.01
32323	10.00	

Figure 3.20
Output from Program 8.

Exercise 8. Write a program to the following specifications:

Input
Use the input data shown in Figure 3.6 whose format is shown in Figure 3.3.

Output
Use the output format shown in Figure 3.E8.

Processing
Print one line of output for each input card. If the quantity on hand is more than 300, move zero into the column headed Quantity on Hand, so that only blanks will appear. If the quantity is 300 or less, print the actual quantity in the column headed Quantity on Hand, and print the word REORDER in the print positions shown in the format.

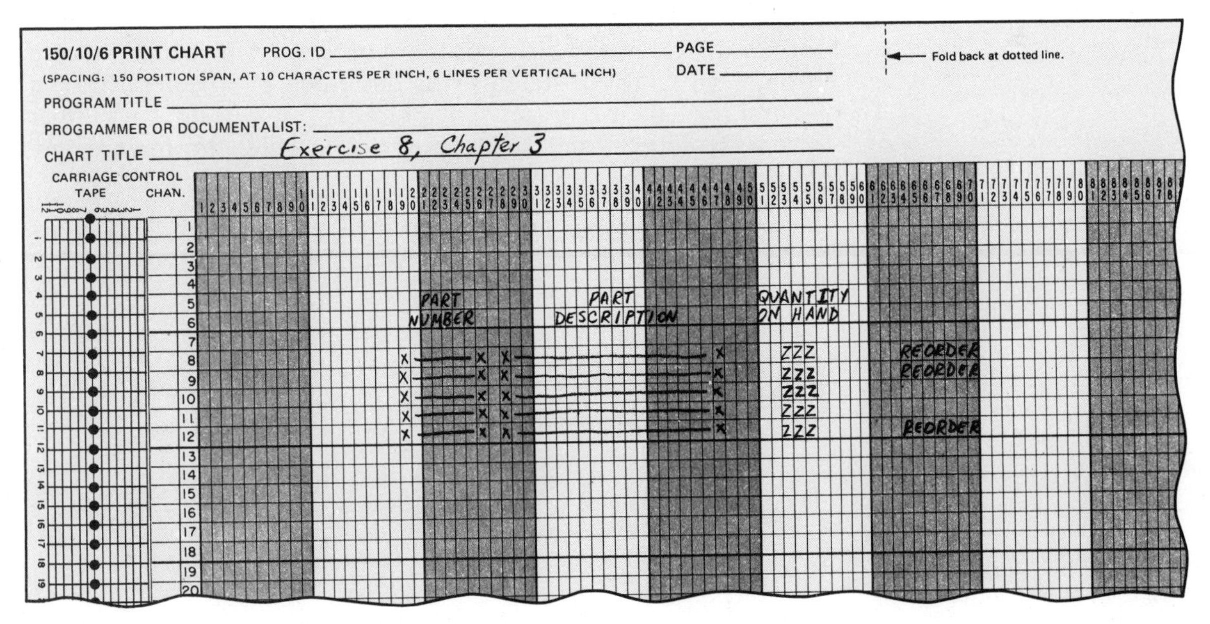

Figure 3.E8 Output format for Exercise 8.

Processing in the No Path Only

A little used form of the IF statement is one where there is no processing in the Yes path but only one or more steps in the No path. An example of such a situation is shown in the flowchart in Figure 3.21. The Procedure Division coding for that logic uses the reserved words **NEXT SENTENCE,** as follows:

```
IF HOURS-WORKED-THIS-WEEK IS EQUAL TO HOURS-WORKED-LAST-WEEK
    NEXT SENTENCE
ELSE
    PERFORM VARYING-HOURS-PROCESS.
```

In COBOL, a **sentence** in the Procedure Division is something that ends with a period, so the clause NEXT SENTENCE sends the processing beyond the next period, which is also the end of the IF statement. The desired processing is thus obtained.

General Formats of COBOL Statements

In this chapter we have seen several forms of the IF statement; we have also seen more than one form of other statements and entries. In order for the programmer to be able to keep track of the different legal forms and to know which words are optional, which are reserved, and so on, each COBOL statement is described by a general format. The general formats use braces { }, brackets [ ], ellipses . . . , upper- and lower-case letters, and underlining to describe the features and

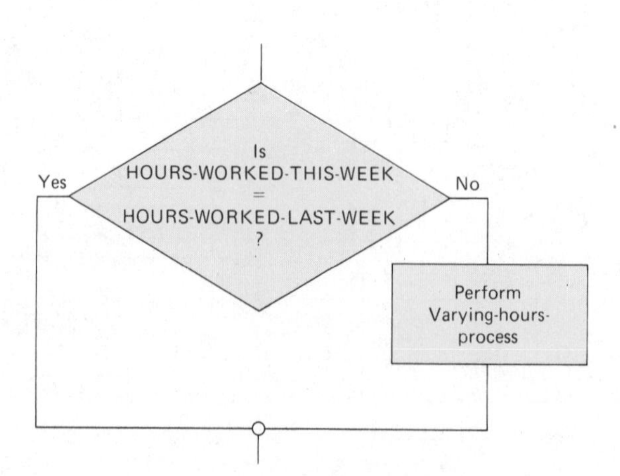

Figure 3.21 Flowchart of logic requiring processing in the No path only.

options of each entry or statement. For example, the format of the IF statement is:

```
IF condition { statement-1     } { ELSE statement-2     }
             { NEXT SENTENCE   } { ELSE NEXT SENTENCE   }
```

Words in capital letters are reserved words; items in lower-case letters are programmer-supplied. Underlined reserved words are required and must be written exactly as shown. The format shows, among other things, that an IF statement must begin with the reserved word IF, spelled as shown, followed by a condition supplied by the programmer. Some of the conditions we have used so far include:

```
QUANTITY-IN IS LESS THAN 400
INDICATOR IS EQUAL TO '-'
HOURS-WORKED-THIS-WEEK IS EQUAL TO HOURS-WORKED-LAST-WEEK
```

Following the condition we see braces, which mean that one of the items within must be chosen to follow the condition. We have used both forms, for in most of our IF statements we have had some processing statement or statements following the condition, and in one of the IF statements in this chapter we had the reserved words NEXT SENTENCE following the condition. The format shows that if NEXT SENTENCE is chosen to be included in the IF statement, it must be spelled as shown. The next set of braces indicates that an ELSE clause is always required. The braces show that the word ELSE must always appear, followed by either a statement or the words NEXT SENTENCE; however, the rules accompanying the format in COBOL manuals, not shown here, say that the words ELSE NEXT SENTENCE may be omitted if they immediately precede the period in the IF statement. That means that the statement

```
IF QUANTITY-IN IS LESS THAN 400
    MOVE 'REORDER' TO MESSAGE-SPACE.
```

is the same as

```
IF QUANTITY-IN IS LESS THAN 400
    MOVE 'REORDER' TO MESSAGE-SPACE
ELSE
    NEXT SENTENCE.
```

The general format says nothing about indenting in the IF statement because COBOL does not examine the indenting. Indenting is only for the programmer's convenience and should be used to improve the readability of programs.

The general format for relational operators is:

```
IS [NOT] GREATER THAN
IS [NOT] LESS THAN
IS [NOT] EQUAL TO
IS [NOT] >
IS [NOT] <
IS [NOT] =
```

Here we have reserved words (in capitals) which are not underlined, which means that they are optional words and may be omitted. If used, however, they must be spelled correctly as shown. So in

```
IF QUANTITY-IN IS LESS THAN 400 . . .
```

both IS and THAN are optional words, and the statement could have begun with:

```
IF QUANTITY-IN LESS THAN 400 . . .
```

or

```
IF QUANTITY-IN IS LESS 400 . . .
```

or

```
IF QUANTITY-IN LESS 400 . . .
```

but the optional words were used to improve the readability. In the EQUAL relation some old COBOL systems still require the word TO, though most modern systems treat it as an optional word as shown in the format. For safety, we will use the word TO whenever we use EQUAL.

Contents of brackets are optional, so the brackets around the words NOT show that the programmer may choose to use NOT or to omit it. The underlining shows that if the programmer chooses to use the contents of the brackets, the word NOT must be used and spelled exactly as shown. The signs >, <, and = may be used instead of the corresponding words. The signs, if used, are required, and they are not underlined because underlining might make them confusing. We will see more general formats in following chapters.

Exercise 9. Here is one general format of the LABEL RECORDS clause:

```
LABEL  { RECORD IS    }  { STANDARD }
       { RECORDS ARE  }  { OMITTED  }
```

Tell whether each of the following is or is not a legal form of the clause:

a. LABEL RECORDS OMITTED
b. LABELS ARE OMITTED
c. LABEL ARE OMITTED
d. LABEL RECORD IS STANDARD
e. RECORDS ARE STANDARD
f. LABEL RECORDS IS OMITTED

Summary

The printer spacing chart can be used to plan printer output formats. The IF statement enables a COBOL program to make decisions based on data. The decision determines which of two paths will be executed. Either or both of the paths may contain as many processing steps as needed. When the Yes path contains no processing, the reserved words NEXT SENTENCE must be used, but when the No path contains no processing they may be omitted. If the No path contains processing, the reserved word ELSE must be used to show where the No path begins. A single period terminates the IF statement and shows where the Yes and No paths rejoin. Indenting the coding of the Yes and No paths is used to improve program readability and is not examined by the COBOL system.

Numeric input data and numeric literals are two ways of introducing numbers into COBOL programs. The rules for data and for literals are different. The most frequently encountered difference is that in input data a decimal point is never punched, while in numeric literals a decimal point may appear anywhere except at the right end. The PICTURE character V tells the COBOL system the location of the assumed decimal point in numeric input data.

Edit PICTUREs may be used to prepare numbers for printing. Commas and a decimal point may be inserted and leading zeros suppressed. The PICTURE character for inserting an actual printing decimal point is the dot.

Every COBOL statement has a general format which tells the programmer what options are available in the statement. A word in capitals is a reserved word. If underlined, it is required in the statement; if not underlined, it is an optional word. If used, it must be spelled correctly. Items in braces require the programmer to choose one, and items in brackets are optional.

Column headings may be defined as constants in the Working Storage Section. They may be written to the output file using a record name in the File Section and the WRITE statement with the FROM option.

Fill-in Exercises

1. The two classes of data on which COBOL operates are ________ and ________.
2. A numeric literal may be up to ________ digits long.
3. The PICTURE character for numeric data is ________.
4. The six relational operators are ________, ________, ________, ________, ________, and ________.
5. A ________ or a ________ may be used to show decision logic in diagram form.
6. The PICTURE character for zero suppression is ________.
7. The ________ clause must be used whenever an IF statement contains processing in the No path but no processing in the Yes path.
8. The phrase ________ may be omitted from an IF statement if it appears immediately before the period.
9. The PICTURE character ________ is used to obtain a printed decimal point in output; the PICTURE character ________ is used for all other representations of the decimal point.
10. A capitalized underlined word in a general format of a COBOL statement is a ________ word and is ________; a capitalized word not underlined is an ________ word.

Review Exercises

1. Using a printer spacing chart, show the following output specification:

Pr. Pos.	Field
1–15	State
21–40	County
41–49	Beginning cubic feet (to one decimal place, comma after the third integer digit)
52–59	Growth in cubic feet (to one decimal place, comma after the second integer digit)
62–70	Final cubic feet (to one decimal place, comma after the third integer digit)

Show zero suppression in all numeric output fields up to one place before the decimal point. Show suitable column headings on the printer spacing chart.

2. Which of the following is (are) valid as numeric input data, which as numeric literals, and which as neither?

   a. 675.0   b. +675.0   c. 675.   d. 675.0−

   e. 6750   f. $675\overset{+}{0}$ g. 6,750

3. Using the printer spacing chart shown in Figure 3.RE3, write the working storage entries to describe the column headings.

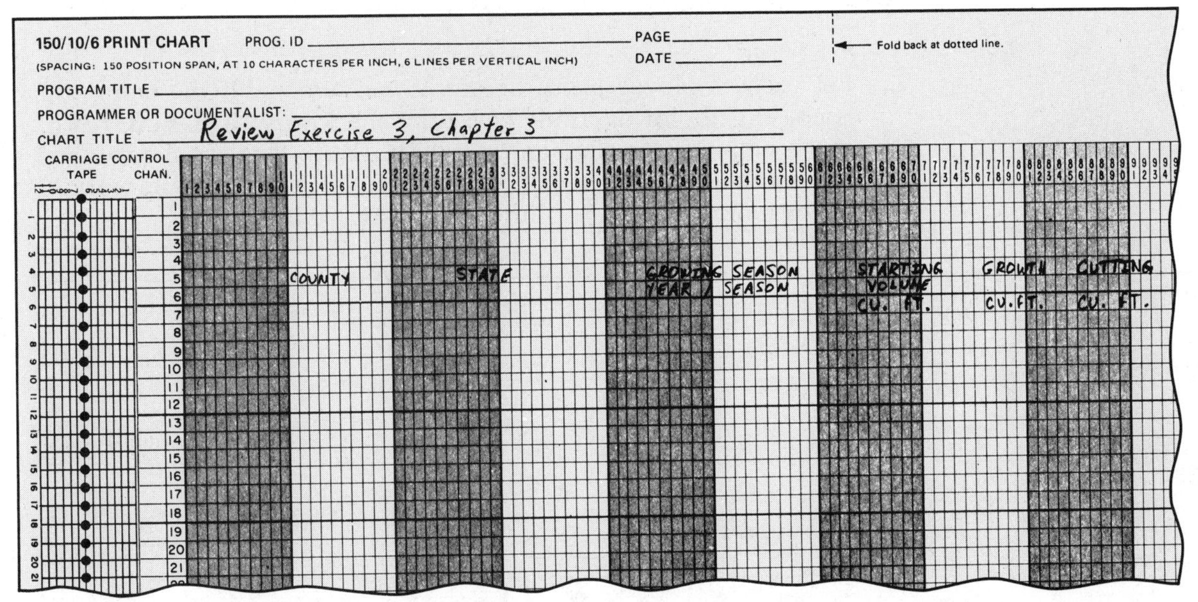

Figure 3.RE3 Output format for Review Exercise 3.

4. Show how the following numbers would print if edited by each of the PICTUREs 9,999.9; Z,ZZZ.9; Z,ZZZ.Z:

   a. 4756‸8   b. 0350‸9   c. 0020‸0

   d. 0005‸0   e. 0000‸0

5. One of the formats of the PERFORM statement is

   $$\underline{\text{PERFORM}}\ \text{procedure-name-1}\left[\left\{\begin{array}{l}\underline{\text{THROUGH}}\\ \underline{\text{THRU}}\end{array}\right\}\text{procedure-name-2}\right]$$

   $\underline{\text{UNTIL}}$ condition-1

   Which part of the format indicates that the use of the word THROUGH is optional?

*6. Write a program to the following specifications:

Input

A deck of data cards in the following format:

Col.	Field
1–5	Employee number
6	Overtime eligibility indicator
	E—not eligible for overtime pay
	N—eligible for overtime pay
7–9	Hours worked, to one decimal place

Output

Use the output format shown in Figure 3.RE6.

Processing

For each employee print the employee number. If the employee is not eligible for overtime pay, print the words NOT ELIGIBLE. If the employee is eligible, print ELIGIBLE. If the hours worked is more than 37.5, print OVERTIME HOURS. If the hours worked is not more than 37.5, print NO OVERTIME HOURS.

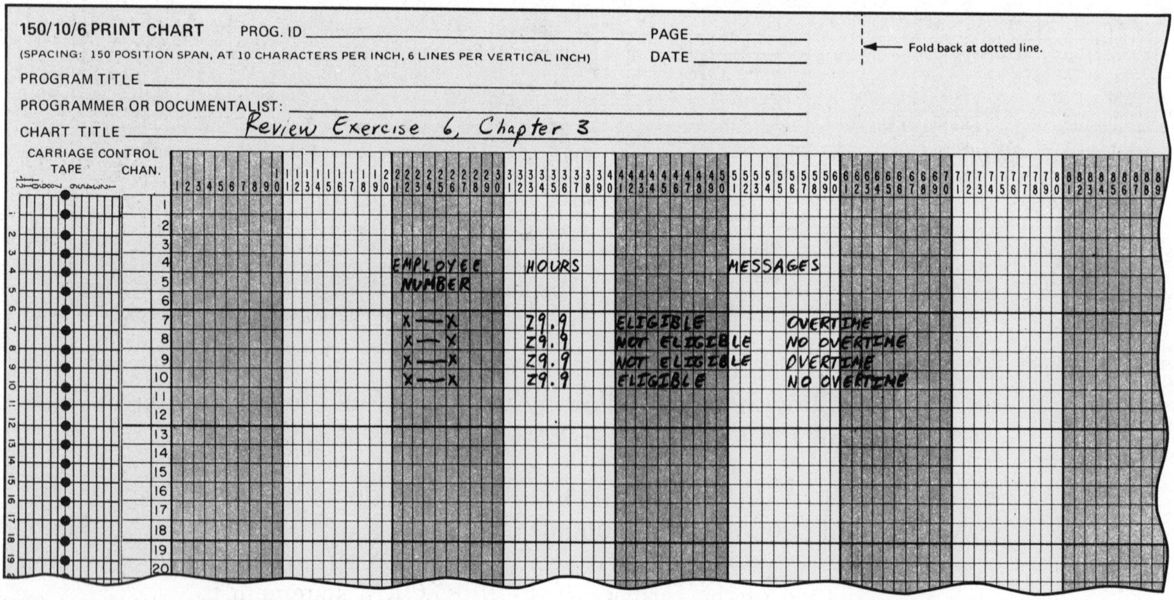

150/10/6 PRINT CHART PROG. ID PAGE

(SPACING: 150 POSITION SPAN, AT 10 CHARACTERS PER INCH, 6 LINES PER VERTICAL INCH) DATE

Fold back at dotted line.

PROGRAM TITLE

PROGRAMMER OR DOCUMENTALIST:

CHART TITLE Review Exercise 6, Chapter 3

CARRIAGE CONTROL TAPE CHAN.

Line	Employee number	Hours	Messages	
4	EMPLOYEE	HOURS	MESSAGES	
5	NUMBER			
7	X—X	Z9.9	ELIGIBLE	OVERTIME
8	X—X	Z9.9	NOT ELIGIBLE	NO OVERTIME
9	X—X	Z9.9	NOT ELIGIBLE	OVERTIME
10	X—X	Z9.9	ELIGIBLE	NO OVERTIME

Figure 3.RE6 Output format for Review Exercise 6.

Arithmetic Verbs

4

KEY POINTS

Here are the key points you should learn from this chapter:

1. the formats of the verbs ADD, SUBTRACT, MULTIPLY, and DIVIDE;
2. top-down design, hierarchy diagrams, and pseudocode;
3. how to use the arithmetic verbs in some sample programs.

KEY WORDS

Key words to recognize and learn:

ADD	identifier	top-down design
SUBTRACT	ROUNDED	hierarchy diagram
MULTIPLY	BY	accumulate
DIVIDE	REMAINDER	pseudocode
GIVING	TO	

In this chapter we will study the arithmetic operations **ADD, SUBTRACT, MULTIPLY,** and **DIVIDE** and some sample programs using these operations. Of the four, ADD is the most difficult and will be left for last. In this chapter we will also see additional tools for attacking a problem in preparation for coding.

SUBTRACT

Some legal subtractions in COBOL are:

```
SUBTRACT 37.5 FROM HOURS-WORKED.
SUBTRACT PAYMENT FROM BALANCE-DUE.
SUBTRACT 40 FROM HOURS-WORKED GIVING OVERTIME-HOURS.
SUBTRACT PAYMENT FROM OLD-BALANCE GIVING NEW-BALANCE.
```

In the first SUBTRACT statement, the resulting difference is assigned to HOURS–WORKED and the original value of HOURS–WORKED is lost. In the second statement, the difference that results from the subtraction is assigned to BALANCE–DUE and the original value of BALANCE–DUE is lost. To preserve the original values in a subtraction, the **GIVING** option may be used, as in the last two SUBTRACT statements. In these, the original values assigned to HOURS–WORKED and OLD–BALANCE remain after the subtraction as they were before the subtraction, and the differences are assigned to OVERTIME–HOURS and NEW–BALANCE, respectively. The fields in SUBTRACT statements need not all have the same definitions in the Data Division. In SUBTRACT statements without the GIVING option, all the fields must be defined as purely numeric (without any actual decimal points, commas, or zero suppression), but in SUBTRACTs with the GIVING option, fields after the word GIVING may be defined as numeric or edited numeric (with actual decimal points, commas, and zero suppression). If necessary, COBOL automatically aligns decimal points in all computations. One of the general formats of the SUBTRACT statement is:

```
SUBTRACT {identifier-1} [identifier-2] ...
         {literal-1   } [literal-2   ]
    FROM identifier-m [ROUNDED]
    [identifier-n [ROUNDED]] . . .
    [ON SIZE ERROR imperative-statement]
```

This is the form of the SUBTRACT without the GIVING option. There is an entirely separate format, which we will get to shortly, for SUBTRACT with the GIVING option. In the format just shown, we see braces after the word SUBTRACT, showing that either a literal or an **identifier** must follow the word SUBTRACT. So far in this book, the only kinds of identifiers we know are data names; that is, the names of fields defined in the Data Division. The numbers attached to the words "identifier" and "literal," as in identifier–1 and literal–1, are used only to distinguish them from other identifers and/or literals in the format.

The brackets around identifier–2 and literal–2 in the format show that a second identifier or literal may optionally appear before the word FROM. We did not have a second identifier or literal in either of the two sample SUBTRACT statements for this format. The ellipsis after the brackets indicates that the contents of the brackets may be repeated as many times as desired by the programmer (within the limits of the size of the COBOL system being used); that is, there could optionally be a third, fourth, fifth, and so on identifier and/or literal if desired.

Following the required word FROM, there is a required identifier. It is here called identifier–m because any number of identifiers might have appeared before the FROM. Following identifier–m is the optional word **ROUNDED,** which we did not use in either of our sample statements. The format then shows additional, optional components, none of which was used in the samples. The format tells us that the following statement is legal:

```
SUBTRACT
   JUNE–PAYMENT,
   JULY–PAYMENT
      FROM BALANCE–DUE.
```

In that statement the values of JUNE–PAYMENT and JULY–PAYMENT would first be added, and then the sum would be subtracted from BALANCE–DUE. The following statement is also legal:

```
SUBTRACT
   BASE–HOURS,
   2.5
      FROM HOURS–WORKED ROUNDED.
```

Here the value of BASE–HOURS would first be added to 2.5, and then the sum subtracted from HOURS–WORKED. If necessary, the difference would be ROUNDED to fit into HOURS–WORKED. If the ROUNDED option is not specified, then truncation of low-order digits occurs if necessary for the result to fit into the result field. The format shows that several identifiers may optionally appear after the word FROM, so the following statement is legal:

```
SUBTRACT 5.00
   FROM
      TEACHER–PAY–RATE,
      JANITOR–PAY–RATE,
      SWEEPER–PAY–RATE.
```

In that statement, 5.00 would be subtracted from the value in each of the result fields. The SIZE ERROR option will be discussed in Chapter 8.

The format of the SUBTRACT with the GIVING option is:

```
SUBTRACT { identifier–1 }  [ identifier–2 ]  . . .
         { literal–1    }  [ literal–2    ]
```

```
FROM {identifier-m } GIVING identifier-n [ROUNDED]
     {literal-m    }
[identifier-o [ROUNDED]] . . .
[ON SIZE ERROR imperative-statement]
```

The format shows that any number of identifiers are permitted after the word GIVING. If more than one identifier appears after GIVING, the difference from the subtraction is assigned to each of the identifiers. Some COBOL systems restrict the programmer to only one identifier after GIVING. The two sample SUBTRACT statements to which this format applies each had only one such identifier.

Exercise 1. Write a COBOL statement which will subtract DEDUCTIONS from PAY and assign the difference to PAY.

Exercise 2. Write a COBOL statement which will subtract DEDUCTIONS from GROSS–PAY and assign the difference to NET–PAY.

Exercise 3. Write a COBOL statement which will subtract CREDITS–EARNED from 66 and assign the difference to CREDITS–REMAINING.

MULTIPLY

Multiplication, like subtraction, can be done with or without a GIVING option. The format for MULTIPLY without the GIVING option is:

```
MULTIPLY {identifier-1} BY identifier-2 [ROUNDED]
         {literal-1   }
  [identifier-3 [ROUNDED]] . . .
  [ON SIZE ERROR imperative-statement]
```

Some COBOL systems permit only one identifier after the word **BY.** Valid MULTIPLYs under this format are:

```
MULTIPLY RATE-OF-PAY BY HOURS-WORKED.
MULTIPLY 1.5 BY OVERTIME-HOURS.
```

In each case the product is assigned to the field named after the word BY. In systems which permit more than one identifier after BY, a valid MULTIPLY is:

```
MULTIPLY 2
  BY
    WHEEL-SPINS,
    PAYOFFS.
```

In this case the values assigned to both WHEEL–SPINS and PAYOFFS would be doubled.

Multiplication can also be written with the GIVING option, so that the product is assigned to a field specified by the programmer. The format of the MULTIPLY with the GIVING option is:

```
MULTIPLY {identifier-1} BY {identifier-2}
         {literal-1   }    {literal-2   }
    GIVING identifier-3 [ROUNDED][identifier-4 [ROUNDED]] . . .
    [ON SIZE ERROR imperative-statement]
```

Some COBOL systems permit only one identifier after the word GIVING. Valid MULTIPLYs using GIVING are:

```
MULTIPLY RATE-OF-PAY BY HOURS-WORKED GIVING GROSS-PAY.
MULTIPLY 1.5 BY OVERTIME-HOURS GIVING OVERTIME-EQUIVALENT.
```

All fields in MULTIPLY statements must be defined as numeric, except that fields after the word GIVING may be defined as numeric or edited numeric.

Exercise 4. Write a COBOL statement that will triple the value assigned to the field GROSS.

Exercise 5. Write a COBOL statement that will multiply together the values assigned to GRADE–POINTS and CREDITS and assign the product to HONOR–POINTS.

DIVIDE

The DIVIDE verb is a little strange, because with the GIVING option come other options. First, the DIVIDE without the GIVING:

```
DIVIDE {identifier-1} INTO identifier-2 [ROUNDED]
       {literal-1   }
    [identifier-3 [ROUNDED]] . . .
    [ON SIZE ERROR imperative-statement]
```

As usual, some COBOL systems permit only one identifier after the word INTO. Here is a sample DIVIDE:

```
DIVIDE NUMBER-OF-EXAMS INTO TOTAL-GRADE.
```

With the GIVING option, the programmer also has a choice of writing the division in either direction (that is, divide *a* into *b*, or divide *b* by *a*) and also the option of having the remainder from the division assigned to a separate field. One widely used format of the DIVIDE with GIVING is shown here. For the format for your particular COBOL system see your COBOL manual:

```
DIVIDE { identifier-1 } { INTO } { identifier-2 }
       { literal-1    } { BY   } { literal-2    }
    GIVING identifier-3 [ROUNDED] [REMAINDER identifier-4]
    [ON SIZE ERROR imperative-statement]
```

Under this format, the following are legal:

```
DIVIDE NUMBER-OF-EXAMS INTO TOTAL-GRADE GIVING AVERAGE-GRADE ROUNDED.
DIVIDE TOTAL-GRADE BY NUMBER-OF-EXAMS GIVING AVERAGE-GRADE ROUNDED.
DIVIDE TOTAL-SCORE BY 3
   GIVING HANDICAP
   REMAINDER H-REM.
```

In the last division, the value assigned to HANDICAP would be the quotient of the division, truncated (not rounded), and the value of the REMAINDER would be assigned to the field H-REM. The REMAINDER from a division is computed by COBOL by multiplying the quotient by the divisor and subtracting that product from the dividend. The use of the REMAINDER option can become quite involved. For example, the remainder field can never have the ROUNDED option and so is truncated if necessary. If the ROUNDED option is used on the quotient field, then the quotient is first truncated for purposes of computing the remainder and is then rounded before being assigned to the quotient field. The quotient and/or remainder fields may have decimal places. Because of these complications some COBOL systems do not always compute the remainder correctly. If you plan to use the REMAINDER option run a small scale test on it first.

All fields in DIVIDE statements must be defined as numeric, except that fields after the words GIVING and REMAINDER may be numeric or edited numeric. In Chapter 8 we discuss methods for handling cases where the divisor may be 0.

ADD

Two formats for the ADD statement will be given. One of them contains the GIVING option. The other contains a required word **TO.** Whenever you write an ADD statement, it must contain either the word TO or the word GIVING; it must never contain both TO and GIVING. The first format is:

```
ADD {identifier-1} [identifier-2] . . . TO identifier-m [ROUNDED]
    {literal-1   } [literal-2   ]
   [identifier-n [ROUNDED]] . . .
   [ON SIZE ERROR imperative-statement]
```

In this format only one identifier is required before the word TO, although there may be as many as desired by the programmer. After the word TO, one identifier is also required, and there may be as many as desired. Sample ADDs are:

```
ADD NEW-PURCHASES TO BALANCE.
ADD
   SALESPERSON-NET,
   SALESPERSON-BONUS
      TO REGION-TOTAL.
ADD
   OVERTIME-HOURS,
   40
      TO
         DEPARTMENT-TOTAL,
         FACTORY-TOTAL.
```

In the first ADD, the value of NEW–PURCHASES is added to the existing value of BALANCE and the sum is assigned to BALANCE. In the second, the values of SALESPERSON–NET, SALESPERSON–BONUS, and REGION–TOTAL are added together and the sum assigned to REGION–TOTAL. In the third ADD, OVERTIME–HOURS is added to 40 and the sum is added to the DEPARTMENT–TOTAL and to the FACTORY–TOTAL.

The ADD with GIVING is:

```
ADD {identifier-1} {identifier-2} [identifier-3] . . .
    {literal-1   } {literal-2   } [literal-3   ]
   GIVING identifier-m [ROUNDED] [identifier-n [ROUNDED]] . . .
   [ON SIZE ERROR imperative-statement]
```

In this form, at least two terms are required before the word GIVING, although there may be as many as desired. The format shows that any number of identifiers are allowed after the GIVING, but some COBOL systems permit only one. Notice that there is no TO in this form of the ADD. Some ADDs in this format are:

```
ADD
   FIELD-A,
   FIELD-B
      GIVING FIELD-C.
ADD
   JUNE-HOURS,
   JULY-HOURS,
   AUGUST-HOURS
      GIVING SUMMER-HOURS.
ADD
   OVERTIME-HOURS,
   40
      GIVING TOTAL-HOURS.
```

In these ADDs, the values of the fields before the word GIVING are added together and the sum is assigned to the field after the word GIVING.

In ADD statements, all fields must be defined as numeric except that fields after the word GIVING may be numeric or edited numeric. The following ADD is invalid because it contains both the words TO and GIVING:

```
ADD FIELD-1 TO FIELD-2 GIVING FIELD-3.
```

Exercise 6. Write a COBOL statement which will add together the values assigned to FIELD–1 and FIELD–2 and assign the sum to FIELD–3.

Exercise 7. Write a COBOL statement which will add the value of CASH–ADVANCES to BALANCE–DUE and assign the sum to BALANCE–DUE.

A Program with Arithmetic

We will now study a program using addition and multiplication. The input format for Program 9 is shown in Figure 4.1. Each record represents a purchase of some parts by a customer. The record shows the quantity purchased, the price per unit, and a handling charge for the order. The program is to read each card and compute the total cost of the merchandise (by multiplying the quantity by the unit price) and

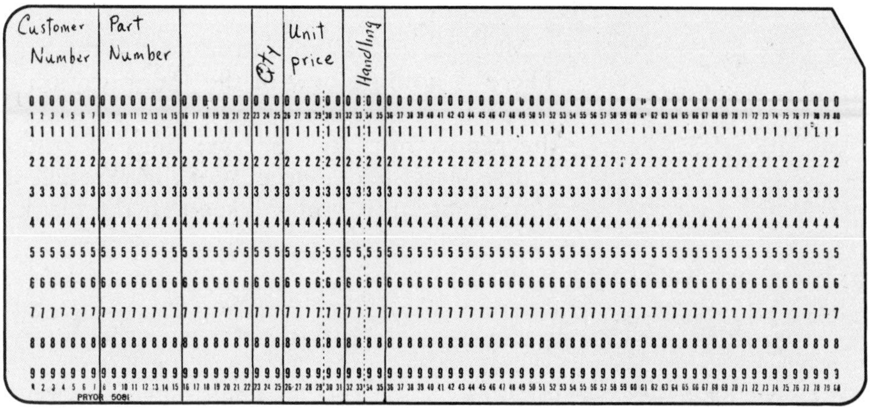

Customer Number | Part Number | Qty | Unit price | Handling

Figure 4.1 Data card format for Program 9.

a tax at 7 percent of the merchandise total. Then the program is to add together the merchandise total, the tax, and the handling charge to arrive at a total due for the order. The information for each order is to be printed on one line as shown in Figure 4.2. Also shown in Figure 4.2 are column headings and a title for the listing, "Daily Order Report."

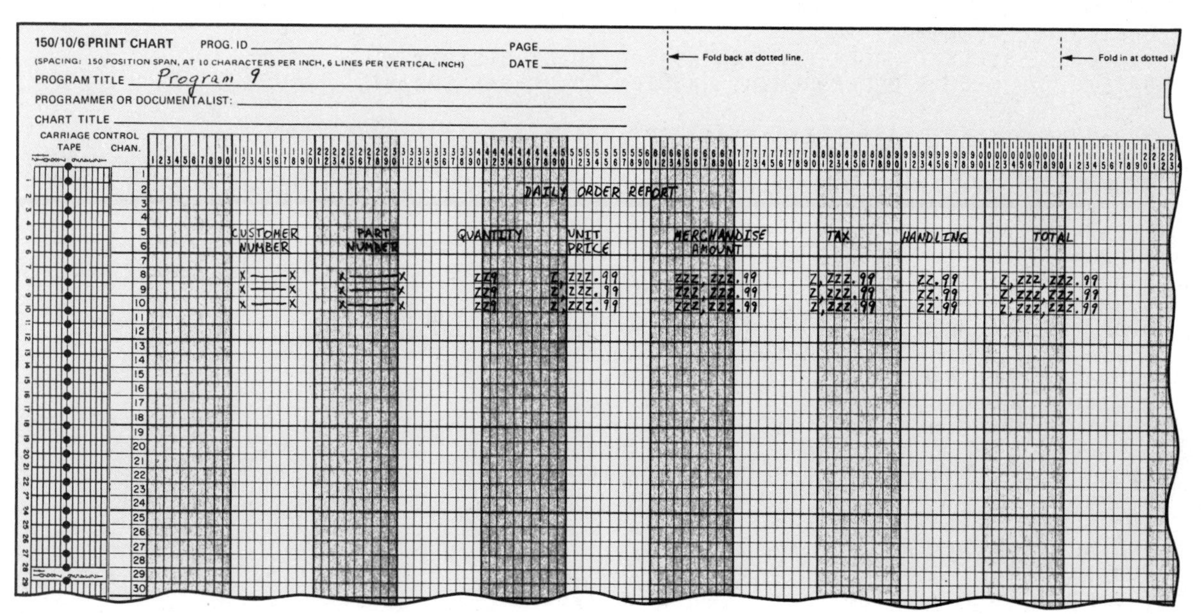

Figure 4.2 Output format for Program 9.

Many new ideas are introduced in the program shown in Figure 4.3. There is nothing new in the Identification or Environment Divisions. In the Data Division we now have several constants to define, namely, the report title and the two lines of column headings. They are of course described in the Working Storage Section. If you study the entries carefully, you will see that the blank spaces between the words

Figure 4.3 Program 9, Part 1 of 2.

```
4-CB2 V4 RELEASE 1.5 10NOV77          IBM OS AMERICAN NATIONAL STANDARD COBOL



000010 IDENTIFICATION DIVISION.
000020 PROGRAM-ID. PROG09.
000030*AUTHOR. PHIL RUBENSTEIN.
000040*
000060*       THIS PROGRAM READS CUSTOMER ORDERS PUNCHED IN
000070*       CARDS AND PRODUCES A DAILY ORDER REPORT.
000075*
000080*****************************************************
000090
000100 ENVIRONMENT DIVISION.
000110 CONFIGURATION SECTION.
000120 SPECIAL-NAMES.
000130     C01 IS TO-NEW-PAGE.
000140 INPUT-OUTPUT SECTION.
000150 FILE-CONTROL.
000160     SELECT ORDER-FILE-IN ASSIGN TO UT-S-SYSIN.
000170     SELECT ORDER-REPORT  ASSIGN TO UT-S-SYSPRINT.
000180
000190*****************************************************
000200
000210 DATA DIVISION.
000220 FILE SECTION.
000230 FD  ORDER-FILE-IN
000240     LABEL RECORDS ARE OMITTED.
000250
000260 01  ORDER-RECORD-IN.
000270     05 CUSTOMER-NUMBER-IN PIC X(7).
000280     05 PART-NUMBER-IN     PIC X(8).
000290     05 FILLER             PIC X(7).
000300     05 QUANTITY-IN        PIC 9(3).
000310     05 UNIT-PRICE-IN      PIC 9(4)V99.
000320     05 HANDLING-IN        PIC 99V99.
000330     05 FILLER             PIC X(45).
000340
000350 FD  ORDER-REPORT
000360     LABEL RECORDS ARE OMITTED.
000370
000380 01  REPORT-LINE           PIC X(115).
000390
```

Figure 4.3 Program 9, Part 1 (continued).

```
000400 WORKING-STORAGE SECTION.
000410 01  MORE-INPUT                 PIC X(3) VALUE 'YES'.
000420
000430 01  REPORT-TITLE.
000440     05 FILLER                  PIC X.
000450     05 FILLER                  PIC X(45) VALUE SPACES.
000460     05 FILLER                  PIC X(18) VALUE 'DAILY ORDER REPORT'.
000470
000480 01  COLUMN-HEADS-1.
000490     05 FILLER                  PIC X.
000500     05 FILLER                  PIC X(10) VALUE SPACES.
000510     05 FILLER                  PIC X(15) VALUE 'CUSTOMER'.
000520     05 FILLER                  PIC X(12) VALUE 'PART'.
000530     05 FILLER                  PIC X(13) VALUE 'QUANTITY'.
000540     05 FILLER                  PIC X(13) VALUE 'UNIT'.
000550     05 FILLER                  PIC X(18) VALUE 'MERCHANDISE'.
000560     05 FILLER                  PIC X(9)  VALUE 'TAX'.
000570     05 FILLER                  PIC X(16) VALUE 'HANDLING'.
000580     05 FILLER                  PIC X(5)  VALUE 'TOTAL'.
000590
000600 01  COLUMN-HEADS-2.
000610     05 FILLER                  PIC X.
000620     05 FILLER                  PIC X(11) VALUE SPACES.
000630     05 FILLER                  PIC X(13) VALUE 'NUMBER'.
000640     05 FILLER                  PIC X(26) VALUE 'NUMBER'.
000650     05 FILLER                  PIC X(15) VALUE 'PRICE'.
000660     05 FILLER                  PIC X(6)  VALUE 'AMOUNT'.
000670
000680 01  BODY-LINE.
000690     05 FILLER                      PIC X.
000700     05 FILLER                      PIC X(11) VALUE SPACES.
000710     05 CUSTOMER-NUMBER-OUT         PIC X(7).
000720     05 FILLER                      PIC X(5) VALUE SPACES.
000730     05 PART-NUMBER-OUT             PIC X(8).
000740     05 FILLER                      PIC X(8) VALUE SPACES.
000750     05 QUANTITY-OUT                PIC ZZ9.
000760     05 FILLER                      PIC X(6) VALUE SPACES.
000770     05 UNIT-PRICE-OUT              PIC Z,ZZZ.99.
000780     05 FILLER                      PIC X(7) VALUE SPACES.
000790     05 MERCHANDISE-AMOUNT-OUT      PIC ZZZ,ZZZ.99.
000800     05 FILLER                      PIC X(6) VALUE SPACES.
000810     05 TAX-OUT                     PIC Z,ZZZ.99.
000820     05 FILLER                      PIC X(5) VALUE SPACES.
000830     05 HANDLING-OUT                PIC ZZ.99.
000840     05 FILLER                      PIC X(5) VALUE SPACES.
000850     05 ORDER-TOTAL-OUT             PIC Z,ZZZ,ZZZ.99.
000860
000870 01  MERCHANDISE-AMOUNT-W           PIC 9(6)V99.
000880 01  TAX-W                          PIC 9(4)V99.
000890
000900*****************************************************
```

in the column headings are handled a little differently from before. For example, the word CUSTOMER in the column heading is only eight letters long, yet its FILLER is described as X(15), 15 characters long. Whenever an alphanumeric constant is described as having more spaces than it needs, COBOL inserts blanks to the right of the constant. In this case, the 15 character positions end up consisting of the word CUSTOMER followed by seven blanks, which is just what is required by the output format in Figure 4.2. Similarly, some other words in the column headings have FILLERS that are too large, and COBOL inserts blanks to the right of the constant.

Another change in the Data Division is that the body line of the report is described in working storage instead of in the File Section as before. The most important reason for doing this will become clear only later on, when we start dealing with much larger volumes of output data. For now, notice that the FILLERs between fields in BODY–LINE contain the VALUE SPACES clause, whereas FILLERs didn't when output lines were defined in the File Section. That is because the VALUE clause is not permitted in the File Section unless it is used with an 88-level entry. The 88 level is a special entry which we will discuss in Chapter 7. Now that we have the VALUE SPACES clauses in the report body line, we will not have to MOVE SPACES to the output line each time before preparing it for printing. There still must be an output record of some sort defined in the File Section, though, so that the WRITE verb will have some object. Since all the real lines to be written are defined in the Working Storage Section, in the File Section we define just a dummy 01-level entry that is as large as the largest of the lines to be written. The longest line on the report is the body line, which comes in at 115 characters including the carriage control character. The dummy output record in the File Section, which in this program is called REPORT–LINE, is then defined with PICTURE X(115). Now any and all lines can be written by saying WRITE REPORT–LINE FROM REPORT–TITLE or WRITE REPORT–LINE FROM COLUMN–HEADS–1, or WRITE REPORT–LINE FROM BODY–LINE and so forth.

The input record definition in the File Section, ORDER–RECORD–IN, has two fields defined with Vs to show the locations of the assumed decimal points. The locations of these points agree with the locations shown in the input format in Figure 4.1.

In the MAIN–PROCESS loop, we do some arithmetic before moving the results to the output fields in BODY–LINE. We first multiply QUANTITY–IN by UNIT–PRICE–IN. This of course yields the merchandise amount, but you notice that the GIVING field is not MERCHANDISE–AMOUNT–OUT in the BODY–LINE but some other field called MERCHANDISE–AMOUNT–W, which is defined in working storage. What is the difference between the definitions of MERCHANDISE–AMOUNT–OUT and MERCHANDISE–

AMOUNT–W? We see that MERCHANDISE–AMOUNT–OUT is defined as edited numeric in agreement with its format in Figure 4.2. But MERCHANDISE–AMOUNT–W is defined as purely numeric, with no actual decimal point or any commas or zero suppression. We need a merchandise-amount field defined as numeric because later in the program we will need the field for arithmetic. Remember that we will be using the merchandise amount to compute the tax, and we will also need it to add to the handling charge. In all arithmetic operations the source fields, the fields named before the words GIVING and REMAINDER, must be numeric; only fields after GIVING and

Figure 4.3 Program 9, Part 2 of 2.

```
000910
000920 PROCEDURE DIVISION.
000930     OPEN INPUT  ORDER-FILE-IN,
000935          OUTPUT ORDER-REPORT.
000940     WRITE REPORT-LINE FROM REPORT-TITLE
000950                                      AFTER ADVANCING TO-NEW-PAGE.
000960     WRITE REPORT-LINE FROM COLUMN-HEADS-1 AFTER ADVANCING 3.
000970     WRITE REPORT-LINE FROM COLUMN-HEADS-2 AFTER ADVANCING 1.
000980     MOVE SPACES TO REPORT-LINE.
000990     WRITE REPORT-LINE AFTER ADVANCING 1.
001000     READ ORDER-FILE-IN
001002         AT END
001004             MOVE 'NO' TO MORE-INPUT.
001010     PERFORM MAIN-PROCESS UNTIL MORE-INPUT IS EQUAL TO 'NO'.
001020     CLOSE ORDER-FILE-IN,
001025           ORDER-REPORT.
001030     STOP RUN.
001040
001050 MAIN-PROCESS.
001060     MULTIPLY QUANTITY-IN BY UNIT-PRICE-IN
001070                                      GIVING MERCHANDISE-AMOUNT-W.
001080     MULTIPLY MERCHANDISE-AMOUNT-W BY .07 GIVING TAX-W ROUNDED.
001090     ADD MERCHANDISE-AMOUNT-W,
001092         TAX-W,
001094         HANDLING-IN
001096             GIVING ORDER-TOTAL-OUT.
001110     MOVE CUSTOMER-NUMBER-IN    TO CUSTOMER-NUMBER-OUT.
001120     MOVE PART-NUMBER-IN        TO PART-NUMBER-OUT.
001130     MOVE QUANTITY-IN           TO QUANTITY-OUT.
001140     MOVE UNIT-PRICE-IN         TO UNIT-PRICE-OUT.
001150     MOVE MERCHANDISE-AMOUNT-W  TO MERCHANDISE-AMOUNT-OUT.
001160     MOVE TAX-W                 TO TAX-OUT.
001170     MOVE HANDLING-IN           TO HANDLING-OUT.
001180     WRITE REPORT-LINE FROM BODY-LINE AFTER ADVANCING 1.
001190     READ ORDER-FILE-IN
001192         AT END
001194             MOVE 'NO' TO MORE-INPUT.
```

REMAINDER are allowed to be edited numeric. Here, even though MERCHANDISE–AMOUNT–W appears after a GIVING in one arithmetic operation, it must still be numeric, because it is going to appear as a source field in later ones. MERCHANDISE–AMOUNT–W is defined as containing as many digits as MERCHANDISE–AMOUNT–OUT but with only the V for the assumed decimal point instead of an actual point, commas, and zero suppression.

The tax is then computed. Since the tax will also be needed later (for adding), we assign the result to TAX–W, a numeric field, instead of to TAX–OUT; TAX–W is defined as containing as many digits as TAX–OUT. In the program the tax rate, 7 percent, is written into the MULTIPLY statement as a numeric literal. In actual practice, a better programming technique would be to define the tax rate as a constant in working storage and use the name of the constant in the multiplication. Then, should the tax rate change while this program is still being used, it would be easier to modify. In all the remaining programs in this book, constants will be defined in working storage.

Now in the addition, when MERCHANDISE–AMOUNT–W, TAX–W, and HANDLING–IN are all added together, the sum can be assigned directly to ORDER–TOTAL–OUT. We don't need a purely numeric field for the total, because the total will not be used in any later arithmetic. Then the rest of the BODY–LINE can be filled by a series of MOVE statements, and the line written with single spacing. Input data for Program 9 is shown in Figure 4.4, and the output is shown in Figure 4.5.

```
ABC1234F2365-09     9000000100005
0968239836-7YT7     8000010500050
ADGH784091AN-07     0500250000500
967547323S-1287     0067000295000
```

Figure 4.4 Input data for Program 9.

Figure 4.5 Output from Program 9.

DAILY ORDER REPORT

CUSTOMER NUMBER	PART NUMBER	QUANTITY	UNIT PRICE	MERCHANDISE AMOUNT	TAX	HANDLING	TOTAL
ABC1234	F2365-09	900	.10	90.00	6.30	.05	96.35
0968239	836-7YT7	800	10.50	8,400.00	588.00	.50	8,988.50
ADGH784	091AN-07	50	250.00	12,500.00	875.00	5.00	13,380.00
9675473	23S-1287	6	7,000.29	42,001.74	2,940.12	50.00	44,991.86

***Exercise 8.** Using the same input data as for Program 8 in Figure 3.19 (format of input data is shown in Figure 3.15), write a program to read each record and compute a 2 percent discount on the amount shown in each record and a net amount (the original amount minus the discount). For each card read, print the Account Number, the amount, the computed discount, and the computed net amount. Print column headings "Acct. No.," "Amount," "Discount," and "Net Amount." Print a title "Discount Report."

Use a printer spacing chart to plan the spacing of the output, the placement of decimal points, commas, zero suppression, and the alignment of column headings and columns of data before you begin coding the Data Division.

A Program with Totals

We will now study Program 10, which produces all the output produced by Program 9 and also prints, at the end of the report, the total of all the merchandise amounts for all the orders and totals of all the tax and handling charge amounts and a grand total of all the order totals. The output should have the format shown in Figure 4.6.

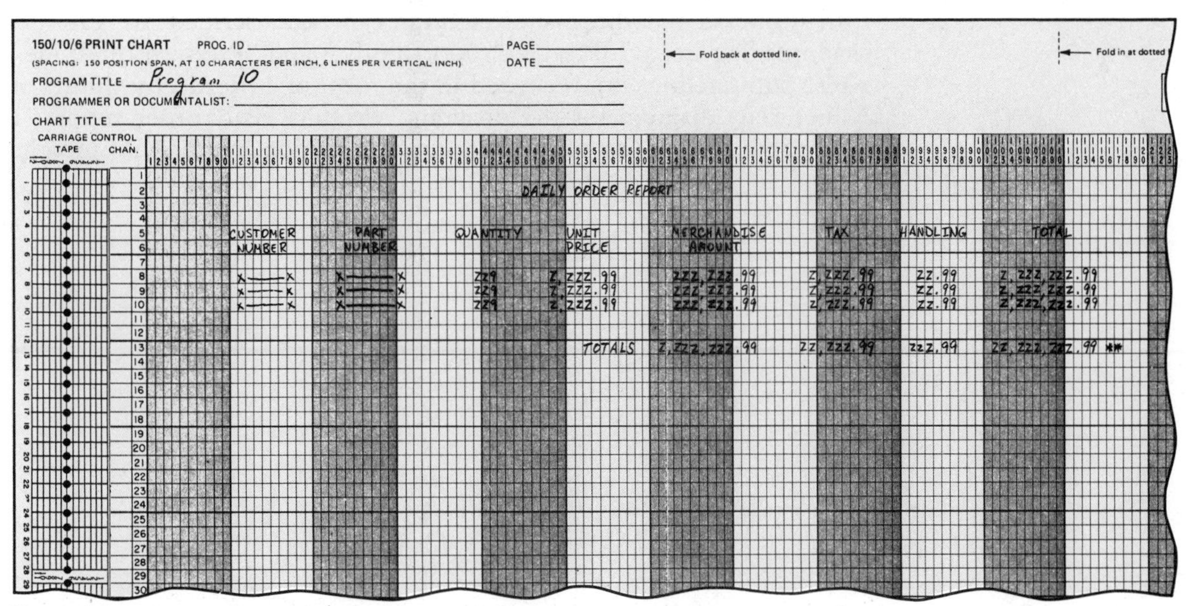
150/10/6 PRINT CHART PROG. ID PAGE
(SPACING: 150 POSITION SPAN, AT 10 CHARACTERS PER INCH, 6 LINES PER VERTICAL INCH) DATE
PROGRAM TITLE Program 10
PROGRAMMER OR DOCUMENTALIST:
CHART TITLE
CARRIAGE CONTROL TAPE CHAN.
Fold back at dotted line.
Fold in at dotted

DAILY ORDER REPORT
CUSTOMER NUMBER PART NUMBER QUANTITY UNIT PRICE MERCHANDISE AMOUNT TAX HANDLING TOTAL
X——X X——X ZZ9 Z,ZZZ.99 ZZZ,ZZZ.99 Z,ZZZ.99 ZZ.99 Z,ZZZ,ZZZ.99
X——X X——X ZZ9 Z,ZZZ.99 ZZZ,ZZZ.99 Z,ZZZ.99 ZZ.99 Z,ZZZ,ZZZ.99
X——X X——X ZZ9 Z,ZZZ.99 ZZZ,ZZZ.99 Z,ZZZ.99 ZZ.99 Z,ZZZ,ZZZ.99
TOTALS Z,ZZZ,ZZZ.99 ZZ,ZZZ.99 ZZZ.99 ZZ,ZZZ,ZZZ.99 **

Figure 4.6 Output format for Program 10.

This program is somewhat more difficult than any we have done so far, so it is worthwhile to plan the program before trying to code it.

Program Design

There are a variety of program design techniques in use now. For any particular program, the choice of which design technique to use often depends on the complexity and/or the size of the program, personal preference, and which technique your instructor tells you to use. We have already seen how a flowchart or a control chart can picture a logical process before it is coded in COBOL. We will now look at two techniques which may be useful for a more complex program, such as we are now attempting.

TOP-DOWN DESIGN

In **top-down design** the programmer avoids trying to think about the whole program at once and instead first decides what the major function of the program is. In our case we can say that the major function is "Produce daily order report." The programmer then decides what subfunctions the program must carry out in order to accomplish the main function. Experience has shown that the coding will come out straightforward if you divide the program into three subfunctions:

a. what the program has to do before entering the main loop;
b. the main loop;
c. what the program has to do after end-of-file.[1]

In our case the three subfunctions can be described as "Produce page heading lines," "Produce report body," and "Produce total line." These subfunctions are recorded in the form of a **hierarchy diagram.** A hierarchy diagram for the program "Produce daily order report" is shown in Figure 4.7 as it reflects the main function and the subfunctions

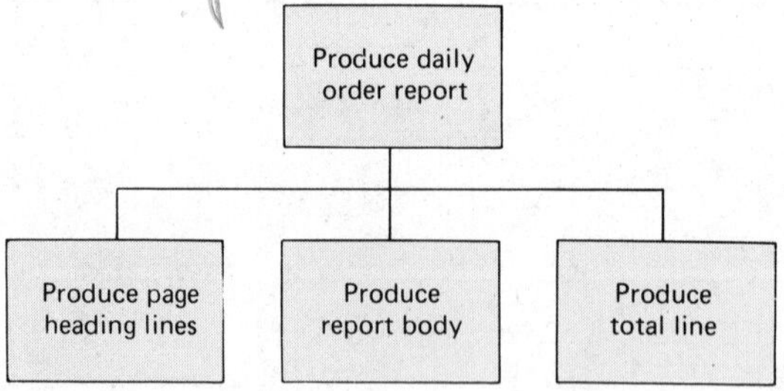

Figure 4.7 Hierarchy diagram with one level of subfunctions.

1. The theory of top-down design dictates otherwise, but experience has shown the theory to be wrong. The best example of the divergence of theory from practice is in Mike Murach and Paul Noll, *Structured ANS COBOL Part 1: A Course for Novices,* Mike Murach and Associates, Inc., Fresno, 1979. On page 165 the authors suggest that breaking a program into the three categories *a, b,* and *c* above is not the best design procedure; on page 160 they correctly design their program by breaking it into the categories *a, b,* and *c.*

thus far. Notice that the main function of the program is shown in a single box at the top, and the subfunctions shown beneath it. Now if any of the subfunctions can be further broken down, we would show that breakdown on the hierarchy diagram also. The subfunction "Produce report body" can indeed be detailed a little further. Figure 4.8 shows the complete hierarchy diagram. Let us examine the subfunction "Accumulate totals." Why is that shown under "Produce report body" instead of under "Produce total line"? Remember that the

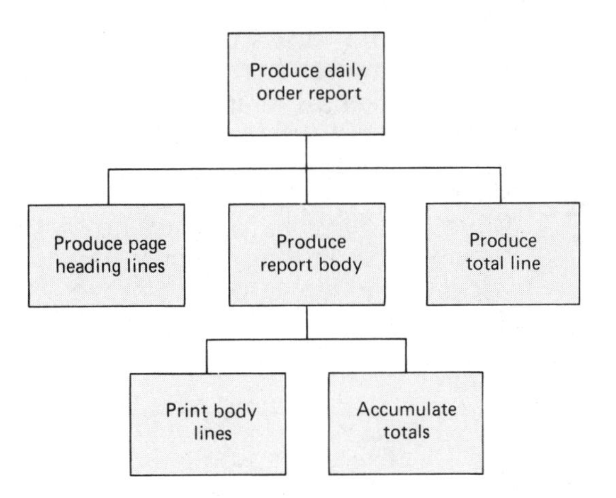

Figure 4.8 Complete hierarchy diagram for Program 10.

program can work on only one input card at a time. The program must do all required processing on each input card before reading in the next. So in this program, as we compute the several money amounts that we need for printing each line, we will also have to **accumulate** those amounts in some way as if we were adding them into a calculator. As each line is printed, therefore, we also add the money amounts into ever growing subtotals, so that, after all the body lines of the report have been printed we will have the totals of the money amounts all ready to be printed in the total line. COBOL provides an easy way for the programmer to do this, and we will see how shortly.

PSEUDOCODE

Another program design technique that has some of the same advantages as top-down design is called **pseudocode.** One of the advantages of pseudocode over all other methods of program design is that it can be written in plain English and does not require drawing any diagrams or charts. Pseudocode is written in a form that resembles COBOL code, but since pseudocode is English, it has none of the language rules of

COBOL (or any programming language). As an example, here is the complete pseudocode for Program 9:

```
Open files
Write page headings and column headings
Read Order-file-in at end move 'No' to More-input
Perform until More-input is equal to 'No'
   Multiply Quantity-in by Unit-price-in giving Merchandise-amount-w
   Multiply Merchandise-amount-w by .07 giving Tax-w
   Add Merchandise-amount-w, Tax-w, Handling-in giving Order-total-out
   Move Customer-number-in to Customer-number-out
   Move Part-number-in to Part-number-out
   Move Quantity-in to Quantity-out
   Move Merchandise-amount-w to Merchandise-amount-out
   Move Tax-w to Tax-out
   Move Handling-in to Handling-out
   Write Report-line from Body-line
   Read Order-file-in at end move 'No' to More-input
End-perform
Close files
Stop run
```

There are some important differences between pseudocode and COBOL. First, in pseudocode only procedural steps are written; there is no reference in pseudocode to data descriptions or identification or environment details. Second, in pseudocode a performed procedure is written directly under the Perform statement that controls it instead of in a paragraph by itself as in COBOL; this makes it unnecessary to ever use paragraph names in pseudocode. The scope of the Perform statement is shown by indenting the performed procedure and ending it with the statement End–perform. Pseudocode may be written in as much or as little detail as desired. In the example given, the performed procedure was written in complete detail but the Open and Close statements and the statement to write the page and column headings were not.

Pseudocode for Program 10 may be approached in the same top-down manner that we used in drawing the hierarchy diagram. That is, we may first write pseudocode at a summary level, showing only the major subfunctions of the program, and fill in the details later. For example, in writing pseudocode for Program 10 we could first write only as much detail as is shown in Figure 4.7. The pseudocode might look like this:

```
Open files
Perform page heading and column heading routine
Read Order-file-in at end move 'No' to More-input
Perform report body routine until More-input is equal to 'No'
```

```
Perform total line routine
Close files
Stop run
```

In actual practice, a programmer would probably not use both hierarchy diagrams and pseudocode to design a program; either one or the other would be used. Here, the pseudocode above was made to correspond to the first hierarchy diagram just so you could see what the first level of subfunctions looks like both ways. Now we can fill in the details of the Perform statements.

The routine to print the page and column headings contains nothing new:

```
Write page heading after advancing to new page
Write column headings
```

Referring to the hierarchy diagram in Figure 4.8, you can see that the routine to produce the report body lines can be expressed in pseudocode as:

```
Perform until More–input is equal to 'No'
   Print body lines
   Accumulate totals
   Read Order–file–in at end move 'No' to More–input
End–perform
```

The pseudocode for each of the two routines "Print body lines" and "Accumulate totals" may now be written. "Print body lines" is right out of Program 9:

```
Multiply Quantity–in by Unit–price–in giving Merchandise–amount–w
Multiply Merchandise–amount–w by .07 giving Tax–w
Add Merchandise–amount–w, Tax–w, Handling–in giving Order–total–out
Move Customer–number–in to Customer–number–out
Move Part–number–in to Part–number–out
Move Quantity–in to Quantity–out
Move Merchandise–amount–w to Merchandise–amount–out
Move Tax–w to Tax–out
Move Handling–in to Handling–out
Write Report–line from Body–line
```

In the routine "Accumulate totals" we must accumulate the money amounts so that their totals will be available for printing the total line after all body lines have been printed. There are four money amounts that must be accumulated for the total line—the merchandise amount, the tax, the handling charge, and the order total. In COBOL we can set up four fields in working storage which we can use to accumulate the four money amounts (when we get to the program you will see how it is done). The four fields might be called Merchandise–amount–tot–w, Tax–tot–w, Handling–tot–w, and Grand–tot–w. The pseudocode for "Accumulate totals" is then:

Add Merchandise–amount–w to Merchandise–amount–tot–w
Add Tax–w to Tax–tot–w
Add Handling–in to Handling–tot–w
Add Order-total–w to Grand–tot–w

The last Perform statement in the pseudocode for Program 10 is the one to perform the total line routine. Since the totals that are needed for the total line have been accumulated in four fields in working storage, all that is needed to print the total line is to move the totals to it and write it:

Move Merchandise–amount–tot–w to Merchandise–amount–tot–out
Move Tax–tot–w to Tax–tot–out
Move Handling–tot–w to Handling–tot–out
Move Grand–tot–w to Grand–tot–out
Write Report–line from Total–line

Program 10 in COBOL

The complete program is shown in Figure 4.9. The coding in the Procedure Division resembles the pseudocode. Writing and understanding the pseudocode was a lot of work, but once pseudocode is completed, it is very easy to code the Procedure Division. Let us now look at some of the things in the Data Division, particularly the Working Storage Section.

Figure 4.9 Program 10, Part 1 of 2.

```
4-CB2 V4 RELEASE 1.5 10NOV77          IBM OS AMERICAN NATIONAL STANDARD COBOL

  000010 IDENTIFICATION DIVISION.
  000020 PROGRAM-ID. PROG10.
  000030*
  000040*       THIS PROGRAM READS CUSTOMER ORDERS PUNCHED IN
  000050*       CARDS AND PRODUCES A DAILY ORDER REPORT.
  000060*
  000070*****************************************************
```

Figure 4.9 Program 10, Part 1 (continued).

```
000080
000090 ENVIRONMENT DIVISION.
000100 CONFIGURATION SECTION.
000110 SPECIAL-NAMES.
000120     C01 IS TO-NEW-PAGE.
000130 INPUT-OUTPUT SECTION.
000140 FILE-CONTROL.
000150     SELECT ORDER-FILE-IN ASSIGN TO UT-S-SYSIN.
000160     SELECT ORDER-REPORT  ASSIGN TO UT-S-SYSPRINT.
000170
000180*******************************************************
000190
000200 DATA DIVISION.
000210 FILE SECTION.
000220 FD  ORDER-FILE-IN
000230     LABEL RECORDS ARE OMITTED.
000240
000250 01  ORDER-RECORD-IN.
000260     05 CUSTOMER-NUMBER-IN PIC X(7).
000270     05 PART-NUMBER-IN     PIC X(8).
000280     05 FILLER             PIC X(7).
000290     05 QUANTITY-IN        PIC 9(3).
000300     05 UNIT-PRICE-IN      PIC 9(4)V99.
000310     05 HANDLING-IN        PIC 99V99.
000320     05 FILLER             PIC X(45).
000330
000340 FD  ORDER-REPORT
000350     LABEL RECORDS ARE OMITTED.
000360
000370 01  REPORT-LINE           PIC X(118).
000380
000390 WORKING-STORAGE SECTION.
000400 01  MORE-INPUT            PIC X(3) VALUE 'YES'.
000410
000420 01  REPORT-TITLE.
000430     05 FILLER             PIC X.
000440     05 FILLER             PIC X(45) VALUE SPACES.
000450     05 FILLER             PIC X(18) VALUE 'DAILY ORDER REPORT'.
000460
000470 01  COLUMN-HEADS-1.
000480     05 FILLER             PIC X.
000490     05 FILLER             PIC X(10) VALUE SPACES.
000500     05 FILLER             PIC X(15) VALUE 'CUSTOMER'.
000510     05 FILLER             PIC X(12) VALUE 'PART'.
000520     05 FILLER             PIC X(13) VALUE 'QUANTITY'.
000530     05 FILLER             PIC X(13) VALUE 'UNIT'.
000540     05 FILLER             PIC X(18) VALUE 'MERCHANDISE'.
000550     05 FILLER             PIC X(9)  VALUE 'TAX'.
000560     05 FILLER             PIC X(16) VALUE 'HANDLING'.
000570     05 FILLER             PIC X(5)  VALUE 'TOTAL'.
000580
000590 01  COLUMN-HEADS-2.
000600     05 FILLER             PIC X.
000610     05 FILLER             PIC X(11) VALUE SPACES.
000620     05 FILLER             PIC X(13) VALUE 'NUMBER'.
000630     05 FILLER             PIC X(26) VALUE 'NUMBER'.
000640     05 FILLER             PIC X(15) VALUE 'PRICE'.
000650     05 FILLER             PIC X(6)  VALUE 'AMOUNT'.
```

Figure 4.9 Program 10, Part 2 of 2.

```
000660
000670 01  BODY-LINE.
000680     05 FILLER                       PIC X.
000690     05 FILLER                       PIC X(11) VALUE SPACES.
000700     05 CUSTOMER-NUMBER-OUT          PIC X(7).
000710     05 FILLER                       PIC X(5) VALUE SPACES.
000720     05 PART-NUMBER-OUT              PIC X(8).
000730     05 FILLER                       PIC X(8) VALUE SPACES.
000740     05 QUANTITY-OUT                 PIC ZZ9.
000750     05 FILLER                       PIC X(6) VALUE SPACES.
000760     05 UNIT-PRICE-OUT               PIC Z,ZZZ.99.
000770     05 FILLER                       PIC X(7) VALUE SPACES.
000780     05 MERCHANDISE-AMOUNT-OUT       PIC ZZZ,ZZZ.99.
000790     05 FILLER                       PIC X(6) VALUE SPACES.
000800     05 TAX-OUT                      PIC Z,ZZZ.99.
000810     05 FILLER                       PIC X(5) VALUE SPACES.
000820     05 HANDLING-OUT                 PIC ZZ.99.
000830     05 FILLER                       PIC X(5) VALUE SPACES.
000840     05 ORDER-TOTAL-OUT              PIC Z,ZZZ,ZZZ.99.
000850
000860 01  TOTAL-LINE.
000870     05 FILLER                       PIC X.
000880     05 FILLER                       PIC X(52) VALUE SPACES.
000890     05 FILLER                       PIC X(9) VALUE 'TOTALS'.
000900     05 MERCHANDISE-AMOUNT-TOT-OUT PIC Z,ZZZ,ZZZ.99.
000910     05 FILLER                       PIC X(5) VALUE SPACES.
000920     05 TAX-TOT-OUT                  PIC ZZ,ZZZ.99.
000930     05 FILLER                       PIC X(4) VALUE SPACES.
000940     05 HANDLING-TOT-OUT             PIC ZZZ.99.
000950     05 FILLER                       PIC X(4) VALUE SPACES.
000960     05 GRAND-TOT-OUT                PIC ZZ,ZZZ,ZZZ.99.
000970     05 FILLER                       PIC X(3) VALUE ' **'.
000980
000990 01  TAX-RATE                        PIC V99 VALUE .07.
001000 01  MERCHANDISE-AMOUNT-W            PIC 9(6)V99.
001010 01  TAX-W                           PIC 9(4)V99.
001020 01  MERCHANDISE-AMOUNT-TOT-W        PIC 9(7)V99 VALUE 0.
001030 01  TAX-TOT-W                       PIC 9(5)V99 VALUE 0.
001040 01  HANDLING-TOT-W                  PIC 999V99 VALUE 0.
001050 01  GRAND-TOT-W                     PIC 9(8)V99 VALUE 0.
001060 01  ORDER-TOTAL-W                   PIC 9(7)V99.
001070
001080*****************************************************************
001090
001100 PROCEDURE DIVISION.
001110     OPEN INPUT  ORDER-FILE-IN,
001120          OUTPUT ORDER-REPORT.
001130     PERFORM PRODUCE-PAGE-HEADING-LINES.
001140     READ ORDER-FILE-IN
001150         AT END
001160             MOVE 'NO' TO MORE-INPUT.
001170     PERFORM PRODUCE-REPORT-BODY
001180         UNTIL
001190             MORE-INPUT IS EQUAL TO 'NO'.
001200     PERFORM PRODUCE-TOTAL-LINE.
001210     CLOSE ORDER-FILE-IN,
001220           ORDER-REPORT.
001230     STOP RUN.
```

Figure 4.9 Program 10, Part 2 (continued).

```
001240
001250 PRODUCE-PAGE-HEADING-LINES.
001260     WRITE REPORT-LINE FROM REPORT-TITLE
001270                                       AFTER ADVANCING TO-NEW-PAGE.
001280     WRITE REPORT-LINE FROM COLUMN-HEADS-1 AFTER ADVANCING 3.
001290     WRITE REPORT-LINE FROM COLUMN-HEADS-2 AFTER ADVANCING 1.
001300     MOVE SPACES TO REPORT-LINE.
001310     WRITE REPORT-LINE AFTER ADVANCING 1.
001320
001330 PRODUCE-REPORT-BODY.
001340     PERFORM PRINT-BODY-LINES.
001350     PERFORM ACCUMULATE-TOTALS.
001360     READ ORDER-FILE-IN
001370         AT END
001380             MOVE 'NO' TO MORE-INPUT.
001390
001400 PRODUCE-TOTAL-LINE.
001410     MOVE MERCHANDISE-AMOUNT-TOT-W TO MERCHANDISE-AMOUNT-TOT-OUT.
001420     MOVE TAX-TOT-W                TO TAX-TOT-OUT.
001430     MOVE HANDLING-TOT-W           TO HANDLING-TOT-OUT.
001440     MOVE GRAND-TOT-W              TO GRAND-TOT-OUT.
001450     WRITE REPORT-LINE FROM TOTAL-LINE AFTER ADVANCING 3.
001460
001470 PRINT-BODY-LINES.
001480     MULTIPLY QUANTITY-IN BY UNIT-PRICE-IN
001490                                  GIVING MERCHANDISE-AMOUNT-W.
001500     MULTIPLY MERCHANDISE-AMOUNT-W BY TAX-RATE
001510                                         GIVING TAX-W ROUNDED.
001520     ADD MERCHANDISE-AMOUNT-W,
001530         TAX-W,
001540         HANDLING-IN
001550             GIVING ORDER-TOTAL-W.
001560     MOVE CUSTOMER-NUMBER-IN    TO CUSTOMER-NUMBER-OUT.
001570     MOVE PART-NUMBER-IN        TO PART-NUMBER-OUT.
001580     MOVE QUANTITY-IN           TO QUANTITY-OUT.
001590     MOVE UNIT-PRICE-IN         TO UNIT-PRICE-OUT.
001600     MOVE MERCHANDISE-AMOUNT-W  TO MERCHANDISE-AMOUNT-OUT.
001610     MOVE TAX-W                 TO TAX-OUT.
001620     MOVE HANDLING-IN           TO HANDLING-OUT.
001630     MOVE ORDER-TOTAL-W         TO ORDER-TOTAL-OUT.
001640     WRITE REPORT-LINE FROM BODY-LINE AFTER ADVANCING 1.
001650
001660 ACCUMULATE-TOTALS.
001670     ADD MERCHANDISE-AMOUNT-W TO MERCHANDISE-AMOUNT-TOT-W.
001680     ADD TAX-W                TO TAX-TOT-W.
001690     ADD HANDLING-IN          TO HANDLING-TOT-W.
001700     ADD ORDER-TOTAL-W        TO GRAND-TOT-W.
```

In this program the page and column heading lines and the body line have been defined in working storage as before. You can see also how the total line is defined in working storage under the name TOTAL–LINE. You can also find the four accumulator fields under the names MERCHANDISE–AMOUNT–TOT–W, TAX–TOT–W,

HANDLING–TOT–W, and GRAND–TOT–W. The fields are each defined as holding as many digits as the corresponding output field in the total line but as purely numeric so that they can be used in arithmetic. They are all defined as starting with VALUE 0 so that the accumulation will be correct.

Another new field, ORDER–TOTAL–W, is defined in working storage. Why is this field needed? Study the three statements in the Procedure Division that use ORDER–TOTAL–W and see if you can determine why ORDER–TOTAL–W and ORDER–TOTAL–OUT are both needed. The output from Program 10 is shown in Figure 4.10.

Figure 4.10 Output from Program 10.

DAILY ORDER REPORT

CUSTOMER NUMBER	PART NUMBER	QUANTITY	UNIT PRICE	MERCHANDISE AMOUNT	TAX	HANDLING	TOTAL
ABC1234	F2365-09	900	.10	90.00	6.30	.05	96.35
0968239	836-7YT7	800	10.50	8,400.00	588.00	.50	8,988.50
ADGH784	091AN-07	50	250.00	12,500.00	875.00	5.00	13,380.00
9675473	23S-1287	6	7,000.29	42,001.74	2,940.12	50.00	44,991.86
			TOTALS	62,991.74	4,409.42	55.55	67,456.71

In Chapter 5 we will study other aspects of taking totals in COBOL, including the computation of averages.

***Exercise 9.** Modify your solution to Exercise 8 so that your program produces all the output produced by the original program and also prints, at the end of the report, a total line showing the sum of all the input amounts, the sum of all the discount amounts, and the sum of all the net amounts. Use the printer spacing chart that you used for Exercise 8 to plan the total line before you begin coding.

Summary

Four of the arithmetic verbs in COBOL are ADD, SUBTRACT, MULTIPLY, and DIVIDE. All four can be written without the GIVING option, as ADD A TO B, SUBTRACT A FROM B, MULTIPLY A BY B, and DIVIDE A INTO B. In each such form the result is assigned to one of the terms in the operation, destroying the original value of that term. The result of arithmetic can optionally be assigned to a particular field designated by the programmer through the use of the GIVING option, as SUBTRACT A FROM B GIVING C and MULTIPLY A BY B GIVING C. When the GIVING option is used with an ADD, the word TO must not also be used, as in the correct statement ADD A, B GIVING C. When the GIVING option is used with DIVIDE, then INTO or BY may be used, as DIVIDE A INTO B GIVING C, or DIVIDE B BY A GIVING C. The DIVIDE statement also has the optional capability of assigning a REMAINDER to a field specified by the programmer. In all four operations COBOL can be directed to round the result if necessary, and some COBOL systems permit more than one source field and/or more than one destination field in certain arithmetic operations.

Fields in the four operations must be defined as numeric, except that after the words GIVING and REMAINDER, the fields may be numeric or edited numeric. This limitation means that results of arithmetic which are going to be both used in further arithmetic and printed as output must be defined as both purely numeric and as edited numeric. Then the numeric definition can be used for any necessary arithmetic, and the edited numeric definition can be used for printing.

It is sometimes convenient to define all the lines of a report in working storage and use WRITE . . . FROM to print them all. In such a case there must be a dummy output record defined in the File Section whose length is at least as large as the longest of the lines to be printed.

It is good programming practice to define and name all program constants in working storage, rather than write them as literals in the Procedure Division. The resulting program is usually easier to understand and to modify.

In top-down design, a programmer produces a hierarchy diagram showing the major function of a program broken down into lower and lower levels of subfunctions. Top-down design enables the programmer to design the whole program by concentrating on only one small portion of it at one time.

Pseudocode has many of the advantages of top-down design and does not require drawing any diagrams. Pseudocode resembles COBOL coding in many ways. Once a program has been designed in pseudocode, writing the Procedure Division of a COBOL program is very easy.

In accumulating totals from input data, the individual amounts must be added as each input item is processed. Then the totals will be available for printing after the entire input file has been processed.

Fill-in Exercises

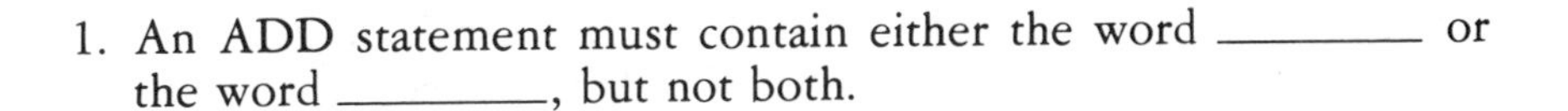

1. An ADD statement must contain either the word ________ or the word ________, but not both.

2. Whenever an alphanumeric constant is defined with a PICTURE having room for more characters than there are in the constant, COBOL inserts ________ to the right of the constant.
3. In arithmetic statements without the GIVING option, all identifiers must be described as ________.
4. In arithmetic statements with the GIVING option, fields after the word GIVING may be described as ________ or as ________.
5. In a top-down design, the programmer draws a ________ diagram.
6. An advantage of pseudocode over other methods of program design is that no ________ or ________ have to be drawn.
7. The language used in this book for writing pseudocode is ________.
8. It is good programming practice to define and name all constants in ________
9. In a DIVIDE statement without the GIVING option, the quotient is assigned to the field named after the word ________.
10. In a SUBTRACT statement without the GIVING option, the difference is assigned to the field named after the word ________.

Review Exercises

1. Refer to the statement formats given in the chapter and explain why each of the following statements is illegal:
   a. ADD A, B TO C GIVING D.
   b. MULTIPLY RATE BY 2.
   c. DIVIDE A BY B.
   d. DIVIDE A INTO B REMAINDER C.
2. Write COBOL statements to accomplish the following:
   a. add the value of DEPOSITS to the value of CURRENT–BALANCE and assign the sum to CURRENT–BALANCE.
   b. subtract the value of WITHDRAWAL from the value of CURRENT–BALANCE and assign the difference to CURRENT–BALANCE.
   c. multiply the value of INTEREST–RATE by the value of CURRENT–BALANCE and assign the result to CURRENT–INTEREST.
3. Write the Procedure Division of Program 5 in pseudocode form. Write the pseudocode statements at the same level of detail as the pseudocode examples in this chapter.
4. Write the working storage entries for the column headings shown in Figure 3.RE3, Chapter 3, page 63.

5. Write a program to read and process a deck of input cards in the format shown in Figure 4.RE5.1. Each card contains the hours worked, Monday through Saturday, for one employee. Print a report in the format shown in Figure 4.RE5.2. For each employee show the total hours worked during the week. Also show, at the

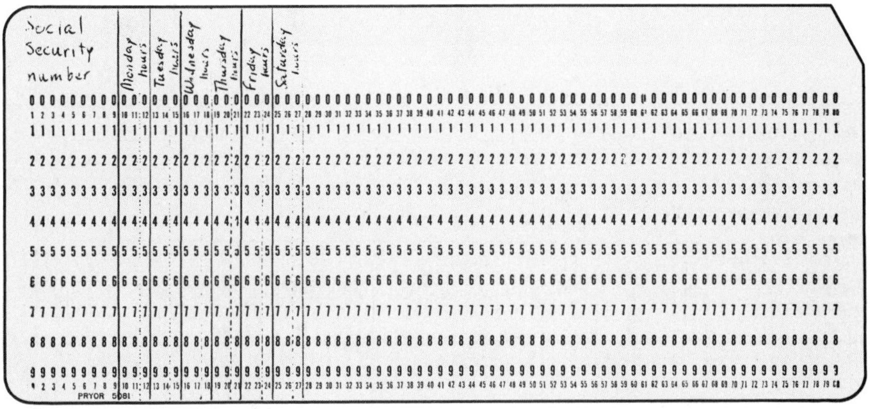

Social Security number | Monday hours | Tuesday hours | Wednesday hours | Thursday hours | Friday hours | Saturday hours

Figure 4.RE5.1 Data card format for Review Exercise 5.

150/10/6 PRINT CHART PROG. ID PAGE
(SPACING: 150 POSITION SPAN, AT 10 CHARACTERS PER INCH, 6 LINES PER VERTICAL INCH) DATE
Fold back at dotted line.
PROGRAM TITLE
PROGRAMMER OR DOCUMENTALIST:
CHART TITLE Review Exercise 5, Chapter 4

WEEKLY LABOR HOURS REPORT

SOC. SEC. NUMBER	MON.	TUE.	WED.	THU.	FRI.	SAT.	TOTAL FOR WEEK
X———X	Z9.9	Z9.9	Z9.9	Z9.9	Z9.9	Z9.9	ZZ9.9
X———X	Z9.9	Z9.9	Z9.9	Z9.9	Z9.9	Z9.9	ZZ9.9
X———X	Z9.9	Z9.9	Z9.9	Z9.9	Z9.9	Z9.9	ZZ9.9
TOTALS	ZZ9.9	ZZ9.9	ZZ9.9	ZZ9.9	ZZ9.9	ZZ9.9	Z,ZZ9.9

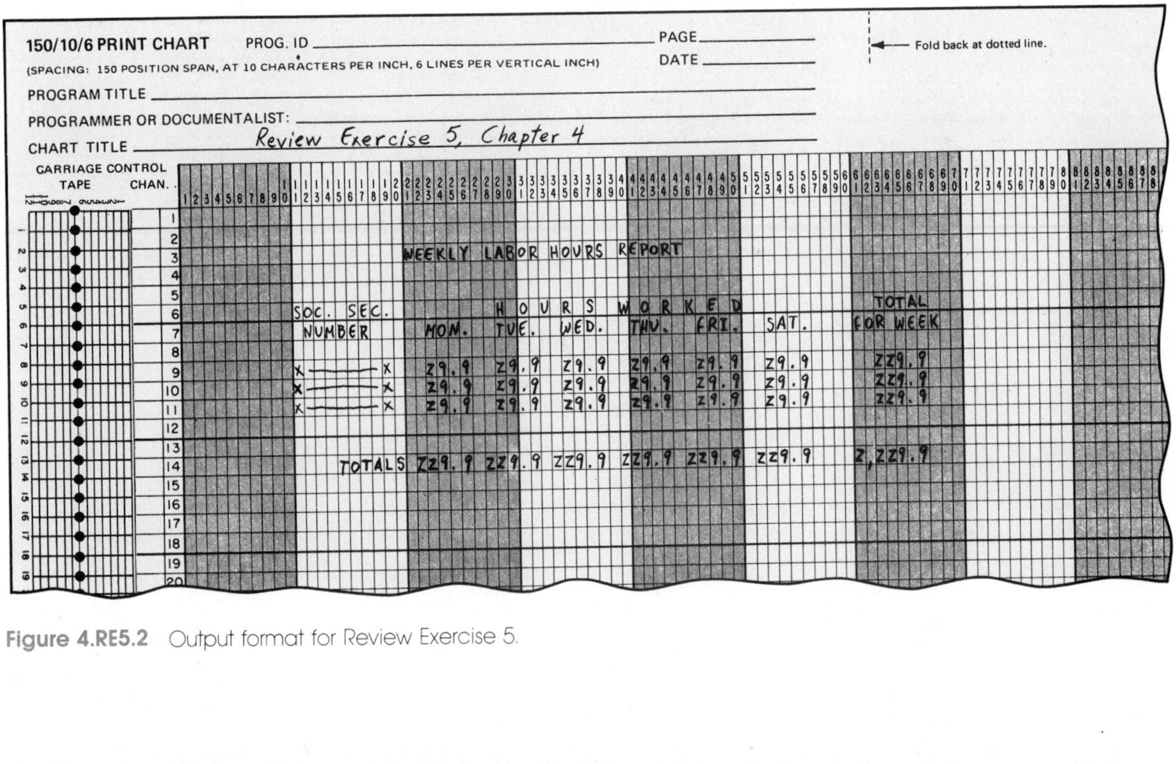

Figure 4.RE5.2 Output format for Review Exercise 5.

end of the report, the total hours worked by all employees on Monday, the total hours worked on Tuesday, and so on, and the grand total of all hours worked during the week.

*6. Using the same input as for Review Exercise 5, produce a report in the format shown in Figure 4.RE6. On the report, list only those employees who worked more than 40 hours during the week. Compute each employee's number of overtime hours (hours worked during the week minus 40). Select a program design technique to use for this program, and use it before you begin coding.

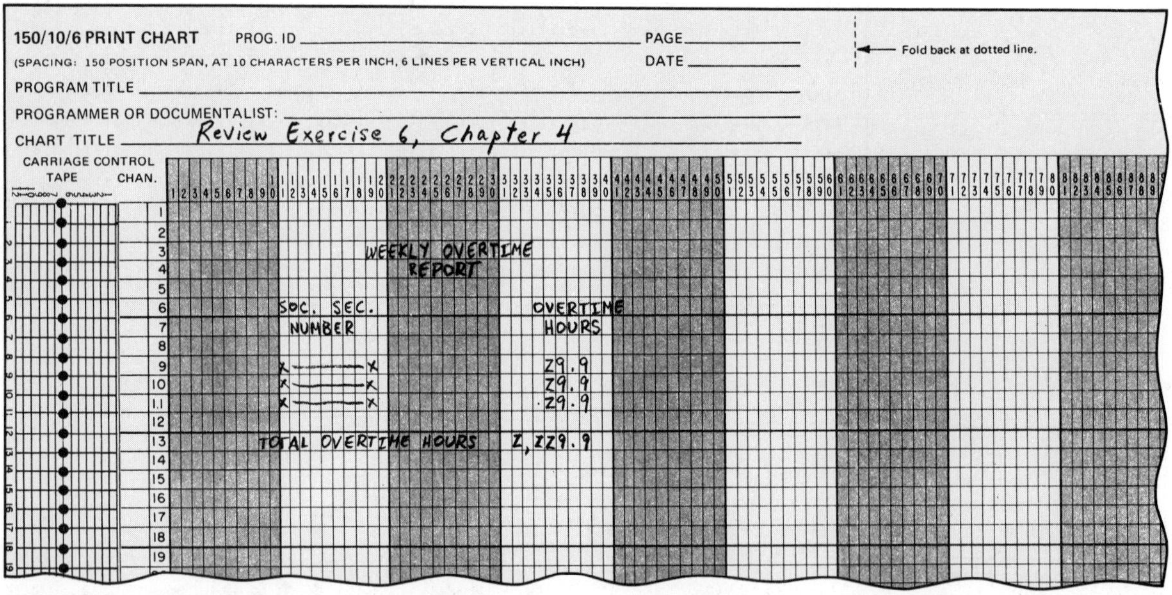

150/10/6 PRINT CHART PROG. ID PAGE
(SPACING: 150 POSITION SPAN, AT 10 CHARACTERS PER INCH, 6 LINES PER VERTICAL INCH) DATE
Fold back at dotted line.
PROGRAM TITLE
PROGRAMMER OR DOCUMENTALIST:
CHART TITLE Review Exercise 6, Chapter 4
CARRIAGE CONTROL TAPE CHAN.

000000000111111111122222222223333333333444444444455555555556666666666777777777788888888889
123456789012345678901234567890123456789012345678901234567890123456789012345678901234567890

1
2
3 WEEKLY OVERTIME
4 REPORT
5
6 SOC. SEC. OVERTIME
7 NUMBER HOURS
8
9 X———X Z9.9
10 X———X Z9.9
11 X———X Z9.9
12
13 TOTAL OVERTIME HOURS Z,ZZ9.9
14
15
16
17
18
19

Figure 4.RE6 Output format for Review Exercise 6.

More on Editing and Other COBOL Features

5

KEY POINTS

Here are the key points you should learn from this chapter:

1. how to use the COMPUTE verb;
2. how to form arithmetic expressions and how such expressions are used;
3. the use of top-down design;
4. how to use the ACCEPT verb;
5. how to use the MOVE verb with the CORRESPONDING option;
6. the COBOL features dealing with signed fields;
7. output editing features of COBOL.

KEY WORDS

Key words to recognize and learn:

COMPUTE	CORRESPONDING	signed numbers
arithmetic expression	data name qualification	SIGN
unary operator	qualified	default
ACCEPT	qualifiers	floating
		check protection

In this chapter we will study arithmetic a little further and then look at a number of COBOL features that make programming easier and improve the usefulness of output. First, in Program 11, we will see a fairly easy problem involving the computation of averages.

Program 11 reads input data in the format shown in Figure 5.1. Each record contains a student number and the grades that the student got on each of four exams. The program is to read the data and print a list showing each student number, the four grades, and the average of the four grades. The output should have the format shown in Figure 5.2.

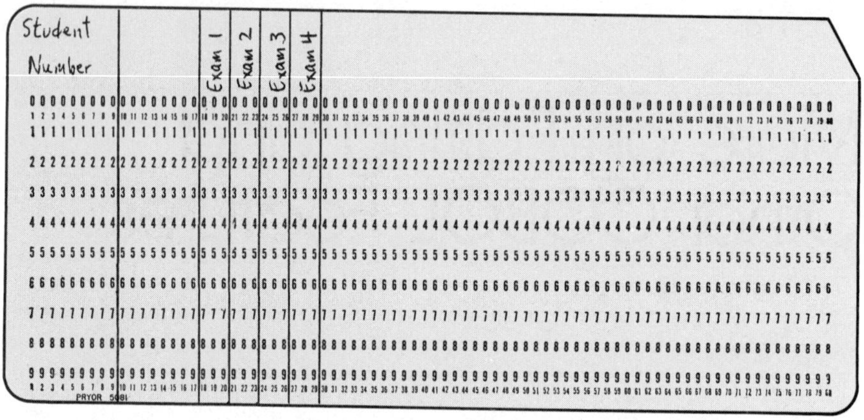

Student Number | Exam 1 | Exam 2 | Exam 3 | Exam 4

0 0 0 0 0 0 0 0 0 0 0 0 0 0 0 0 0 0 0 0 0 0 0 0 0 0 0 0 0 0 0 0 0 0 0 0 0 0 0 0 0 0 0 0 0 0 0 0 0 0 0 0 0 0 0 0 0 0 0 0 0 0 0 0 0 0 0 0 0 0 0 0 0 0 0 0 0 0 0 0
1 1 1 1 1 1 1 1 1 1 1 1 1 1 1 1 1 1 1 1 1 1 1 1 1 1 1 1 1 1 1 1 1 1 1 1 1 1 1 1 1 1 1 1 1 1 1 1 1 1 1 1 1 1 1 1 1 1 1 1 1 1 1 1 1 1 1 1 1 1 1 1 1 1 1 1 1 1 1 1
2 2 2 2 2 2 2 2 2 2 2 2 2 2 2 2 2 2 2 2 2 2 2 2 2 2 2 2 2 2 2 2 2 2 2 2 2 2 2 2 2 2 2 2 2 2 2 2 2 2 2 2 2 2 2 2 2 2 2 2 2 2 2 2 2 2 2 2 2 2 2 2 2 2 2 2 2 2 2 2
3 3 3 3 3 3 3 3 3 3 3 3 3 3 3 3 3 3 3 3 3 3 3 3 3 3 3 3 3 3 3 3 3 3 3 3 3 3 3 3 3 3 3 3 3 3 3 3 3 3 3 3 3 3 3 3 3 3 3 3 3 3 3 3 3 3 3 3 3 3 3 3 3 3 3 3 3 3 3 3
4 4 4 4 4 4 4 4 4 4 4 4 4 4 4 4 4 4 4 4 4 4 4 4 4 4 4 4 4 4 4 4 4 4 4 4 4 4 4 4 4 4 4 4 4 4 4 4 4 4 4 4 4 4 4 4 4 4 4 4 4 4 4 4 4 4 4 4 4 4 4 4 4 4 4 4 4 4 4 4
5 5 5 5 5 5 5 5 5 5 5 5 5 5 5 5 5 5 5 5 5 5 5 5 5 5 5 5 5 5 5 5 5 5 5 5 5 5 5 5 5 5 5 5 5 5 5 5 5 5 5 5 5 5 5 5 5 5 5 5 5 5 5 5 5 5 5 5 5 5 5 5 5 5 5 5 5 5 5 5
6 6 6 6 6 6 6 6 6 6 6 6 6 6 6 6 6 6 6 6 6 6 6 6 6 6 6 6 6 6 6 6 6 6 6 6 6 6 6 6 6 6 6 6 6 6 6 6 6 6 6 6 6 6 6 6 6 6 6 6 6 6 6 6 6 6 6 6 6 6 6 6 6 6 6 6 6 6 6 6
7 7 7 7 7 7 7 7 7 7 7 7 7 7 7 7 7 7 7 7 7 7 7 7 7 7 7 7 7 7 7 7 7 7 7 7 7 7 7 7 7 7 7 7 7 7 7 7 7 7 7 7 7 7 7 7 7 7 7 7 7 7 7 7 7 7 7 7 7 7 7 7 7 7 7 7 7 7 7 7
8 8 8 8 8 8 8 8 8 8 8 8 8 8 8 8 8 8 8 8 8 8 8 8 8 8 8 8 8 8 8 8 8 8 8 8 8 8 8 8 8 8 8 8 8 8 8 8 8 8 8 8 8 8 8 8 8 8 8 8 8 8 8 8 8 8 8 8 8 8 8 8 8 8 8 8 8 8 8 8
9 9 9 9 9 9 9 9 9 9 9 9 9 9 9 9 9 9 9 9 9 9 9 9 9 9 9 9 9 9 9 9 9 9 9 9 9 9 9 9 9 9 9 9 9 9 9 9 9 9 9 9 9 9 9 9 9 9 9 9 9 9 9 9 9 9 9 9 9 9 9 9 9 9 9 9 9 9 9 9

PRYOR 5081

Figure 5.1 Data card format for Program 11.

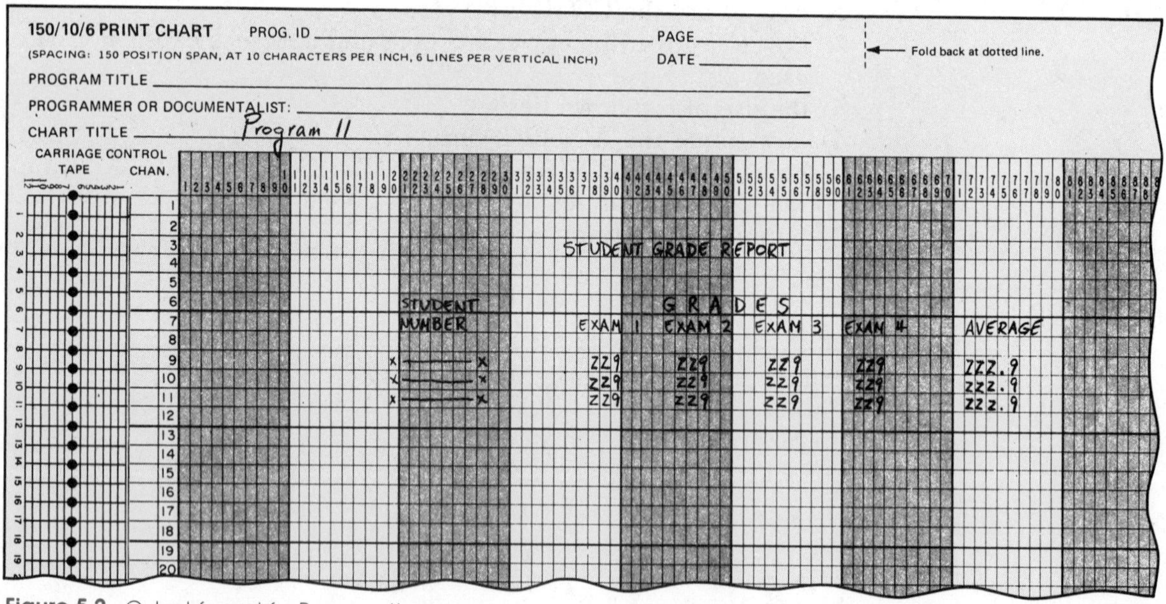

150/10/6 PRINT CHART PROG. ID ______ PAGE ______
(SPACING: 150 POSITION SPAN, AT 10 CHARACTERS PER INCH, 6 LINES PER VERTICAL INCH) DATE ______
Fold back at dotted line.
PROGRAM TITLE ______
PROGRAMMER OR DOCUMENTALIST: ______
CHART TITLE Program 11
CARRIAGE CONTROL TAPE CHAN.

STUDENT GRADE REPORT

G R A D E S

STUDENT NUMBER	EXAM 1	EXAM 2	EXAM 3	EXAM 4	AVERAGE
X———X	ZZ9	ZZ9	ZZ9	ZZ9	ZZZ.9
X———X	ZZ9	ZZ9	ZZ9	ZZ9	ZZZ.9
X———X	ZZ9	ZZ9	ZZ9	ZZ9	ZZZ.9

Figure 5.2 Output format for Program 11.

There is very little that is new in this program, shown in Figure 5.3. The field SUM–OF–GRADES is needed only so that the ADD statement has a place to assign its result in preparation for the DIVIDE. The input data for Program 11 is shown in Figure 5.4 and the output in Figure 5.5.

Figure 5.3 Program 11, Part 1 of 2.

```
4-CB2 V4 RELEASE 1.5 10NOV77          IBM OS AMERICAN NATIONAL STANDARD COBOL



000010 IDENTIFICATION DIVISION.
000020 PROGRAM-ID. PROG11.
000025*AUTHOR. PHIL RUBENSTEIN.
000030*
000040*      THIS PROGRAM READS A FILE OF EXAM GRADES AND
000050*      COMPUTES EACH STUDENT'S AVERAGE.
000060*
000070**********************************************************
000080
000090 ENVIRONMENT DIVISION.
000100 CONFIGURATION SECTION.
000110 SPECIAL-NAMES.
000120     C01 IS TO-NEW-PAGE.
000130 INPUT-OUTPUT SECTION.
000140 FILE-CONTROL.
000150     SELECT EXAM-GRADE-FILE-IN    ASSIGN TO UT-S-SYSIN.
000160     SELECT STUDENT-GRADE-REPORT  ASSIGN TO UT-S-SYSPRINT.
000170
000180**********************************************************
000190
000200 DATA DIVISION.
000210 FILE SECTION.
000220 FD  EXAM-GRADE-FILE-IN
000230     LABEL RECORDS ARE OMITTED.
000240
000250 01  EXAM-GRADE-RECORD.
000260     05 STUDENT-NUMBER-IN  PIC X(9).
000270     05 FILLER             PIC X(8).
000280     05 GRADE-1-IN         PIC 9(3).
000290     05 GRADE-2-IN         PIC 9(3).
000300     05 GRADE-3-IN         PIC 9(3).
000310     05 GRADE-4-IN         PIC 9(3).
000320     05 FILLER             PIC X(51).
000330
000340 FD  STUDENT-GRADE-REPORT
000350     LABEL RECORDS ARE OMITTED.
000360
000370 01  REPORT-LINE           PIC X(79).
000380
000390 WORKING-STORAGE SECTION.
000400 01  MORE-INPUT            PIC X(3) VALUE 'YES'.
000410
000420 01  REPORT-TITLE.
000430     05 FILLER             PIC X.
000440     05 FILLER             PIC X(35) VALUE SPACES.
000450     05 FILLER             PIC X(20) VALUE 'STUDENT GRADE REPORT'.
000460
000470 01  COLUMN-HEADS-1.
000480     05 FILLER             PIC X.
```

Figure 5.3 Program 11, Part 2 of 2.

```
000490        05 FILLER                    PIC X(20) VALUE SPACES.
000500        05 FILLER                    PIC X(24) VALUE 'STUDENT'.
000510        05 FILLER                    PIC X(12) VALUE 'G R A D E S '.
000580
000590 01   COLUMN-HEADS-2.
000600        05 FILLER                    PIC X.
000610        05 FILLER                    PIC X(20) VALUE SPACES.
000620        05 FILLER                    PIC X(16) VALUE 'NUMBER'.
000630        05 FILLER                    PIC X(8)  VALUE 'EXAM 1'.
000640        05 FILLER                    PIC X(8)  VALUE 'EXAM 2'.
000650        05 FILLER                    PIC X(8)  VALUE 'EXAM 3'.
000653        05 FILLER                    PIC X(11) VALUE 'EXAM 4'.
000656        05 FILLER                    PIC X(7)  VALUE 'AVERAGE'.
000660
000670 01   BODY-LINE.
000680        05 FILLER                        PIC X.
000690        05 FILLER                        PIC X(19) VALUE SPACES.
000700        05 STUDENT-NUMBER-OUT            PIC X(9).
000710        05 FILLER                        PIC X(9)  VALUE SPACES.
000720        05 GRADE-1-OUT                   PIC ZZ9.
000730        05 FILLER                        PIC X(5)  VALUE SPACES.
000740        05 GRADE-2-OUT                   PIC ZZ9.
000750        05 FILLER                        PIC X(5)  VALUE SPACES.
000760        05 GRADE-3-OUT                   PIC ZZ9.
000770        05 FILLER                        PIC X(5)  VALUE SPACES.
000780        05 GRADE-4-OUT                   PIC ZZ9.
000790        05 FILLER                        PIC X(7)  VALUE SPACES.
000800        05 AVERAGE-GRADE                 PIC ZZZ.9.
000805
000810 01   SUM-OF-GRADES                      PIC 9(3).
000820 01   NUMBER-OF-EXAMS                    PIC 9 VALUE 4.
000830
000850*************************************************************
001090
001100 PROCEDURE DIVISION.
001110      OPEN INPUT  EXAM-GRADE-FILE-IN,
001115           OUTPUT STUDENT-GRADE-REPORT.
001120      PERFORM REPORT-HEADINGS-ROUTINE.
001130      READ EXAM-GRADE-FILE-IN
001132          AT END
001134              MOVE 'NO' TO MORE-INPUT.
001140      PERFORM MAIN-PROCESS UNTIL MORE-INPUT IS EQUAL TO 'NO'.
001160      CLOSE EXAM-GRADE-FILE-IN,
001165            STUDENT-GRADE-REPORT.
001170      STOP RUN.
001180
001190 REPORT-HEADINGS-ROUTINE.
001200      WRITE REPORT-LINE FROM REPORT-TITLE
001210                                 AFTER ADVANCING TO-NEW-PAGE.
001220      WRITE REPORT-LINE FROM COLUMN-HEADS-1 AFTER ADVANCING 3.
001230      WRITE REPORT-LINE FROM COLUMN-HEADS-2 AFTER ADVANCING 1.
001240      MOVE SPACES TO REPORT-LINE.
001250      WRITE REPORT-LINE AFTER ADVANCING 1.
001260
001270 MAIN-PROCESS.
001280      MOVE STUDENT-NUMBER-IN TO STUDENT-NUMBER-OUT.
001290      MOVE GRADE-1-IN        TO GRADE-1-OUT.
```

Figure 5.3 Part 2 (continued).

```
001300        MOVE GRADE-2-IN          TO GRADE-2-OUT.
001315        MOVE GRADE-3-IN          TO GRADE-3-OUT.
001325        MOVE GRADE-4-IN          TO GRADE-4-OUT.
001330        ADD GRADE-1-IN,
001332            GRADE-2-IN,
001334            GRADE-3-IN,
001336            GRADE-4-IN
001338                GIVING SUM-OF-GRADES.
001340        DIVIDE SUM-OF-GRADES BY NUMBER-OF-EXAMS
001350                                     GIVING AVERAGE-GRADE ROUNDED.
001360        WRITE REPORT-LINE FROM BODY-LINE AFTER ADVANCING 1.
001370        READ EXAM-GRADE-FILE-IN
001380            AT END
001390                MOVE 'NO' TO MORE-INPUT.
```

Figure 5.4
Input data for Program 11.

```
070507324        100078098084
093456544        090085098000
987654677        022067076057
546787655        076083082092
```

Figure 5.5
Output from Program 11.

```
                      STUDENT GRADE REPORT

STUDENT                        G R A D E S
NUMBER           EXAM 1    EXAM 2    EXAM 3    EXAM 4        AVERAGE

070507324          100        78        98        84           90.0
093456544           90        85        98         0           68.3
987654677           22        67        76        57           55.5
546787655           76        83        82        92           83.3
```

Exercise 1. The data card format in Figure 5.E1 shows a Salesperson Number and five sale amounts. Write a program to read a deck of such cards and print, for each card, the Salesperson Number, the five sale amounts, and the average sale amount, rounded to two decimal places. Design the output on a printer spacing chart using a suitable report title and appropriate column headings.

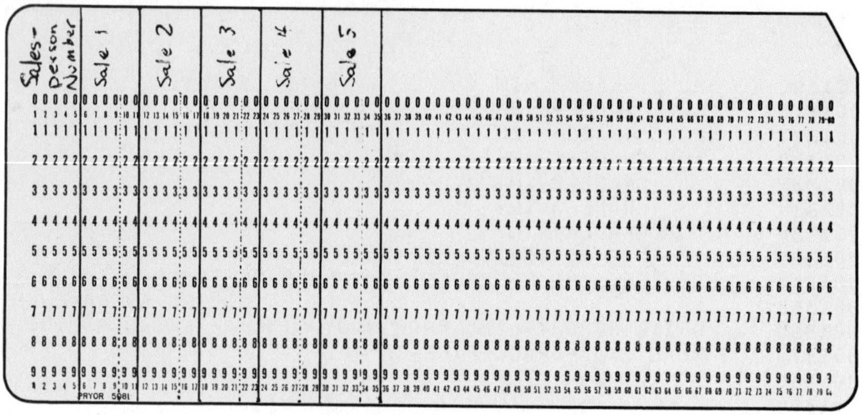

Figure 5.E1
Data card format for Exercise 1.

COMPUTE

COBOL provides a convenient way to avoid having to define fields for intermediate results. The fifth and last arithmetic verb **COMPUTE** permits the programmer to perform more than one arithmetic operation in a single statement. In Program 11, the ADD and DIVIDE statements could have been replaced with:

```
COMPUTE AVERAGE-GRADE ROUNDED =
   (GRADE-1-IN +
    GRADE-2-IN +
    GRADE-3-IN +
    GRADE-4-IN) / NUMBER-OF-EXAMS.
```

and the field SUM–OF–GRADES could be entirely omitted. In doing the computation, COBOL would set up its own field for the intermediate result. In COMPUTE statements, COBOL performs the computation indicated on the right side of the equal sign and assigns the result of the computation to the field named on the left side. One general format of the COMPUTE statement is

```
COMPUTE identifier-1 [ROUNDED] [identifier-2 [ROUNDED]] . . .
   = arithmetic-expression
   [ON SIZE ERROR imperative-statement]
```

The format shows that one identifier is required after the word COMPUTE and there can be as many as desired. Some COBOL systems permit only one identifier after the word COMPUTE. One **arithmetic expression** is required to the right of the equal sign.

An arithmetic expression may consist of one numeric literal or one identifier defined as a numeric field or any combination of literals and/or identifiers connected by arithmetic operators. The arithmetic operators are:

+	**addition**
−	**subtraction**
*	**multiplication**
/	**division**
**	**exponentiation**

Whenever arithmetic operators are used, they must be preceded and followed by a space. The COMPUTE statement earlier in this chapter shows the plus signs and the division sign each preceded and followed by a space. The expression:

```
VOLTAGE ** 2
```

would compute the square of the value assigned to VOLTAGE.

Expressions may be enclosed in parentheses to indicate the order in which the arithmetic should be performed. Some COBOL systems permit parentheses to be preceded and followed by a space, but in some systems left parentheses must be preceded by a space but not followed by a space, and right parentheses must not be preceded by a space but followed by a space. For safety, we can use the spacing shown in the COMPUTE statement earlier in this chapter.

Arithmetic is performed in the following order: contents of parentheses are evaluated first; then exponentiation is done; then multiplication and division; then addition and subtraction. If any ambiguity remains, then arithmetic is performed from left to right. So the expression:

```
A * B / C * D
```

is evaluated as if it were written:

```
A * (B / C) * D
```

A different result would be obtained if the expression were written:

```
(A * B) / (C * D)
```

Notice that the spacing of the arithmetic operators and the parentheses follows the rules given.

The **unary operators,** + and −, are also permitted. They may precede any expression. A + before an expression is only decorative, and a − before an expression has the same effect as multiplying the expression by −1. When unary operators are used, they must be preceded and followed by a space, except that in some COBOL systems the unary operator may not be preceded by a space if it follows a left parenthesis; in some systems the space following the unary operator may be omitted.) Check with your instructor for the spacing rules that apply to your particular COBOL system.

A More Difficult Program Using Averaging

Program 12 uses the same exam-grade input as Program 11, but Program 12, in addition to producing all of the same output as Program 11, will compute the average of all the grades on exam 1, the average of the grades on exams 2, 3, and 4, and the grand class average of all the individual student averages. The output format is shown in Figure 5.6

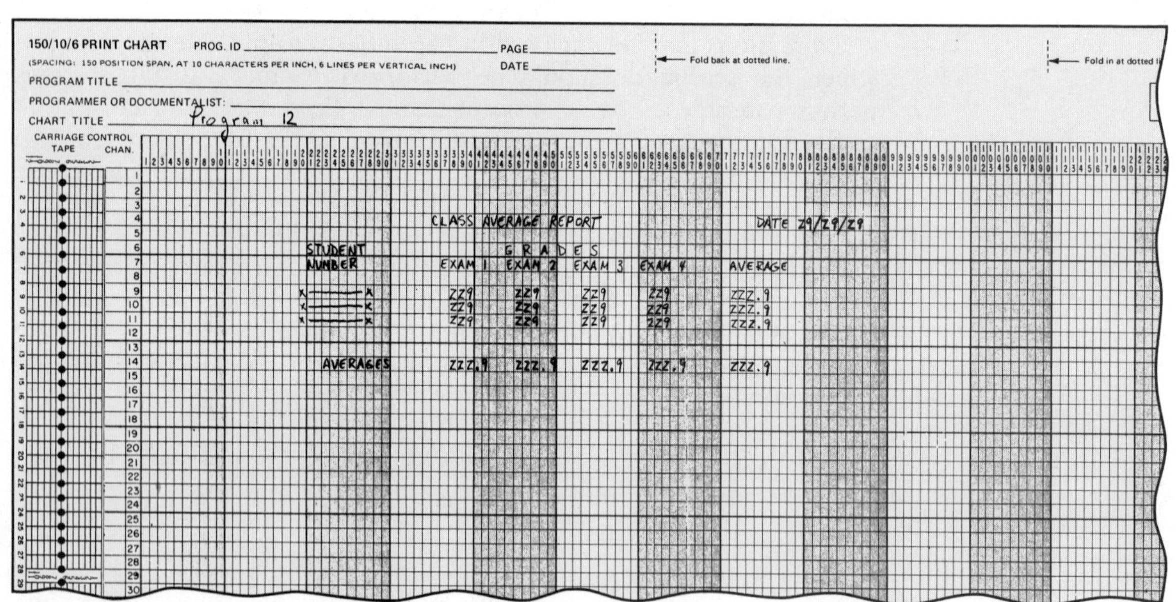

Figure 5.6 Output format for Program 12.

This program is sufficiently complicated so that a little top-down design may be in order. The main function of this program, "Produce class average report," and its main subfunctions are shown in Figure 5.7. Let us consider the subfunction "Produce final line of averages" for a moment. What are the subfunctions of "Produce final line of

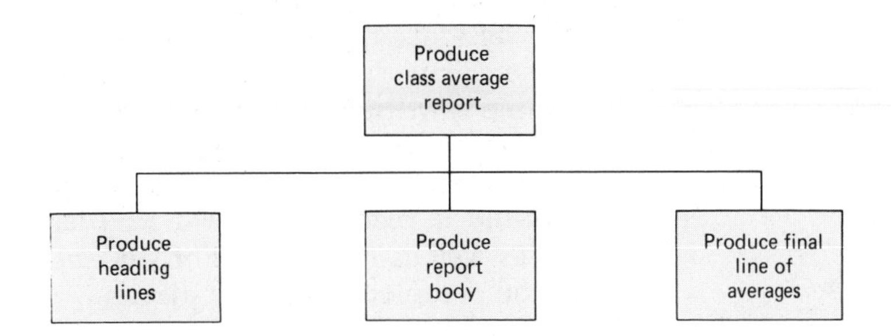

Figure 5.7
First-level hierarchy diagram for Program 12.

averages"? We know that all the subfunctions of "Produce final line of averages" are things that are done after end-of-file is detected in the input. This includes computing the final averages, moving the final averages to the final line and printing the final line. We place those subfunctions on the diagram as shown in Figure 5.8.

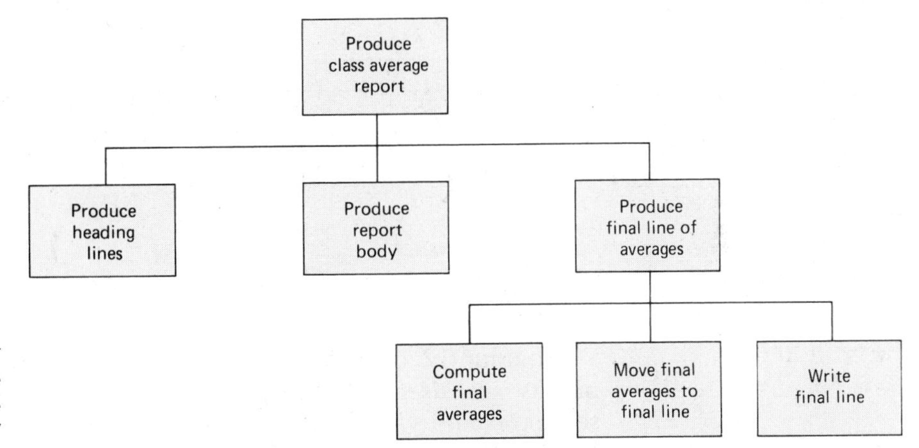

Figure 5.8
Hierarchy diagram for Program 12 showing the subfunctions of "Produce final line of averages."

This leaves everything else to be subfunctions of "Produce report body." One of these subfunctions is "Produce a body line," which we already know how to do from Program 11. As part of "Produce report body," we will have to generate the data that will be needed at end-of-file for "Compute final averages" to do its work. What data will "Compute final averages" need to compute, let's say, the average grade on exam 1? It will need the sum of all the grades on exam 1 and also the number of students who took the exam. "Compute final averages" will have to have similar information about exams 2, 3, and 4. To provide this information, we show the subfunctions "Accumulate grades" and "Count number of students" as subfunctions of "Produce report body," as in Figure 5.9. These two new subfunctions are added under "Produce report body" instead of somewhere else in the diagram because this is the only place where they can be carried out. As each body line is written onto the report, we will add the grades on the four exams into four accumulators and also the number 1 to another accumulator that will serve as a count of the number of students. When end-of-file is reached, we will have all the totals of grades that we need, as well as the number of students. The diagram in Figure 5.9 is then the complete hierarchy diagram.

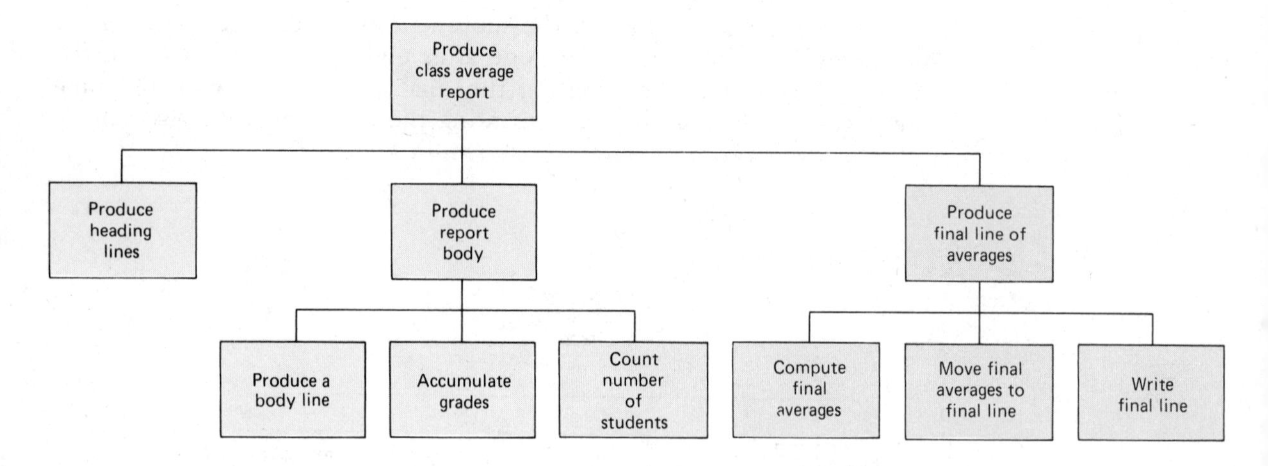

Figure 5.9 Complete hierarchy diagram for Program 12.

Program 12 is shown in Figure 5.10. Although several new techniques are used in this program, much of the coding should be self-explanatory. As shown in the hierarchy diagram, the loop PRODUCE–REPORT–BODY carries out the three functions of producing a report body line, accumulating the grades for each exam, and counting the number of students. Grade totals are accumulated in the fields EXAM–1–SUM, EXAM–2–SUM, EXAM–3–SUM, and EXAM–4–SUM. Each of the accumulators is defined as purely numeric and large enough to hold the largest possible expected sum of all the student grades on a single exam.

The routine PRODUCE–FINAL–LINE–OF–AVERAGES also carries out its functions as indicated in the hierarchy diagram. Among other things, it computes the class average grade on each of the four exams by using the values of the fields EXAM–1–SUM, EXAM–2–SUM, EXAM–3–SUM, EXAM–4–SUM, and NUMBER–OF–STUDENTS. How do those fields happen to have the correct values at the time PRODUCE–FINAL–LINE–OF–AVERAGES is executed? Well, PRO-

Figure 5.10 Program 12, Part 1 of 3.

```
4-CB2 V4 RELEASE 1.5 10NOV77        IBM OS AMERICAN NATIONAL STANDARD COBOL



000010 IDENTIFICATION DIVISION.
000020 PROGRAM-ID. PROG12.
000030*AUTHOR. PHIL RUBENSTEIN.
000040*
000050*      THIS PROGRAM READS A FILE OF EXAM GRADES AND
000060*      COMPUTES EACH STUDENT'S AVERAGE AND THE CLASS AVERAGE.
000070*
000080*******************************************************
000090
000100 ENVIRONMENT DIVISION.
000110 CONFIGURATION SECTION.
000120 SPECIAL-NAMES.
000130      C01 IS TO-NEW-PAGE.
000140 INPUT-OUTPUT SECTION.
000150 FILE-CONTROL.
000160      SELECT EXAM-GRADE-FILE-IN     ASSIGN TO UT-S-SYSIN.
000170      SELECT STUDENT-GRADE-REPORT   ASSIGN TO UT-S-SYSPRINT.
000180
000190*******************************************************
000200
000210 DATA DIVISION.
000220 FILE SECTION.
000230 FD   EXAM-GRADE-FILE-IN
000240      LABEL RECORDS ARE OMITTED.
000250
000260 01   EXAM-GRADE-RECORD.
000270      05 STUDENT-NUMBER-IN   PIC X(9).
000280      05 FILLER              PIC X(8).
000290      05 GRADE-1-IN          PIC 9(3).
000300      05 GRADE-2-IN          PIC 9(3).
000310      05 GRADE-3-IN          PIC 9(3).
000320      05 GRADE-4-IN          PIC 9(3).
000330      05 FILLER              PIC X(51).
000340
000350 FD   STUDENT-GRADE-REPORT
000360      LABEL RECORDS ARE OMITTED.
000370
000380 01   REPORT-LINE            PIC X(88).
000390
000400 WORKING-STORAGE SECTION.
000410 01   MORE-INPUT             PIC X(3) VALUE 'YES'.
000420
000430 01   RUN-DATE.
000440      05 RUN-YEAR            PIC 99.
000450      05 RUN-MONTH           PIC 99.
000460      05 RUN-DAY             PIC 99.
000470
000480 01   REPORT-TITLE.
000490      05 FILLER              PIC X.
```

Figure 5.10 Program 12, Part 2 of 3.

```
000500        05 FILLER                PIC X(35) VALUE SPACES.
000510        05 FILLER                PIC X(39) VALUE 'CLASS AVERAGE REPORT'.
000520        05 FILLER                PIC X(5)  VALUE 'DATE'.
000530        05 RUN-MONTH             PIC Z9.
000540        05 FILLER                PIC X     VALUE '/'.
000550        05 RUN-DAY               PIC 99.
000560        05 FILLER                PIC X     VALUE '/'.
000570        05 RUN-YEAR              PIC Z9.
000580
000590 01     COLUMN-HEADS-1.
000600        05 FILLER                PIC X.
000610        05 FILLER                PIC X(20) VALUE SPACES.
000620        05 FILLER                PIC X(24) VALUE 'STUDENT'.
000630        05 FILLER                PIC X(12) VALUE 'G R A D E S '.
000640
000650 01     COLUMN-HEADS-2.
000660        05 FILLER                PIC X.
000670        05 FILLER                PIC X(20) VALUE SPACES.
000680        05 FILLER                PIC X(16) VALUE 'NUMBER'.
000690        05 FILLER                PIC X(8)  VALUE 'EXAM 1'.
000700        05 FILLER                PIC X(8)  VALUE 'EXAM 2'.
000710        05 FILLER                PIC X(8)  VALUE 'EXAM 3'.
000720        05 FILLER                PIC X(11) VALUE 'EXAM 4'.
000730        05 FILLER                PIC X(7)  VALUE 'AVERAGE'.
000740
000750 01     BODY-LINE.
000760        05 FILLER                      PIC X.
000770        05 FILLER                      PIC X(19) VALUE SPACES.
000780        05 STUDENT-NUMBER-OUT          PIC X(9).
000790        05 FILLER                      PIC X(9)  VALUE SPACES.
000800        05 GRADE-1-OUT                 PIC ZZ9.
000810        05 FILLER                      PIC X(5)  VALUE SPACES.
000820        05 GRADE-2-OUT                 PIC ZZ9.
000830        05 FILLER                      PIC X(5)  VALUE SPACES.
000840        05 GRADE-3-OUT                 PIC ZZ9.
000850        05 FILLER                      PIC X(5)  VALUE SPACES.
000860        05 GRADE-4-OUT                 PIC ZZ9.
000870        05 FILLER                      PIC X(7)  VALUE SPACES.
000880        05 AVERAGE-GRADE               PIC ZZZ.9.
000890
000900 01     FINAL-AVERAGE-LINE.
000910        05 FILLER                      PIC X.
000920        05 FILLER                      PIC X(22) VALUE SPACES.
000930        05 FILLER                      PIC X(15) VALUE 'AVERAGES'.
000940        05 EXAM-1-AVERAGE-OUT          PIC ZZZ.9.
000950        05 FILLER                      PIC X(3)  VALUE SPACES.
000960        05 EXAM-2-AVERAGE-OUT          PIC ZZZ.9.
000970        05 FILLER                      PIC X(3)  VALUE SPACES.
000980        05 EXAM-3-AVERAGE-OUT          PIC ZZZ.9.
000990        05 FILLER                      PIC X(3)  VALUE SPACES.
001000        05 EXAM-4-AVERAGE-OUT          PIC ZZZ.9.
001010        05 FILLER                      PIC X(5)  VALUE SPACES.
001020        05 CLASS-AVERAGE-OUT           PIC ZZZ.9.
001030
001040 01     NUMBER-OF-EXAMS                PIC 9     VALUE 4.
001050 01     EXAM-1-SUM                     PIC 9(5)  VALUE 0.
001060 01     EXAM-2-SUM                     PIC 9(5)  VALUE 0.
```

Figure 5.10 Part 2 (continued).

```
001070 01  EXAM-3-SUM                         PIC 9(5)  VALUE 0.
001080 01  EXAM-4-SUM                         PIC 9(5)  VALUE 0.
001090 01  NUMBER-OF-STUDENTS                 PIC 9(3)  VALUE 0.
001100
001110*********************************************************
001120
001130 PROCEDURE DIVISION.
001140     OPEN INPUT  EXAM-GRADE-FILE-IN,
001150          OUTPUT STUDENT-GRADE-REPORT.
001160     PERFORM PRODUCE-HEADING-LINES.
001170     READ EXAM-GRADE-FILE-IN
001180         AT END
001190             MOVE 'NO' TO MORE-INPUT.
001200     PERFORM PRODUCE-REPORT-BODY
001210         UNTIL
001220             MORE-INPUT IS EQUAL TO 'NO'.
001230     PERFORM PRODUCE-FINAL-LINE-OF-AVERAGES.
001240     CLOSE EXAM-GRADE-FILE-IN,
001250           STUDENT-GRADE-REPORT.
001260     STOP RUN.
001270
001280 PRODUCE-HEADING-LINES.
001290     ACCEPT RUN-DATE FROM DATE.
001300     MOVE CORRESPONDING RUN-DATE TO REPORT-TITLE.
001310     WRITE REPORT-LINE FROM REPORT-TITLE
001320                                AFTER ADVANCING TO-NEW-PAGE.
001330     WRITE REPORT-LINE FROM COLUMN-HEADS-1 AFTER ADVANCING 3.
001340     WRITE REPORT-LINE FROM COLUMN-HEADS-2 AFTER ADVANCING 1.
001350     MOVE SPACES TO REPORT-LINE.
001360     WRITE REPORT-LINE AFTER ADVANCING 1.
001370
001380 PRODUCE-REPORT-BODY.
001390     PERFORM PRODUCE-A-LINE.
001400     PERFORM ACCUMULATE-GRADES.
001410     PERFORM COUNT-NUMBER-OF-STUDENTS.
001420     READ EXAM-GRADE-FILE-IN
001430         AT END
001440             MOVE 'NO' TO MORE-INPUT.
001450
001460 PRODUCE-FINAL-LINE-OF-AVERAGES.
001470     DIVIDE EXAM-1-SUM BY NUMBER-OF-STUDENTS
001480                              GIVING EXAM-1-AVERAGE-OUT ROUNDED.
001490     DIVIDE EXAM-2-SUM BY NUMBER-OF-STUDENTS
001500                              GIVING EXAM-2-AVERAGE-OUT ROUNDED.
001510     DIVIDE EXAM-3-SUM BY NUMBER-OF-STUDENTS
001520                              GIVING EXAM-3-AVERAGE-OUT ROUNDED.
001530     DIVIDE EXAM-4-SUM BY NUMBER-OF-STUDENTS
001540                              GIVING EXAM-4-AVERAGE-OUT ROUNDED.
001550     COMPUTE CLASS-AVERAGE-OUT ROUNDED =
001560         (EXAM-1-SUM +
001570          EXAM-2-SUM +
001580          EXAM-3-SUM +
001590          EXAM-4-SUM) /
001600         (NUMBER-OF-EXAMS * NUMBER-OF-STUDENTS).
001610     WRITE REPORT-LINE FROM FINAL-AVERAGE-LINE AFTER ADVANCING 3.
001620
001630 PRODUCE-A-LINE.
```

Figure 5.10 Program 12, Part 3 of 3.

```
001640         MOVE STUDENT-NUMBER-IN TO STUDENT-NUMBER-OUT.
001650         MOVE GRADE-1-IN        TO GRADE-1-OUT.
001660         MOVE GRADE-2-IN        TO GRADE-2-OUT.
001670         MOVE GRADE-3-IN        TO GRADE-3-OUT.
001680         MOVE GRADE-4-IN        TO GRADE-4-OUT.
001690         COMPUTE AVERAGE-GRADE ROUNDED =
001700              (GRADE-1-IN +
001710               GRADE-2-IN +
001720               GRADE-3-IN +
001730               GRADE-4-IN) / NUMBER-OF-EXAMS.
001740         WRITE REPORT-LINE FROM BODY-LINE AFTER ADVANCING 1.
001750
001760     ACCUMULATE-GRADES.
001770         ADD GRADE-1-IN TO EXAM-1-SUM.
001780         ADD GRADE-2-IN TO EXAM-2-SUM.
001790         ADD GRADE-3-IN TO EXAM-3-SUM.
001800         ADD GRADE-4-IN TO EXAM-4-SUM.
001810
001820     COUNT-NUMBER-OF-STUDENTS.
001830         ADD 1 TO NUMBER-OF-STUDENTS.
```

DUCE–FINAL–LINE–OF–AVERAGES is executed only after the end-of-file is reached on the input data. Remember that the PERFORM PRODUCE–REPORT–BODY . . . statement will continue to execute as long as there is input data; once there is no more data the next PERFORM statement, PERFORM PRODUCE–FINAL–LINE–OF–AVERAGES, will execute. In PRODUCE–REPORT–BODY of course we accumulate all the values we would need, so that when we get to execute PRODUCE–FINAL–LINE–OF–AVERAGES the counts and sums are ready and waiting to be used.

The four DIVIDE statements in PRODUCE–FINAL–LINE–OF–AVERAGES compute the averages on the four exams and assign them directly to the output fields EXAM–1–AVERAGE–OUT, EXAM–2–AVERAGE–OUT, EXAM–3–AVERAGE–OUT and EXAM–4–AVERAGE–OUT. The COMPUTE statement computes the final average of all the averages by taking the sum of all the exam grades and dividing them by the number of students and the number of exams. All of the results of the computations in PRODUCE–FINAL–LINE–OF–AVERAGES go directly to edited numeric output fields instead of to purely numeric working fields, because no further arithmetic will be done on any of them.

Several new techniques are involved in printing the date as part of the title line of the report. The **ACCEPT** statement extracts the date from the COBOL system on the day the program is run. One form of the ACCEPT verb,

ACCEPT identifier FROM DATE

enables a COBOL program to move the date into any desired field described anywhere in the Data Division.) The date is defined by COBOL as a six-digit numeric field, and the statement:

```
ACCEPT RUN-DATE FROM DATE.
```

moves the date to the programmer-supplied field RUN–DATE in YYMMDD format (two digits for the year of the century, two for the month, and two for the day of the month). The programmer-supplied field must be large enough to hold six digits: RUN–DATE, defined in working storage, has its six positions broken down into two-digit fields called RUN–YEAR, RUN–MONTH, and RUN–DAY.

The three fields within RUN–DATE must be listed in that order because that is the order in which they are supplied by the ACCEPT verb. But suppose we don't want the date to print as YYMMDD? Suppose we want it to print in the usual American way MM/DD/YY (as we do) or the usual European way DD/MM/YY? COBOL makes it easy to rearrange the order of the fields and perform zero suppression on any or all of the three. You can see how the three fields are defined in REPORT–TITLE in the desired order, with the identical names that were used in RUN–DATE. Notice that no suffixes were used to distinguish the two fields called RUN–MONTH from each other. Similarly, there are two fields whose names are exactly RUN–DAY and two that are named RUN–YEAR. (Using duplicate names in COBOL is useful when the **CORRESPONDING** option is used with verbs.)

Study the MOVE CORRESPONDING statement following the ACCEPT. It tells COBOL to move all fields having identical names from RUN–DATE to REPORT–TITLE. (In executing the MOVE CORRESPONDING, COBOL examines all the fields that make up the source and destination areas and, for any fields having identical names, carries out a MOVE which is independent of any other MOVEs it might do.) So this MOVE CORRESPONDING will carry out three separate MOVEs, of RUN–MONTH, RUN–DAY, and RUN–YEAR, just as if we had written three separate MOVE statements.

The output from Program 12 is shown in Figure 5.11.

Figure 5.11
Output from Program 12.

CLASS AVERAGE REPORT DATE 2/08/80

STUDENT NUMBER	EXAM 1	GRADES EXAM 2	EXAM 3	EXAM 4	AVERAGE
070543242	100	78	98	84	90.0
091020222	90	85	98	0	68.3
075655343	22	67	76	57	55.5
513467845	76	83	82	92	83.3
AVERAGES	72.0	78.3	88.5	58.3	74.3

In future programs we will often use MOVE CORRESPONDING to save writing and keying and to make coding more compact and easier to understand. The ADD and SUBTRACT verbs also have the CORRESPONDING option.

*Exercise 2. Using the same input data as in Exercise 1, write a program to print a report containing all of the output as in Exercise 1, and in addition at the end of the report print a single item showing the average of all sales. Also, print the words BELOW QUOTA next to any salesperson line where the average sale is less than $100.

Plan the output for both the individual salesperson line and the final line on a printer spacing chart before you begin coding.

Data Name Qualification

In the previous section we saw how fields in the Data Division could be given duplicate names. In COBOL, data names need not be unique; that is, the same name may be used for more than one field, as in Program 12. But COBOL still requires that every field be in some way distinguishable from every other field. Although we had two fields called RUN–DAY in Program 12, anyone could tell which was which since one of the fields called RUN–DAY was in RUN–DATE and the other RUN–DAY was in REPORT–TITLE. Distinguishing between two or more fields on the basis of some larger field is called **data name qualification;** that is, if it were necessary for us to distinguish one RUN–DAY from another in a COBOL program, we could call one of the fields RUN–DAY IN RUN–DATE and the other RUN–DAY IN REPORT–TITLE. The name RUN–DAY would in each case be said to be **qualified**, and the names of the larger fields, in this case RUN–DATE and REPORT–TITLE, would be said to be the **qualifiers.** For example, if your particular COBOL system doesn't happen to have the CORRESPONDING option, then the following statements could be used in Program 12 in place of the MOVE CORRESPONDING:

```
MOVE RUN-MONTH IN RUN-DATE TO RUN-MONTH IN REPORT-TITLE.
MOVE RUN-DAY   IN RUN-DATE TO RUN-DAY   IN REPORT-TITLE.
MOVE RUN-YEAR  IN RUN-DATE TO RUN-YEAR  IN REPORT-TITLE.
```

In data name qualification, the word OF may be used interchangeably with IN, as in:

```
MOVE RUN-MONTH OF RUN-DATE TO RUN-MONTH IN REPORT-TITLE.
```

A qualified data name is an identifier and may be used in any statement whose general format calls for an identifier. Now we know two kinds of identifiers—ordinary data names of the sort we have been using all along and now qualified data names.

Additional PICTURE Features

In this section we will cover a number of features that make the PICTURE clause more flexible than it has been so far.

SIGNED FIELDS

Up to now we have been dealing with unsigned numbers in all our programs. We have been using numeric input, numeric output, numeric literals, and numeric intermediate results in working storage, but they have all been positive or zero. COBOL has a number of features dealing with **signed numbers.**

Numeric literals: We have already seen (in Table 3.1) that a numeric literal may be immediately preceded by a plus or a minus sign. A plus sign is only decorative; a minus sign causes COBOL to treat the number as negative. The only way to designate a negative literal is to place a minus sign at its left end with no space between the sign and the number.

Numeric input data: Numeric input data may be signed positive or negative and may have its sign in any one of four places. As shown in Table 3.1, the sign may be immediately to the left or to the right of the number or over the leftmost or rightmost digit of the number. Signed numeric input data is designated in the File Section by the PICTURE character S at the left end of the PICTURE. The character S goes at the left end of the PICTURE regardless of where the sign is keyed in the number. So a signed, three-digit integer input item would be described with PICTURE S999.

The programmer tells COBOL where the sign is located by using the **SIGN** clause. The SIGN clause if needed can be included right after the PICTURE clause in the description of a data item. The possible locations of the sign in numeric input data and the corresponding SIGN clauses are:

Location of sign	SIGN clause
Over leftmost digit	SIGN IS LEADING
Over rightmost digit	SIGN IS TRAILING
Immediately to left of number	SIGN IS LEADING SEPARATE CHARACTER
Immediately to right of number	SIGN IS TRAILING SEPARATE CHARACTER

Some COBOL systems have a more or less standard place where the sign is usually expected to appear in input data. If the sign is in fact located in that place, then the SIGN clause can be omitted. In the COBOL system used to run the programs in this book, the **default** location of the sign in input data is over the rightmost digit of the number; so if a signed, five-digit, dollars-and-cents field with its sign over the rightmost digit is being read, its description might be:

```
05 MONEY-IN     PIC S999V99 SIGN IS TRAILING.
```

or equivalently:

```
05 MONEY-IN     PIC S999V99.
```

Numeric intermediate results in working storage: The PICTURE character S may be used on purely numeric items in working storage. The S is required if the value of the numeric item is expected to be negative. If the value of the item might be either negative or positive, the presence of an S does not in any way interfere with the proper processing of positive values. In fact, some COBOL systems operate more efficiently (that is, programs occupy less computer storage space and execute faster) if all working storage numeric items have S in their PICTURE, regardless of whether their values are expected to be positive, negative, or both. So in all the programs we have done so far, it would do no harm (and possibly some good) if we included an S in the PICTURE of all the intermediate result fields. If used, the PICTURE character S must appear at the left end of the PICTURE, as in:

```
05 MERCHANDISE-AMOUNT-W     PIC S9(6)V99.
```

In all subsequent programs, we will include S in the descriptions of numeric intermediate items.

Numeric output: When signed numbers are printed as output, the programmer may designate how and where the sign of the number is to print. Positive numbers may have a plus sign printed at or near the right or left end of the number, and negative numbers may have either a minus sign at or near the right or left end of the number or the letters CR or DB at or near the right end of the number. Here are some ways that a positive number could be made to print:

```
+675.28
+  675.28
   675.28+
   675.28   +
```

Here are some ways that a negative number can be made to print:

```
  675.28-
 -675.28
-  675.28
  675.28CR
  675.28  DB
```

There can be as many blanks as desired between the number and its sign designation. Each digit, blank, and sign occupies one print position. The sign designations CR and DB each occupy two print positions. The programmer indicates the type of sign designation desired by placing the sign character in the edit PICTURE of the output field; the following is a legal description:

```
05 FIELD-1-OUT          PIC -ZZZ.99.
```

This PICTURE tells COBOL to print a minus sign in the position shown if the printed value happens to be negative. If a positive value is edited under this PICTURE, a blank will print in the position allotted to the sign. Here is how several values would print under the edit PICTURE –ZZZ.99:

Value	–ZZZ.99
–67528	–675.28
–00560	– 5.60
–00050	– .50
+67528	675.28
00000	.00

Similarly, the PICTURE character – may be placed at the right end of the PICTURE, or the letters CR or DB may be placed at the right end of the PICTURE. You may not use the minus sign and the CR or DB designation in the same PICTURE. Here is how some values will print under the PICTUREs ZZ9.99–; ZZZ.99CR; and Z99.99DB:

Value	ZZ9.99–	ZZZ.99CR	Z99.99DB
–675^28	675.28–	675.28CR	675.28DB
–005^60	5.60–	5.60CR	05.60DB
–000^50	0.50–	.50CR	00.50DB
+675^28	675.28	675.28	675.28
000^00	0.00	.00	00.00

To obtain one or more blank spaces between the sign indication and the number, the PICTURE character B may be used to position the sign where it is wanted. Here is how some values will print under several edit PICTUREs:

Value	-BZZZ,ZZ9	-BBZZZ,ZZ9	ZZZ,ZZ9BCR	ZZZ,ZZ9BBDB
-589000	- 589,000	- 589,000	589,000 CR	589,000 DB
-006890	- 6,890	- 6,890	6,890 CR	6,890 DB
-000055	- 55	- 55	55 CR	55 DB
+589000	589,000	589,000	589,000	589,000
000000	0	0	0	0

To obtain a printed plus sign, the PICTURE character + may be used. But the character + works a little differently from the other three sign designators −, CR, and DB. When you use + you will always get some printed sign indication, whether the printed value happens to be positive, negative, or zero, unless all the digits are zero-suppressed, in which case the sign also disappears. The other three sign designators print only if the value happens to be negative, and they give no sign indication when the value is positive. When the PICTURE character + is used, a plus sign will print next to positive or zero values and a minus sign next to negative values. Here is how some values will print under two different PICTUREs:

Value	+ZZZ	ZZ9B+
+520	+520	520 +
-520	-520	520 -
+006	+ 6	6 +
000		0 +

The way to remember when to use S and when to use −, +, CR, and DB is: −, +, CR, and DB are used for printing final output onto paper; S is used for all other signed fields.

Exercise 3. Write 05-level entries for the following input fields:

a. a seven-digit dollars-and-cents field, called COMMISSION–IN, with a sign over the leftmost digit,
b. a six-digit field with three decimal places, called GAMMA–IN, with a sign immediately to the right of the number,
c. a nine-digit integer, called DISTANCE–IN, with a sign immediately to the left of the number.

Exercise 4. Write an 05-level working storage entry for a signed, eight-digit field with four decimal places, called VELOCITY–W.

Exercise 5. Write the PICTURE for each of the following types of output editing:

a. a six-digit number, three decimal places, minus sign to print at the left end of the number if it is negative; no sign to print if the number is positive. Provide zero suppression up to the decimal point.
b. a six-digit dollars-and-cents number, comma after the first digit, the letters CR to print one space from the right end of the number if it is negative; no sign indication to print if the number is positive. Provide for zero suppression of the entire field.
c. a seven-digit integer, comma after the fourth digit only, a plus sign to print at the right end of the number if it is positive; a minus sign to print at the right end of the number if it is negative; zero suppression up to but not including the rightmost digit.

FLOATING INSERTION CHARACTERS

In the previous section we saw how a plus sign or a minus sign could be made to print in a fixed position to the left of a printed number. It is possible to have COBOL print instead a plus or a minus sign (and as we shall see, a dollar sign as well) immediately to the left of the most significant digit in the number as printed. Such an operation is called **floating** because the sign floats to the right until it finds the most significant digit and prints itself there. A floating sign is indicated by a string of two or more such signs in the edit PICTURE of the output field. A floating sign also carries out zero suppression, and Zs must not be used in any PICTURE that specifies a floating sign. Here is how some values would print under one PICTURE containing a fixed minus sign and another containing a floating minus sign:

Value	– ZZZ,ZZ9	----,--9
–589000	–589,000	–589,000
–006809	– 6,809	–6,809
–000050	– 50	–50
–000500	– 500	–500
+589000	589,000	589,000
000000	0	0

Notice that, when a floating sign is used, there must be enough signs to allow for the largest possible number that might print and for the printing sign itself. So a three-digit number could have PICTURE ---- to allow one sign to print, even if digits occupy the other three places. Of course, a three-digit number could also have PICTURE ---9 if the programmer wants zero suppression to stop before the rightmost digit. A PICTURE for a four-digit number would have to allow six places altogether, if a comma were also wanted (four for the digits, one for the sign, one for the comma.) PICTURE --,--- or PICTURE --,--9 would be legal. When a floating sign is used, there must always be at least two signs to the left of the leftmost comma. To see why, try to figure out what size number this illegal PICTURE is trying to represent: PICTURE -,---.

A string of plus signs can be used to obtain floating sign indication to the left of the most significant digit, whether the number happens to be positive or negative.

The PICTURE character $ can be used to insert a dollar sign at the left end of a number in either a fixed or floating position. If the programmer wants the dollar sign to print in a fixed position, a single $ should be inserted in the PICTURE where the dollar sign is to print, as PICTURE $ZZZ,ZZZ.99. If a floating dollar sign is wanted, a string of two or more $ should be used in the PICTURE. No Zs may appear in the PICTURE if a floating dollar sign is used; if a fixed dollar sign is used no Bs may appear to the left of the dollar sign.

Here is how some values will print under one edit PICTURE with a fixed dollar sign and under two with floating dollar signs:

Value	$ZZ,ZZZ.99	$$$,$$$.99	$$$,$$$.$$
5890503	$58,905.03	$58,905.03	$58,905.03
0890503	$ 8,905.03	$8,905.03	$8,905.03
0060050	$ 600.50	$600.50	$600.50
0000529	$ 5.29	$5.29	$5.29
0000006	$ .06	$.06	$.06
0000000	$ .00	$.00	

Notice that when a field is entirely zero-suppressed by a floating dollar sign to the right of the decimal point, the entire field is blanked when the value of the number is zero. The same holds true when a floating minus or plus sign appears to the right of a decimal point.

It is possible to combine fixed minus and plus signs with fixed and/or floating dollar signs in certain combinations. In a PICTURE with a fixed dollar sign, a fixed plus or minus sign may appear to the left of the dollar sign, or a fixed plus or minus sign (or CR or DB) may appear at the right end of the number. In a PICTURE with a floating dollar sign, a fixed plus or minus sign (or CR or DB) may appear at the right end of the number.

CHECK PROTECTION

When COBOL programs are used to print checks it is usually undesirable to leave any blank spaces in the place on the check where the money amount is printed. If the check field has room for, let's say, a seven-digit number, and the particular check amount has only four or five digits, the blanks that result from suppression of the leading zeros are an invitation to forgery. The invitation is especially appealing if the check is for an amount less than $1 and only a decimal point and some pennies are printed. For this reason, COBOL provides a form of zero suppression where leading zeros are replaced not with blanks but with asterisks (*). Replacing leading zeros with * is called **check protection.** To obtain check protection, the programmer uses the PICTURE character * the same way that the PICTURE character Z would be used. * and Z may not appear in the same PICTURE, although * may appear in a PICTURE with a dollar sign, a plus or minus sign, CR, DB, commas, and a decimal point. Here are some examples of money amounts printed with and without check protection:

Value	$ZZ,ZZZ.99	$**,***.99	$**,***.**
5890503	$58,905.03	$58,905.03	$58,905.03
0890503	$ 8,905.03	$*8,905.03	$*8,905.03
0060050	$ 600.50	$***600.50	$***600.50
0000529	$ 5.29	$*****5.29	$*****5.29
0000006	$ .06	$******.06	$******.06
0000000	$ .00	$******.00	* ******.**

Notice that when a field is entirely zero-suppressed by use of asterisks, it is not blanked when zero. Blank when zero occurs only if a field is entirely zero-suppressed through the use of Z, $, +, or −.

OTHER PICTURE FEATURES

There are many more details of output editing which we will not cover here, for example, the use of the PICTURE character B in combination with floating insertion characters. For additional editing details, see your COBOL manual.

Exercise 6. Show how each of the following values would print under each of the edit PICTURES shown:

Value	$ZZ9−	$ZZZ+	$$$$−	***CR	$***DB	$**9+
+590						
−590						
+060						
−060						
+005						
−005						
000						

Summary

The COMPUTE verb can be used to perform more than one arithmetic operation in a single statement and can sometimes eliminate the need for fields to hold intermediate results of arithmetic. A COMPUTE statement always contains an equal sign, and COBOL assigns the results of the computation to the field whose name appears to the left of the sign. There may be an arithmetic expression or just a literal or the name of a single field defined as numeric to the right of the sign.

The order of arithmetic in an expression is exponentiation first, then multiplication and division, then addition and subtraction. Parentheses may be used to change the order of arithmetic. The arithmetic operators are +, −, *, /; for exponentiation, **. The unary operators + and − may also appear before any expression.

Top-down design can be used to organize the functions needed in a program of moderate complexity, such as one to compute averages of values in a file. There is no single best hierarchy diagram that would be agreed to by all programmers, and sometimes a case can be made for placing a certain function in one or another place in the diagram.

The ACCEPT verb and the reserved word DATE may be used to obtain the current date from the COBOL system on the day a program is run. The ACCEPT verb moves the date in six-digit YYMMDD format into a field defined by the programmer.

The MOVE CORRESPONDING statement can be used to move data from fields in one area to fields in another where the sending and receiving fields have duplicate names. Duplicate named fields can be distingushed from one another if necessary by data name qualification. They must be defined so that, through the use of the word IN or OF, every field is capable of being made unique. A qualified data name may be used in any COBOL statement whose general format calls for an identifier.

COBOL has special features that deal with signed fields. Numeric literals may have a plus or minus sign at the left end. Numeric input may have a sign at the left or right end of the number or over the leftmost or rightmost digit. The location of the sign is given in the SIGN clause. The presence of a sign is indicated by the PICTURE character S in the PICTURE of the input data item. Intermediate numeric results in working storage may also be signed, and the presence of a sign is indicated by the PICTURE character S in the description of the working storage item.

Printed output may have its sign designated by a fixed or floating plus or minus sign at the left end or by a fixed +, −, CR, or DB at the right end.

A fixed or floating dollar sign may be printed with numeric output through the use of the PICTURE character $. Fixed positive and negative sign indications may be used with a fixed or floating dollar sign in certain combinations.

COBOL provides a check protection feature through the use of the PICTURE character *. In check protection, asterisks instead of blanks replace leading zeros to discourage inserting forged numbers on checks. The character * can be used in a PICTURE anywhere that a Z would otherwise be.

Fill-in Exercises

1. The ________ verb permits more than one arithmetic operation to be carried out in a single statement.
2. The order of evaluation in an arithmetic expression is contents of parentheses first, then ________, then ________ and ________, then ________ and ________.
3. When the signs + and − appear immediately before an arithmetic expression, they are called ________.
4. The ________ verb can be used to obtain the date from the COBOL system on the day a program is run.
5. The ________ option can do several independent MOVEs in one MOVE statement.
6. Distinguishing between two or more fields having identical names by using the names of some larger fields is called ________ ________.
7. The ________ clause may be included in the description of an input data item to indicate the location of a plus or minus sign.
8. The PICTURE character S when used must always appear at the ________ end of the PICTURE.
9. The PICTURE character ________ causes some sign to print whether the value of the output data item is positive, negative, or zero.
10. The sign designation DB prints only if the output data value is ________.

Review Exercises

1. Using the input data shown in Figure 2.11 (whose format is shown in Figure 2.10), write a program to print the contents of each card read and, at the end of the report, the average annual salary of all employees. Use a printer spacing chart to plan the output.

*2. Using the same input data as in Review Exercise 1, write a program to print the contents of each card. On each line print the words

SALARY OVER $10,000 or SALARY NOT OVER $10,000, as appropriate. At the end of the report, print the number of employees with salary over $10,000, the number of employees with salary not over $10,000, the average of the salaries over $10,000, and the average of the salaries not over $10,000. Plan the output on a printer spacing chart before you begin coding. Include the run date as part of the report heading.

3. Given the following data definitions:

```
01  GROUP-1.
    05 FIELD-1          PIC X(5).
    05 FIELD-2          PIC X(8).
    05 FIELD-3          PIC X(6).
01  GROUP-2.
    05 FIELD-1          PIC X(5).
    05 FIELD-2          PIC X(8).
    05 FIELD-3          PIC X(6).
```

Which of the following is (are) valid COBOL statements?

   a. MOVE FIELD-1 TO FIELD-1.
   b. MOVE FIELD-1 IN GROUP-2 TO FIELD-1 IN GROUP-1.
   c. MOVE CORRESPONDING FIELD-1 TO FIELD-1.
   d. MOVE CORRESPONDING GROUP-1 TO GROUP-2.
   e. MOVE CORRESPONDING FIELD-1 TO FIELD-2.

4. Write 05-level entries for the following input fields:

   a. a nine-digit dollars-and-cents field called GROSS-REVENUES, with a sign over the rightmost digit;
   b. a six-digit integer called VELOCITY, with a sign over the leftmost digit;
   c. an unsigned seven-digit integer called BOXCARS.

5. Write the PICTURE for each of the following types of output editing:

   a. an eight-digit dollars-and-cents number, unsigned, fixed dollar sign to print at the left end of the number, comma after the third digit;
   b. an eight-digit dollars-and-cents number, minus sign to print at the right end if the number is negative; no sign to print if the number is positive, dollar sign to float to two places before the decimal point, no comma;
   c. a six-digit integer, comma after the third digit, zero suppression up to but not including the rightmost digit, plus sign to float up to but not including the rightmost digit if the number is positive; negative sign to float up to but not including the rightmost digit if the number is negative.

6. Show how each of the following values would print under each of the edit PICTUREs shown:

Value	$$$$,$$9.99	$***,*99.99	$***,***.**
389654^90			
003746^23			
000556^75			
000056^75			
000004^98			
000000^29			
000000^05			
000000^00			

Control Breaks

6

KEY POINTS

Here are the key points you should learn from this chapter:

1. what control breaks are and why they are important in data processing;
2. a generalized approach to developing hierarchy diagrams for programs with control breaks;
3. how to program control breaks in COBOL.

KEY WORDS

Key words to recognize and learn:

control break	figurative constant	major control field
control field	ZEROS	minor control field
detail line	ZEROES	intermediate control field
COMPUTATIONAL	SPACE	group indicating
LINE–COUNTER	group item	nested IF
COMP	elementary item	group printing
USAGE	subordinate	summary reporting
DISPLAY	BEFORE ADVANCING	
ZERO	mnemonic name	

In this chapter we will study techniques for printing totals throughout a report as well as at its end. This kind of processing is carried out in almost every COBOL installation in the world and can become quite complicated. By starting at the beginning and using top-down design, we will be able to keep the difficulties under control. In the course of studying these totalling techniques, we will also see some other useful COBOL features.

A Report with One Control Break

The first program in this chapter will use input in the format shown in

Figure 3.15. We are going to assume for Program 13 that there may be any number of input cards for a single account number; that is, in any run, each account number will probably have several credits and debits in the input. The cards for each account number will be grouped together in the input. The program is to read the input file, print the contents of each card, and also print the total debits and the total credits for each account number. The totals for each account number are to be printed immediately following the printing of the group of lines for that account number. The output from Program 13 is to look like Figure 6.1, from input shown in Figure 6.2.

Figure 6.1 Output from Program 13, Part 1 of 3.

```
                         ACCOUNT ACTIVITY REPORT

   DATE  2/13/80                                                      PAGE  1

                   ACCT. NO.          DEBITS         CREDITS

                     02005              30.34
                     02005                                 .00
                     02005                           98,762.01
                     02005                .06
                     02005          89,235.51
                     02005                           34,859.10

TOTALS FOR ACCOUNT NUMBER 02005    $89,265.91     $133,621.11

                     04502                           99,999.99
                     04502              83.99
                     04502          45,672.12

TOTALS FOR ACCOUNT NUMBER 04502    $45,756.11      $99,999.99

                     12121          75,499.00
                     12121                .02
                     12121                .00
                     12121                           23,411.11
                     12121                               66.67
                     12121          66,662.22

TOTALS FOR ACCOUNT NUMBER 12121   $142,161.24      $23,477.78

                     19596                           92,929.29
                     19596          12,547.20
                     19596          23,487.64
                     19596                .20
                     19596                                2.39
                     19596                              213.45

TOTALS FOR ACCOUNT NUMBER 19596    $36,035.04      $93,145.13

                     20023          99,999.99
```

Figure 6.1 Output from Program 13, Part 2 of 3.

```
                              ACCOUNT ACTIVITY REPORT

   DATE  2/13/80                                                          PAGE  2

                     ACCT. NO.           DEBITS          CREDITS

                       20023           87,654.99
                       20023           86,868.68
                       20023                           88,888.80
                       20023                           88,553.32
                       20023                           56,789.23

 TOTALS FOR ACCOUNT NUMBER 20023     $274,523.66     $234,231.35

                       23456                2.11
                       23456                4.53
                       23456               12.00
                       23456                                 .12
                       23456                               18.00
                       23456                                 .54

 TOTALS FOR ACCOUNT NUMBER 23456          $18.64          $18.66

                       30721                1.01
                       30721                               99.12
                       30721               98.44

 TOTALS FOR ACCOUNT NUMBER 30721          $99.45          $99.12

                       40101              342.87
                       40101                              212.45
                       40101           60,002.01
                       40101                               67.00
                       40101           32,343.23
                       40101                           90,023.41

 TOTALS FOR ACCOUNT NUMBER 40101      $92,688.11      $90,302.86

                       67689               11.14
                       67689                            1,010.10
```

For the first time we have so much output that it occupies more than one page. Each page of output has its own heading, and the pages are numbered consecutively. You can see that for a single account number there may be several debits and/or credits. All the lines for an account number print consecutively, and at the end of the group of lines there is a line showing the total of the debits and the total of the credits for the account number. At the end of the next account number are its

Figure 6.1 Part 3 of 3.

```
                          ACCOUNT ACTIVITY REPORT
   DATE   2/13/80                                                  PAGE  3

                   ACCT. NO.         DEBITS         CREDITS

                    67689             33.35
                    67689                           2,203.30
                    67689          3,040.20
                    67689                          10,000.00

TOTALS FOR ACCOUNT NUMBER 67689    $3,084.69      $13,213.40

                    72332               .04
                    72332                              44.44
                    72332            444.44
                    72332                               4.44
                    72332               .44
                    72332                           4,040.40

TOTALS FOR ACCOUNT NUMBER 72332     $444.92       $4,089.28

                    74567             22.21
                    74567             22.11
                    74567         98,777.73
                    74567              3.24
                    74567                          87,654.32
                    74567                              78.93

TOTALS FOR ACCOUNT NUMBER 74567   $98,825.29      $87,733.25
```

totals, and so on for all the account numbers. The input to this program is in account number order; that is, each account number in the input is higher than the account number before it, and all cards for a single account number are together in the input. This program would have worked as well even if the account numbers were not in order, just as long as all cards for a single account come in one after the other in the input. The account numbers were ordered so that it would be easy to find them on the output.

Figure 6.2
Input data for Program 13.

```
-02005   0003034
 02005   0000000
 02005   9876201
-02005   0000006
-02005   8923551
 02005   3485910
 04502   9999999
-04502   0008399
-04502   4567212
-12121   7549900
-12121   0000002
-12121   0000000
 12121   2341111
 12121   0006667
-12121   6666222
 19596   9292929
-19596   1254720
-19596   2348764
-19596   0000020
 19596   0000239
 19596   0021345
-20023   9999999
-20023   8765499
-20023   8686868
 20023   8888880
 20023   8855332
 20023   5678923
-23456   0000211
-23456   0000453
-23456   0001200
 23456   0000012
 23456   0001800
 23456   0000054
-30721   0000101
 30721   0009912
-30721   0009844
-40101   0034287
 40101   0021245
-40101   6000201
 40101   0006700
-40101   3234323
 40101   9002341
-67689   0001114
 67689   0101010
-67689   0003335
 67689   0220330
-67689   0304020
 67689   1000000
-72332   0000004
 72332   0004444
-72332   0044444
 72332   0000444
-72332   0000044
 72332   0404040
-74567   0002221
-74567   0002211
-74567   9877773
-74567   0000324
 74567   8765432
 74567   0007893
```

Before doing a hierarchy diagram for this program, let's try to think about how this program can be done. We already know how to accumulate totals, so there is no problem accumulating the total of the debits and the total of the credits for the first account number as we print each line. But how will the program know when all the cards have been read for the first account number and that it is time to print the total for that account? The program will know that all of the first account is processed when it reads the first card of the second account number. As the program reads cards, it will have to compare the account number on each card read to the account number on the previous, until it detects a change in account number. The change in the account number is called a **control break,** and the account number field itself is called a **control field.**

Once a control break is detected, the program will print the total line for the first account number and prepare to process the next group of input cards, the ones containing the second account number. The preparation consists mainly of zeroing out the accumulator fields that were used to total the debits and credits for the first account number, so that the same fields may be used to total the debits and credits for the second. When the second control break is detected, the total line for the second account number can be printed, and the accumulators zeroed again in preparation for processing the third account number. In this way we can process an indefinite number of account numbers without having to provide endless numbers of accumulator fields.

Printing the very last total line sometimes presents a problem to programmers. Since a total line is printed only after the first card of the next group is detected, what will be printed when there is no next group? When end-of-file is detected on the input file, we have still not printed the totals of the debits and credits for the last account number, for the program has not detected a control break, a change in account number. So when doing the hierarchy diagram, we need only to remember that, after end-of-file is detected, one more total line has to be written, and that is the total line for the last account number.

One thing about the output from this program that we have not seen in earlier programs is that it occupies several pages, with a complete heading at the top of every page. This program will count the number of lines that are printed on a page. As each line is printed, the program will add 1 to an accumulator set aside to serve as a line counter (or 2 if a line is printed with double-spacing). When the line counter gets to some predetermined amount, the program will skip to a new page and print complete headings. Then the program must set the line counter back to the beginning and resume adding into it for every line printed.

THE HIERARCHY DIAGRAM

We can now begin to create the hierarchy diagram for this program. The first level of main subfunctions, as shown in Figure 6.3, is of the same general shape as on other one-level diagrams. We have an initialization box, a main loop, and some process which is performed after end-of-file on the input. In this program, what is carried out after end-of-file is only printing the last total line on the report—just the total line for the last group. All the other total lines on the report are printed by the main loop as we go along.

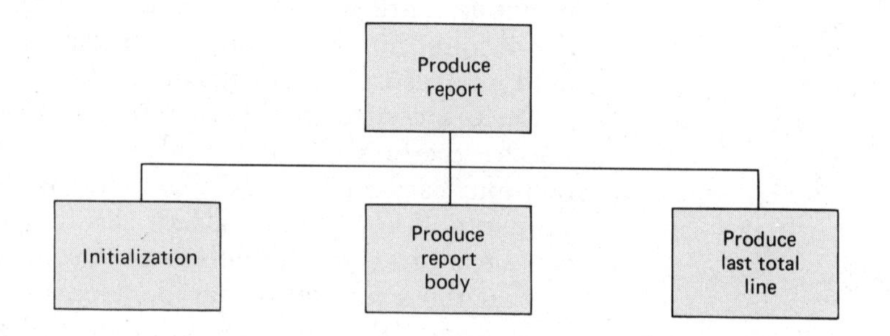

Figure 6.3 First-level hierarchy diagram for Program 13.

And so we next add to the hierarchy diagram two boxes showing that the main loop produces both **detail lines** and total lines. Detail line is just a name given to a line on a report that is derived from a single input record. The hierarchy diagram showing "Produce a detail line" and "Produce a total line" is in Figure 6.4. Of course, the box "Produce a total line" will produce all total lines except the last one on the report, and the box "Produce last total line" will generate the final

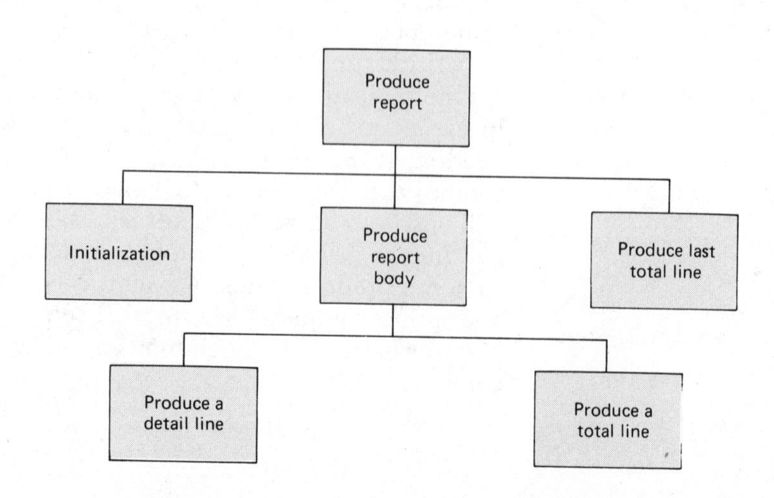

Figure 6.4 Second-level hierarchy diagram for Program 13.

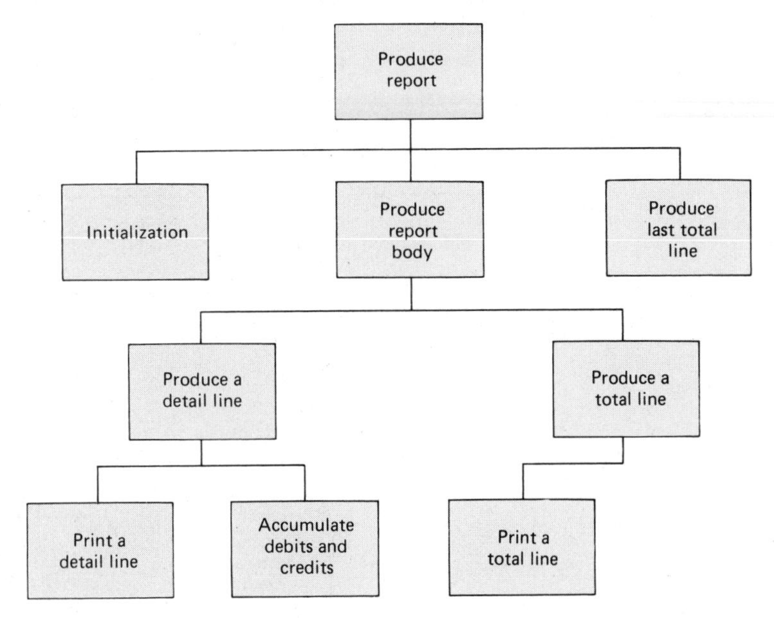

Figure 6.5 Third-level hierarchy diagram for Program 13.

line. Now in Figure 6.4 the relationship between "Produce a detail line" and "Produce a total line" is similar to what we have seen before in previous diagrams, and we can fill in the third level with "Print a detail line," "Accumulate debits and credits," and "Print a total line," as shown in Figure 6.5. But this time things are a little more involved than before. First, some detail line or another will find that the output report page is already full and that there is no room for it to print. The program will have to skip to a new page, print page headings, and then print the detail line. This means we have to include a new subfunction "Print page headings" as a subfunction of "Print a detail line." Also, after a total line is printed, we must then prepare to process the next control group (remember zeroing the accumulators), and so "Set up to process a control group" is shown as a subfunction of "Produce a total line." The hierarchy diagram with four levels is shown in Figure 6.6.

Now we can examine some higher-level boxes for a moment. What does "Initialization" consist of in this program? A case could be made for including in the initialization printing the first page heading and setting up to process the first control group. It is not absolutely necessary for these subfunctions to be included in "Initialization," but you will see that the coding will turn out to be quite convenient if we do include "Print page headings" and "Set up to process a control group" as subfunctions of "Initialization." Remember that a hierarchy diagram is not cast in concrete, and if it turns out to be subject to

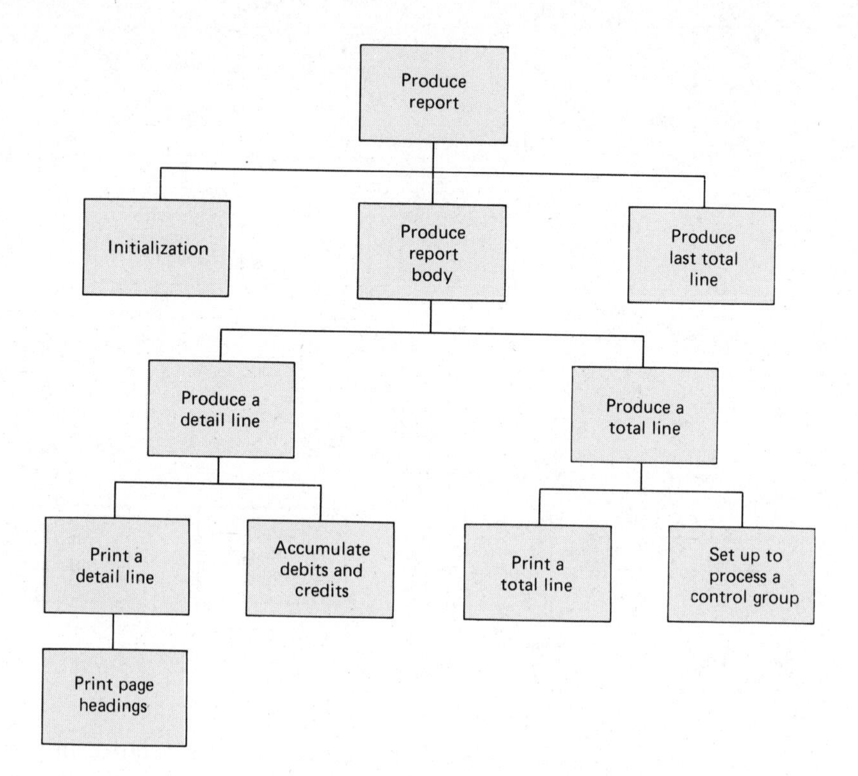

Figure 6.6
Fourth-level hierarchy diagram for Program 13.

improvement, we can always go back and make whatever changes we like. In any case, it certainly doesn't seem unreasonable to include "Print page headings" and "Set up to process a control group" as part of initialization, so it's worth a try. As already noted, it turns out just ducky.

Since these two subfunctions of "Initialization" are already in the hierarchy diagram in Figure 6.6 it would be legal to just connect the "Initialization" box to its two subfunctions with lines. But to do that we would have to cross existing lines on the diagram, and crossed lines make the diagram harder to understand. Line crossings can always be avoided by duplicating one or more boxes, so in Figure 6.7 the box "Set up to process a control group" has been duplicated at the left. The new box still represents the same coding as the original "Set up to process a control group." In the program the coding for the box will appear only once.

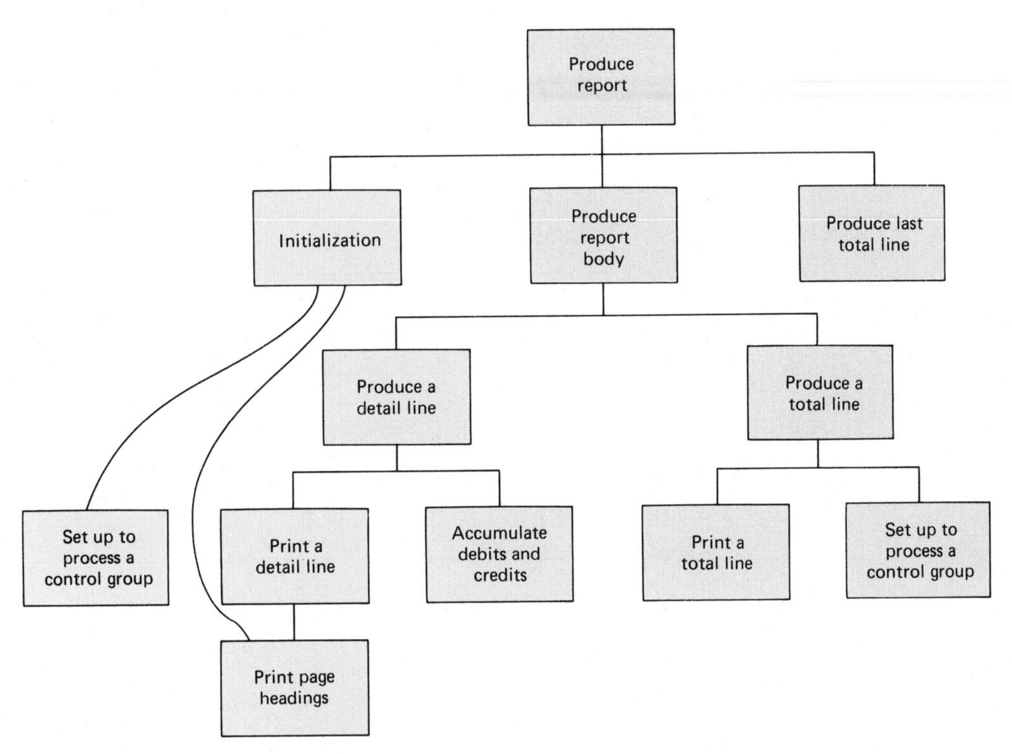

Figure 6.7 Hierarchy diagram for Program 13 showing the subfunctions of "Initialization."

Finally, what are the subfunctions, if any, of "Produce last total line"? The box "Print a total line" is all that is needed to produce the last total line, and so we could connect "Produce last total line" to "Print a total line." But to do that we would have to cross a line, so instead the box "Print a total line" is duplicated at the right of the diagram in Figure 6.8. It of course represents the same coding as the original "Print a total line," and the coding will appear only once in the program.

What is not shown as a subfunction of "Print a total line" is "Print page headings," for even if a total line finds a page already full, we don't want to skip to a new page, print page headings, and then print the total line at the top of the page without the detail group that it belongs to. In fact, in this kind of processing, it is customary never to have total lines print at the top of a new page away from their corresponding detail lines. So sufficient space must be left at the bottom of each page of report output to accommodate any total lines that might appear. Figure 6.8 is the complete hierarchy diagram.

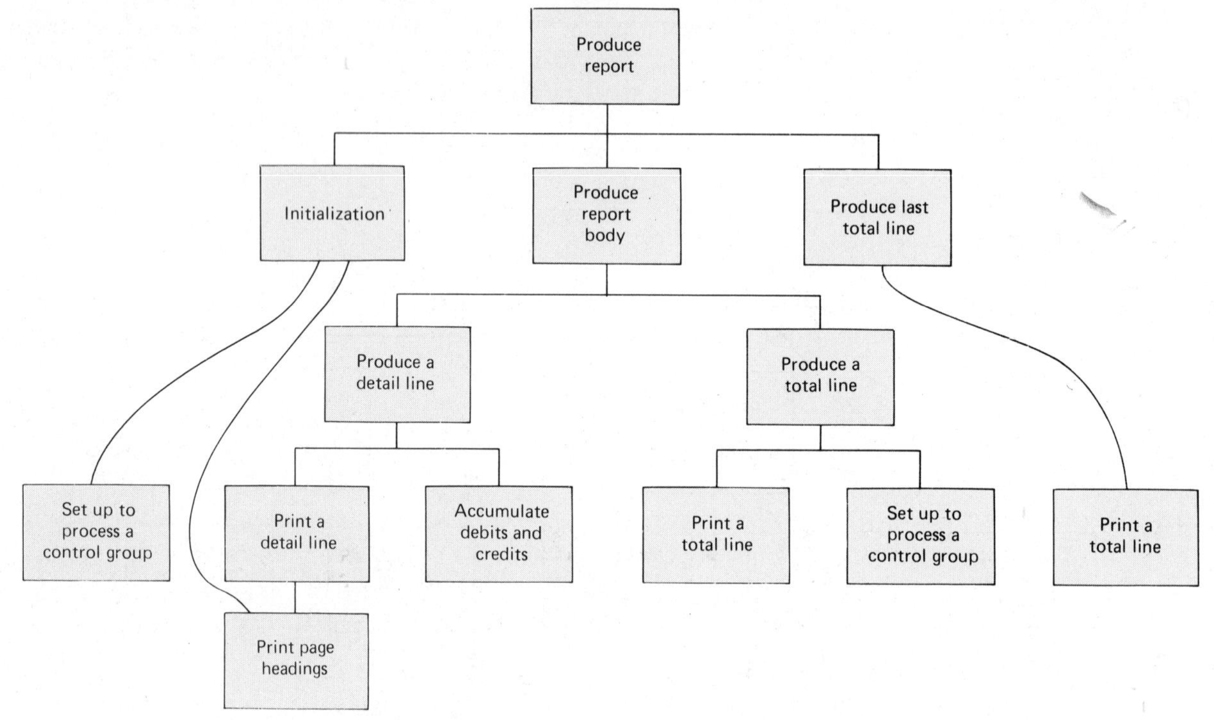

Figure 6.8 Complete hierarchy diagram for Program 13.

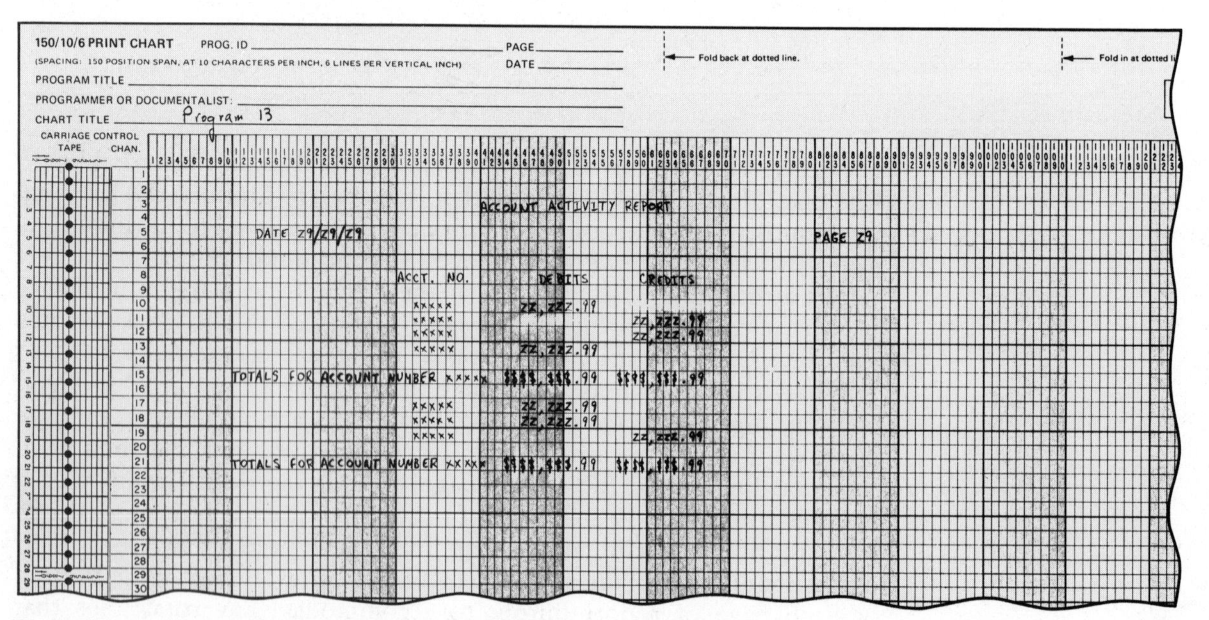

Figure 6.9 Output format for Program 13.

THE FIRST CONTROL BREAK PROGRAM

The output format for program 13 is shown in printer spacing chart form in Figure 6.9. Program 13 is shown in Figure 6.10. The program contains many new programming techniques.

Figure 6.10 Program 13, Part 1 of 4.

```
4-CB2 V4 RELEASE 1.5 10NOV77          IBM OS AMERICAN NATIONAL STANDARD COBOL



000010 IDENTIFICATION DIVISION.
000020 PROGRAM-ID. PROG13.
000030*AUTHOR. PHIL RUBENSTEIN.
000040*
000050*    THIS PROGRAM PRODUCES AN ACCOUNT ACTIVITY REPORT
000060*    WITH TOTALS OF DEBITS AND CREDITS FOR EACH ACCOUNT.
000070*
000080*****************************************************************
000090
000100 ENVIRONMENT DIVISION.
000110 CONFIGURATION SECTION.
000120 SPECIAL-NAMES.
000130     C01 IS TO-NEW-PAGE.
000140 INPUT-OUTPUT SECTION.
000150 FILE-CONTROL.
000160     SELECT ACCOUNT-ACTIVITY-REPORT  ASSIGN TO UT-S-SYSPRINT.
000170     SELECT ACCOUNT-FILE-IN          ASSIGN TO UT-S-SYSIN.
000180
000190*****************************************************************
000200
000210 DATA DIVISION.
000220 FILE SECTION.
000230 FD  ACCOUNT-FILE-IN
000240     LABEL RECORDS ARE OMITTED.
000250
000260 01  ACCOUNT-RECORD-IN.
000270     05 INDICATOR              PIC X.
000280     05 ACCOUNT-NUMBER-IN      PIC X(5).
000290     05 FILLER                 PIC X(3).
000300     05 AMOUNT                 PIC 9(5)V99.
000310     05 FILLER                 PIC X(64).
000320
000330 FD  ACCOUNT-ACTIVITY-REPORT
000340     LABEL RECORDS ARE OMITTED.
000350
000360 01  REPORT-LINE               PIC X(88).
000370
000380 WORKING-STORAGE SECTION.
000390 01  MORE-INPUT                PIC X(3) VALUE 'YES'.
000400
000410 01  PAGE-HEAD-1.
000420     05 FILLER             PIC X.
000430     05 FILLER             PIC X(40) VALUE SPACES.
000440     05 FILLER         PIC X(23) VALUE 'ACCOUNT ACTIVITY REPORT'.
```

Figure 6.10 Program 13, Part 2 of 4.

```
000450
000460 01   PAGE-HEAD-2.
000470      05 FILLER                    PIC X.
000480      05 FILLER                    PIC X(13) VALUE SPACES.
000490      05 FILLER                    PIC X(5)  VALUE 'DATE'.
000500      05 RUN-MONTH                 PIC Z9.
000510      05 FILLER                    PIC X     VALUE '/'.
000520      05 RUN-DAY                   PIC 99.
000530      05 FILLER                    PIC X     VALUE '/'.
000540      05 RUN-YEAR                  PIC 99.
000550      05 FILLER                    PIC X(54) VALUE SPACES.
000560      05 FILLER                    PIC X(5)  VALUE 'PAGE'.
000570      05 PAGE-NUMBER-OUT           PIC Z9.
000580
000590 01   PAGE-HEAD-3.
000600      05 FILLER                    PIC X.
000610      05 FILLER                    PIC X(30) VALUE SPACES.
000620      05 FILLER                    PIC X(17) VALUE 'ACCT. NO.'.
000630      05 FILLER                    PIC X(12) VALUE 'DEBITS'.
000640      05 FILLER                    PIC X(7)  VALUE 'CREDITS'.
000650
000660 01   DETAIL-LINE.
000670      05 FILLER                    PIC X.
000680      05 ACCOUNT-NUMBER-OUT        PIC B(32)X(5).
000690      05 DEBITS                    PIC B(8)ZZ,ZZZ.99.
000700      05 CREDITS                   PIC B(4)ZZ,ZZZ.99.
000710
000720 01   CONTROL-BREAK-LINE.
000730      05 FILLER                    PIC X.
000740      05 FILLER                    PIC X(10) VALUE SPACES.
000750      05 FILLER        PIC X(26) VALUE 'TOTALS FOR ACCOUNT NUMBER'.
000760      05 ACCOUNT-NUMBER-SAVE       PIC X(5).
000770      05 DEBITS-TOTAL              PIC $$$$$$,$$$.99.
000780      05 CREDITS-TOTAL             PIC $$$$$$,$$$.99.
000790
000800 01   ACCUMULATORS-W.
000810      05 DEBITS-TOTAL              PIC S9(6)V99.
000820      05 CREDITS-TOTAL             PIC S9(6)V99.
000830
000840 01   PAGE-NUMBER-W                PIC S99 VALUE ZERO.
000850 01   LINE-COUNT                   PIC S99 COMPUTATIONAL.
000860 01   PAGE-LIMIT                   PIC S99 VALUE 40 COMPUTATIONAL.
000870 01   NO-OF-LINES-IN-PAGE-HEADINGS PIC S9 VALUE 6 COMPUTATIONAL.
000880 01   LINE-SPACING                 PIC S9 COMPUTATIONAL.
000890
000900 01   TODAYS-DATE.
000910      05 RUN-YEAR                  PIC 99.
000920      05 RUN-MONTH                 PIC 99.
000930      05 RUN-DAY                   PIC 99.
```

In the Data Division, you can see the use of the editing PICTURE character B to introduce spaces between fields. We have earlier seen how oversized PICTUREs could generate blanks in FILLER fields, such as in column heading descriptions. Now we can use B to generate blanks in named fields. Study the entries in DETAIL–LINE to see how the Bs generate the blanks as specified in the printer spacing chart.

All fields being used for intermediate working storage results, such as the two accumulator fields in ACCUMULATORS–W, are described with the PICTURE character S for more efficient processing, even though we expect all values to be positive. In addition, some fields are described as being **COMPUTATIONAL.** COMPUTATIONAL is a designation that may obtain more efficient processing of fields which are used solely for operations which are wholly internal to the program; that is, COMPUTATIONAL should always be used on fields that are neither read in nor written out and which have no operations performed on them with fields that are read in or written out. An example of such a field is LINE–COUNT. (Beware! **LINE–COUNTER** is a COBOL reserved word. Its special use will be discussed in Chapter 10. LINE–COUNT is a programmer-supplied word for this program.) LINE–COUNT is the accumulator that will be used to count lines as they are printed to determine when to skip to a new page. The only values that are added to this field are internal to the program; no input data ever interacts with this field. Furthermore, the field is never written out and so should be designated COMPUTATIONAL. The other COMPUTATIONAL fields also do not interact with input or output. Mercifully, COMPUTATIONAL has an authorized abbreviation, and we will use it in later programs. It is the reserved word **COMP.**[1]

It is permissible to use COMPUTATIONAL on fields which do interact with input or output, but improved efficiency is then no longer always likely.[2] When COMPUTATIONAL is used as we have done here on wholly internal fields, it can never do harm and it may do some good. But when it is used on fields which interact with input, its use can in certain cases result in loss of efficiency. The only way to know for sure is to run the program with and without the COMPUTA-

1. In some COBOL systems, the SYNCHRONIZED clause is required in order to guarantee the full efficiency benefits of COMPUTATIONAL. In many systems, though, synchronization is either not required in order to obtain such benefits, or else all necessary synchronization is obtained through the use of the 01 level, or both. Your instructor will show you how to use the SYNCHRONIZED clause if necessary.
2. In systems having the COMPUTATIONAL–3 option, its use on any working storage field on which arithmetic is performed can never result in loss of efficiency, and will usually improve program efficiency considerably. Depending on the kind and amount of arithmetic carried out, COMPUTATIONAL or COMPUTATIONAL–3 may be more efficient. We refrain from using COMPUTATIONAL–3 in this book since it is not part of the 1974 standard.

TIONAL designation and see which version takes less computer time. COMPUTATIONAL is referred to as a **USAGE** of a field. Fields whose USAGE is not specified as COMPUTATIONAL are said to have the USAGE **DISPLAY.**

The field called LINE–SPACING will be used to control the spacing of detail lines as they are printed. The printer spacing chart shows that a detail line should be double-spaced if it appears immediately after the column headings or after a total line and should be single-spaced if it appears after another detail line. When we look at the Procedure Division, you will see how LINE–SPACING controls the variable spacing.

The word **ZERO** is a **figurative constant.** It can be used wherever a numeric or nonnumeric literal of zero would be legal. COBOL interprets ZERO as being numeric or nonnumeric depending on its context. The figurative constants **ZEROS** and **ZEROES** have exactly the same meaning as ZERO. They are provided only for the programmer's convenience.

Another figurative constant that we have been using all along is SPACES. It may be used interchangeably with the word **SPACE,** whose meaning in COBOL is identical to SPACES. SPACE and SPACES are always interpreted as nonnumeric literals consisting of as many blank characters as are required by the context in which they are used.

There is no great advantage in using the word ZERO instead of the numeral 0 when a numeric literal is desired, except perhaps ease of readability. But when a nonnumeric literal consisting of a string of zeros is desired, then ZERO (or ZEROS or ZEROES) will be interpreted by COBOL to give the exact number of characters needed.

The Procedure Division of this program is organized slightly differently from that of earlier programs. The main control paragraph is shorter than before because now essentially everything has been put into PERFORMed paragraphs.

Some of the earlier programs in this book contain a slight error that would be noticeable only if there happened to be no input data whatever in the input file. Some of these earlier programs would print page and/or column headings even if there were no input data, and some would print a total line of all zeros even if no input data were found. In Program 12 where there is a DIVIDE involving the number of students, the program would try to divide by zero if there were no student input. Different COBOL systems might do different things in case of an attempted division by zero, but none of the things would be good. In Program 13 we see one way of being sure that the body of the program is carried out only if there is some input data. We do the priming READ first, right after the OPEN, and only if some input data is found do we PERFORM the three first-level paragraphs INITIALIZATION, PRODUCE–REPORT–BODY, and PRODUCE–LAST–TOTAL–LINE.

Figure 6.10 Program 13, Part 3 of 4.

```
000940
000950*****************************************************************
000960
000970 PROCEDURE DIVISION.
000980     OPEN INPUT  ACCOUNT-FILE-IN,
000990          OUTPUT ACCOUNT-ACTIVITY-REPORT.
001000     READ ACCOUNT-FILE-IN
001010         AT END
001020             MOVE 'NO' TO MORE-INPUT.
001030     IF MORE-INPUT = 'YES'
001040         PERFORM INITIALIZATION
001050         PERFORM PRODUCE-REPORT-BODY UNTIL MORE-INPUT = 'NO'
001060         PERFORM PRODUCE-LAST-TOTAL-LINE.
001070     CLOSE ACCOUNT-FILE-IN,
001080           ACCOUNT-ACTIVITY-REPORT.
001090     STOP RUN.
001100
001110 INITIALIZATION.
001120     ACCEPT TODAYS-DATE FROM DATE.
001130     MOVE CORRESPONDING TODAYS-DATE TO PAGE-HEAD-2.
001140     PERFORM PRINT-PAGE-HEADINGS.
001150     PERFORM SET-UP-TO-PROCESS-A-CNTRL-GRP.
001160
001170 PRODUCE-REPORT-BODY.
001180     IF ACCOUNT-NUMBER-IN NOT = ACCOUNT-NUMBER-SAVE
001190         PERFORM PRODUCE-A-TOTAL-LINE.
001200     PERFORM PRODUCE-A-DETAIL-LINE.
001210
001220 PRODUCE-A-TOTAL-LINE.
001230     PERFORM PRINT-A-TOTAL-LINE.
001240     PERFORM SET-UP-TO-PROCESS-A-CNTRL-GRP.
001250
001260 PRODUCE-A-DETAIL-LINE.
001270     MOVE SPACES TO DETAIL-LINE.
001280     MOVE ACCOUNT-NUMBER-IN TO ACCOUNT-NUMBER-OUT.
001290     IF INDICATOR IS EQUAL TO '-'
001300         MOVE AMOUNT TO DEBITS
001310         ADD  AMOUNT TO DEBITS-TOTAL IN ACCUMULATORS-W
001320     ELSE
001330         MOVE AMOUNT TO CREDITS
001340         ADD  AMOUNT TO CREDITS-TOTAL IN ACCUMULATORS-W.
001350     MOVE ACCOUNT-NUMBER-IN TO ACCOUNT-NUMBER-OUT.
001360     PERFORM PRINT-A-DETAIL-LINE.
001370     READ ACCOUNT-FILE-IN
001380         AT END
001390             MOVE 'NO' TO MORE-INPUT.
001400
001410 PRINT-A-TOTAL-LINE.
001420     MOVE CORRESPONDING ACCUMULATORS-W TO CONTROL-BREAK-LINE.
001430     WRITE REPORT-LINE FROM CONTROL-BREAK-LINE AFTER ADVANCING 2.
001440     ADD 2 TO LINE-COUNT.
001450
001460 SET-UP-TO-PROCESS-A-CNTRL-GRP.
001470     MOVE 2                       TO LINE-SPACING.
001480     MOVE ZEROS                   TO ACCUMULATORS-W.
001490     MOVE ACCOUNT-NUMBER-IN TO ACCOUNT-NUMBER-SAVE.
001500
```

The INITIALIZATION paragraph does a few things: It first fetches the date from the COBOL system and inserts it into the appropriate heading line, PAGE–HEAD–2. Then, in accordance with the hierarchy diagram, it executes the first PRINT–PAGE–HEADINGS and also sets up to process the first account number group. We will look at the routine called PRINT–PAGE–HEADINGS later, but it is worthwhile to look at the paragraph SET–UP–TO–PROCESS–A–CNTRL–GRP now.

SET–UP–TO–PROCESS–A–CNTRL–GRP does only three things: First, it sets LINE–SPACING to 2. This will mean, as you will see, that the first detail line of each new group is to be double-spaced. In fact, throughout the execution of the program, the field LINE–SPACING will always contain either a 2 or a 1, and you will soon see how the 2 and 1 are used to control the spacing of the detail lines.

The statement that zeros out the two accumulators in ACCUMULATORS–W looks innocent and harmless enough but could easily have not worked correctly if we were not careful. Any field that is broken down into smaller fields, as ACCUMULATORS–W is, is called a **group item.** Any field not broken down into smaller fields is called an **elementary item.** The elementary items into which a group item is broken down are said to be **subordinate** to the group item. The accumulators DEBITS–TOTAL and CREDITS–TOTAL are elementary items.

Group items are always considered to be alphanumeric items of USAGE DISPLAY, so whenever you treat a group of accumulators as a single group item, the accumulators must also be defined as having USAGE DISPLAY. Remember that when a USAGE is not specified for a field, it is assumed to be DISPLAY, so the accumulators DEBITS–TOTAL and CREDITS–TOTAL are DISPLAY fields, as required. The only restriction on the programmer is that if you wish to handle accumulators using group-level operations, the accumulators cannot be COMPUTATIONAL. If for some reason you want to make the accumulators COMPUTATIONAL, then they must be processed one at a time by elementary operations.

The last thing that SET–UP–TO–PROCESS–A–CNTRL–GRP does is MOVE the current account number to a field called ACCOUNT–NUMBER–SAVE, which happens to be part of CONTROL–BREAK–LINE but could have been described anywhere in working storage. ACCOUNT–NUMBER–SAVE will be the field to which we assign the account number currently being worked on by the program. As the program reads each card, it will compare the account number on the card to the value of ACCOUNT–NUMBER–SAVE to see whether a control break has occurred. If the account number on the just-read card is the same as ACCOUNT–NUMBER–SAVE, that means that the program is still working on the same account number. If the two are different, it means that all the cards for the previous account have been read and that a control break has occurred.

Let us now look at PRODUCE–REPORT–BODY. The program always enters PRODUCE–REPORT–BODY with a newly-read card waiting to be processed. The IF statement checks to see whether this new card is the first card of a new group. If it is, a control break has occurred, so the program PERFORMs PRODUCE–A–TOTAL–LINE, as indicated in the hierarchy diagram, before proceeding to PRODUCE–A–DETAIL–LINE for this first card of the new group. If not, it goes directly to PRODUCE–A–DETAIL–LINE for the card. The paragraph PRODUCE–A–TOTAL–LINE PERFORMs the two functions indicated for it in the hierarchy diagram, namely PRINT–A–TOTAL–LINE and SET–UP–TO–PROCESS–A–CNTRL–GRP.

The paragraph PRODUCE–A–DETAIL–LINE contains the detailed processing needed to handle a single input card. The program first checks to see whether the input item is a debit or a credit, processes it accordingly, and completes the formatting of the output line by moving ACCOUNT–NUMBER–IN to ACCOUNT–NUMBER–OUT. The program then writes the detail line by using the routine PRINT–A–DETAIL–LINE and READs the next input record.

In the routine PRODUCE–A–TOTAL–LINE, the program is essentially required to print out a total line. The totals it needs are already accumulated in the two ACCUMULATORS–W fields, and those merely have to be MOVEd to the total line, which in this program is called CONTROL–BREAK–LINE. The only other variable field in CONTROL–BREAK–LINE, ACCOUNT–NUMBER–SAVE, had the account number MOVEd to it when the program first started processing the group, so there is no further formatting needed in this line, and it

Figure 6.10 Program 13, Part 4 of 4.

```
001510 PRINT-A-DETAIL-LINE.
001520      IF LINE-COUNT GREATER THAN PAGE-LIMIT
001530          PERFORM PRINT-PAGE-HEADINGS.
001540      WRITE REPORT-LINE FROM DETAIL-LINE AFTER LINE-SPACING.
001550      ADD LINE-SPACING   TO LINE-COUNT.
001560      MOVE 1             TO LINE-SPACING.
001570
001580 PRINT-PAGE-HEADINGS.
001590      ADD 1              TO PAGE-NUMBER-W.
001600      MOVE PAGE-NUMBER-W TO PAGE-NUMBER-OUT.
001610      WRITE REPORT-LINE FROM PAGE-HEAD-1
001620                                  AFTER ADVANCING TO-NEW-PAGE.
001630      WRITE REPORT-LINE FROM PAGE-HEAD-2 AFTER 2.
001640      WRITE REPORT-LINE FROM PAGE-HEAD-3 AFTER 3.
001650      MOVE 2 TO LINE-SPACING.
001660      MOVE NO-OF-LINES-IN-PAGE-HEADINGS TO LINE-COUNT.
001670
001680 PRODUCE-LAST-TOTAL-LINE.
001690      PERFORM PRINT-A-TOTAL-LINE.
```

is written with double-spacing. A 2 is MOVEd to the field LINE–SPACING, because the printer spacing chart shows that double-spacing is required after a control break line.

Let us look at PRINT–A–DETAIL–LINE to see how it handles page overflow and how it uses LINE–SPACING to get most detail lines single-spaced, but double-spaced after page headings and total lines. Page overflow is no problem. The program compares the LINE–COUNT to the constant PAGE–LIMIT and if too many lines have already been printed on the page, PERFORMs PRINT–PAGE–HEADINGS. PAGE–LIMIT has a value set by the programmer, and it can be changed if the programmer decides later that more or fewer lines should be printed on each page.

The WRITE statement in this routine uses the clause AFTER LINE–SPACING. This is a form of the AFTER ADVANCING clause. One widely used format of the AFTER ADVANCING clause is:

```
{BEFORE}               {identifier LINES}
{AFTER }  ADVANCING    {integer LINES   }
                       {mnemonic–name   }
```

The first set of braces shows that either of the words BEFORE or AFTER may be used for printing before or after spacing the paper, respectively. We have never had occasion to print a line **BEFORE ADVANCING** the paper; we have always printed our output AFTER ADVANCING the paper. The second set of braces shows that one of the three forms must be chosen by the programmer. We have used both integer LINES and mnemonic–name.

We have used 2 LINES, and we would have used 1 LINES if it didn't seem wrong. But the format shows that LINES is an optional word, and so we have often omitted it (it is not underlined, and only underlined words are required). We have also used the **mnemonic name** TO–NEW–PAGE. In COBOL, all mnemonic names are defined in the SPECIAL–NAMES paragraph of the Environment Division.

Now it turns out that there is a third way to use the AFTER ADVANCING clause, namely identifier LINES, and furthermore that the word ADVANCING is optional. If identifier LINES is used, the identifier must be defined as an integer. In Program 13 the identifier LINE–SPACING is defined as an integer. If LINE–SPACING has a value of 2 at the time the clause AFTER LINE–SPACING is executed, COBOL will give double-spacing. If LINE–SPACING is 1 at that time, the paper will be single-spaced. Remember that both PRINT–PAGE–HEADINGS and SET–UP–TO–PROCESS–A–CNTRL–GRP set LINE–SPACING to 2 so the first detail line printed after the page headings or at the beginning of a new group will have double-spacing. PRINT–A–DETAIL–LINE itself turns LINE–SPACING to 1, so that subsequent detail lines will be single-spaced.

The last routine that we will look at is PRINT–PAGE–HEADINGS. Its operation is mostly self-evident. The field PAGE–NUMBER–W starts out with VALUE ZERO, and 1 is ADDed to it every time a page heading is printed, including the first time. We see in this routine some examples of the AFTER ADVANCING clause without its ADVANCING. After the headings are printed, the LINE–COUNT is set to reflect the number of lines that have already been printed on the page, so that the subsequent counting of lines as they are printed will be accurate.

Exercise 1. Explain why some working storage fields in Program 13 need a VALUE clause and some do not. These are the fields with a VALUE clause:

PAGE–NUMBER–W
PAGE–LIMIT
NO–OF–LINES–IN–PAGE–HEADING

These fields do not have a VALUE clause:

PAGE–NUMBER–OUT
LINE–COUNT
DEBITS–TOTAL IN ACCUMULATORS–W

*Exercise 2. Write a program to read and process data in the format shown in Figure 4.1. Assume that there are several cards for each customer, each card containing a different part number. For each card, print a line showing the Customer number, Part number, Quantity, Unit price, Handling, and Merchandise amount (compute the Merchandise amount by multiplying the unit price by the Quantity). For each customer, print a total line showing the total Handling and the total Merchandise amount. Design the output on a printer spacing chart before you begin coding.

Multiple Control Breaks

It is not uncommon for input data to have more than one control field and therefore often necessary for reports to have several different kinds of total lines, each kind showing totals for a different control field. For example, consider input data in the format shown in Figure 6.11. Each input record in this format shows the dollar amount of a purchase made by a customer. The record also shows the number of the salesperson who serviced the customer and the store number where the sale took place. In Program 14 we will produce a list of all these sales and show the total amount of sales made by each salesperson, the total amount of sales in each store, and the total amount of all sales. The output should look like Figure 6.12, from input in Figure 6.13. The output format in printer spacing chart form is shown in Figure 6.14.

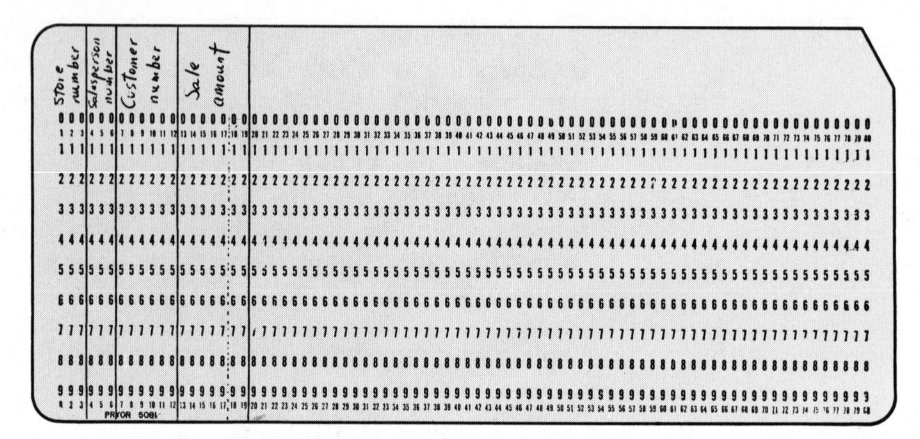

Figure 6.11
Data card format for Program 14.

```
                       SALES REPORT
  DATE   2/19/80                                  PAGE   1

        STORE      SALES-       CUSTOMER        SALE
         NO.       PERSON        NUMBER        AMOUNT

         010        101          003001      1,234.56
                                 007002          2.24
                                 011003      6,665.70
                                 039004      8,439.20
                                 046005      8,448.48
                                 053006         12.34
                                 060006      9,494.93
                                 067007      4,000.03

                                            38,297.48 *

         010        102          074212      5,454.99
                                 081012           .33
                                 088013      5,849.58
                                 095015        393.90

                                            11,698.80 *

         010        103          003234        303.03

                                               303.03 *

          TOTAL FOR STORE NO. 010         $ 50,299.31 **

         020        011          007567      9,999.99
                                 011454        456.00
                                 015231      8,484.39
                                 019345      8,459.44
                                 023345      8,333.33

                                            35,733.15 *

         020        222          027345      4,343.43
```

Figure 6.12
Output from Program 14, Part 1 of 3.

Figure 6.12
Output from Program 14, Part 2 of 3.

```
                    SALES REPORT
DATE  2/19/80                                       PAGE  2

     STORE      SALES-      CUSTOMER        SALE
      NO.       PERSON       NUMBER        AMOUNT

      020        222         031567      9,903.30
                             035001         34.21

                                        14,280.94 *

      020        266         039903      4,539.87
                             043854      5,858.30

                                        10,398.17 *

       TOTAL FOR STORE NO. 020      $ 60,412.26 **

      030        193         047231      9,391.93
                             051342      5,937.43
                             055034      9,383.22
                             059932      5,858.54
                             063419      3,949.49

                                        34,520.61 *

       TOTAL FOR STORE NO. 030      $ 34,520.61 **

      040        045         067333           .00
                             071323      5,959.50

                                         5,959.50 *

      040        403         048399      3,921.47
                             054392           .00

                                         3,921.47 *

      040        412         060111      9,999.99

                                         9,999.99 *

       TOTAL FOR STORE NO. 040      $ 19,880.96 **
```

Figure 6.12
Output from Program 14, Part 3 of 3.

```
                    SALES REPORT
DATE  2/19/80                                        PAGE  3

        STORE       SALES-       CUSTOMER        SALE
         NO.        PERSON        NUMBER        AMOUNT

        046          012          013538             .00
                                  017521          690.78
                                  021504        1,381.56
                                  025487        2,072.34
                                  029470        2,763.12

                                                6,907.80 *

        046          028          033453        3,453.90
                                  037436        4,144.68
                                  041419        4,835.46

                                               12,434.04 *

          TOTAL FOR STORE NO. 046          $ 19,341.84 **

                   GRAND TOTAL      $     184,454.98 ***
```

Obviously, many of the ideas from Program 13 can be used in Program 14. The new ideas that we will need in order to create a report with three levels of totals will be general enough so that you will then be able to apply them yourself to reports of two levels or of four or more levels.

The input data for Program 14 is grouped so that the printed output will be in the correct order for the totals to print. All sales for each store are grouped together in the input. Within each group of sales for a store, all sales made by one salesperson are grouped together; within each group of sales made by one salesperson, the customer numbers are in order. The store number is the most significant field and is called the **major control field.** The customer number is the least significant field and is called the **minor control field.** The salesperson number, whose significance falls between that of the store number and the customer number, is called the **intermediate control field.** If there had been more than three fields used for ordering the data, the most significant would be the major, the least significant the minor, and all those between major and minor would be called intermediate. This input is said to be sorted by store by salesperson by customer;

Figure 6.13 Input data for Program 14.

```
0101010030011 23456
```

alternatively, the input can be described as being sorted by customer number within salesperson number within store number. In this program, control breaks will be taken only on salesperson number and store number and totals printed for those breaks.

As each input card is read, the program will print it and add the sale amount to some accumulator or other. In a multiple-control break program, there must be at least one accumulator for each level of control break. In this program there must be one accumulator to collect the total sales for a salesperson, another for the total sales for a store, and a third for the grand total of all sales. The exact arrangement and use of the accumulators is sometimes a matter of programmer preference, and you will see how the accumulators are used when we look at the coding for Program 14.

The program will have to check each input card to see whether there is a control break on salesperson number and/or on store number. If there is a break on salesperson number, the program will print a total for the previous salesperson, zero the salesperson accumulator, and prepare to process the next salesperson group. But if there is a break on store number, that means that all the data have been read not only for the previous store but also for the last salesperson of the previous store. So a break on store number first requires printing a total line for the last salesperson of the previous store and then printing a total line for the previous store. Also, the accumulators for both salesperson and store must be zeroed in preparation for processing the next store (and the first salesperson in the new store). When end-of-file is reached on the input, there still remain to be printed the total line for the last salesperson in the last store, the total line for the last store, and the grand total line.

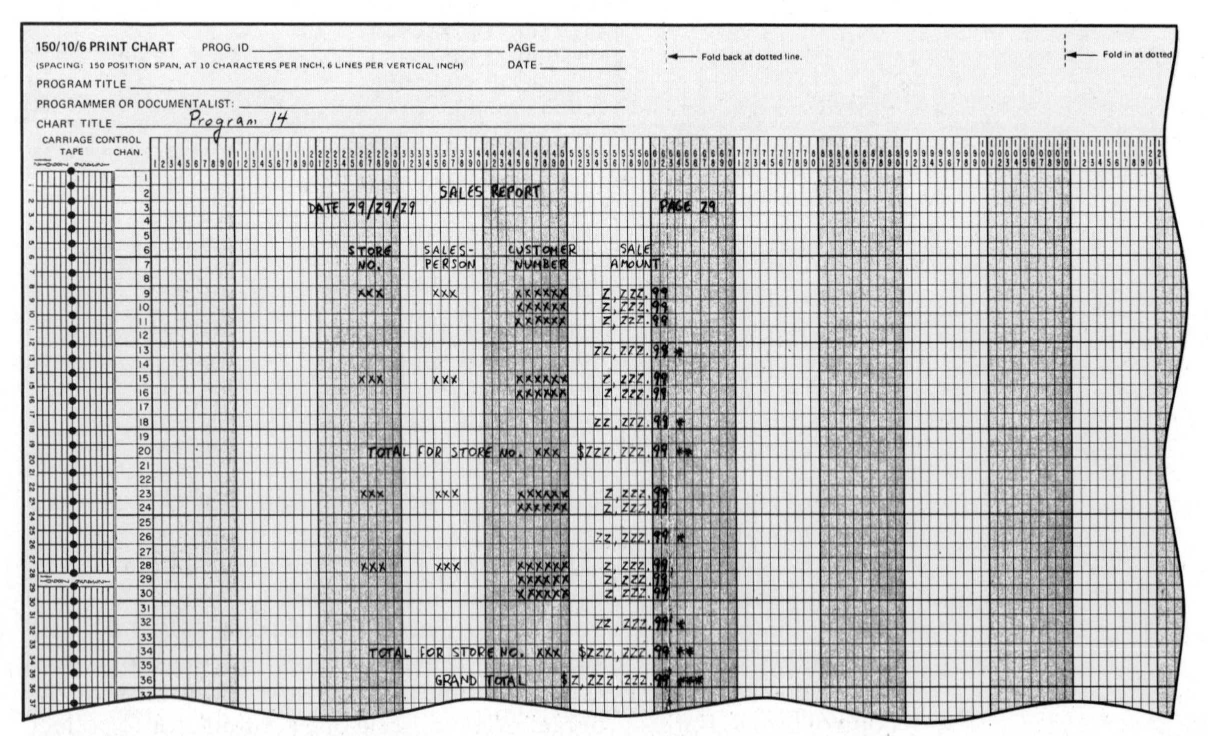

Figure 6.14 Output format for Program 14.

The output in Figure 6.12 also introduces **group indicating.** Authorities disagree on the exact definition of group indicating: Some say that group indicating is printing control fields on only the first detail line on each page and on the first detail line after every control break.[3] You can see in Figure 6.12 that store number and salesperson number fields print on only the first detail line after every control break and on the first detail line of each page, instead of printing on every line. In Program 13 we did not have group indicating; there we printed the control field, account number, on every line.

HIERARCHY DIAGRAM FOR PROGRAM 14

Using the ideas and rationale from the hierarchy diagram for Program 13, we can start with the diagram for Program 14 shown in Figure 6.15. Here the box "Produce report body" has three subfunctions

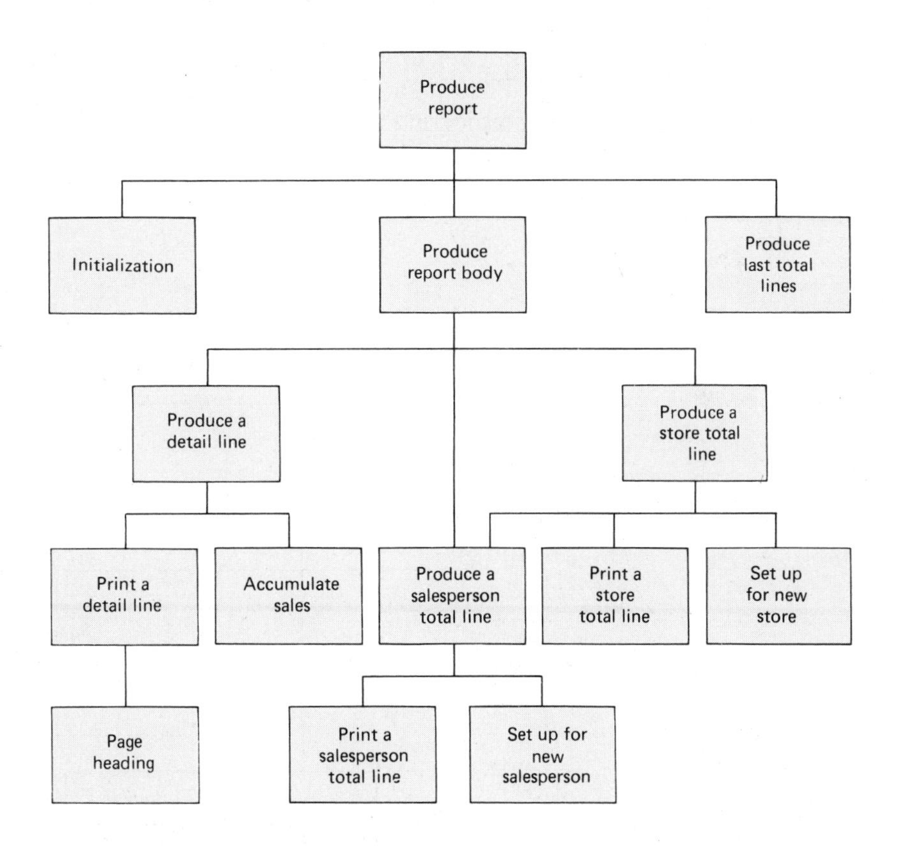

Figure 6.15
Partial hierarchy diagram for Program 14.

3. American National Standards Institute, *American National Standard Programming Language COBOL,* 1974, p. VIII-31. An alternative definition of group indicating as printing each control field on only the first line of its group and on the first detail line of every page is given in Gary B. Shelly and Thomas J. Cashman, *Introduction to Computer Programming Structured COBOL,* Anaheim Publishing Co., 1977, p. 9.17.

instead of two. Now its subfunctions are "Produce a detail line," "Produce a salesperson total line," and "Produce a store total line." The box "Produce a salesperson total line" is shown at a lower level than the other two because it is a subfunction also of "Produce a store total line." Remember that whenever we get a break on store number, the program must first execute the break on salesperson number. If you study the structure of the hierarchy diagram, you will see how it could be generalized to handle more than three levels of totals.

The diagram is not complete, though, for we have not yet filled in the subfunctions of "Initialization" and "Produce last total lines." For "Initialization" we must print the first page headings and set up for the first store and the first salesperson. For "Produce last total lines" we must print the total line for the last salesperson in the last store, the total line for the last store, and the grand total line. The hierarchy diagram can be completed by duplicating the boxes "Set up for new store," "Set up for new salesperson," "Print a salesperson total line," and "Print a store total line," adding the box "Print grand total line," and connecting all the boxes properly. This is done in Figure 6.16.

Figure 6.16 Complete hierarchy diagram for Program 14.

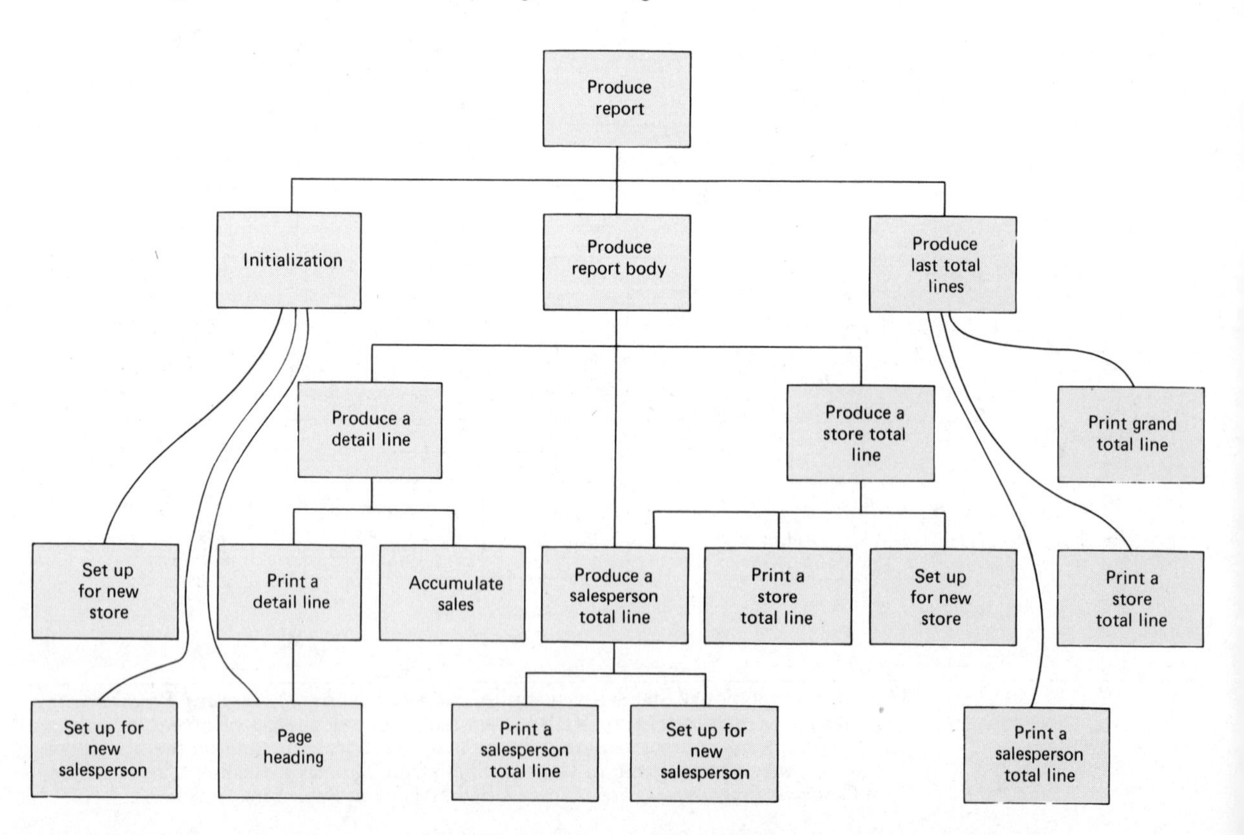

A PROGRAM WITH THREE LEVELS OF TOTALS

Program 14 is shown in Figure 6.17. As in Program 13, the Procedure Division is made up of many small paragraphs, some having only three statements. The structure and content of the paragraphs were made to correspond exactly to the hierarchy diagram. But some programmers

Figure 6.17 Program 14, Part 1 of 3.

```
4-CB2 V4 RELEASE 1.5 10NOV77          IBM OS AMERICAN NATIONAL STANDARD COBOL



000010 IDENTIFICATION DIVISION.
000020 PROGRAM-ID. PROG14.
000030*AUTHOR. PHIL RUBENSTEIN.
000040*
000050*     THIS PROGRAM PRODUCES A SALES REPORT
000060*     WITH THREE LEVELS OF TOTALS.
000070*
000080*****************************************************************
000090
000100 ENVIRONMENT DIVISION.
000110 CONFIGURATION SECTION.
000120 SPECIAL-NAMES.
000130      C01 IS NEW-PAGE.
000140 INPUT-OUTPUT SECTION.
000150 FILE-CONTROL.
000160      SELECT SALES-REPORT  ASSIGN TO UT-S-SYSPRINT.
000170      SELECT SALES-FILE-IN ASSIGN TO UT-S-SYSIN.
000180
000190*****************************************************************
000200
000210 DATA DIVISION.
000220 FILE SECTION.
000230 FD   SALES-FILE-IN
000240      LABEL RECORDS ARE OMITTED.
000250
000260 01   SALES-RECORD-IN.
000270      05 STORE-NUMBER-IN        PIC 9(3).
000280      05 SALESPERSON-NUMBER-IN  PIC 9(3).
000290      05 CUSTOMER-NUMBER        PIC 9(6).
000300      05 SALE-AMOUNT            PIC 9(4)V99.
000310      05 FILLER                 PIC X(62).
000320
000330 FD   SALES-REPORT
000340      LABEL RECORDS ARE OMITTED.
000350
000360 01   REPORT-LINE               PIC X(69).
000370
000380 WORKING-STORAGE SECTION.
000390 01   MORE-INPUT                PIC X(3) VALUE 'YES'.
000400
000410 01   PAGE-HEAD-1.
000420      05 FILLER                 PIC X.
000430      05 FILLER                 PIC X(35) VALUE SPACES.
000440      05 FILLER                 PIC X(12) VALUE 'SALES REPORT'.
```

Figure 6.17 Program 14, Part 2 of 3.

```
000450
000460 01  PAGE-HEAD-2.
000470     05 FILLER                 PIC X.
000480     05 FILLER                 PIC X(19) VALUE SPACES.
000490     05 FILLER                 PIC X(5)  VALUE 'DATE'.
000500     05 RUN-MONTH              PIC Z9.
000510     05 FILLER                 PIC X     VALUE '/'.
000520     05 RUN-DAY                PIC 99.
000530     05 FILLER                 PIC X     VALUE '/'.
000540     05 RUN-YEAR               PIC Z9.
000550     05 FILLER                 PIC X(29) VALUE SPACES.
000560     05 FILLER                 PIC X(5)  VALUE 'PAGE'.
000570     05 PAGE-NUMBER-OUT        PIC Z9.
000580
000590 01  PAGE-HEAD-3.
000600     05 FILLER                 PIC X.
000610     05 FILLER                 PIC X(24) VALUE SPACES.
000620     05 FILLER                 PIC X(9)  VALUE 'STORE'.
000630     05 FILLER                 PIC X(10) VALUE 'SALES-'.
000640     05 FILLER                 PIC X(13) VALUE 'CUSTOMER'.
000650     05 FILLER                 PIC X(4)  VALUE 'SALE'.
000660
000670 01  PAGE-HEAD-4.
000680     05 FILLER                 PIC X.
000690     05 FILLER                 PIC X(25) VALUE SPACES.
000700     05 FILLER                 PIC X(8)  VALUE 'NO.'.
000710     05 FILLER                 PIC X(11) VALUE 'PERSON'.
000720     05 FILLER                 PIC X(11) VALUE 'NUMBER'.
000730     05 FILLER                 PIC X(6)  VALUE 'AMOUNT'.
000740
000750 01  DETAIL-LINE.
000760     05 FILLER                 PIC X.
000770     05 STORE-NUMBER-OUT       PIC B(25)9(3).
000780     05 SALESPERSON-NUMBER-OUT    PIC B(6)9(3).
000790     05 CUSTOMER-NUMBER        PIC B(7)9(6).
000800     05 SALE-AMOUNT            PIC B(4)Z,ZZZ.99.
000810
000820 01  SALESPERSON-TOTAL-LINE.
000830     05 FILLER                 PIC X.
000840     05 SALESPERSON-TOTAL-OUT  PIC B(53)ZZ,ZZZ.99B.
000850     05 FILLER                 PIC X     VALUE '*'.
000860
000870 01  STORE-TOTAL-LINE.
000880     05 FILLER                 PIC X.
000890     05 FILLER                 PIC X(26) VALUE SPACES.
000900     05 FILLER       PIC X(20) VALUE 'TOTAL FOR STORE NO.'.
000910     05 STORE-NUMBER-SAVE      PIC 9(3).
000920     05 FILLER                 PIC X(2)  VALUE SPACES.
000930     05 STORE-TOTAL-OUT        PIC $ZZZ,ZZZ.99B.
000940     05 FILLER                 PIC X(2)  VALUE '**'.
000950
000960 01  GRAND-TOTAL-LINE.
000970     05 FILLER                 PIC X.
000980     05 FILLER                 PIC X(34) VALUE SPACES.
000990     05 FILLER                 PIC X(15) VALUE 'GRAND TOTAL'.
001000     05 GRAND-TOTAL-OUT        PIC $Z,ZZZ,ZZZ.99B.
001010     05 FILLER                 PIC X(3)  VALUE '***'.
```

Figure 6.17 Part 2 (continued).

```
001020
001030 01  PAGE-NUMBER-W               PIC 99 VALUE 0.
001040 01  LINE-COUNT                  PIC 99 COMP.
001050 01  PAGE-LIMIT                  PIC 99 VALUE 40 COMP.
001060 01  NO-OF-LINES-IN-PAGE-HEADING PIC 9 VALUE 6 COMP.
001070 01  LINE-SPACING                PIC 9 COMP.
001080 01  SALESPERSON-TOTAL-W         PIC 9(5)V99.
001090 01  STORE-TOTAL-W               PIC 9(6)V99.
001100 01  GRAND-TOTAL-W               PIC 9(7)V99 VALUE 0.
001110 01  SALESPERSON-NUMBER-SAVE     PIC 9(3).
001120
001130 01  TODAYS-DATE.
001140     05 RUN-YEAR                 PIC 99.
001150     05 RUN-MONTH                PIC 99.
001160     05 RUN-DAY                  PIC 99.
001170
001180*****************************************************************
001190
001200 PROCEDURE DIVISION.
001210     OPEN INPUT  SALES-FILE-IN,
001220          OUTPUT SALES-REPORT.
001230     READ SALES-FILE-IN
001240         AT END
001250             MOVE 'NO' TO MORE-INPUT.
001260     IF MORE-INPUT = 'YES'
001270         PERFORM INITIALIZATION
001280         PERFORM PRODUCE-REPORT-BODY UNTIL MORE-INPUT = 'NO'
001290         PERFORM PRODUCE-FINAL-LINES.
001300     CLOSE SALES-FILE-IN,
001310           SALES-REPORT.
001320     STOP RUN.
001330
001340 INITIALIZATION.
001350     ACCEPT TODAYS-DATE FROM DATE.
001360     MOVE CORRESPONDING TODAYS-DATE TO PAGE-HEAD-2.
001370     PERFORM SET-UP-FOR-NEW-STORE.
001380     PERFORM SET-UP-FOR-NEW-SALESPERSON.
001390     PERFORM PAGE-HEADING.
001400
001410 PRODUCE-REPORT-BODY.
001420     IF STORE-NUMBER-IN NOT = STORE-NUMBER-SAVE
001430         PERFORM PRODUCE-STORE-TOTAL-LINE
001440     ELSE
001450         IF SALESPERSON-NUMBER-IN NOT = SALESPERSON-NUMBER-SAVE
001460             PERFORM PRODUCE-SALESPERSON-TOTAL-LINE.
001470     PERFORM PRODUCE-A-DETAIL-LINE.
001480     READ SALES-FILE-IN
001490         AT END
001500             MOVE 'NO' TO MORE-INPUT.
001510
001520 SET-UP-FOR-NEW-STORE.
001530     MOVE STORE-NUMBER-IN TO STORE-NUMBER-SAVE,
001540                             STORE-NUMBER-OUT.
001550     MOVE 0 TO STORE-TOTAL-W.
001560     MOVE 3 TO LINE-SPACING.
001570
001580 SET-UP-FOR-NEW-SALESPERSON.
```

Figure 6.17 Program 14, Part 3 of 3.

```
001590      MOVE SALESPERSON-NUMBER-IN TO SALESPERSON-NUMBER-OUT,
001600                                    SALESPERSON-NUMBER-SAVE.
001610      MOVE STORE-NUMBER-IN          TO STORE-NUMBER-OUT.
001620      MOVE 0                        TO SALESPERSON-TOTAL-W.
001630      MOVE 2                        TO LINE-SPACING.
001640
001650 PRODUCE-STORE-TOTAL-LINE.
001660      PERFORM PRODUCE-SALESPERSON-TOTAL-LINE.
001670      PERFORM PRINT-A-STORE-TOTAL-LINE.
001680      PERFORM SET-UP-FOR-NEW-STORE.
001690
001700 PRODUCE-SALESPERSON-TOTAL-LINE.
001710      PERFORM PRINT-A-SLSPSN-TOTAL-LINE.
001720      PERFORM SET-UP-FOR-NEW-SALESPERSON.
001730
001740 PRODUCE-FINAL-LINES.
001750      PERFORM PRINT-A-SLSPSN-TOTAL-LINE.
001760      PERFORM PRINT-A-STORE-TOTAL-LINE.
001770      MOVE GRAND-TOTAL-W TO GRAND-TOTAL-OUT.
001780      WRITE REPORT-LINE FROM GRAND-TOTAL-LINE AFTER 2.
001790
001800 PRINT-A-STORE-TOTAL-LINE.
001810      MOVE STORE-TOTAL-W TO STORE-TOTAL-OUT.
001820      WRITE REPORT-LINE FROM STORE-TOTAL-LINE AFTER 2.
001830      ADD STORE-TOTAL-W TO GRAND-TOTAL-W.
001840      ADD 2 TO LINE-COUNT.
001850
001860 PRINT-A-SLSPSN-TOTAL-LINE.
001870      MOVE SALESPERSON-TOTAL-W TO SALESPERSON-TOTAL-OUT.
001880      WRITE REPORT-LINE FROM SALESPERSON-TOTAL-LINE AFTER 2.
001890      ADD SALESPERSON-TOTAL-W TO STORE-TOTAL-W.
001900      ADD 2 TO LINE-COUNT.
001910
001920 PRODUCE-A-DETAIL-LINE.
001930      MOVE CORRESPONDING SALES-RECORD-IN TO DETAIL-LINE.
001940      ADD SALE-AMOUNT IN SALES-RECORD-IN TO SALESPERSON-TOTAL-W.
001960      IF LINE-COUNT GREATER THAN PAGE-LIMIT
001970          PERFORM PAGE-HEADING.
001980      WRITE REPORT-LINE FROM DETAIL-LINE AFTER LINE-SPACING.
001990      ADD LINE-SPACING TO LINE-COUNT.
002000      MOVE 1           TO LINE-SPACING.
002010      MOVE SPACES      TO DETAIL-LINE.
002020
002030 PAGE-HEADING.
002040      ADD 1 TO PAGE-NUMBER-W.
002050      MOVE PAGE-NUMBER-W TO PAGE-NUMBER-OUT.
002060      WRITE REPORT-LINE FROM PAGE-HEAD-1 AFTER NEW-PAGE.
002070      WRITE REPORT-LINE FROM PAGE-HEAD-2 AFTER 1.
002080      WRITE REPORT-LINE FROM PAGE-HEAD-3 AFTER 3.
002090      WRITE REPORT-LINE FROM PAGE-HEAD-4 AFTER 1.
002100      MOVE 2 TO LINE-SPACING.
002110      MOVE NO-OF-LINES-IN-PAGE-HEADING TO LINE-COUNT.
002120      MOVE STORE-NUMBER-SAVE           TO STORE-NUMBER-OUT.
002130      MOVE SALESPERSON-NUMBER-SAVE     TO SALESPERSON-NUMBER-OUT.
```

feel that paragraphs containing fewer than five statements are more confusing than helpful and that paragraphs carrying out such tiny functions do not contribute to the readability of the program. It is possible to push a very small paragraph up into the paragraph that PERFORMs it, if the small paragraph is PERFORMed from only one place in the program. For example, consider the two paragraphs PRODUCE–SALESPERSON–TOTAL–LINE and PRODUCE–FINAL–LINES, each of them quite small. The first, PRODUCE–SALESPERSON–TOTAL–LINE, is PERFORMed from two places in the program and cannot be pushed up. The other, PRODUCE–FINAL–LINES, is PERFORMed from only one place, in the main control paragraph at the beginning of the Procedure Division. So it would be possible to replace, in the main control paragraph, the statement PERFORM PRODUCE–FINAL–LINES with the entire PRODUCE–FINAL–LINES paragraph itself. If that were done the IF statement in the main control paragraph would be:

```
IF MORE–INPUT = 'YES'
    PERFORM INITIALIZATION
    PERFORM PRODUCE–REPORT–BODY UNTIL MORE–INPUT = 'NO'
    PERFORM PRINT–A–SLSPSN–TOTAL–LINE
    PERFORM PRINT–A–STORE–TOTAL–LINE
    MOVE GRAND–TOTAL–W TO GRAND–TOTAL–OUT
    WRITE REPORT–LINE FROM GRAND–TOTAL–LINE AFTER 2.
```

Even though the logic of the program would remain unchanged it would no longer reflect the hierarchy chart. It is a matter of personal opinion whether the IF statement is more easily understood as it appears above or as it appears in Program 14.

The IF statement in the paragraph PRODUCE–REPORT–BODY carries out the logic shown in the flowchart in Figure 6.18. As each input card is read, it is tested first to see whether a major control break has occurred. If so, the major control break routine is PERFORMed; if not, it is then necessary to see whether the lower-level break has occurred. So we have here in the No path of an IF test another IF statement. The whole thing is referred to as a **nested IF** and usually presents no problem and requires no special techniques. We will see in the next chapter, however, some rather involved uses of nested IF statements and a kind of logic that cannot be coded by using only nested IFs because of a limitation in the COBOL language.

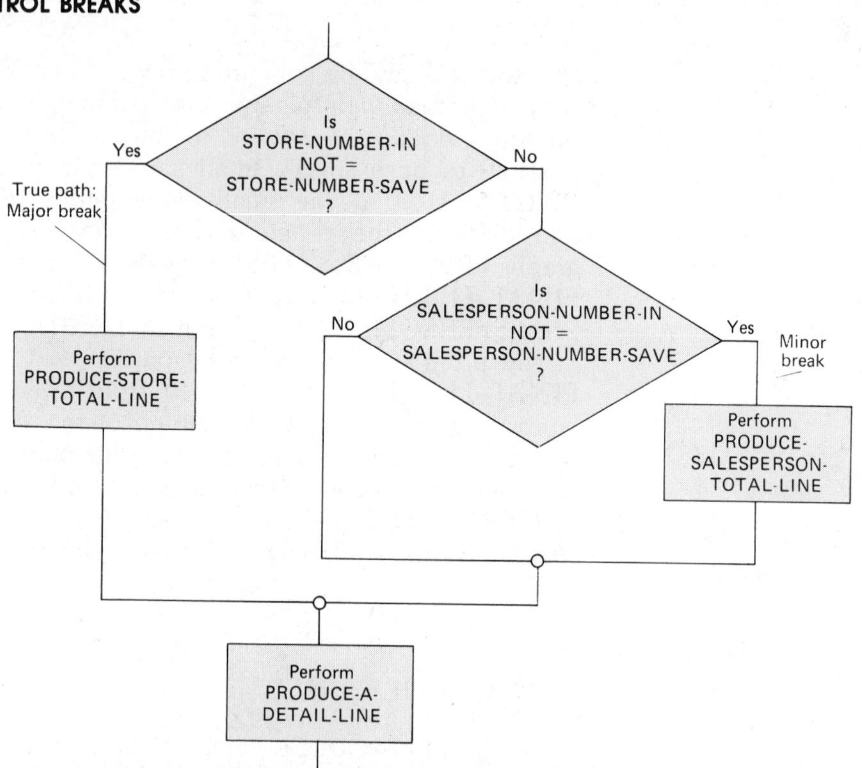

Figure 6.18
Flowchart of the control break testing logic.

***Exercise 3.** Rewrite the program in Exercise 2 with group indication of the customer number; that is, have each customer number print on only the first detail line after a control break and on the first detail line of a page.

***Exercise 4.** Using the input of Program 14, write a program that will print a line for each card showing the customer number, the sales amount, a tax amount on the sale (at 8 per cent), and the total of the sale amount and the tax. On each total line for salesperson, store, and grand total, show the total of all the appropriate sale amounts, tax amounts, and the totals of the sale amounts and the tax amounts. Group indicate the store numbers and salesperson numbers.

***Exercise 5.** Modify your solution to Exercise 4 so that no detail lines print; only the various levels of total lines should print. This is called **group printing** or **summary reporting.** On salesperson total lines, show the store number and the salesperson number. On store total lines, show the store number. Arrange the page overflow logic so that a salesperson total line may be the first line on a page (after the page headings) but that a store total line will never print at the top of a page right after the headings.

Exercise 6. Modify your solution to Exercise 5 so that on each total line there also prints the number of sale amounts that make up the total.

Summary

COBOL programs must often produce reports which have total lines throughout, as well as at the end; there may also be one or more levels of totals. There must be at least one accumulator for each level of total, including one accumulator for the final total. One way of handling several levels of accumulators is shown in Program 14. The input data are added into the minor accumulator. When a minor total line is printed, the contents of the minor accumulator are added into the next higher-level accumulator. When a total line for that accumulator is printed, its contents are added into the next higher-level accumulator, and so on.

COBOL provides convenient facilities for handling page overflow when output occupies more than one page. An accumulator may be set up to serve as a line counter, and, as each line is printed, an appropriate amount added into the accumulator. The accumulator may be tested to see whether the output page is full and a new page heading printed on a new page if necessary. The line counter accumulator must then be reset to count lines on the new page.

COMPUTATIONAL may be designated as a USAGE of a field. Designating as COMPUTATIONAL those fields used entirely for operations internal to the program can improve program efficiency. When COMPUTATIONAL is designated for other fields, such as those which interact with input or output fields, it may improve or worsen program efficiency. The abbreviation of COMPUTATIONAL is COMP. Fields whose USAGE is not COMPUTATIONAL are DISPLAY fields.

The figurative constants ZERO, ZEROS, and ZEROES may be used wherever a numeric or nonnumeric constant of zero would otherwise be legal. COBOL interprets the figurative constant as numeric or nonnumeric, depending on the context. If a nonnumeric literal of zeros is called for, COBOL provides as many zeros as are needed, depending on the context.

SPACE and SPACES are equivalent, and both may be used wherever a nonnumeric constant would be legal. When SPACE or SPACES is used, COBOL provides a string of blanks, the exact number of blank characters being determined by the context.

Any item that is broken down into smaller fields is called a group item; any item not broken down is called an elementary item. Group items are always treated as though they were alphanumeric fields of USAGE DISPLAY, even if their component fields are defined as COMPUTATIONAL.

The BEFORE or AFTER ADVANCING clause may be used with either an integer, a mnemonic name, or an identifier to indicate the desired spacing or skipping in the output report. Mnemonic names are defined in the SPECIAL–NAMES paragraph. ADVANCING is an optional word, as is LINES.

Group indicating is printing each control field on only the first detail line after every control break and on the first detail line of each page. Group printing, also called summary reporting, is printing total lines only but not any detail lines. A detail line is a line on a report that is derived from just one input record.

A nested IF statement is an IF statement whose paths contain one or more IFs.

Fill-in Exercises

1. A change in the value of a control ________ is called a control ________.
2. When end-of-file is detected on the input file, the last ________ line(s) still remain to be printed.
3. On a hierarchy diagram, boxes may be duplicated to prevent ________ of lines.
4. A line on a report that is attributable to a single input record is called a ________.
5. Designating fields which do not interact with input or output fields as ________. can improve program efficiency.
6. The two field USAGEs we have studied so far are ________ and ________.
7. SPACE, SPACES, ZERO, ZEROS, and ZEROES are called ________ ________.
8. A field that is broken down into smaller fields is called a ________ ________.
9. A field that is not broken down into smaller fields is called an ________ ________.
10. In the BEFORE/AFTER ADVANCING clause, ________ and ________ are optional words.

Review Exercises

1. Modify your solution to Exercise 3, page 152, to provide a grand total line in addition to the customer total lines.
2. Which of the following clauses is (are) legal:

   a. BEFORE ADVANCING 2 LINES
   b. BEFORE 2 LINES
   c. BEFORE 2

d. AFTER ADVANCING 1
e. AFTER 1
f. AFTER

*3. Using the input for Program 13, write a program that will print and accumulate only the credits. Print one line for each input card containing a credit. Print a total line for each account, showing the total of only the credits for that account. Also show a final total line. Group indicate the account number.
4. Modify your solution to Exercise 4, page 152, so that on each salesperson total line there also prints the average amount of sales (before tax) and, if the average sale amount is less than $100, the words BELOW QUOTA.
5. Modify your solution to Exercise 4, page 152, so that only sale amounts of $100 or greater are printed and added into the totals. Sale amounts of less than $100 should be ignored by the program.
*6. Modify your solution to Exercise 4, page 152, so that on each total line there prints the number of sales of less than $100 which are included in the total and the number of sales of $100 or greater which are included in the total.

More About IF Statements

7

KEY POINTS

Here are the key points you should learn from this chapter:

1. the rules for forming nested IF statements;
2. the kind of logic that cannot be coded using only nested IF statements and how to code it properly;
3. how to code IF statements with complex conditions;
4. how to write abbreviated relation conditions;
5. how to use condition names.

KEY WORDS

Key words to recognize and learn:

complex condition	inner IF statement	object
abbreviated relation condition	compound condition	CORR
condition name	subject	case
outer IF statement	88-level	subscripting
	update	indexing

In this chapter you will study features related to the IF statement that make it more flexible than it has been. These are the nested IF, the **complex condition, abbreviated relation conditions,** and **condition names.**

Nested IF Statements

A nested IF statement was first used in Program 14 in the previous chapter. A nested IF statement is one where one or more IF statements appear in the True path or the False path, or both, of an IF statement. In Program 14 the nested IF consisted of an **outer IF statement** containing an **inner IF statement** in its No path. The flowchart in Figure 7.1 shows a slightly more complex nested IF. In Figure 7.1 both the Yes and No paths of the outer IF contain an IF; in addition, there

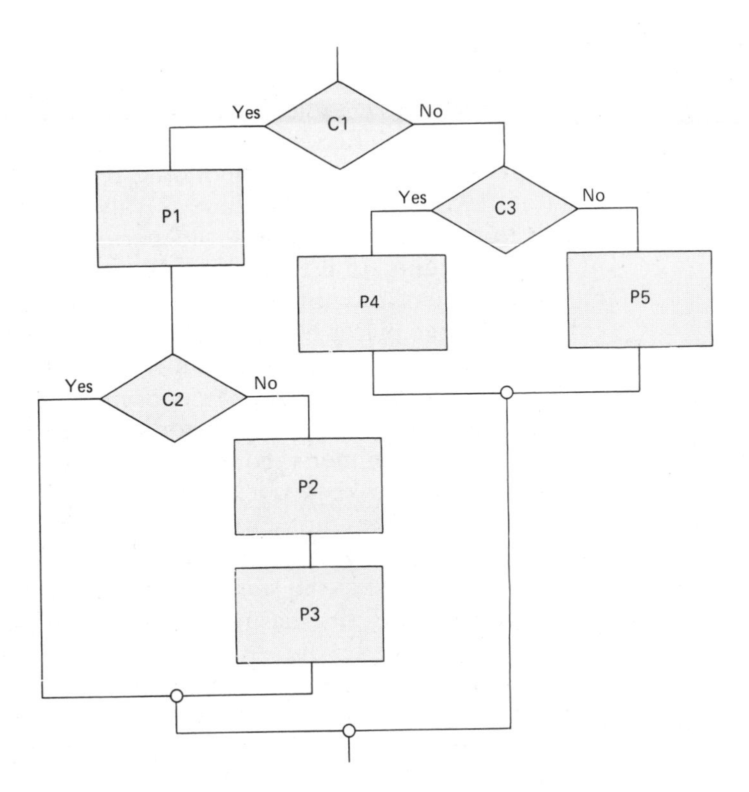

Figure 7.1
A nested IF statement.

is more than one processing step in one of the paths. Figure 7.1 may be coded in COBOL as follows:

```
IF C1
    PERFORM P1
    IF C2
        NEXT SENTENCE
    ELSE
        PERFORM P2
        PERFORM P3
ELSE
    IF C3
        PERFORM P4
    ELSE
        PERFORM P5.
```

Nested IF statements should always be written with proper indenting so that the programmer can see which ELSE belongs to which IF and where the Yes and No paths of each of the IF tests are. In the IF statement above, the first IF tests the condition C1 and starts the Yes path for that condition. The ELSE aligned with the first IF starts the No path for that same condition, C1. Indented further are other IFs and ELSEs, each ELSE aligned with its corresponding IF. Each ELSE ends the Yes path for its corresponding IF and starts the No path.

Most importantly, there is only one period at the end of the whole thing. Remember that the period rejoins Yes and No paths in IF statements. In flowcharts of IF statements the rejoining of Yes and No paths is shown by a little circle. You can see in Figure 7.1 that there are three places where Yes and No paths rejoin, and in the IF statement the single period rejoins all three pairs of paths.

It has already been mentioned that COBOL does not examine indenting in IF statements or any others and that indenting is only for the programmer's convenience. Here the programmer would use indenting to keep track of which ELSE belongs to which IF. But since COBOL does not examine indenting, it uses the following rule to determine for itself which ELSE belongs to which IF: Any ELSE encountered in the statement is considered to apply to the immediately preceding IF that has not already been paired with an ELSE. You need never concern yourself with this rule if you are careful to use indenting properly. Properly indented IFs and ELSEs will always give the same logical result as the rule that COBOL uses for pairing IFs with their corresponding ELSEs.

Here is an example of a nested IF where there is a little less processing than in the previous nested IF. Figure 7.2 shows a flowchart where there are two paths with no processing. The coding for the flowchart in Figure 7.2 is:

```
IF C1
    IF C2
        PERFORM P1
        PERFORM P2
        PERFORM P3
    ELSE
        NEXT SENTENCE
ELSE
    IF C3
        NEXT SENTENCE
    ELSE
        PERFORM P4.
```

Notice that in this statement the reserved words NEXT SENTENCE must be used whenever there is no processing in a path, and of course the single period at the end of the sentence rejoins all three pairs of Yes and No paths.

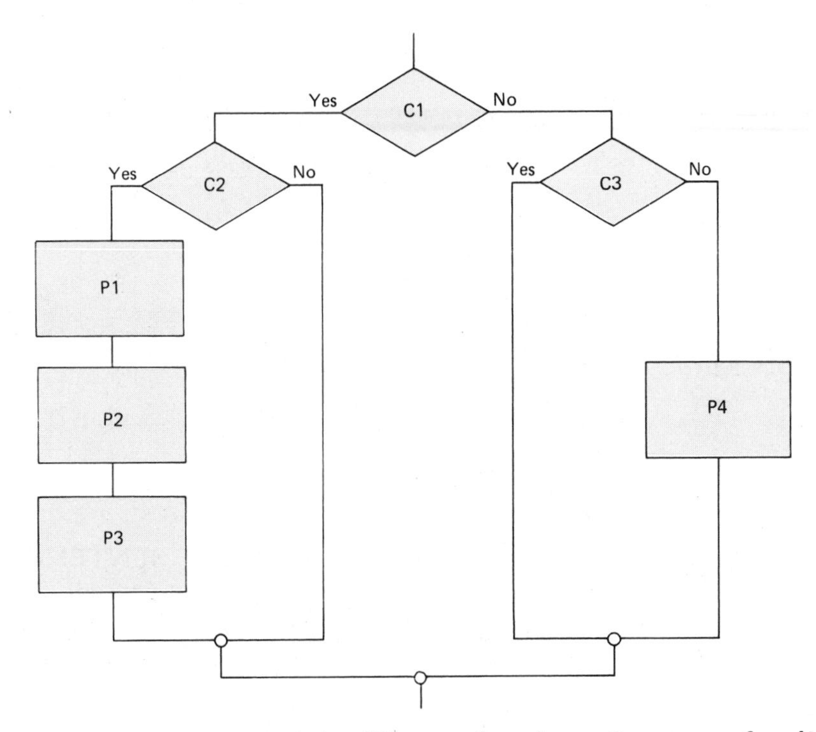

Figure 7.2
A nested IF statement in which two paths contain no processing.

Figure 7.3 shows a slightly different flowchart. One way of coding Figure 7.3 is:

```
IF C1
    IF C2
        PERFORM P1
    ELSE
        NEXT SENTENCE
ELSE
    IF C3
        PERFORM P2
    ELSE
        NEXT SENTENCE.
```

But the rules of COBOL say that the words ELSE NEXT SENTENCE may be omitted if they appear immediately before a period, so Figure 7.3 may be equally well coded as:

```
IF C1
    IF C2
        PERFORM P1
    ELSE
        NEXT SENTENCE
ELSE
    IF C3
        PERFORM P2.
```

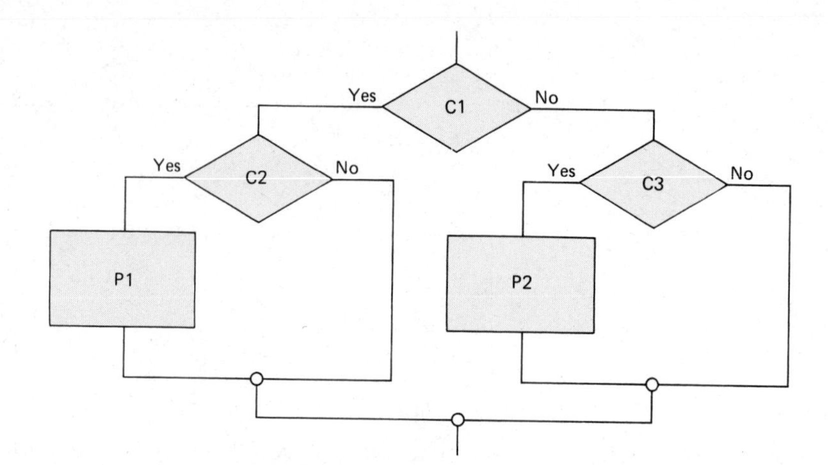

Figure 7.3 Another flowchart in which two paths contain no processing.

The remaining **ELSE NEXT SENTENCE** clause cannot be removed since it does not immediately precede the period.

Figure 7.4 shows a flowchart of nested logic which cannot be implemented on most COBOL systems by using just a nested IF.[1]

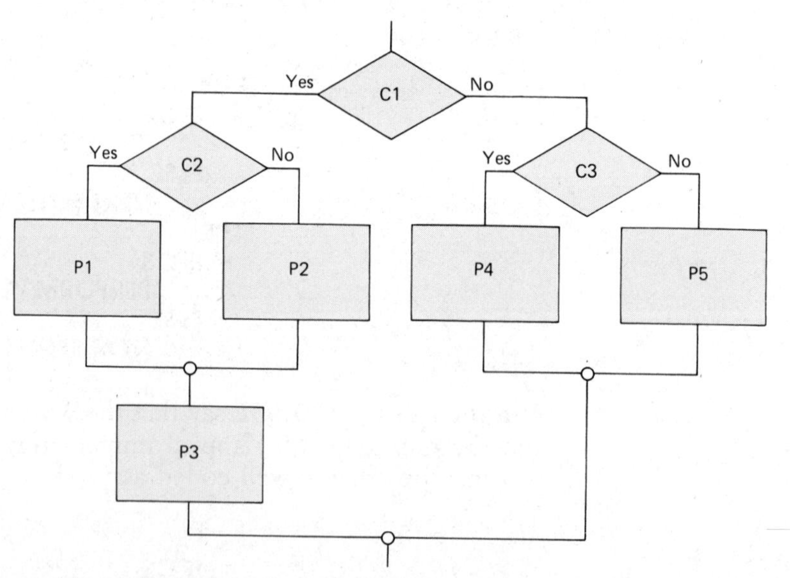

Figure 7.4 Flowchart of logic which cannot be implemented in COBOL using only a nested IF statement. The joining of a Yes path with a No path before box "P3" cannot be coded.

1. The logic in Figure 7.4 can be implemented using only a nested IF on some of the newest COBOL systems having the END–IF statement.

What is it about Figure 7.4 that makes it impossible to code using only a nested IF? It is the paths from the condition C2: They rejoin before the end of the logic. In COBOL, the only way to join Yes and No paths is with a period. There can be only one period in an IF statement, and so the statement can be implemented with just a nested IF construction only if all the Yes and No pairs join at the end. An attempt to write the logic of Figure 7.4 might look like this:

```
IF C1
    IF C2
        PERFORM P1
    ELSE
        PERFORM P2
PERFORM P3
ELSE
    IF C3
        PERFORM P4
    ELSE
        PERFORM P5.
```

But aligning the statement PERFORM P3 with the first IF doesn't fool anyone but the programmer. It certainly doesn't fool COBOL, which does not examine the indenting and interprets the IF statement as if it were written:

```
IF C1
    IF C2
        PERFORM P1
    ELSE
        PERFORM P2
        PERFORM P3
ELSE
    IF C3
        PERFORM P4
    ELSE
        PERFORM P5.
```

The logic implied by the IF statement above is given in Figure 7.5.

The correct way to handle a situation like the logic in Figure 7.4 is to remove from the flowchart the entire portion that is causing the trouble and put it in a separate paragraph. The exact portion of the flowchart that must be removed in such cases can be located as follows: Find a little circle that appears anywhere before the end of the logic. In this case it's the circle before the box "P3." Then trace back from that circle to its corresponding condition; in this case the condition that corresponds to the troublesome circle is C2. Now remove everything between the condition and the circle from this flowchart,

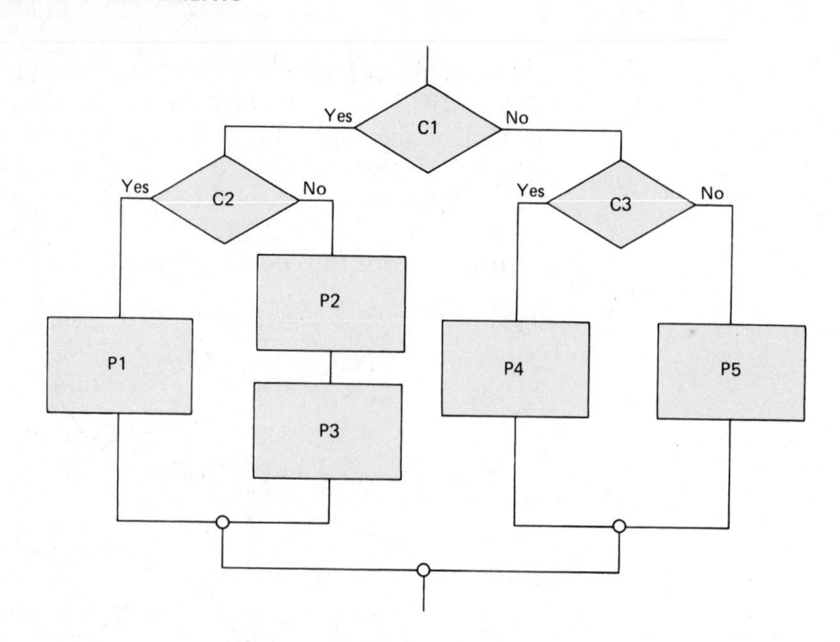

Figure 7.5
Logic implied by an IF statement trying to implement the logic shown in Figure 7.4.

including the condition and the circle, and place them all in a separate paragraph. Let's call that new paragraph C2–CONDITION–TEST, and redraw Figure 7.4 as Figure 7.6.

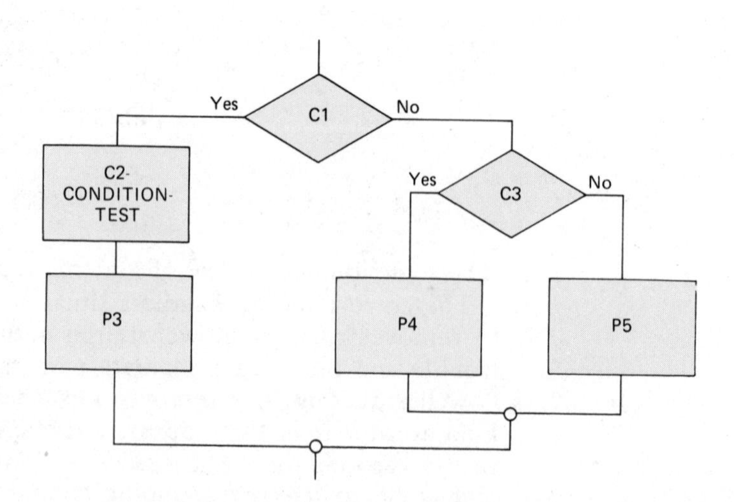

Figure 7.6
Figure 7.4 redrawn so that it can be coded in COBOL.

Now there are no little circles except at the end of the logic, and Figure 7.6 can be coded as:

```
IF C1
    PERFORM C2–CONDITION–TEST
    PERFORM P3
ELSE
    IF C3
        PERFORM P4
    ELSE
        PERFORM P5.
```

The paragraph C2–CONDITION–TEST, which could be placed anywhere in the Procedure Division, would be:

```
C2–CONDITION–TEST.
    IF C2
        PERFORM P1
    ELSE
        PERFORM P2.
```

IF statements may be nested to any level within the limits of the size of the computer; that is, an IF statement in one of the paths of an outer IF statement may itself have an IF statement in one of its own paths, as shown in Figure 7.7. Figure 7.7 may be coded as:

```
IF C1
    PERFORM P1
    IF C2
        NEXT SENTENCE
    ELSE
        PERFORM P2
        IF C3
            PERFORM P3
            PERFORM P4
        ELSE
            NEXT SENTENCE
ELSE
    NEXT SENTENCE.
```

Or, by removing the unneeded ELSE NEXT SENTENCE clauses:

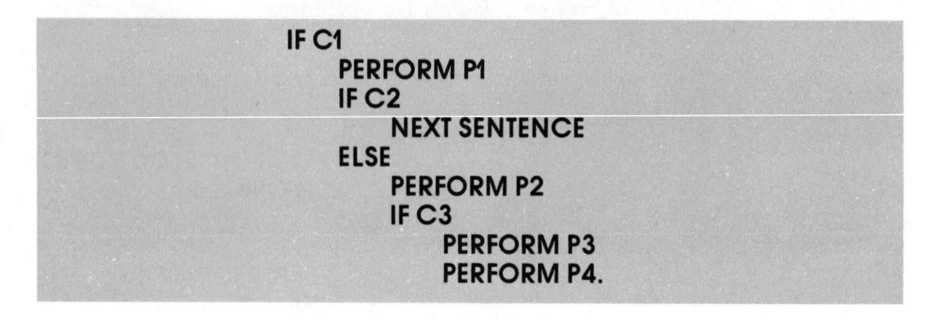

```
IF C1
    PERFORM P1
    IF C2
        NEXT SENTENCE
    ELSE
        PERFORM P2
        IF C3
            PERFORM P3
            PERFORM P4.
```

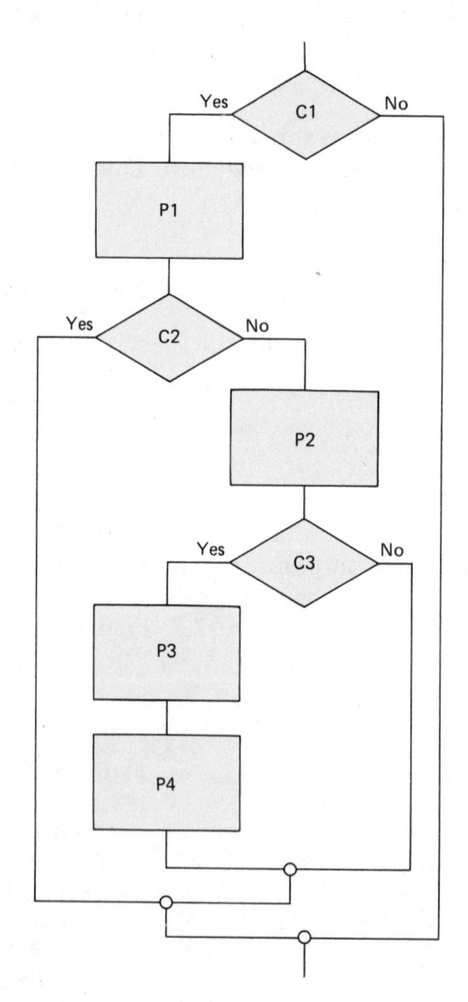

Figure 7.7
Three levels of nesting in an IF statement.

Exercise 1. Code the logic shown in Figure 7.E1 using only nested IF statements.

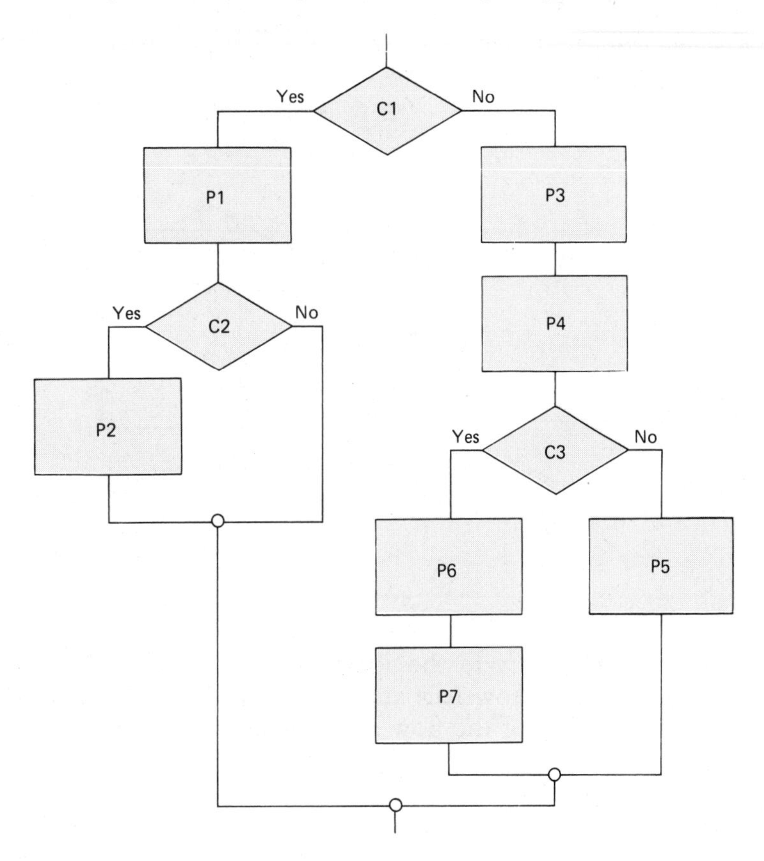

Figure 7.E1(a)
Flowcharts for Exercise 1.

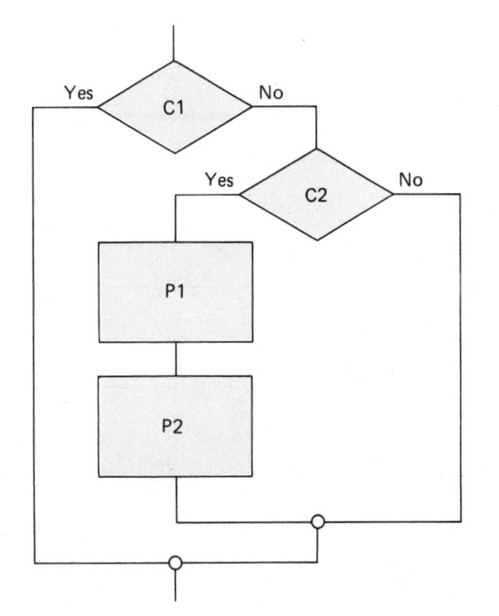

Figure 7.E1(b)

Figure 7.E1(c)

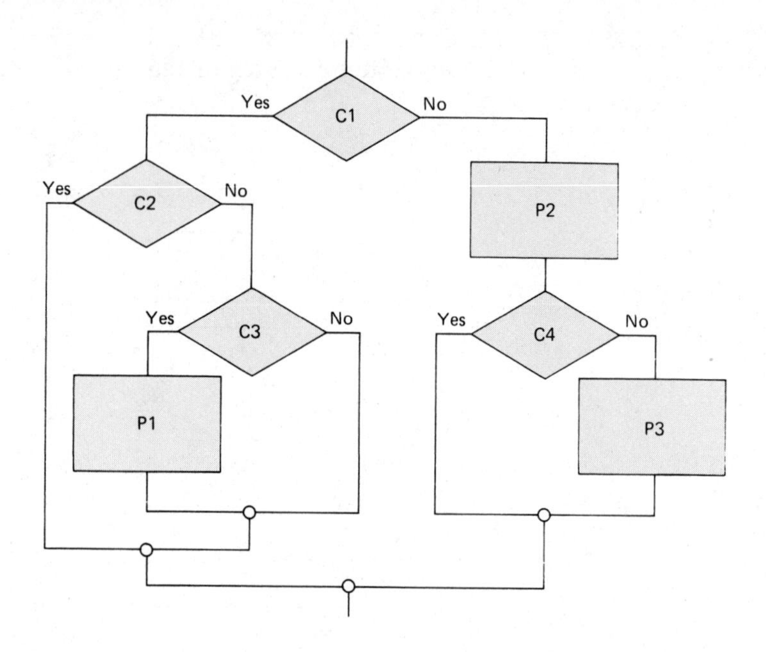

Exercise 2. Code the logic shown in Figure 7.E2 by first removing from each flowchart any segments whose Yes and No paths rejoin before the end of the flow and placing them into a separate paragraph.

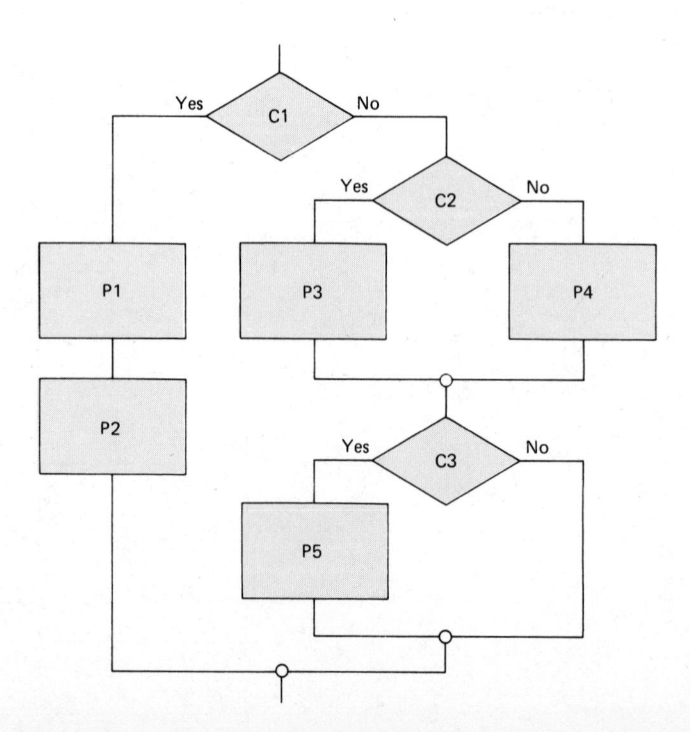

Figure 7.E2(a)
Flowcharts for Exercise 2.

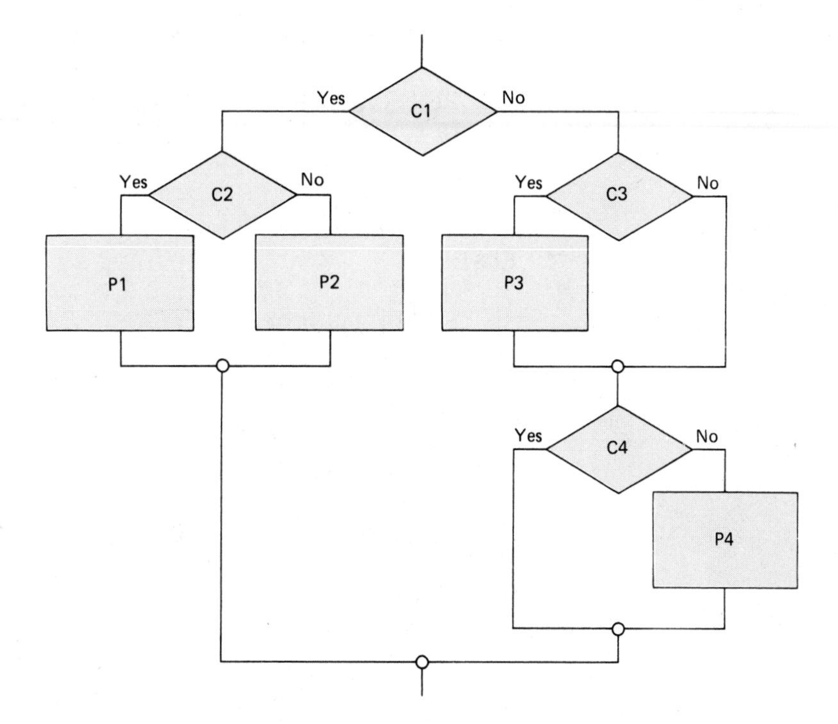

Figure 7.E2(b)

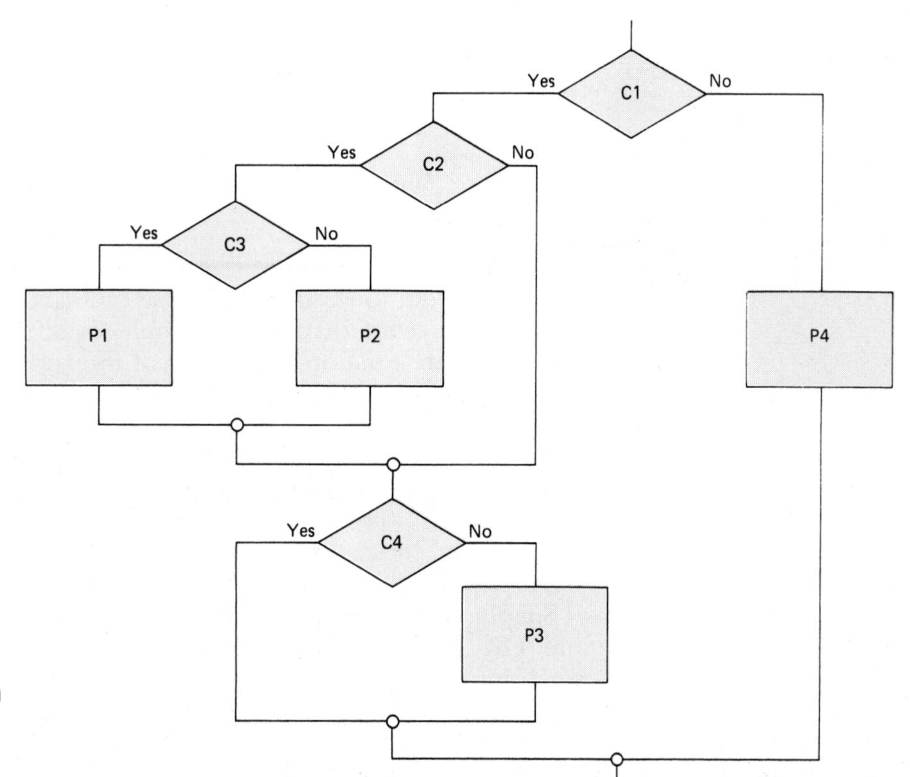

Figure 7.E2(c)

Exercise 3. Draw a flowchart showing the logic of each of the following statements:

a.
```
IF C1
    PERFORM P1
    IF C2
        NEXT SENTENCE
    ELSE
        PERFORM P2
ELSE
    IF C3
        PERFORM P3.
```

b.
```
IF C1
    IF C2
        PERFORM P1
        PERFORM P2
    ELSE
        NEXT SENTENCE
ELSE
    IF C3
        NEXT SENTENCE
    ELSE
        PERFORM P3
        PERFORM P4.
```

c.
```
IF C1
    IF C2
        IF C3
            PERFORM P1.
```

Complex Conditions

Complex conditions, sometimes also called **compound conditions,** are certain legal combinations of simple conditions. Among the simple conditions are relation conditions of the sort that we have been using in IF statements so far in this book; for example:

```
QUANTITY-IN IS LESS THAN 400
STORE-NUMBER-IN NOT = STORE-NUMBER-SAVE
```

are relation conditions. There are other kinds of simple conditions which we will discuss later.

Simple conditions may be connected by the reserved words AND and OR to form complex conditions. An example of a complex condition used in an IF statement might be:

```
IF QUANTITY-IN IS LESS THAN 400 AND
   UNIT-PRICE IS GREATER THAN 10.00
      PERFORM REORDER-HIGH-PRICE
ELSE
      PERFORM TEST-2.
```

In the preceding IF statement both simple conditions would have to be true in order for the True path of the IF to be executed. In the statement:

```
IF A = B OR
   C = D
      PERFORM FOUND-A-MATCH
ELSE
      PERFORM NO-MATCH.
```

if either A equals B or C equals D the True path will be executed.

ANDs and ORs may be combined in a complex condition. Ordinarily, ANDs are evaluated first, followed by ORs. So the condition:

```
IF A = B OR C = D AND E = F . . .
```

would be evaluated as if it were written:

```
IF A = B OR (C = D AND E = F) . . .
```

Parentheses may be used in complex conditions to change the order of evaluation. Unnecessary parentheses may be used for readability and will not interfere with proper execution of the statement. In any particular COBOL system, the rules regarding spacing before and after parentheses in complex conditions are the same as the rules regarding spacing before and after parentheses in arithmetic expressions.

As we have seen earlier, a relation condition may contain the word NOT, as in:

```
IF STORE-NUMBER-IN NOT = STORE-NUMBER-SAVE . . .
```

Such conditions may be used in complex conditions, as in:

```
IF A NOT GREATER THAN B OR C = D . . .
```

or:

```
IF A NOT LESS THAN B AND C NOT GREATER THAN D . . .
```

Also, complex conditions may be negated, as in:

```
IF NOT (A = B AND C = D)
    PERFORM NOT-BOTH-EQUAL
ELSE
    PERFORM BOTH-EQUAL.
```

Complex conditions can become extremely complicated; see your COBOL manual for a full discussion of them. Complex conditions should be used only when their use makes the meaning of the program clearer than it would be without the complex condition.

Abbreviated Relation Conditions

In a complex condition made up of simple relation conditions, more than one of the relation conditions sometimes have the same **subject** and/or the same relational operator. For example, in the condition:

```
IF A = B OR A IS GREATER THAN C . . .
```

there are two simple relation conditions and they both have A as their subject. Such a condition could be abbreviated by leaving out the repetition of the subject, so the following would be legal:

```
IF A = B OR IS GREATER THAN C
    PERFORM IN-RANGE.
```

If the simple conditions in a complex condition have the same subject and the same relational operator, then repetition of the subject and operator may be omitted. So the condition:

```
IF A NOT = B AND A NOT = C . . .
```

could be abbreviated to yield the following legal statement:

```
IF A NOT = B AND C
    PERFORM EQUALS-NEITHER.
```

If you use NOT in a complex condition in the relational operator, as in the IF statement just given, and also to negate a condition, you may confuse someone trying to read the program. You will almost certainly confuse yourself, and you may even confuse the COBOL system. For example, different COBOL systems might interpret the condition:

```
IF A IS GREATER THAN B AND NOT LESS THAN C AND D . . .
```

either as:

```
IF A IS GREATER THAN B AND A IS NOT LESS THAN C AND A
   IS NOT LESS THAN D . . .
```
[2]

or as:

```
IF A IS GREATER THAN B AND NOT A IS LESS THAN C AND A
   IS LESS THAN D . . .
```
[3]

Since such constructions have very little meaning, if any, to someone reading them, they should be avoided. (Furthermore, parentheses are not permitted with an abbreviated condition, so they cannot be used even to improve readability. Parentheses are permitted to completely surround a condition, however.)

A word of warning about a common error made even by experienced programmers. In testing whether the value of some data name is not equal to any of several values, the incorrect connective is sometimes used. For example, if a field called CODE–IN is always supposed to be either a 1, a 2, or a 3, we might want to test to see whether it is and PERFORM an error routine if it is not. Unfortunately, COBOL does not permit us to say:

```
IF CODE-IN IS NEITHER 1 NOR 2 NOR 3
     PERFORM ERROR-ROUTINE.
```

Whenever you find this NEITHER ... NOR situation, there are three correct ways to code it, and one commonly used incorrect way (which of course must eventually be corrected by the programmer if the program is to work). You may write it as an OR and negate the whole condition, as in:

```
IF NOT (CODE-IN = 1 OR 2 OR 3)
     PERFORM ERROR-ROUTINE.
```

you may reverse the Yes and No paths, as:

```
IF CODE-IN = 1 OR 2 OR 3
     NEXT SENTENCE
ELSE
     PERFORM ERROR-ROUTINE.
```

2. American National Standards Institute, *American National Standard Programming Language COBOL,* 1974, p. II-48.
3. International Business Machines Corporation, *IBM OS Full American National Standard COBOL, sixth ed.,* 1975, p. 165.

or, in the form which is closest to the original wording:

```
IF CODE-IN NOT = 1 AND 2 AND 3
    PERFORM ERROR-ROUTINE.
```

This last form may look a little strange, but it makes sense when you realize that in its fully expanded, unabbreviated form, the statement would be:

```
IF CODE-IN NOT = 1 AND
   CODE-IN NOT = 2 AND
   CODE-IN NOT = 3
     PERFORM ERROR-ROUTINE.
```

For the common wrong way of coding this condition, see Exercise 4.

Some situations involving multiple testing of this sort can be handled best by using an **88-level** entry in the Data Division. We will discuss the 88 level after these exercises.

Exercise 4. Given the following COBOL statement:

```
IF CODE-IN NOT = 1 OR 2 OR 3
    PERFORM ERROR-ROUTINE.
```

How would the statement execute if CODE-IN is equal to:

a. 1 b. 2 c. 3 d. 4

Exercise 5. Given the following statement:

```
IF AMOUNT IS GREATER THAN 1000000
    PERFORM ERROR-ROUTINE
ELSE
    IF CHARGE-CODE = 'H'
        NEXT SENTENCE
    ELSE
        PERFORM ERROR-ROUTINE.
```

Rewrite this statement using a complex condition so that the logic is made clearer.

Exercise 6. Rewrite the following conditions using abbreviated subjects and/or operators, as appropriate:

a. IF FIELD-1 GREATER THAN FIELD-2 AND FIELD-1 LESS THAN FIELD-3 . . .
b. IF FIELD-1 NOT GREATER THAN FIELD-2 AND FIELD-1 NOT GREATER THAN FIELD-3 . . .

The 88 Level

An 88-level entry in the Data Division can be used by the programmer to give a name to one or more values that might be assigned to a field. 88-level entries are most useful when used on input data fields in the File Section. Many programmers use them on fields in the Working Storage Section, but their use there must often be carefully examined to see whether they make the program easier to understand or more difficult.

When an 88-level entry is used on a field, the field must still be described in the usual way. That is, the field must have an ordinary level number (from 01 to 49), a PICTURE, and may have other optional clauses, such as SIGN, VALUE, and/or COMPUTATIONAL. As an example, we might want to use an 88-level entry on the CODE–IN field mentioned in the example in the previous section. Since CODE–IN can have valid values of 1, 2, or 3, it might be described as:

```
05 CODE-IN          PIC 9.
```

We can use an 88-level entry on this field to give a condition name to the valid codes. The general format of the 88-level entry is:

```
88 condition-name { VALUE IS   }
                  { VALUES ARE }

   literal-1 [ { THROUGH } literal-2 ]
               { THRU    }

 [ literal-3 [ { THROUGH } literal-4 ] ] ...
               { THRU    }
```

An 88-level entry, when used, must immediately follow the description of the field to which it applies. If we want to give the name CODE–IS–OK to the values 1, 2, and 3 we can write:

```
05 CODE-IN              PIC 9.
   88 CODE-IS-OK        VALUES ARE 1 THROUGH 3.
```

or:

```
05 CODE-IN              PIC 9.
   88 CODE-IS-OK        VALUES 1, 2, 3.
```

The words THROUGH and THRU are equivalent. The rules for making up condition names are the same as for making up data names. You can see from the general format that one VALUE clause is required in an 88-level entry and that no other clauses are permitted. Note especially that the PICTURE clause is not permitted in an 88-level entry.

In the Procedure Division the condition name may be tested in an IF statement; for example:

```
IF CODE-IS-OK
    NEXT SENTENCE
ELSE
    PERFORM ERROR-ROUTINE.
```

or

```
IF NOT CODE-IS-OK
    PERFORM ERROR-ROUTINE.
```

Notice that there is no relational operator in these IF statements. The condition name alone constitutes the entire condition in the IF.

The use of an 88-level entry in connection with coded fields like CODE-IN makes it very convenient to see at a glance what all the valid codes are. The program is also very easy to change if the set of valid codes should change, and the Procedure Division coding is clearer when condition names are properly used.

Exercise 7. Write an 05-level entry and an 88-level entry for the following field: a three-digit, unsigned integer field called WORK-STATION. The condition name VALID-WORK-STATION should be given to the valid work stations, whose numbers lie in the range 100 to 999, inclusive.

FICA Tax Calculation

The FICA tax is a payroll tax that is applied to the paychecks and pay envelopes of most American workers. The letters FICA stand for Federal Insurance Contributions Act, but the contributions made by workers to the Federal government under the Act are like the contributions you give to a robber when he or she has a lethal weapon pointed at your head. The common name for the FICA tax is Social Security tax. It is computed as a certain percentage of the worker's gross pay, with some limit on the maximum tax that may be extracted from a worker in one calendar year. In 1979 the maximum tax "contribution" was $1,403.77. After an employee had paid $1,403.77, no further FICA deductions were made from gross pay.

Let us now try to write the FICA tax portion of a payroll program. Assume that in the Data Division are fields with the following meanings:

CURRENT–GROSS	**the employee's gross earnings for the current period (week, half-month, month, and so on); this is the amount on which the tax is to be computed**
TAX–RATE	**the tax percentage**
YEAR–TO–DATE–TAX	**the amount of FICA tax that has already been paid by the employee so far this year, excluding any tax that might be due because of earnings in this pay period**
MAXIMUM–TAX	**the maximum tax that an employee is required to pay in one year; in 1979 it was $1,403.77**
CURRENT–TAX	**the program is to assign the FICA tax for this period to this field when the computation is complete**

The tax computation routine is also to **update** the YEAR–TO–DATE–TAX field; that is, after computing the tax for this period, the routine is to change the value of the YEAR–TO–DATE–TAX field so that it will reflect the tax paid through the current pay period.

There are a number of ways to approach this problem. One way is shown in the flowchart of Figure 7.8. The flowchart first tests to see whether the YEAR–TO–DATE–TAX is already at the maximum; if so, there is no more tax to be paid, and CURRENT–TAX is zero. If not, a calculation and test are done to see whether tax on this period's gross amount will put the YEAR–TO–DATE–TAX over the maximum. The outcome of that test then causes the CURRENT–TAX to be computed in one of two ways to insure that the CURRENT–TAX does not put the YEAR–TO–DATE–TAX over the top.

Figure 7.9 shows the coding for the flowchart in Figure 7.8. One COBOL feature not used before can be found in the inner IF statement. The condition is written as:

```
IF CURRENT-TAX + YEAR-TO-DATE-TAX LESS THAN MAXIMUM-TAX . . .
```

showing that an arithmetic expression may appear as the subject of a relation condition. An arithmetic expression may also appear in a relation condition following the relational operator as the **object** of the relation condition. The general format for a relation condition is:

$$\left\{\begin{array}{l}\textbf{identifier-1}\\ \textbf{literal-1}\\ \textbf{arithmetic-expression-1}\end{array}\right\} \textbf{ relational-operator } \left\{\begin{array}{l}\textbf{identifier-2}\\ \textbf{literal-2}\\ \textbf{arithmetic-expression-2}\end{array}\right\}$$

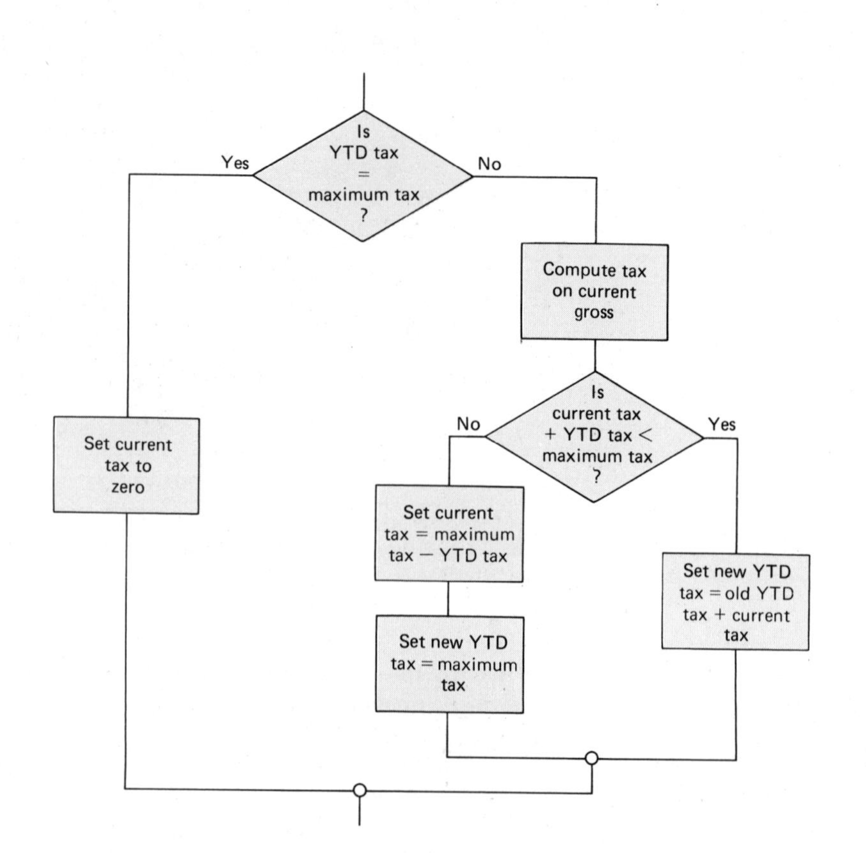

Figure 7.8
Flowchart of one way to carry out the FICA tax computation.

Figure 7.9 Coding for the flowchart in Figure 7.8.

```
000010     IF YEAR-TO-DATE-TAX = MAXIMUM-TAX
000020         MOVE 0 TO CURRENT-TAX
000030     ELSE
000040         MULTIPLY CURRENT-GROSS BY TAX-RATE
000050                                 GIVING CURRENT-TAX ROUNDED
000060         IF CURRENT-TAX + YEAR-TO-DATE-TAX LESS THAN MAXIMUM-TAX
000070             ADD CURRENT-TAX TO YEAR-TO-DATE-TAX
000080         ELSE
000090             COMPUTE CURRENT-TAX = MAXIMUM-TAX - YEAR-TO-DATE-TAX
000100             MOVE MAXIMUM-TAX TO YEAR-TO-DATE-TAX.
```

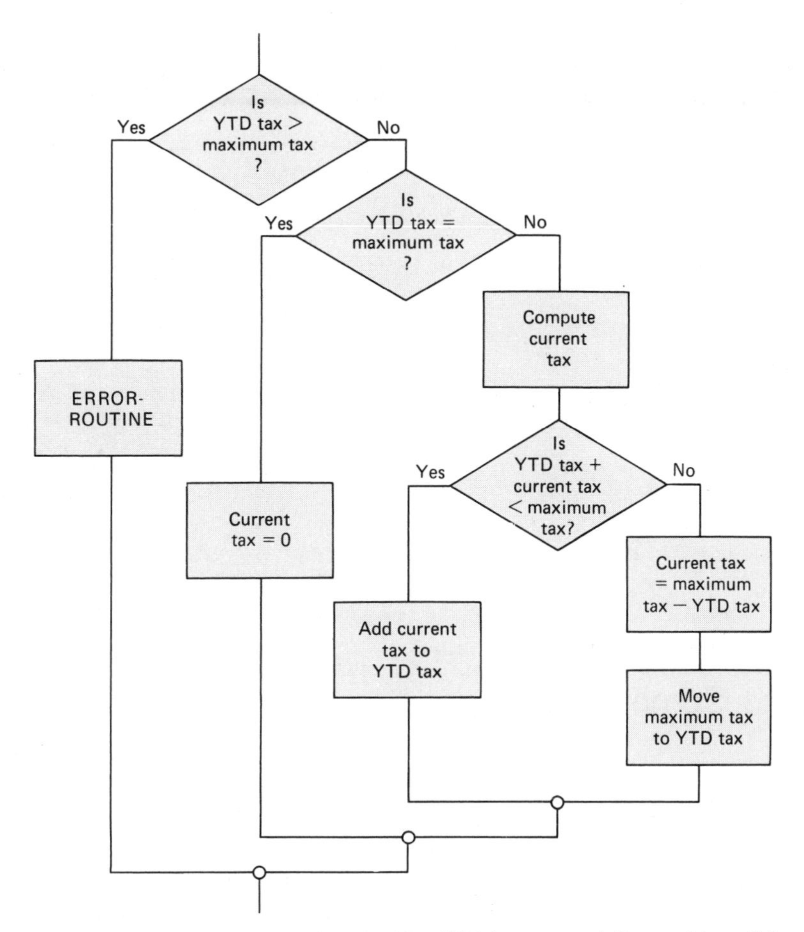

Figure 7.10 A more elaborate treatment of the FICA tax computation problem. This flowchart tests whether the YEAR–TO–DATE–TAX has erroneously exceeded the MAXIMUM–TAX.

The general format for the relational operators was given in Chapter 3. The subject and object of a relation condition cannot both be literals.

A somewhat more elaborate treatment of the FICA tax computation problem is shown in Figure 7.10. This flowchart allows for the possibility that somehow, due to errors in input data or in the program or in the functioning of the computer, the YEAR–TO–DATE–TAX has become larger than the MAXIMUM–TAX. If this condition is found, an error routine is executed. The coding for the flowchart is shown in Figure 7.11.

Figure 7.11 Coding for the flowchart in Figure 7.10.

```
000010     IF YEAR-TO-DATE-TAX GREATER THAN MAXIMUM-TAX
000020         PERFORM ERROR-ROUTINE
000030     ELSE
000040         IF YEAR-TO-DATE-TAX = MAXIMUM-TAX
000050             MOVE 0 TO CURRENT-TAX
000060         ELSE
000070             MULTIPLY CURRENT-GROSS BY TAX-RATE
000080                                 GIVING CURRENT-TAX ROUNDED
000090             IF CURRENT-TAX + YEAR-TO-DATE-TAX LESS THAN
000100                                                 MAXIMUM-TAX
000110                 ADD CURRENT-TAX TO YEAR-TO-DATE-TAX
000120             ELSE
000130                 COMPUTE CURRENT-TAX =
000140                             MAXIMUM-TAX - YEAR-TO-DATE-TAX
000150                 MOVE MAXIMUM-TAX TO YEAR-TO-DATE-TAX.
```

Sales Commissions

We now do a program for computing commissions on sales, using a somewhat involved schedule of commission rates. This program shows the use of a nested IF statement, complex conditions, and 88-level entries in context.

In this problem, the salespeople are each assigned to some class; the salespeople classes are A through H. All salespeople assigned to classes A through F are considered to be junior salespeople; salespeople in class G are associate salespeople; and class H salespeople are senior. Salespeople in different titles have different quotas, and their commission rates depend on whether or not they meet their quota. The quotas are: for junior salespeople, none; for associates, an average sale amount of $10,000; for seniors, an average sale amount of $50,000. The commission rates for each of the salespeople titles are shown in Table 7.1.

For example, suppose that an associate salesperson has three sales of $5,000, $10,000, and $20,000. The average of these is greater than the quota for associates, so the commission would be 20 percent of the $5,000 sale and 30 percent of each of the other two.

The input format is shown in Figure 7.12 and has space for three sales. The program will always consider the average of the three sales in determining which commission schedule to use. The output format is shown in Figure 7.13. Two types of error lines are shown. This program will check the salesperson class in each input record before processing it, and, if the class is missing or invalid, the program will print an appropriate error message. The program, Program 15, is shown in Figure 7.14.

Title	commission schedule if quota is not met	commission schedule if quota is met
junior	10% of sale amount on all sales	10% of sale amount on all sales
associate	5% of sale amount on any sales less than quota amount; 20% of sale amount on other sales	20% of sale amount on any sales less than quota amount; 30% of sale amount on other sales
senior	20% of sale amount on any sales less than quota amount; 30% of sale amount on other sales	30% of sale amount on any sales less than quota amount; 40% of sale amount on other sales

Table 7.1. Commission schedules for three salespeople titles for Program 15.

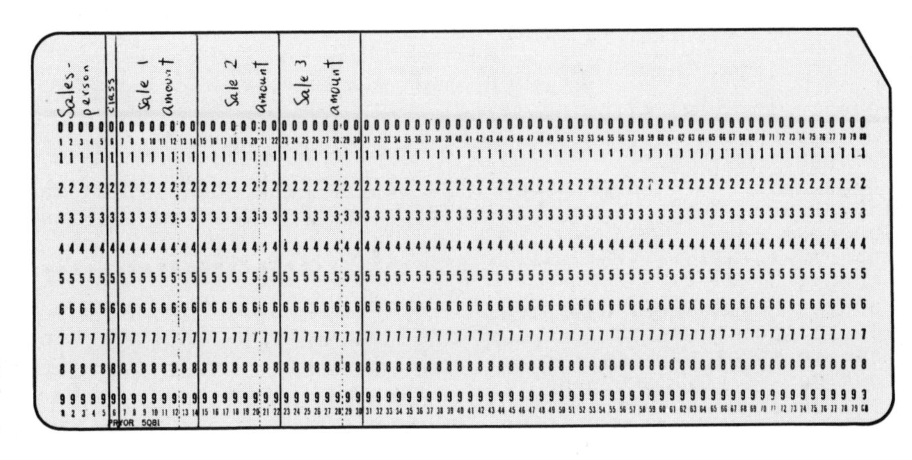

Figure 7.12
Data card format for Program 15.

Figure 7.13 Output format for Program 15.

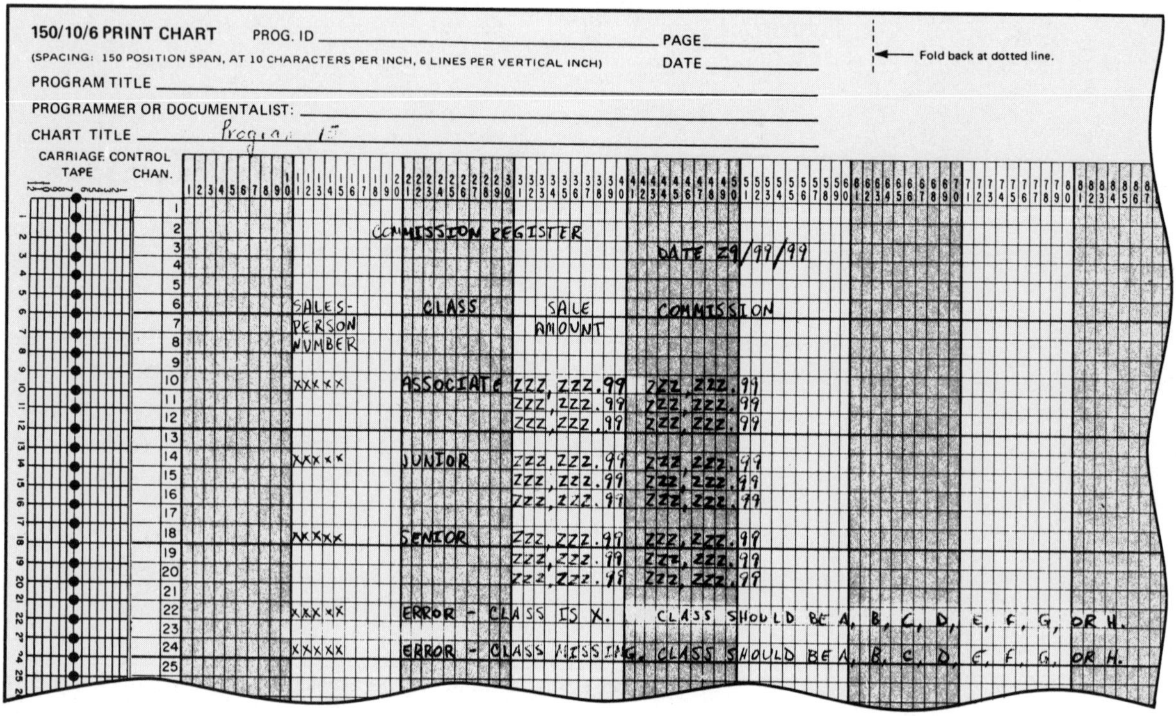

150/10/6 PRINT CHART PROG. ID ______ PAGE ______

(SPACING: 150 POSITION SPAN, AT 10 CHARACTERS PER INCH, 6 LINES PER VERTICAL INCH) DATE ______

← Fold back at dotted line.

PROGRAM TITLE ______

PROGRAMMER OR DOCUMENTALIST: ______

CHART TITLE Program 15

CARRIAGE CONTROL TAPE CHAN.

```
 2                          COMMISSION REGISTER
 3                                                 DATE Z9/99/99
 6             SALES-        CLASS          SALE          COMMISSION
 7             PERSON                       AMOUNT
 8             NUMBER
10             XXXXX         ASSOCIATE  ZZZ,ZZZ.99    ZZZ,ZZZ.99
11                                      ZZZ,ZZZ.99    ZZZ,ZZZ.99
12                                      ZZZ,ZZZ.99    ZZZ,ZZZ.99
14             XXXXX         JUNIOR     ZZZ,ZZZ.99    ZZZ,ZZZ.99
15                                      ZZZ,ZZZ.99    ZZZ,ZZZ.99
16                                      ZZZ,ZZZ.99    ZZZ,ZZZ.99
18             XXXXX         SENIOR     ZZZ,ZZZ.99    ZZZ,ZZZ.99
19                                      ZZZ,ZZZ.99    ZZZ,ZZZ.99
20                                      ZZZ,ZZZ.99    ZZZ,ZZZ.99
22             XXXXX         ERROR - CLASS IS X.    CLASS SHOULD BE A, B, C, D, E, F, G, OR H.
24             XXXXX         ERROR - CLASS MISSING. CLASS SHOULD BE A, B, C, D, E, F, G, OR H.
```

Figure 7.14 Program 15. Part 1 of 3.

```
4-CB2 V4 RELEASE 1.5 10NOV77          IBM OS AMERICAN NATIONAL STANDARD COBOL

000010 IDENTIFICATION DIVISION.
000020 PROGRAM-ID. PROG15.
000030*
000040*     THIS PROGRAM COMPUTES SALESPERSON COMMISSIONS FOR JUNIOR
000050*     SALESPERSONS, ASSOCIATE SALESPERSONS, AND SENIOR SALESPERSONS
000060*
000070*****************************************************************
000080
000090 ENVIRONMENT DIVISION.
000100 CONFIGURATION SECTION.
000110 SPECIAL-NAMES.
000120     C01 IS NEW-PAGE.
000130 INPUT-OUTPUT SECTION.
000140 FILE-CONTROL.
000150     SELECT COMMISSION-REPORT ASSIGN TO UT-S-SYSPRINT.
000160     SELECT SALES-FILE-IN     ASSIGN TO UT-S-SYSIN.
000170
000180*****************************************************************
```

Figure 7.14 Part 1 (continued).

```
000190
000200 DATA DIVISION.
000210 FILE SECTION.
000220 FD  SALES-FILE-IN
000230     LABEL RECORDS ARE OMITTED.
000240
000250 01  SALES-RECORD-IN.
000260     05 SALESPERSON-NUMBER-IN      PIC 9(5).
000270     05 CLASS-IN                   PIC X.
000280        88 SALESPERSON-IS-JUNIOR      VALUES 'A' THRU 'F'.
000290        88 SALESPERSON-IS-ASSOCIATE VALUE 'G'.
000300        88 SALESPERSON-IS-SENIOR      VALUE 'H'.
000310        88 CLASS-CODE-IS-MISSING      VALUE SPACE.
000320     05 SALE-1-IN                  PIC 9(6)V99.
000330     05 SALE-2-IN                  PIC 9(6)V99.
000340     05 SALE-3-IN                  PIC 9(6)V99.
000350     05 FILLER                     PIC X(50).
000360
000370 FD  COMMISSION-REPORT
000380     LABEL RECORDS ARE OMITTED.
000390
000400 01  REPORT-LINE                   PIC X(89).
000410
000420 WORKING-STORAGE SECTION.
000430 01  MORE-INPUT                    PIC X(3) VALUE 'YES'.
000440
000450 01  PAGE-HEAD-1.
000460     05 FILLER                     PIC X.
000470     05 FILLER                     PIC X(17) VALUE SPACES.
000480     05 FILLER              PIC X(19) VALUE 'COMMISSION REGISTER'.
000490
000500 01  PAGE-HEAD-2.
000510     05 FILLER                     PIC X.
000520     05 FILLER                     PIC X(43) VALUE SPACES.
000530     05 FILLER                     PIC X(5)  VALUE 'DATE'.
000540     05 RUN-MONTH                  PIC Z9.
000550     05 FILLER                     PIC X     VALUE '/'.
000560     05 RUN-DAY                    PIC 99.
000570     05 FILLER                     PIC X     VALUE '/'.
000580     05 RUN-YEAR                   PIC Z9.
000590
000600 01  PAGE-HEAD-3.
000610     05 FILLER                     PIC X.
000620     05 FILLER                     PIC X(10) VALUE SPACES.
000630     05 FILLER                     PIC X(12) VALUE 'SALES-'.
000640     05 FILLER                     PIC X(11) VALUE 'CLASS'.
000650     05 FILLER                     PIC X(10) VALUE 'SALE'.
000660     05 FILLER                     PIC X(10) VALUE 'COMMISSION'.
000670
000680 01  PAGE-HEAD-4.
000690     05 FILLER                     PIC X.
000700     05 FILLER                     PIC X(10) VALUE SPACES.
000710     05 FILLER                     PIC X(22) VALUE 'PERSON'.
000720     05 FILLER                     PIC X(6)  VALUE 'AMOUNT'.
000730
```

Figure 7.14 Program 15, Part 2 of 3.

```
000740 01  PAGE-HEAD-5.
000750     05 FILLER                     PIC X.
000760     05 FILLER                     PIC X(10) VALUE SPACES.
000770     05 FILLER                     PIC X(6)  VALUE 'NUMBER'.
000780
000790 01  DETAIL-LINE.
000800     05 FILLER                     PIC X.
000810     05 SALESPERSON-NUMBER-OUT     PIC B(10)X(5).
000820     05 CLASS-TITLE-OUT            PIC B(5)X(9)B.
000830     05 SALE-AMOUNT-OUT            PIC ZZZ,ZZZ.99BBB.
000840     05 COMMISSION-OUT             PIC ZZZ,ZZZ.99.
000850
000860 01  ERROR-LINE.
000870     05 FILLER                     PIC X.
000880     05 SALESPERSON-NUMBER-ERR     PIC B(10)X(5)B(5).
000890     05 FILLER                     PIC X(8)  VALUE 'ERROR -'.
000900     05 ERROR-MESSAGE              PIC X(15).
000910     05 FILLER PIC X(29) VALUE 'CLASS SHOULD BE A, B, C, D, E'.
000920     05 FILLER PIC X(13) VALUE ', F, G, OR H.'.
000930
000940 01  ERROR-MESSAGES.
000950     05 CLASS-MISSING-MESSAGE      PIC X(15) VALUE 'CLASS MISSING.'.
000960     05 INVALID-CLASS-MESSAGE.
000970        10 FILLER                  PIC X(9) VALUE 'CLASS IS'.
000980        10 INVALID-CLASS           PIC X.
000990        10 FILLER                  PIC X    VALUE '.'.
001000
001010 01  AVERAGE-SALE-AMOUNT           PIC S9(6)V99.
001020 01  NUMBER-OF-SALES               PIC S9    VALUE 3.
001030 01  SALE-QUOTA                    PIC S9(5).
001040
001050 01  SALE-QUOTAS.
001060     05 JUNIOR-SALE-QUOTA          PIC S9(5) VALUE 0.
001070     05 ASSOCIATE-SALE-QUOTA       PIC S9(5) VALUE 10000.
001080     05 SENIOR-SALE-QUOTA          PIC S9(5) VALUE 50000.
001090
001100 01  COMMISSION-RATES.
001110     05 LOW-RATE                   PIC SV99.
001120     05 HIGH-RATE                  PIC SV99.
001130     05 COMMISSION-RATE-1          PIC SV99  VALUE .10.
001140     05 COMMISSION-RATE-2          PIC SV99  VALUE .10.
001150     05 COMMISSION-RATE-3          PIC SV99  VALUE .05.
001160     05 COMMISSION-RATE-4          PIC SV99  VALUE .20.
001170     05 COMMISSION-RATE-5          PIC SV99  VALUE .20.
001180     05 COMMISSION-RATE-6          PIC SV99  VALUE .30.
001190     05 COMMISSION-RATE-7          PIC SV99  VALUE .30.
001200     05 COMMISSION-RATE-8          PIC SV99  VALUE .40.
001210
001220 01  TODAYS-DATE.
001230     05 RUN-YEAR                   PIC 99.
001240     05 RUN-MONTH                  PIC 99.
001250     05 RUN-DAY                    PIC 99.
001260
001270*****************************************************************
```

Figure 7.14 Part 2 (continued).

```
001280
001290 PROCEDURE DIVISION.
001300     OPEN INPUT  SALES-FILE-IN,
001310          OUTPUT COMMISSION-REPORT.
001320     READ SALES-FILE-IN
001330         AT END
001340             MOVE 'NO' TO MORE-INPUT.
001350     IF MORE-INPUT = 'YES'
001360         PERFORM INITIALIZATION
001370         PERFORM MAIN-PROCESS UNTIL MORE-INPUT = 'NO'.
001380     CLOSE SALES-FILE-IN,
001390           COMMISSION-REPORT.
001400     STOP RUN.
001410
001420 INITIALIZATION.
001430     ACCEPT TODAYS-DATE FROM DATE.
001440     MOVE CORR TODAYS-DATE TO PAGE-HEAD-2.
001450     WRITE REPORT-LINE FROM PAGE-HEAD-1 AFTER NEW-PAGE.
001460     WRITE REPORT-LINE FROM PAGE-HEAD-2 AFTER 1.
001470     WRITE REPORT-LINE FROM PAGE-HEAD-3 AFTER 3.
001480     WRITE REPORT-LINE FROM PAGE-HEAD-4 AFTER 1.
001490     WRITE REPORT-LINE FROM PAGE-HEAD-5 AFTER 1.
```

In the Data Division we see for the first time the use of a level number other than 01 or 05. The entries for INVALID–CLASS–MESSAGE show that an item which is part of another can itself still be broken down. The two fields CLASS–MISSING–MESSAGE and INVALID–CLASS–MESSAGE could each have been made independent 01-level items, and the program would still execute the same way. But program efficiency can often be improved if DISPLAY items[4] are grouped together under a single 01 entry (the program may occupy less computer storage space), and so the two error messages were placed under a single 01-level entry. The commission rates, which are all defined as constants in working storage so that if the rates change the program will be easy to modify, are also grouped under a single 01 entry.

In the INITIALIZATION paragraph we see for the first time the abbreviation **CORR** for CORRESPONDING, which is another of the few authorized abbreviations in COBOL. In the MAIN–PROCESS loop the program determines which commission schedule applies to the current input record and sets up the fields SALE–QUOTA, LOW–RATE, and HIGH–RATE, accordingly. Then the routine WRITE–A–SALESPERSON–GROUP can compute and print the commission.

4. And COMPUTATIONAL–3 items, for COBOL systems with that USAGE.

Figure 7.14 Program 15, Part 3 of 3.

```
001500
001510 MAIN-PROCESS.
001520     IF CLASS-CODE-IS-MISSING
001530         PERFORM CLASS-MISSING-ROUTINE
001540     ELSE
001550         IF SALESPERSON-IS-JUNIOR
001560             PERFORM JUNIOR-SALESPERSON-COMMISSION
001570         ELSE
001580             COMPUTE AVERAGE-SALE-AMOUNT ROUNDED =
001590                 (SALE-1-IN +
001600                  SALE-2-IN +
001610                  SALE-3-IN) / NUMBER-OF-SALES
001620             IF SALESPERSON-IS-ASSOCIATE AND
001630                AVERAGE-SALE-AMOUNT LESS THAN ASSOCIATE-SALE-QUOTA
001640                 PERFORM LEVEL-2-COMMISSION
001650             ELSE
001660                 IF SALESPERSON-IS-ASSOCIATE
001670                     PERFORM LEVEL-3-COMMISSION
001680                 ELSE
001690                     IF SALESPERSON-IS-SENIOR AND
001700                        AVERAGE-SALE-AMOUNT LESS THAN
001710
001720                                              SENIOR-SALE-QUOTA
001730                         PERFORM LEVEL-4-COMMISSION
001740                     ELSE
001750                         IF SALESPERSON-IS-SENIOR
001760                             PERFORM LEVEL-5-COMMISSION
001770                         ELSE
001780                             PERFORM INVALID-CLASS-ROUTINE.
001790     READ SALES-FILE-IN
001800         AT END
001810             MOVE 'NO' TO MORE-INPUT.
001820
001830 JUNIOR-SALESPERSON-COMMISSION.
001840     MOVE 'JUNIOR'            TO CLASS-TITLE-OUT.
001850     MOVE JUNIOR-SALE-QUOTA   TO SALE-QUOTA.
001860     MOVE COMMISSION-RATE-1   TO LOW-RATE.
001870     MOVE COMMISSION-RATE-2   TO HIGH-RATE.
001880     PERFORM WRITE-A-SALESPERSON-GROUP.
001890
001890 LEVEL-2-COMMISSION.
001900     MOVE 'ASSOCIATE'             TO CLASS-TITLE-OUT.
001910     MOVE ASSOCIATE-SALE-QUOTA    TO SALE-QUOTA.
001920     MOVE COMMISSION-RATE-3       TO LOW-RATE.
001930     MOVE COMMISSION-RATE-4       TO HIGH-RATE.
001940     PERFORM WRITE-A-SALESPERSON-GROUP.
001950
001960 LEVEL-3-COMMISSION.
001970     MOVE 'ASSOCIATE'             TO CLASS-TITLE-OUT.
001980     MOVE ASSOCIATE-SALE-QUOTA    TO SALE-QUOTA.
001990     MOVE COMMISSION-RATE-5       TO LOW-RATE.
002000     MOVE COMMISSION-RATE-6       TO HIGH-RATE.
002010     PERFORM WRITE-A-SALESPERSON-GROUP.
002020
```

Figure 7.14 Part 3 (continued).

```
002030 LEVEL-4-COMMISSION.
002040     MOVE 'SENIOR'            TO CLASS-TITLE-OUT.
002050     MOVE SENIOR-SALE-QUOTA TO SALE-QUOTA.
002060     MOVE COMMISSION-RATE-5 TO LOW-RATE.
002070     MOVE COMMISSION-RATE-6 TO HIGH-RATE.
002080     PERFORM WRITE-A-SALESPERSON-GROUP.
002090
002100 LEVEL-5-COMMISSION.
002110     MOVE 'SENIOR'            TO CLASS-TITLE-OUT.
002120     MOVE SENIOR-SALE-QUOTA TO SALE-QUOTA.
002130     MOVE COMMISSION-RATE-7 TO LOW-RATE.
002140     MOVE COMMISSION-RATE-8 TO HIGH-RATE.
002150     PERFORM WRITE-A-SALESPERSON-GROUP.
002160
002170 WRITE-A-SALESPERSON-GROUP.
002180     MOVE SALESPERSON-NUMBER-IN TO SALESPERSON-NUMBER-OUT.
002190     IF SALE-1-IN LESS THAN SALE-QUOTA
002200         MULTIPLY SALE-1-IN BY LOW-RATE
002210                                 GIVING COMMISSION-OUT ROUNDED
002220     ELSE
002230         MULTIPLY SALE-1-IN BY HIGH-RATE
002240                                 GIVING COMMISSION-OUT ROUNDED.
002250     MOVE SALE-1-IN TO SALE-AMOUNT-OUT.
002260     WRITE REPORT-LINE FROM DETAIL-LINE AFTER 2.
002270     MOVE SPACES TO DETAIL-LINE.
002280     IF SALE-2-IN LESS THAN SALE-QUOTA
002290         MULTIPLY SALE-2-IN BY LOW-RATE
002300                                 GIVING COMMISSION-OUT ROUNDED
002310     ELSE
002320         MULTIPLY SALE-2-IN BY HIGH-RATE
002330                                 GIVING COMMISSION-OUT ROUNDED.
002340     MOVE SALE-2-IN TO SALE-AMOUNT-OUT.
002350     WRITE REPORT-LINE FROM DETAIL-LINE AFTER 1.
002360     IF SALE-3-IN LESS THAN SALE-QUOTA
002370         MULTIPLY SALE-3-IN BY LOW-RATE
002380                                 GIVING COMMISSION-OUT ROUNDED
002390     ELSE
002400         MULTIPLY SALE-3-IN BY HIGH-RATE
002410                                 GIVING COMMISSION-OUT ROUNDED.
002420     MOVE SALE-3-IN TO SALE-AMOUNT-OUT.
002430     WRITE REPORT-LINE FROM DETAIL-LINE AFTER 1.
002440
002450 CLASS-MISSING-ROUTINE.
002460     MOVE CLASS-MISSING-MESSAGE TO ERROR-MESSAGE.
002470     PERFORM WRITE-ERROR-LINE.
002480
002490 INVALID-CLASS-ROUTINE.
002500     MOVE CLASS-IN              TO INVALID-CLASS.
002510     MOVE INVALID-CLASS-MESSAGE TO ERROR-MESSAGE.
002520     PERFORM WRITE-ERROR-LINE.
002530
002540 WRITE-ERROR-LINE.
002550     MOVE SALESPERSON-NUMBER-IN TO SALESPERSON-NUMBER-ERR.
002560     WRITE REPORT-LINE FROM ERROR-LINE AFTER 2.
```

The nested IF statement in MAIN–PROCESS in this program and the IF statement in Figure 7.11 are of a special type. In both statements the words IF and ELSE alternate. In this kind of IF statement it is very easy to see which ELSE belongs to which IF, since there are no unrelated IFs and ELSEs in the way. This form of nested IF statement is called a **case.** In a case only one or none of the True paths will be executed. Since it is so easy to know which ELSE belongs to which IF, it is not important to indent the levels of the nest as we have done. In fact, some programmers feel that since in a case one of the True paths at most will be executed, it is clearer to write a case as shown in Figure 7.15.

Figure 7.15 The nested IF statement from Program 15 rewritten with case indenting.

```
000010     IF CLASS-CODE-IS-MISSING
000020         PERFORM CLASS-MISSING-ROUTINE
000030     ELSE
000040     IF SALESPERSON-IS-JUNIOR
000050         PERFORM JUNIOR-SALESPERSON-COMMISSION
000060     ELSE
000070         COMPUTE AVERAGE-SALE-AMOUNT =
000080             (SALE-1-IN +
000090              SALE-2-IN +
000100              SALE-3-IN) / NUMBER-OF-SALES
000110     IF SALESPERSON-IS-ASSOCIATE AND
000120        AVERAGE-SALE-AMOUNT LESS THAN ASSOCIATE-SALE-QUOTA
000130         PERFORM LEVEL-2-COMMISSION
000140     ELSE
000150     IF SALESPERSON-IS-ASSOCIATE
000160         PERFORM LEVEL-3-COMMISSION
000170     ELSE
000180     IF SALESPERSON-IS-SENIOR AND
000190        AVERAGE-SALE-AMOUNT LESS THAN SENIOR-SALE-QUOTA
000200         PERFORM LEVEL-4-COMMISSION
000210     ELSE
000220     IF SALESPERSON-IS-SENIOR
000230         PERFORM LEVEL-5-COMMISSION
000240     ELSE
000250         PERFORM INVALID-CLASS-ROUTINE.
```

In Chapter 9 you will see how to use **subscripting** and **indexing** to make Program 15 shorter and more understandable by eliminating a lot of nearly identical repetitive code. Input data used to run Program 15 are shown in Figure 7.16, and the output produced is shown in Figure 7.17.

```
09834A00198500015550000085/580
02784F01475300009850000200 5050
26374G00198500015550000085/580
27634G01475300009850000200 5050
36472H035080600405000006070080
34782H055689000460980056700000
23478 678909096859409834789657
847530456738574632837564728374
```

Figure 7.16 Input data for Program 15.

Figure 7.17 Output from Program 15.

```
          COMMISSION REGISTER
                                        DATE  2/13/80


SALES-          CLASS          SALE         COMMISSION
PERSON                        AMOUNT
NUMBER

09834         JUNIOR          1,985.00        198.50
                             15,550.00      1,555.00
                              8,575.80        857.58

02784         JUNIOR         14,753.00      1,475.30
                              9,850.00        985.00
                             20,050.50      2,005.05

26374         ASSOCIATE       1,985.00         99.25
                             15,550.00      3,110.00
                              8,575.80        428.79

27634         ASSOCIATE      14,753.00      4,425.90
                              9,850.00      1,970.00
                             20,050.50      6,015.15

36472         SENIOR         35,080.60      7,016.12
                             40,500.00      8,100.00
                             60,700.80     18,210.24

34782         SENIOR         55,689.00     22,275.60
                             46,098.00     13,829.40
                            567,000.00    226,800.00

23478         ERROR - CLASS MISSING. CLASS SHOULD BE A, B, C, D, E, F, G, OR H.

84753         ERROR - CLASS IS 0.    CLASS SHOULD BE A, B, C, D, E, F, G, OR H.
```

Exercise 8. Rewrite the nested IF statement shown in Figure 7.11 using case indenting.

Exercise 9. Rewrite the MAIN–PROCESS loop of Program 15 without using any complex conditions. The resulting IF statement will no longer be a case. Which of the two forms of the IF statement do you think is easier to understand? What are the good and bad features of each form of the statement?

Exercise 10. Modify Program 15 to allow for page overflow. Change the report heading to include space for page number, and modify the Procedure Division to print the page number on each page. All three sales for any salesperson should appear on the same page together.

Summary

A nested IF statement consists of one or more IF statements in the True and/or False paths of another IF statement. An IF statement may be nested to essentially any level, and there is only one period at the end of the entire nest of IFs. In order for the programmer to keep track of which ELSE belongs to which IF, conventional indenting should be adhered to. There may be as many processing steps as desired in any of the paths at any level of the nest, and any path containing no processing must use the words NEXT SENTENCE. The clause ELSE NEXT SENTENCE may be omitted if it appears immediately before the period.

Some nested logic cannot be implemented on most COBOL systems using just a nested IF statement. This will occur whenever a Yes and No path of one of the inner IFs rejoin before the end of the outer IF. In such a case, a portion of the logic must be written as a separate paragraph and PERFORMed from the proper place in the IF statement.

Complex conditions are formed of simple conditions connected by the words AND and OR. The types of simple conditions discussed in this chapter are relation conditions, which we have used since Chapter 3, and condition name conditions, which depend on 88-level entries. Relation conditions may contain the word NOT, as in IF A NOT GREATER THAN B; simple conditions may be negated, as in IF NOT CODE–IS–OK; and complex conditions may be negated, as in IF NOT (A = B AND C = D). Care should be taken to avoid writing IF statements whose meaning is not clear to a human reader.

Relation conditions may be abbreviated when two or more of them are used in a complex condition. The subject of all but the first relation condition may be omitted if it is the same for all the relation conditions in the complex condition. If the subject and the relational operator are the same for all the relation conditions, they may both be omitted from all but the first.

An 88-level entry may be used to give a condition name to particular values of fields. When an 88-level entry is used, the field must still be defined in the usual way. Condition names are most useful in the File Section, but many programmers also use them in the Working Storage Section. When an 88-level entry is used, it must immediately follow the description of the field to which it applies. An 88-level entry must contain exactly one VALUE clause and no other clauses, especially no PICTURE clause.

An arithmetic expression may be used as the subject and/or object of a relation condition. A literal may also be used as either the subject or object of a relation condition, but not both.

Fields in the Data Division may be broken down to more than one level. In working storage, you can often improve program efficiency by grouping DISPLAY (and COMPUTATIONAL–3) items under a single 01-level entry.

A special kind of nested IF statement where IFs and ELSEs alternate and there are no intervening IFs or ELSEs is called a case. In a case, only one or none of the True paths is executed. Special indenting conventions different from the indenting used for regular nested IF statements are often used in writing a case.

Fill-in Exercises

1. In a nested IF statement the ________ appearance of the word IF begins the outer IF.
2. Use of proper ________ enables the programmer to see which ELSE belongs to which IF.
3. The words ELSE NEXT SENTENCE may not be removed from an IF statement if they do not appear immediately before the ________.
4. IF statements may be ________ to any level.
5. Simple conditions may be connected by the words ________ and ________ to form complex conditions.
6. The order of evaluation of complex conditions is ________ first and then ________.
7. Complex conditions should be used only to make program logic more ________.
8. If the same ________ appears more than once in a complex condition, the repetitions may be omitted.
9. An 88-level entry can be used to give a ________ to one or more values that might be assigned to a field.
10. Exactly one ________ clause is required in an 88-level entry.

Review Exercises

1. Draw a flowchart showing the logic of the following statement:

```
IF C1
    PERFORM P1
ELSE
    IF C2
        PERFORM P2
        PERFORM P3
```

```
        IF C3
            IF C4
                PERFORM P4
            ELSE
                NEXT SENTENCE
        ELSE
            NEXT SENTENCE
    ELSE
        PERFORM P5.
```

2. Rewrite the statement in Review Exercise 1 by combining C3 and C4 into a complex condition. Reduce the depth of nesting by one level.
3. Expand the following abbreviated condition so that all subjects and all relational operators are stated explicitly:

   IF A IS GREATER THAN B AND C OR D . . .

4. Write an 05-level entry and 88-level entries for the following field: a six-digit, dollars-and-cents input amount called PURCHASE–AMOUNT–IN. The name PURCHASE–AMOUNT–IS–TOO–LOW should be given to the range 0.00–999.99; the name PURCHASE–AMOUNT–IS–LOW to the range 1,000.00–3,999.99; the name PURCHASE–AMOUNT–IS–NORMAL to 4,000.00–8,000.00; the name PURCHASE–AMOUNT–IS–HIGH to 8,000.01–9,999.98; and the name PURCHASE–AMOUNT–IS–TOP to the value 9,999.99.

*5. Write a program to process input data in the format shown in Figure 7.RE5.1. Each input record is for one employee and shows

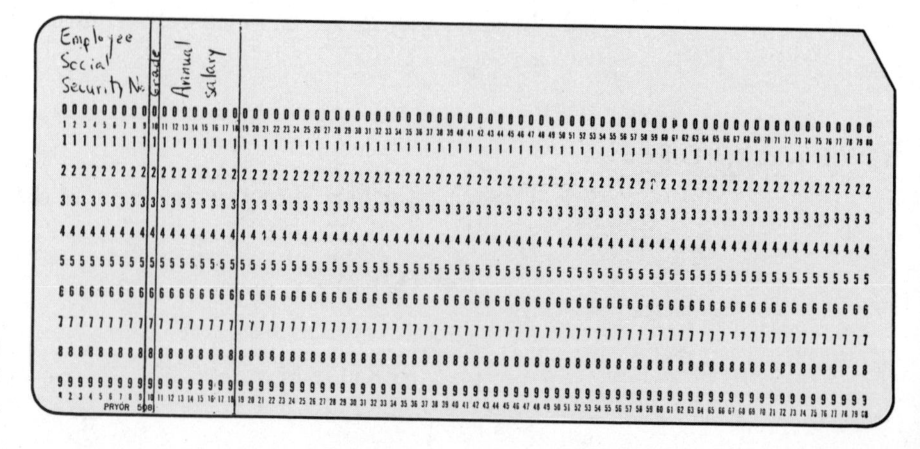

Figure 7.RE5.1
Data card format for Review Exercise 5.

the employee's grade and current annual salary. The grades are L through P and 1 through 6. The titles for each grade are:

L	Programmer trainee	P	Group leader
M	Junior programmer	1	Manager
N	Programmer	2–6	Assistant vice-president
O	Project leader		

The program is required to compute a cost-of-living raise for each employee based on the employee's grade and current annual salary and print the results in the format shown in Figure 7.RE5.2.

150/10/6 PRINT CHART PROG. ID ______ PAGE ______
(SPACING: 150 POSITION SPAN, AT 10 CHARACTERS PER INCH, 6 LINES PER VERTICAL INCH) DATE ______
Fold back at dotted line.
PROGRAM TITLE ______
PROGRAMMER OR DOCUMENTALIST: ______
CHART TITLE Review Exercise 5, Chapter 7
CARRIAGE CONTROL TAPE CHAN.

```
                       COST-OF-LIVING INCREASES SCHEDULE

                                                               DATE Z9/99/99


   EMPLOYEE SOCIAL    GRADE    CURRENT       INCREASE           NEW
   SECURITY NUMBER              SALARY                        SALARY

     XXX-XX-XXXX        X    ZZZ,ZZZ.99     Z,ZZZ.99       ZZZ,ZZZ.99
     XXX-XX-XXXX        X    ZZZ,ZZZ.99     Z,ZZZ.99       ZZZ,ZZZ.99
     XXX-XX-XXXX        X    ZZZ,ZZZ.99     Z,ZZZ.99       ZZZ,ZZZ.99
 **  XXX-XX-XXXX        X  ERROR - INVALID GRADE.
 **  XXX-XX-XXXX           ERROR - GRADE MISSING.
```

Figure 7.RE5.2 Output format for Review Exercise 5.

The schedule for cost-of-living increases is:

Grade	Salary	Increase
Any	Less than $10,000	10% of current salary
L	Any	10% of current salary
M–P	$10,000–$14,999.99	7% of current salary, plus $200
M–P	$15,000 and over	5% of current salary, plus $500
1	$10,000–$19,999.99	5% of current salary
1	$20,000 and over	$1,400
2–6	Any	$2,000

6. Modify your solution to Review Exercise 5 in order to count the number of employees in each grade and also to compute the total current salaries, the total increases, and the total salaries after the increase. The program should produce output in the format shown in Figure 7.RE6.

150/10/6 PRINT CHART PROG. ID ________ PAGE ________

(SPACING: 150 POSITION SPAN, AT 10 CHARACTERS PER INCH, 6 LINES PER VERTICAL INCH) DATE ________

Fold back at dotted line.

PROGRAM TITLE ________

PROGRAMMER OR DOCUMENTALIST: ________

CHART TITLE Review Exercise 6, Chapter 7

CARRIAGE CONTROL TAPE CHAN.

```
                           COST-OF-LIVING INCREASES SCHEDULE
                                                                      DATE Z9/99/99

               EMPLOYEE SOCIAL     GRADE     CURRENT          INCREASE          NEW
               SECURITY NUMBER                SALARY                           SALARY

                 XXX-XX-XXXX         X      ZZZ,ZZZ.99      Z,ZZZ.99       ZZZ,ZZZ.99
                 XXX-XX-XXXX         X      ZZZ,ZZZ.99      Z,ZZZ.99       ZZZ,ZZZ.99
                 XXX-XX-XXXX         X      ZZZ,ZZZ.99      Z,ZZZ.99       ZZZ,ZZZ.99
            **   XXX-XX-XXXX         X  ERROR - INVALID GRADE.
            **   XXX-XX-XXXX            ERROR - GRADE MISSING.

                               TOTALS Z,ZZZ,ZZZ.99     ZZ,ZZZ.99    Z,ZZZ,ZZZ.99

               NUMBER OF EMPLOYEES IN GRADE L     ZZ9
                                            N     ZZ9
                                            M     ZZ9
                                            O     ZZ9
                                            P     ZZ9
                                            1     ZZ9
                                            2-6   ZZ9
```

Figure 7.RE6 Output format for Review Exercise 6.

Validity Checking

8

KEY POINTS

Here are the key points you should learn from this chapter:

1. the importance of checking the validity of input data;
2. how to check for presence of data, class of data, valid codes, and reasonableness;
3. how to program for arithmetic overflow in COBOL.

KEY WORDS

Key words to recognize and learn:

REDEFINES   class condition   SIZE ERROR

COBOL has several features that permit a program to check the validity of the data it is working on. Invalid data can arise essentially in two ways, the most common of which is invalid data in program input. Most input data are prepared by people, usually on key machines such as a keypunch or a terminal. Even though such data is proofread and verified in other ways before being processed by the program, it is usual to expect that the input data will still contain errors, and the variety of errors will be literally impossible to imagine. After you think you have seen all the kinds of errors that key operators can make, you ain't seen nothing yet.

Obviously programs should not execute on incorrect or invalid data. The worst thing that can happen if a program processes incorrect data is that the output will be incorrect and no one will notice the error until it is too late; another is that incorrect data will cause the program to behave in such an obviously nonsensical way that the error becomes apparent to all. The best thing, though, is for the program itself to be able to detect errors and to handle them in a rational, planned way.

Aside from program input, the other source of invalid data is the program itself. It sometimes happens that in the course of execution the program generates unexpected intermediate results which cannot be further processed properly. COBOL provides ways for a program to protect itself against certain kinds of invalid input data and certain kinds of internally generated invalid data.

Validity Check of Input Data

Since the variety of possible invalid inputs is infinite, it is customary for COBOL programs to make just a few kinds of checks on the data before proceeding. The program can check that fields which are required in the input are in fact present; that fields which are supposed to contain only numbers are in fact purely numeric (and that fields which are supposed to contain only letters are in fact alphabetic); and that the specific contents of fields are reasonable values. Since the last of these is the most difficult, we will look at the others first.

CHECKING FOR PRESENCE OF DATA

In all the programs we have done so far, every field of input data was required to be present. For example, in Figure 7.12 each input record had to contain some salesperson number, some class, and three sales amounts. If any of those fields had been accidentally left blank on any card, the program would not have been able to process the card. Program 15 did check that the class field was present, but the absence of any of the other fields would have also been a fatal error.

In real data processing in industry, not all input records always need to have all of their fields present, but, when presence of a field is required, there are ways that COBOL can check that the field is filled in. One technique may be used for fields defined as alphanumeric and other techniques used for fields defined as integers or as fractions or mixed numbers.

Checking for the presence of data in alphanumeric fields: As in Program 15, the programmer may attach an 88-level entry to the field. A suitable condition name may be given and the VALUE SPACE or VALUE SPACES clause used, as in:

```
05 CLASS-IN            PIC X.
    88 CLASS-CODE-IS-MISSING   VALUE SPACE.
```

Then in the Procedure Division, the test:

```
IF CLASS-CODE-IS-MISSING . . .
```

may be used. The following relation test would work as well:

```
IF CLASS-IN = SPACE . . .
```

The choice of which form to use is a matter of programmer preference.

Checking for the presence of integers: It is illegal in COBOL to use the VALUE SPACES clause with an item described as numeric. So an 88-level entry cannot be used to test for a missing integer. (88-level entries may be used with numeric data in other contexts, however; see for example Review Exercise 4, Chapter 7.) The easiest way to check for the absence of an integer is with a relation condition in an IF statement, as for example:

```
IF STORE-NUMBER-IN = SPACES . . .
```

Notice that it is legal to use SPACES in an IF statement with an integer field.

Checking for the presence of fractions or mixed numbers: If your particular COBOL system permits using SPACES in an IF statement with a field containing a fractional part, then the easiest way to test for the absence of such a number is as with an integer, for example:

```
IF MONEY-AMOUNT-IN = SPACES . . .
```

But some COBOL systems do not permit SPACES to be used in an IF statement with non-integer data. In such a case, you can take advantage of the COBOL feature which permits a single field to have more than one PICTURE and more than one name if necessary.

The **REDEFINES** clause permits the programmer to give as many different PICTUREs as desired to a single field, so for purposes of testing the field the programmer can assign a nonnumeric PICTURE to the field and use the techniques given earlier for testing for the presence of data in an alphanumeric field. For example, the following entries:

```
05  MONEY-AMOUNT-IN                    PIC  9(6)V99.
05  MONEY-AMOUNT-IN-X
    REDEFINES MONEY-AMOUNT-IN          PIC  X(8).
```

will let the programmer refer to the same field by either of the two names, MONEY-AMOUNT-IN or MONEY-AMOUNT-IN-X. The choice of which name the programmer will use in any particular place in the Procedure Division depends on how the field is used: If the program is to do arithmetic with the field, the name MONEY-AMOUNT-IN would be used. But if the program is to use the field in some context where only alphanumeric fields are allowed, then the name MONEY-AMOUNT-IN-X would be used.[1] So to test MONEY-AMOUNT-IN for absence of data, the following would be legal:

1. Names used to redefine fields are ignored by the CORRESPONDING option.

```
IF MONEY-AMOUNT-IN-X = SPACES . . .
```

But it is permissible to attach an 88-level entry to a REDEFINES statement, so the following would be legal:

```
05 MONEY-AMOUNT-IN                      PIC 9(6)V99.
05 MONEY-AMOUNT-IN-X
   REDEFINES MONEY-AMOUNT-IN            PIC X(8).
   88 MONEY-AMOUNT-IS-MISSING               VALUE SPACES.
```

Then in the Procedure Division, the programmer may write:

```
IF MONEY-AMOUNT-IS-MISSING . . .
```

The presence of the REDEFINES clause does not interfere with other uses of the field MONEY-AMOUNT-IN. For example, MONEY-AMOUNT-IN could still have its own numeric 88-level entries if desired, as follows:

```
05 MONEY-AMOUNT-IN                          PIC 9(6)V99.
   88 MONEY-AMOUNT-IS-LOW   VALUES 0 THRU 4999.99.
   88 MONEY-AMOUNT-IS-HIGH  VALUES 5000.00 THRU 999999.99.
05 MONEY-AMOUNT-IN-X
   REDEFINES MONEY-AMOUNT-IN                PIC X (8).
   88 MONEY-AMOUNT-IS-MISSING               VALUE SPACES.
```

This REDEFINES technique can of course also be used with integer fields as well as with non-integer fields. In Chapter 9 we will see an entirely different use of the REDEFINES clause.

Exercise 1. Given the following COBOL statement:

```
05 PART-DESCRIPTION-IN        PIC X(20).
```

a. without using an 88-level entry, write an IF test to check for the absence of data in the field;
b. write an 88-level entry using a VALUE SPACES clause and write an IF test to check for the absence of data in the field.

Exercise 2. Given the following COBOL statement:

```
05 NUMBER-OF-SHEEP-IN        PIC 9(5).
```

a. without using a REDEFINES entry or an 88-level entry, write an IF test to check for absence of data in the field;
b. using a REDEFINES entry and an 88-level entry with a VALUE SPACES clause, write an IF test to check for the absence of data in the field.

CHECKING THE CLASS OF DATA

COBOL provides two **class condition** tests which allow the program to check whether a particular field does or does not contain only numbers, or whether it does or does not contain only the characters A through Z and the character space. The general format of the class condition is:

```
identifier IS [NOT] { NUMERIC    }
                    { ALPHABETIC }
```

The test IF MONEY–AMOUNT–IN IS NOT NUMERIC . . . would execute the True path if any of the characters in MONEY–AMOUNT–IN were other than the digits 0 through 9. If the description of a field being tested contains the PICTURE character S, then the contents of the field would be considered numeric if the field contained the digits 0 through 9 and a valid plus or minus sign. Our field MONEY–AMOUNT–IN, which has no S in its PICTURE, would be considered not numeric if it were found to contain a sign.

The results of not testing a numeric field for absence or invalidity of data can range from insignificant to disastrous. If the field is merely printed on a report, then its absence or invalidity will appear in the report and cause just that one item to be unreadable. If, instead, the missing or invalid field is a control field, then the control breaks for that group will not operate properly and a whole section of the output may be useless. The remainder of the output may be useable, however.

If a field used for numeric comparisons or arithmetic is missing or invalid, any of several very unpleasant things can happen. Depending on the nature of the error in the numeric field and on the COBOL system being used, the program might just make up its own number and go merrily along using that number in processing; of course, the output will be totally wrong and the error might not be noticed until too late. Another thing that might happen is that the program terminates execution. At least that way everyone knows that an error has occurred, but none of the input records following the bad one will be processed. It is best to have the program check all numeric input fields and take care of the erroneous ones before they take care of you.

The ALPHABETIC class condition permits a program to test whether all the characters in a field are the alphabetic characters, which are defined in COBOL as the letters A through Z and the character space. So for a field defined as follows:

```
05  PART-NUMBER-IN        PIC X(12).
```

the test IF PART–NUMBER–IN IS NOT ALPHABETIC . . . would execute the True path if any of the characters in PART–NUMBER–IN were not the letters A through Z or space.

It is legal and sometimes useful to be able to determine whether a field described with Xs contains all numbers. The PART–NUMBER–IN field, for example, could be tested as follows:

```
IF PART–NUMBER–IN IS NUMERIC . . .
```

In this IF statement, the True path would be executed only if all the characters in PART–NUMBER–IN were the digits 0 through 9. The ALPHABETIC class test may not be executed on a field defined as numeric.

CHECKING DATA FOR REASONABLENESS

Incorrect input data can often slip past tests for presence and tests of class. For example, a field described as:

```
05 HOURS–WORKED        PIC 99V9.
```

with room for three digits, might accidentally contain 93.0 instead of the correct value of 39.0. This kind of error, which would not be detected by a class test or a presence test, can be detected by the program because it is so much larger than what one would expect. The test could be coded this way:

```
05 HOURS–WORKED          PIC 99V9.
   88 HOURS–WORKED–IS–VERY–HIGH   VALUES 60.1 THRU 99.9.
```

and:

```
IF HOURS–WORKED–IS–VERY–HIGH . . .
```

Fields which are supposed to contain only certain valid codes can be checked to see that one of the valid codes is present:

```
05  MARITAL–STATUS                      PIC X.
    88  MARITAL–STATUS–IS–SINGLE            VALUE 'S'.
    88  MARITAL–STATUS–IS–MARRIED           VALUE 'M'.
    88  LEGALLY–SEPARATED                   VALUE 'L'.
    88  LIVING–APART                        VALUE 'V'.
    88  LIVING–TOGETHER                     VALUE 'T'.
    88  DONT–KNOW–MARITAL–STATUS            VALUE 'U'.
    88  WIDOW–OR–WIDOWER                    VALUE 'W'.
    88  MARITAL–STATUS–IS–DIVORCED          VALUE 'D'.
    88  MARITAL–STATUS–IS–MISSING           VALUE SPACE.
    88  MARITAL–STATUS–CODE–IS–OK           VALUES 'S', 'M',
                                            'L', 'T' THRU 'W',
                                            'D'.
```

Then for processing the field the program may refer to any of the 88-level names and for checking the field we can have:

```
IF MARITAL-STATUS-IS-MISSING . . .
```

and

```
IF NOT MARITAL-STATUS-CODE-IS-OK . . .
```

Sometimes individual fields cannot be tested for reasonableness in cases where the data appears invalid only because of an unlikely combination of fields in the input. For example, in a certain payroll application it may be reasonable to have annual salaries in the range $5,000 to $90,000, because everyone from janitor to vice-president is processed by the one program. Complex conditions can sometimes be used to detect unlikely combinations of data, as in:

```
IF WORK-CODE-IS-JANITOR AND SALARY GREATER THAN
   JANITOR-SALARY-LIMIT . . .
```

It is usually not wise to go to great lengths to include complicated and extensive reasonableness testing in a program. For no matter how thorough the validity checking might be, some creative key operator will make an error that gets through it. The easiest tests to make, like those for presence, class, and valid codes, and some straightforward reasonableness tests, are the ones that catch the greatest number of errors.

Exercise 3. Given the following field:

05 HOURS–WORKED PIC 99V9.

Write an 88-level entry and an IF test to determine whether the contents of the field are outside the reasonable range. Fewer than 5 hours worked, or more than 65, is outside the reasonable range. Write the IF test so that the True path will be executed if the field contents are found unreasonable.

Exercise 4. Given the following field:

05 TYPE–OF–SALE PIC X.

Write 88-level entries giving suitable names to the following codes for the different types of sales; wholesale, W; retail, R; return, N; preferred customer, P. Also write an 88-level entry giving a name to a missing code. Write IF tests to check for the absence of data in the field and for an invalid code.

A Program with Validity Checking of Input Data

In Program 16 we will see how to program validity checking of input data. Program 16 uses the same input data format as Program 15, shown in Figure 7.12. In Program 16 each input record is supposed to contain a valid salesperson class; the valid classes are the letters A through H as before. Three sales amounts are also to appear in the record as before. The program is to check the input records for validity. If the record contains no validity errors, the program is to add the three sales amounts together and print the sum, as shown in Figure 8.1.

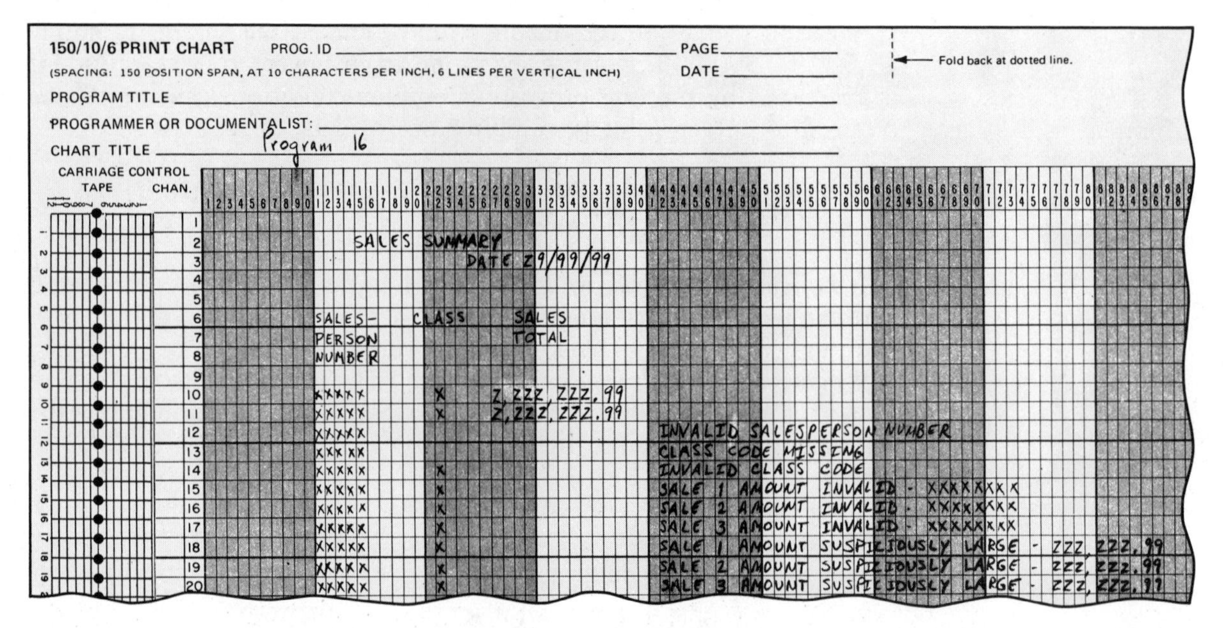

Figure 8.1 Output format for Program 16.

Figure 8.1 also shows the different kinds of error messages that the program might produce and thus implies the kinds of errors the program is supposed to check for. The error message about the sale amount being suspiciously large refers to sales of more than $500,000. In this sales application it is assumed that any sale amount that comes into the program so large is probably an error.

The program is to check for all possible errors in each input record. If more than one error appears in a single record, the program is to print an error line for each error found. The line is to be printed showing the sum of the sales only if the record contains no errors. If any errors are found in an input record, only the error lines are to print for that record. A hierarchy diagram for Program 16 is shown in

Figure 8.2. The subfunctions of "Produce detail lines" are the different validity checks that the program makes on each input record. For each type of error that may be found, the hierarchy chart shows that the program will set up an appropriate error line and PERFORM a common routine called "Write an error line." If no errors are found in an input record, the program will carry out the step "Set up to write a good line." The error-line flag referred to in the hierarchy diagram is used to indicate to the program whether any errors have been found in the input record being processed. Its use will become clear when we look at the coding for Program 16.

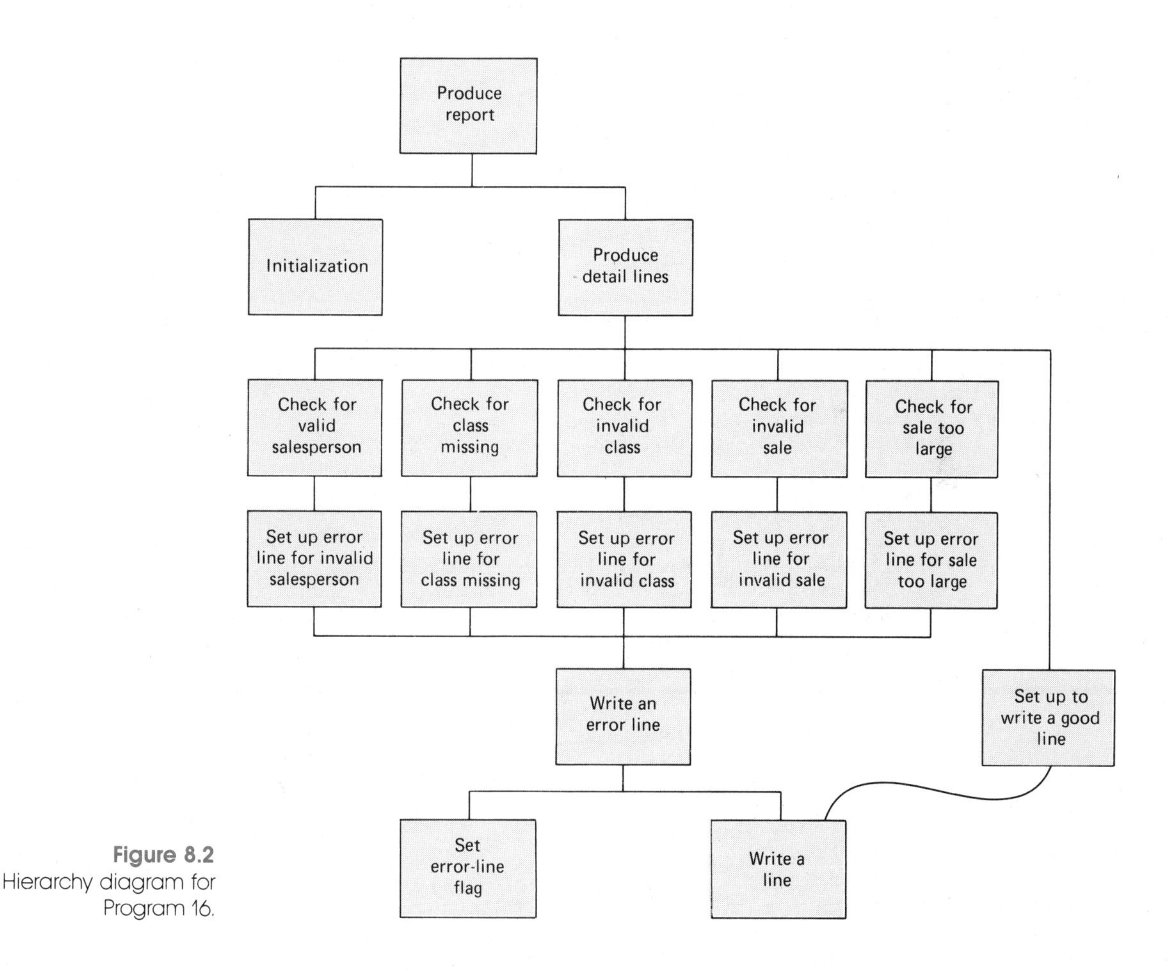

Figure 8.2
Hierarchy diagram for Program 16.

Program 16 is shown in Figure 8.3. In the description of SALES–RECORD–IN, 88-level entries have been assigned to suspiciously large sale amounts. The three sale amounts have also been redefined with alphanumeric PICTUREs. This is part of the coding

Figure 8.3 Program 16, Part 1 of 3.

```
4-CB2 V4 RELEASE 1.5 10NOV77        IBM OS AMERICAN NATIONAL STANDARD COI



  000010 IDENTIFICATION DIVISION.
  000020 PROGRAM-ID. PROG16.
  000030*AUTHOR. PHIL RUBENSTEIN.
  000040*
  000050*    THIS PROGRAM CHECKS THE VALIDITY OF FIELDS
  000060*    IN THE INPUT AND PRODUCES A SALES SUMMARY.
  000070*
  000080*****************************************************************
  000090
  000100 ENVIRONMENT DIVISION.
  000110 CONFIGURATION SECTION.
  000120 SPECIAL-NAMES.
  000130     C01 IS NEW-PAGE.
  000140 INPUT-OUTPUT SECTION.
  000150 FILE-CONTROL.
  000160     SELECT SALES-SUMMARY-REPORT ASSIGN TO UT-S-SYSPRINT.
  000170     SELECT SALES-FILE-IN        ASSIGN TO UT-S-SYSIN.
  000180
  000190*****************************************************************
  000200
  000210 DATA DIVISION.
  000220 FILE SECTION.
  000230 FD  SALES-FILE-IN
  000240     LABEL RECORDS ARE OMITTED.
  000250
  000260 01  SALES-RECORD-IN.
  000270     05 SALESPERSON-NUMBER-IN     PIC X(5).
  000280     05 CLASS-IN                  PIC X.
  000290        88 CLASS-CODE-IS-MISSING     VALUE SPACE.
  000300        88 CLASS-CODE-IS-VALID       VALUES 'A' THRU 'H'.
  000310     05 SALE-1-IN                 PIC 9(6)V99.
  000320        88 SALE-1-TOO-LARGE  VALUES 500000.01 THRU 999999.99.
  000330     05 SALE-1-IN-X
  000340        REDEFINES SALE-1-IN       PIC X(8).
  000350     05 SALE-2-IN                 PIC 9(6)V99.
  000360        88 SALE-2-TOO-LARGE  VALUES 500000.01 THRU 999999.99.
  000370     05 SALE-2-IN-X
  000380        REDEFINES SALE-2-IN       PIC X(8).
  000390     05 SALE-3-IN                 PIC 9(6)V99.
  000400        88 SALE-3-TOO-LARGE  VALUES 500000.01 THRU 999999.99.
  000410     05 SALE-3-IN-X
  000420        REDEFINES SALE-3-IN       PIC X(8).
  000430     05 FILLER                    PIC X(50).
  000440
  000450 FD  SALES-SUMMARY-REPORT
  000460     LABEL RECORDS ARE OMITTED.
  000470
  000480 01  REPORT-LINE                  PIC X(87).
  000490
```

Figure 8.3 Part 1 (continued).

```
000500 WORKING-STORAGE SECTION.
000510 01  MORE-INPUT                      PIC X(3) VALUE 'YES'.
000520
000530 01  PAGE-HEAD-1.
000540     05 FILLER                       PIC X.
000550     05 FILLER                       PIC X(14) VALUE SPACES.
000560     05 FILLER                PIC X(13) VALUE 'SALES SUMMARY'.
000570
000580 01  PAGE-HEAD-2.
000590     05 FILLER                       PIC X.
000600     05 FILLER                       PIC X(24) VALUE SPACES.
000610     05 FILLER                       PIC X(5)  VALUE 'DATE'.
000620     05 RUN-MONTH                    PIC Z9.
000630     05 FILLER                       PIC X     VALUE '/'.
000640     05 RUN-DAY                      PIC 99.
000650     05 FILLER                       PIC X     VALUE '/'.
000660     05 RUN-YEAR                     PIC Z9.
000670
000680 01  PAGE-HEAD-3.
000690     05 FILLER                       PIC X.
000700     05 FILLER                       PIC X(10) VALUE SPACES.
000710     05 FILLER                       PIC X(9)  VALUE 'SALES-'.
000720     05 FILLER                       PIC X(9)  VALUE 'CLASS'.
000730     05 FILLER                       PIC X(5)  VALUE 'SALES'.
000740
000750 01  PAGE-HEAD-4.
000760     05 FILLER                       PIC X.
000770     05 FILLER                       PIC X(10) VALUE SPACES.
000780     05 FILLER                       PIC X(18) VALUE 'PERSON'.
000790     05 FILLER                       PIC X(5)  VALUE 'TOTAL'.
000800
000810 01  PAGE-HEAD-5.
000820     05 FILLER                       PIC X.
000830     05 FILLER                       PIC X(10) VALUE SPACES.
000840     05 FILLER                       PIC X(6)  VALUE 'NUMBER'.
```

needed to properly carry out validity checking on the sales fields. The sale amount fields will be tested in the Procedure Division to see whether they contain only numbers. If they do not, they are invalid and are to be MOVEd to the output area as part of the error message. Once the field is found to be not numeric, we don't know what kinds of characters it contains, and the safest thing then is to do an alphanumeric MOVE from one PICTURE X field to another. COBOL does not permit moving a non-integer field to an alphanumeric field.

Somewhat the same reasoning applies to the field SALESPERSON–NUMBER–IN. It is defined as alphanumeric even though the salesperson number is supposed to be just that, a number. The definition was made alphanumeric so that, if a SALESPERSON–NUMBER–IN is found to contain not numbers, it can be MOVEd to the output line regardless of what characters there might be.

The formatting of the error lines is handled a little differently in this program than in previous programs. Here, all detail lines on the report, whether they contain an error message or the sum of the three sales, are printed from the same area in working storage, BODY–LINE. BODY–LINE has place for both the sum of the sales and the longest possible error message, in fields called SALES–TOTAL–OUT and ERROR–MESSAGE–OUT. For any single printed line of course either the sum of the sales or an error message will print, but never both. Some of the different kinds of error messages that may be MOVEd to ERROR–MESSAGE–OUT in the course of processing are defined in the 01-level fields SALE–ERROR–MESSAGE and ERROR–TYPES, both also in working storage. Their use will become clear when we look at the Procedure Division. The field HAVE–ANY–ERRORS–BEEN–FOUND will be used in the Procedure Division to determine whether a line containing the sum of the three sales should be printed for any particular input record. Remember, only if an input record is completely free of errors do we want to print the sum of the sales.

Figure 8.3 Program 16, Part 2 of 3.

```
000850
000860 01  BODY-LINE.
000870     05 FILLER                    PIC X.
000880     05 SALESPERSON-NUMBER-OUT    PIC B(10)X(5).
000890     05 CLASS-OUT                 PIC B(6)XB(4).
000900     05 SALES-TOTAL-OUT           PIC Z,ZZZ,ZZZ.99BBB.
000910     05 ERROR-MESSAGE-OUT         PIC X(45).
000920
000930 01  SALE-ERROR-MESSAGE.
000940     05 FILLER                    PIC X(5)  VALUE 'SALE'.
000950     05 SALE-NUMBER               PIC 9B.
000960     05 FILLER                    PIC X(7)  VALUE 'AMOUNT'.
000970     05 ERROR-TYPE                PIC X(31).
000980
000990 01  ERROR-TYPES.
001000     05 SALE-INVALID.
001010        10 FILLER                 PIC X(10) VALUE 'INVALID -'.
001020        10 INVALID-SALE-AMOUNT    PIC X(8).
001030     05 SALE-TOO-LARGE.
001040        10 FILLER           PIC X(21) VALUE 'SUSPICIOUSLY LARGE -'.
001050        10 LARGE-SALE-AMOUNT      PIC 999,999.99.
001060
001070 01  HAVE-ANY-ERRORS-BEEN-FOUND   PIC X(8).
001080
001090 01  TODAYS-DATE.
001100     05 RUN-YEAR                  PIC 99.
001110     05 RUN-MONTH                 PIC 99.
001120     05 RUN-DAY                   PIC 99.
001130
001140*****************************************************************
```

Figure 8.3 Part 2 (continued).

```
001150
001160 PROCEDURE DIVISION.
001170     OPEN INPUT  SALES-FILE-IN,
001180          OUTPUT SALES-SUMMARY-REPORT.
001190     READ SALES-FILE-IN
001200         AT END
001210             MOVE 'NO' TO MORE-INPUT.
001220     IF MORE-INPUT = 'YES'
001230         PERFORM INITIALIZATION
001240         PERFORM PRODUCE-DETAIL-LINES UNTIL MORE-INPUT = 'NO'.
001250     CLOSE SALES-FILE-IN,
001260           SALES-SUMMARY-REPORT.
001270     STOP RUN.
001280
001290 INITIALIZATION.
001300     ACCEPT TODAYS-DATE FROM DATE.
001310     MOVE CORR TODAYS-DATE TO PAGE-HEAD-2.
001320     WRITE REPORT-LINE FROM PAGE-HEAD-1 AFTER NEW-PAGE.
001330     WRITE REPORT-LINE FROM PAGE-HEAD-2 AFTER 1.
001340     WRITE REPORT-LINE FROM PAGE-HEAD-3 AFTER 3.
001350     WRITE REPORT-LINE FROM PAGE-HEAD-4 AFTER 1.
001360     WRITE REPORT-LINE FROM PAGE-HEAD-5 AFTER 1.
001370     MOVE SPACES TO REPORT-LINE.
001380     WRITE REPORT-LINE AFTER 1.
001390
```

The Procedure Division in this program does not follow the structure of the hierarchy diagram. To follow the structure of the diagram would have resulted in a large number of very tiny paragraphs, and so those tiny paragraphs have simply been pushed up into the paragraphs that would otherwise have PERFORMed them. The logic implied by the diagram is adhered to, however.

The paragraph PRODUCE–DETAIL–LINES starts by moving the SALESPERSON–NUMBER–IN to the output line, for the salesperson number is to print regardless of whether or not the line is an error line and regardless of whether or not the salesperson number is valid. The flag HAVE–ANY–ERRORS–BEEN–FOUND is set to 'NONE YET,' because the input record has not yet been checked for validity, and indeed no errors have yet been found in it. PRODUCE–DETAIL–LINES checks the input record for all the possible types of errors, sets up the appropriate error line, and PERFORMs the routine to write the error line on the report. Whenever a field must be checked to see that it is both numeric and reasonable, the NUMERIC test must be done first. When the paragraph WRITE–AN–ERROR–

Figure 8.3 Program 16, Part 3 of 3.

```
001400 PRODUCE-DETAIL-LINES.
001410     MOVE SPACES TO BODY-LINE.
001420     MOVE SALESPERSON-NUMBER-IN TO SALESPERSON-NUMBER-OUT.
001430     MOVE 'NONE YET' TO HAVE-ANY-ERRORS-BEEN-FOUND.
001440     IF SALESPERSON-NUMBER-IN NOT NUMERIC
001450         MOVE 'INVALID SALESPERSON NUMBER' TO ERROR-MESSAGE-OUT
001460         PERFORM WRITE-AN-ERROR-LINE.
001470     IF CLASS-CODE-IS-MISSING
001480         MOVE 'CLASS CODE MISSING' TO ERROR-MESSAGE-OUT
001490         PERFORM WRITE-AN-ERROR-LINE
001500     ELSE
001510         MOVE CLASS-IN TO CLASS-OUT
001520         IF NOT CLASS-CODE-IS-VALID
001530             MOVE 'INVALID CLASS CODE' TO ERROR-MESSAGE-OUT
001540             PERFORM WRITE-AN-ERROR-LINE.
001550     IF SALE-1-IN NOT NUMERIC
001560         MOVE 1 TO SALE-NUMBER
001570         MOVE SALE-1-IN-X TO INVALID-SALE-AMOUNT
001580         PERFORM WRITE-AN-INVALID-SALE-LINE
001590     ELSE
001600     IF SALE-1-TOO-LARGE
001610         MOVE 1 TO SALE-NUMBER
001620         MOVE SALE-1-IN TO LARGE-SALE-AMOUNT
001630         PERFORM WRITE-A-LARGE-SALE-LINE.
001640     IF SALE-2-IN NOT NUMERIC
001650         MOVE 2 TO SALE-NUMBER
001660         MOVE SALE-2-IN-X TO INVALID-SALE-AMOUNT
001670         PERFORM WRITE-AN-INVALID-SALE-LINE
001680     ELSE
001690     IF SALE-2-TOO-LARGE
001700         MOVE 2 TO SALE-NUMBER
001710         MOVE SALE-2-IN TO LARGE-SALE-AMOUNT
001720         PERFORM WRITE-A-LARGE-SALE-LINE.
001730     IF SALE-3-IN NOT NUMERIC
001740         MOVE 3 TO SALE-NUMBER
001750         MOVE SALE-3-IN-X TO INVALID-SALE-AMOUNT
001760         PERFORM WRITE-AN-INVALID-SALE-LINE
001770     ELSE
001780     IF SALE-3-TOO-LARGE
001790         MOVE 3 TO SALE-NUMBER
001800         MOVE SALE-3-IN TO LARGE-SALE-AMOUNT
001810         PERFORM WRITE-A-LARGE-SALE-LINE.
001820     IF HAVE-ANY-ERRORS-BEEN-FOUND = 'NONE YET'
001830         ADD SALE-1-IN,
001840             SALE-2-IN,
001850             SALE-3-IN
001860                 GIVING SALES-TOTAL-OUT
001870         PERFORM WRITE-A-LINE.
001880     READ SALES-FILE-IN
001890         AT END
001900             MOVE 'NO' TO MORE-INPUT.
001910
001920 WRITE-AN-INVALID-SALE-LINE.
001930     MOVE SALE-INVALID        TO ERROR-TYPE.
001940     MOVE SALE-ERROR-MESSAGE TO ERROR-MESSAGE-OUT.
001950     PERFORM WRITE-AN-ERROR-LINE.
```

Figure 8.3 Part 3 (continued).

```
001960
001970 WRITE-A-LARGE-SALE-LINE.
001980      MOVE SALE-TOO-LARGE      TO ERROR-TYPE.
001990      MOVE SALE-ERROR-MESSAGE TO ERROR-MESSAGE-OUT.
002000      PERFORM WRITE-AN-ERROR-LINE.
002010
002020 WRITE-AN-ERROR-LINE.
002030      MOVE 'YES' TO HAVE-ANY-ERRORS-BEEN-FOUND.
002040      PERFORM WRITE-A-LINE.
002050
002060 WRITE-A-LINE.
002070      WRITE REPORT-LINE FROM BODY-LINE AFTER 1.
```

LINE executes, it MOVEs 'YES' to HAVE–ANY–ERRORS–BEEN–FOUND, indicating that at least one error has been found in the current input record. The paragraphs WRITE–AN–INVALID–SALE–LINE and WRITE–A–LARGE–SALE–LINE are included in the Procedure Division, even though they are not shown in the hierarchy diagram. Their use eliminates some duplicate coding that would otherwise be required if those paragraphs were not used.

The last IF statement in PRODUCE–DETAIL–LINES tests to see whether any errors have been found in the input record. If none have been found even after all the error checks have been made, then normal processing may be done and the sum of the three sales computed and printed.

The input data used to run Program 16 are shown in Figure 8.4, and the output produced by the program is shown in Figure 8.5.

```
09834A001985000155500000857580
02784F014753000098500002005050
26374G001985000155500000857580
27634G014753000098500002005050
36472H035080600405000000607080
34782H055689000460980056700000
23478 678909096859409834789657
847530456738574632837564728374
12345D    33331234333350000002
23456#0000Q0004343434930306043
34567I000033330303335445555U77
45678]   43433000000033900000 0
67890J4444444R7770523E1234567A
09876A3212322   65566000000003
     S343343344432112321123211
AFVERD939202223311101229221191
12343H191919191111101012831018
```

Figure 8.4
Input data for Program 16.

Figure 8.5 Output from Program 16.

```
        SALES SUMMARY
                   DATE  2/19/80


     SALES-     CLASS     SALES
     PERSON               TOTAL
     NUMBER

     09834        A        26,110.80
     02784        F        44,653.50
     26374        G        26,110.80
     27634        G        44,653.50
     36472        H       136,281.40
     34782        H                       SALE 3 AMOUNT SUSPICIOUSLY LARGE - 567,000.00
     23478                                CLASS CODE MISSING
     23478                                SALE 1 AMOUNT SUSPICIOUSLY LARGE - 678,909.09
     23478                                SALE 2 AMOUNT SUSPICIOUSLY LARGE - 685,940.98
     84753        O                       INVALID CLASS CODE
     84753        O                       SALE 3 AMOUNT SUSPICIOUSLY LARGE - 647,283.74
     12345        D                       SALE 1 AMOUNT INVALID -     3333
     12345        D                       SALE 3 AMOUNT SUSPICIOUSLY LARGE - 500,000.02
     23456        #                       INVALID CLASS CODE
     23456        #                       SALE 1 AMOUNT INVALID - 0000Q000
     34567        I                       INVALID CLASS CODE
     34567        I                       SALE 3 AMOUNT INVALID - 45555U77
     45678        ]                       INVALID CLASS CODE
     45678        ]                       SALE 1 AMOUNT INVALID -    43433
     45678        ]                       SALE 3 AMOUNT INVALID - 900000 0
     67890        J                       INVALID CLASS CODE
     67890        J                       SALE 1 AMOUNT INVALID - 4444444R
     67890        J                       SALE 2 AMOUNT INVALID - 7770523E
     67890        J                       SALE 3 AMOUNT INVALID - 1234567A
     09876        A                       SALE 1 AMOUNT INVALID - 3212322
     09876        A                       SALE 2 AMOUNT INVALID -   655660
                                          INVALID SALESPERSON NUMBER
                  S                       INVALID CLASS CODE
     AFVER                                INVALID SALESPERSON NUMBER
     AFVER        D                       SALE 1 AMOUNT SUSPICIOUSLY LARGE - 939,202.22
     12343        H       431,339.47
```

Exercise 5.

Write a program to process a deck of input cards in the format shown in Figure 7.RE5.1, page 190. Have the program make the following validity checks on each input record:

a. Social Security number numeric
b. Grade present
c. Grade valid (L through P, 1 through 6)
d. Salary numeric
e. Salary not greater than $300,000.

Check each input record for all possible types of errors. Print an error line for each error found. If a record is completely free of errors, print its contents on one line (formatted so that it can be read easily). At the end of the report, print a total of all the salaries printed (that is, a total of all the salaries that were found in the error-free records).

Design a report format with a suitable title and suitable column headings. Include the date in the heading.

Arithmetic Overflow

Sometimes invalid data in the form of arithmetic overflow can be generated during execution of a program. An arithmetic result is considered to overflow if there are more places to the left of the decimal point in the answer than places to the left of the point in the field provided by the programmer. Excess places to the right of the decimal point are never considered to be an overflow condition. Excess places to the right of the point are either truncated or rounded, depending on whether the ROUNDED option is absent or present.

The programmer can test for arithmetic overflow by using the **SIZE ERROR** clause. The SIZE ERROR clause is optional with all five arithmetic verbs ADD, SUBTRACT, MULTIPLY, DIVIDE and COMPUTE. The format of the SIZE ERROR clause has been seen as early as Chapter 4, and is:

```
ON SIZE ERROR imperative-statement
```

Following the words SIZE ERROR in an arithmetic operation, there may be one or more COBOL statements, as in:

```
ADD A TO B
    SIZE ERROR
        MOVE 'YES' TO IS-THERE-A-SIZE-ERROR.
```

or

```
ADD A TO B ROUNDED
    SIZE ERROR
        MOVE SIZE-ERROR-MESSAGE TO ERROR-MESSAGE-OUT
        WRITE ERROR-LINE AFTER 1
        MOVE 0 to B.
```

In these examples, there is only one period at the end of the complete SIZE ERROR processing. Later in this chapter we will see that when arithmetic is done inside an IF statement, the SIZE ERROR processing steps can be terminated with an ELSE. The period or the ELSE ends the last of the steps that are to be carried out in case of arithmetic overflow. Indenting is of course only for the programmer's convenience and is not examined by COBOL.

No conditional statements may follow the words SIZE ERROR up to the place where the SIZE ERROR processing ends (at the period or the ELSE); this means that no IF statements may appear, no READ statements (for READ statements contain the conditional AT END clause), and no arithmetic statements containing a SIZE ERROR clause. If any such statements are needed for the program to properly process a SIZE ERROR, the statements can be placed in a separate paragraph. Then the SIZE ERROR clause can be written as something like:

```
ADD A TO B
    SIZE ERROR
        PERFORM SIZE-ERROR-ROUTINE.
```

If the ROUNDED option is specified in an arithmetic operation, rounding is carried out before the field is tested for SIZE ERROR. Division by zero always causes a SIZE ERROR condition. If there is more than one result field given in an arithmetic statement, a SIZE ERROR on one of the fields will not interfere with normal execution of arithmetic on the others. If a size error occurs during execution of a statement that contains a SIZE ERROR clause, the result field is left unchanged from before the operation. If a size error occurs in a statement where the SIZE ERROR clause is not specified, the results may be unpredictable.

A Program with a SIZE ERROR

We now do a program which shows the SIZE ERROR clause in context. Program 17 uses input data in the format shown in Figure 4.RE5.1 (page 91). Each record shows the number of hours worked by an employee on Monday, the number of hours worked on Tuesday, on Wednesday, and so on through Saturday. The program is to read each card and compute the total number of hours worked by each employee during the week. Since we would not expect the number of hours worked in a week to exceed 99.9, we can allow just that size for the output field. Then during processing, the program can check whether the total hours for the week exceeds 99.9 by using a SIZE ERROR clause and handle the error appropriately. The program will also check that each hours-worked field contains only numbers.

The program is also to compute the total hours worked on Monday by all employees, the total worked on Tuesday by all employees, and so on through Saturday, except that if any employee record is found to be in error, its hours are not to be included in the totals. The output format for Program 17, showing the types of error messages that are to appear, is given in Figure 8.6.

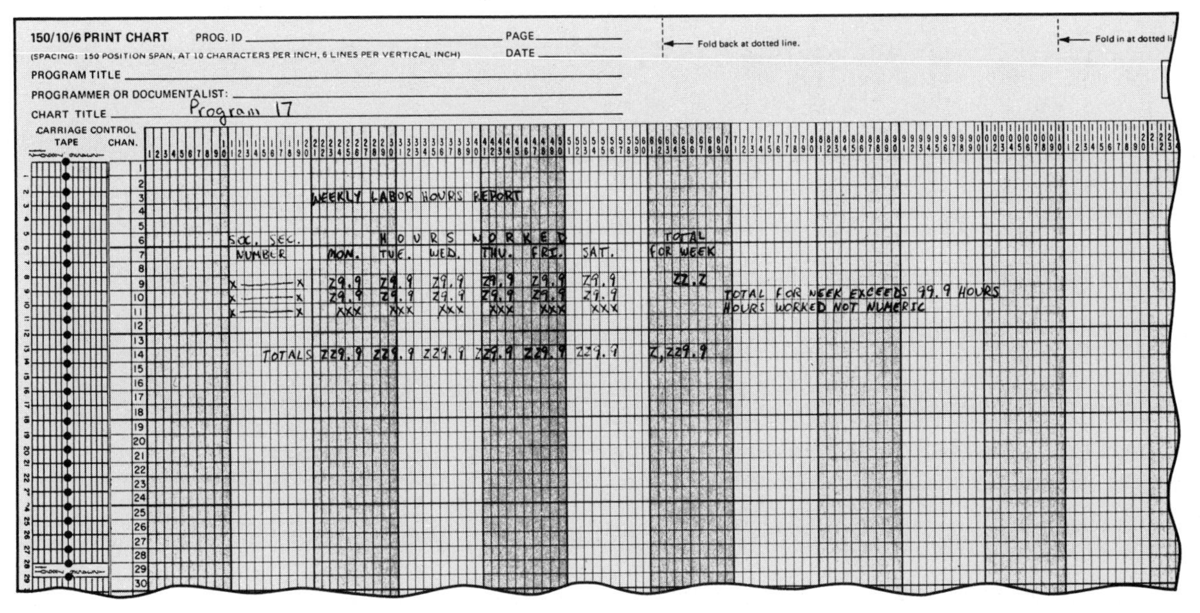

Figure 8.6 Output format for Program 17.

Notice that next to the error message HOURS WORKED NOT NUMERIC the hours worked appear in alphanumeric format. This is done so that if any of the six hours-worked fields in an input record is found not to be numeric, all six can be MOVEd to the output area in alphanumeric form.

Program 17 is shown in Figure 8.7. In LABOR–RECORD–IN, the numeric fields are all redefined as alphanumeric, so if any of the numeric fields are found not to contain all numbers, they can be handled as alphanumeric regardless of what characters they contain. Two different formats for detail lines are described in working storage. One of them, REGULAR–BODY–LINE, is used for printing correct lines showing the total hours worked for the week and also for printing numerical values of the hours worked when the total for the week exceeds 99.9 hours. The other, called ALPHA–BODY–LINE, is used for printing the input values of hours worked when one or more of them is found to be not numeric. You will see in the Procedure Division how the program selects which line to use and how it fills the appropriate line with output data.

Figure 8.7 Program 17, Part 1 of 2.

```
4-CB2 V4 RELEASE 1.5 10NOV77        IBM OS AMERICAN NATIONAL STANDARD COBOL



000010 IDENTIFICATION DIVISION.
000020 PROGRAM-ID. PROG17.
000030*AUTHOR. PHIL RUBENSTEIN.
000040*
000050*     THIS PROGRAM READS AN EMPLOYEE LABOR FILE
000060*     AND PRODUCES A WEEKLY LABOR HOURS REPORT.
000070*
000080*****************************************************************
000090
000100 ENVIRONMENT DIVISION.
000110 CONFIGURATION SECTION.
000120 SPECIAL-NAMES.
000130     C01 IS NEW-PAGE.
000140 INPUT-OUTPUT SECTION.
000150 FILE-CONTROL.
000160     SELECT LABOR-HOURS-REPORT       ASSIGN TO UT-S-SYSPRINT.
000170     SELECT EMPLOYEE-LABOR-FILE-IN   ASSIGN TO UT-S-SYSIN.
000180
000190*****************************************************************
000200
000210 DATA DIVISION.
000220 FILE SECTION.
000230 FD  EMPLOYEE-LABOR-FILE-IN
000240     LABEL RECORDS ARE OMITTED.
000250
000260 01  LABOR-RECORD-IN.
000270     05 SOCIAL-SECURITY-NUMBER       PIC X(9).
000280     05 MONDAY-HOURS                 PIC 99V9.
000290     05 MONDAY-HOURS-XX
000300        REDEFINES MONDAY-HOURS       PIC X(3).
000310     05 TUESDAY-HOURS                PIC 99V9.
000320     05 TUESDAY-HOURS-XX
000330        REDEFINES TUESDAY-HOURS      PIC X(3).
000340     05 WEDNESDAY-HOURS              PIC 99V9.
000350     05 WEDNESDAY-HOURS-XX
000360        REDEFINES WEDNESDAY-HOURS    PIC X(3).
000370     05 THURSDAY-HOURS               PIC 99V9.
000380     05 THURSDAY-HOURS-XX
000390        REDEFINES THURSDAY-HOURS     PIC X(3).
000400     05 FRIDAY-HOURS                 PIC 99V9.
000410     05 FRIDAY-HOURS-XX
000420        REDEFINES FRIDAY-HOURS       PIC X(3).
000430     05 SATURDAY-HOURS               PIC 99V9.
000440     05 SATURDAY-HOURS-XX
000450        REDEFINES SATURDAY-HOURS     PIC X(3).
000460     05 FILLER                       PIC X(53)
000470
000480 FD  LABOR-HOURS-REPORT
000490     LABEL RECORDS ARE OMITTED.
000500
000510 01  REPORT-LINE                 PIC X(103).
```

Figure 8.7 Part 1 (continued).

```
000520
000530 WORKING-STORAGE SECTION.
000540 01  MORE-INPUT                     PIC X(3) VALUE 'YES'.
000550
000560 01  PAGE-HEAD-1.
000570     05 FILLER                      PIC X.
000580     05 FILLER                      PIC X(20) VALUE SPACES.
000590     05 FILLER        PIC X(25) VALUE 'WEEKLY LABOR HOURS REPORT'.
000600
000610 01  PAGE-HEAD-2.
000620     05 FILLER                      PIC X.
000630     05 FILLER                      PIC X(10) VALUE SPACES.
000640     05 FILLER                      PIC X(18) VALUE 'SOC. SEC.'.
000650     05 FILLER        PIC X(34) VALUE 'H O U R S  W O R K E D'.
000660     05 FILLER                      PIC X(5)  VALUE 'TOTAL'.
000670
000680 01  PAGE-HEAD-3.
000690     05 FILLER                      PIC X.
000700     05 FILLER                      PIC X(11) VALUE SPACES.
000710     05 FILLER                      PIC X(11) VALUE 'NUMBER'.
000720     05 FILLER                      PIC X(6)  VALUE 'MON.'.
000730     05 FILLER                      PIC X(6)  VALUE 'TUE.'.
000740     05 FILLER                      PIC X(6)  VALUE 'WED.'.
000750     05 FILLER                      PIC X(6)  VALUE 'THU.'.
000760     05 FILLER                      PIC X(6)  VALUE 'FRI.'.
000770     05 FILLER                      PIC X(8)  VALUE 'SAT.'.
000780     05 FILLER                      PIC X(8)  VALUE 'FOR WEEK'.
000790
000800 01  REGULAR-BODY-LINE.
000810     05 FILLER                      PIC X.
000820     05 SOCIAL-SECURITY-NUMBER      PIC B(10)X(9).
000830     05 MONDAY-HOURS                PIC B(3)Z9.9.
000840     05 TUESDAY-HOURS               PIC B(2)Z9.9.
000850     05 WEDNESDAY-HOURS             PIC B(2)Z9.9.
000860     05 THURSDAY-HOURS              PIC B(2)Z9.9.
000870     05 FRIDAY-HOURS                PIC B(2)Z9.9.
000880     05 SATURDAY-HOURS              PIC B(2)Z9.9.
000890     05 TOTAL-FOR-WEEK              PIC B(7)ZZ.ZBB.
000900     05 ERROR-MESSAGE-OUT           PIC X(33).
000910
000920 01  ALPHA-BODY-LINE.
000930     05 FILLER                      PIC X.
000940     05 SOCIAL-SECURITY-NUMBER      PIC B(10)X(9).
000950     05 MONDAY-HOURS-X              PIC B(4)X(3).
000960     05 TUESDAY-HOURS-X             PIC B(3)X(3).
000970     05 WEDNESDAY-HOURS-X           PIC B(3)X(3).
000980     05 THURSDAY-HOURS-X            PIC B(3)X(3).
000990     05 FRIDAY-HOURS-X              PIC B(3)X(3).
001000     05 SATURDAY-HOURS-X            PIC B(3)X(3)B(13).
001010     05 FILLER        PIC X(24) VALUE 'HOURS WORKED NOT NUMERIC'.
001020
001030 01  ACCUMULATORS-W.
001040     05 MONDAY-HOURS                   PIC 9(3)V9 VALUE 0.
001050     05 TUESDAY-HOURS                  PIC 9(3)V9 VALUE 0.
001060     05 WEDNESDAY-HOURS                PIC 9(3)V9 VALUE 0.
001070     05 THURSDAY-HOURS                 PIC 9(3)V9 VALUE 0.
001080     05 FRIDAY-HOURS                   PIC 9(3)V9 VALUE 0.
```

The flag IS–THERE–AN–ERROR is used to signal the program when any sort of error is found in an input record. As each record is read, and before the validity checks on it begin, IS–THERE–AN–ERROR is set to 'NO', for no error has yet been found. If and when an error is found in the record, IS–THERE–AN–ERROR is set to 'YES'.

Figure 8.7 Program 17, Part 2 of 2.

```
001090         05 SATURDAY-HOURS                      PIC 9(3)V9 VALUE 0.
001100         05 TOTAL-FOR-WEEK                      PIC 9(4)V9 VALUE 0.
001110
001120 01    TOTAL-LINE.
001130         05 FILLER                              PIC X.
001140         05 FILLER                              PIC X(14) VALUE SPACES.
001150         05 FILLER                              PIC X(7)  VALUE 'TOTALS'.
001160         05 MONDAY-HOURS                        PIC ZZ9.9.
001170         05 TUESDAY-HOURS                       PIC BZZ9.9.
001180         05 WEDNESDAY-HOURS                     PIC BZZ9.9.
001190         05 THURSDAY-HOURS                      PIC BZZ9.9.
001200         05 FRIDAY-HOURS                        PIC BZZ9.9.
001210         05 SATURDAY-HOURS                      PIC BZZ9.9.
001220         05 TOTAL-FOR-WEEK                      PIC B(4)Z,ZZ9.9.
001230
001240 01    IS-THERE-AN-ERROR                        PIC X(3).
001250
001260 01    INTERMEDIATE-RESULTS.
001270         05 TOTAL-FOR-WEEK                      PIC 99V9.
001280
001290******************************************************************
001300
001310 PROCEDURE DIVISION.
001320         OPEN INPUT  EMPLOYEE-LABOR-FILE-IN,
001330              OUTPUT LABOR-HOURS-REPORT.
001340         READ EMPLOYEE-LABOR-FILE-IN
001350             AT END
001360                 MOVE 'NO' TO MORE-INPUT.
001370         IF MORE-INPUT = 'YES'
001380             PERFORM INITIALIZATION
001390             PERFORM PRODUCE-DETAIL-LINES UNTIL MORE-INPUT = 'NO'
001400             PERFORM PRODUCE-TOTAL-LINE.
001410         CLOSE EMPLOYEE-LABOR-FILE-IN,
001420               LABOR-HOURS-REPORT.
001430         STOP RUN.
001440
001450 INITIALIZATION.
001460         WRITE REPORT-LINE FROM PAGE-HEAD-1 AFTER NEW-PAGE.
001470         WRITE REPORT-LINE FROM PAGE-HEAD-2 AFTER 3.
001480         WRITE REPORT-LINE FROM PAGE-HEAD-3 AFTER 1.
001490         MOVE SPACES TO REPORT-LINE.
001500         WRITE REPORT-LINE AFTER 1.
001510
001520 PRODUCE-DETAIL-LINES.
001530         MOVE 'NO' TO IS-THERE-AN-ERROR.
```

Figure 8.7 Part 2 (continued).

```
001540     IF MONDAY-HOURS      IN LABOR-RECORD-IN NOT NUMERIC OR
001550        TUESDAY-HOURS     IN LABOR-RECORD-IN NOT NUMERIC OR
001560        WEDNESDAY-HOURS   IN LABOR-RECORD-IN NOT NUMERIC OR
001570        THURSDAY-HOURS    IN LABOR-RECORD-IN NOT NUMERIC OR
001580        FRIDAY-HOURS      IN LABOR-RECORD-IN NOT NUMERIC OR
001590        SATURDAY-HOURS    IN LABOR-RECORD-IN NOT NUMERIC
001600         MOVE 'YES' TO IS-THERE-AN-ERROR
001610         MOVE CORR LABOR-RECORD-IN TO ALPHA-BODY-LINE
001620         MOVE MONDAY-HOURS-XX        TO MONDAY-HOURS-X
001630         MOVE TUESDAY-HOURS-XX       TO TUESDAY-HOURS-X
001640         MOVE WEDNESDAY-HOURS-XX     TO WEDNESDAY-HOURS-X
001650         MOVE THURSDAY-HOURS-XX      TO THURSDAY-HOURS-X
001660         MOVE FRIDAY-HOURS-XX        TO FRIDAY-HOURS-X
001670         MOVE SATURDAY-HOURS-XX      TO SATURDAY-HOURS-X
001680         WRITE REPORT-LINE FROM ALPHA-BODY-LINE AFTER 1
001690     ELSE
001700         MOVE SPACES TO REGULAR-BODY-LINE
001710         MOVE CORR LABOR-RECORD-IN TO REGULAR-BODY-LINE
001720         ADD MONDAY-HOURS     IN LABOR-RECORD-IN,
001730             TUESDAY-HOURS    IN LABOR-RECORD-IN,
001740             WEDNESDAY-HOURS  IN LABOR-RECORD-IN,
001750             THURSDAY-HOURS   IN LABOR-RECORD-IN,
001760             FRIDAY-HOURS     IN LABOR-RECORD-IN,
001770             SATURDAY-HOURS   IN LABOR-RECORD-IN
001780                 GIVING TOTAL-FOR-WEEK IN INTERMEDIATE-RESULTS
001790                 SIZE ERROR
001800                     MOVE 'YES' TO IS-THERE-AN-ERROR
001810                     MOVE 'TOTAL FOR WEEK EXCEEDS 99.9 HOURS'
001820                                              TO ERROR-MESSAGE-OUT
001830                     WRITE REPORT-LINE FROM REGULAR-BODY-LINE
001835                                                          AFTER 1.
001840     IF IS-THERE-AN-ERROR = 'NO'
001850         MOVE CORR INTERMEDIATE-RESULTS TO REGULAR-BODY-LINE
001860         WRITE REPORT-LINE FROM REGULAR-BODY-LINE AFTER 1
001870         ADD CORR LABOR-RECORD-IN       TO ACCUMULATORS-W
001880         ADD CORR INTERMEDIATE-RESULTS TO ACCUMULATORS-W.
001890     READ EMPLOYEE-LABOR-FILE-IN
001900         AT END
001910             MOVE 'NO' TO MORE-INPUT.
001920
001930 PRODUCE-TOTAL-LINE.
001940     MOVE CORR ACCUMULATORS-W TO TOTAL-LINE.
001950     WRITE REPORT-LINE FROM TOTAL-LINE AFTER 3.
```

The main loop, PRODUCE–DETAIL–LINES, consists of only four sentences, one of them very long. PRODUCE–DETAIL–LINES is entered with a newly-read card waiting to be processed. The paragraph begins by moving 'NO' to IS–THERE–AN–ERROR, since no error has yet been found. It then tests whether any of the hours-worked fields do not contain only numbers. If any nonnumeric values are found, an error line is printed and 'YES' is MOVEd to IS–THERE–AN–ERROR. If the hours-worked fields are found to correctly contain only numbers, then the six hours-worked fields are ADDed to determine the total hours worked for the week. If that total is greater than 99.9 (generating a SIZE ERROR), an error line is printed and IS–THERE–AN–ERROR is set to 'YES'. Only after all these validity checks are completed does the last IF statement test to see whether any errors were found in this input record. Remember that only if the record is error-free is the program to print a line showing the total hours worked for the week. And only hours from error-free records are to be added into ACCUMULATORS–W in order to be printed in the total line at the end of the report.

For the first time we see the use of the CORRESPONDING option with an arithmetic verb. The CORRESPONDING option is available in ADD and SUBTRACT statements when the GIVING option is not used. In the statement:

```
ADD CORR LABOR–RECORD–IN TO ACCUMULATORS–W.
```

COBOL finds all the fields in LABOR–RECORD–IN having names that match fields in ACCUMULATORS–W. Those fields are MONDAY–HOURS, TUESDAY–HOURS, WEDNESDAY–HOURS, THURSDAY–HOURS, FRIDAY–HOURS, and SATURDAY–HOURS. The values of those fields in LABOR–RECORD–IN are added to their corresponding fields in ACCUMULATORS–W and the sums are stored in ACCUMULATORS–W. Of course, the data names in LABOR–RECORD–IN must be identical to those in ACCUMULATORS–W for the CORRESPONDING option to work correctly.

Notice that a SIZE ERROR clause can appear within an IF statement, but remember that an IF statement cannot appear in a SIZE ERROR clause. The general rule is that anything at all can appear in the True or False paths of IF statements but that there must be no conditional statements in a SIZE ERROR clause. The input data for Program 17 are shown in Figure 8.8, and the output in Figure 8.9.

```
123344543085084070060080090
123321202085045036120004060
020556234209023400200309306
123456789805085850032073036
987654321095085075065055045
098765432125115106093084072
345678901803021032043054076
456789012 85   6 45 65 77 77
765432198123098094094095090
```

Figure 8.8
Input data for Program 17.

Figure 8.9 Output from Program 17.

```
                WEEKLY LABOR HOURS REPORT

SOC. SEC.              H O U R S  W O R K E D                 TOTAL
 NUMBER       MON.  TUE.  WED.  THU.  FRI.  SAT.            FOR WEEK

123344543      8.5   8.4   7.0   6.0   8.0   9.0               46.9
123321202      8.5   4.5   3.6  12.0   0.4   6.0               35.0
020556234      209   023   400   200   309   303                     HOURS WORKED NOT NUMERIC
123456789     80.5   8.5  85.0   3.2   7.3   3.3                     TOTAL FOR WEEK EXCEEDS 99.9 HOURS
987654321      9.5   8.5   7.5   6.5   5.5   4.5               42.0
098765432     12.5  11.5  10.6   9.3   8.4   7.2               59.5

345678901     80.3   2.1   3.2   4.3   5.4   7.6                     TOTAL FOR WEEK EXCEEDS 99.9 HOURS
456789012       85     6    45    65    77    77                     HOURS WORKED NOT NUMERIC
765432198      123   098   094   094   095   090                     HOURS WORKED NOT NUMERIC

     TOTALS   39.0  32.9  28.7  33.8  22.3  26.7              183.4
```

***Exercise 6.** Write a program to read the same input data used in Exercise 5. The program is to compute a 15 percent Christmas bonus based on the employee's annual salary. For each card read, print the employee's Social Security number, the annual salary, and the bonus amount, except if the bonus amount exceeds 99,999.99. If the bonus amount exceeds 99,999.99, print no bonus amount, but instead print the message BONUS AMOUNT SUSPICIOUSLY LARGE. Accumulate the total of the bonus amounts printed and print the total at the end of the report.

SIZE ERROR Processing Terminated by ELSE

Figure 8.10 shows a flowchart of a situation where the steps of the SIZE ERROR processing are terminated not by a period but by ELSE. The flowchart could be coded in COBOL as follows:

```
IF C1
    arithmetic-statement ON SIZE ERROR
        PERFORM SIZE-ERROR-ROUTINE
ELSE
    PERFORM NO-PATH-PROCESSING.
```

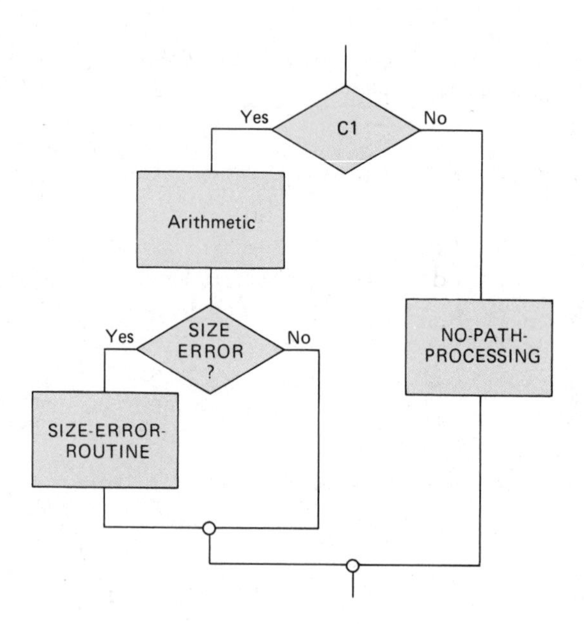

Figure 8.10
Flowchart showing SIZE ERROR processing within an IF statement where the SIZE ERROR steps are terminated by ELSE.

Here, the SIZE ERROR processing is contained between the words SIZE ERROR and ELSE. As before, no conditional statements may appear among the SIZE ERROR processing steps. Of course, the paragraph SIZE–ERROR–ROUTINE may contain any legal statements.

There can never be processing steps in the No path of a SIZE ERROR test. If the logic of a problem demands that certain steps be carried out only if there is no SIZE ERROR, then a SIZE ERROR flag must be used as in Program 17.

Summary

Whenever a program reads input data that have been prepared by people, the program should check the validity of the data before proceeding to process it. The program should check that data fields which are required to be in the data are in fact filled in. Different techniques may be used to check for the presence of alphanumeric data and for integers and fractions. Among the COBOL features that can be used to check one or another kind of field are an 88-level entry with a VALUE SPACES clause, a relation condition where the field is compared to SPACES, and a REDEFINES clause.

The REDEFINES clause permits the programmer to assign more than one name and more than one PICTURE to a single field. Then the programmer may use any of the names of the field in the Procedure Division, depending on the requirements of each context.

If a program reads a coded field as input data, the program should check that the field contains one of the valid codes. If numeric fields are being read, the program should use the class test to check that the numeric fields contain only numbers. Sometimes there can be disastrous results if a program tries to do numeric comparisons or arithmetic with fields containing characters other than numbers. Alphabetic fields can be checked for all letters and spaces.

Input data should be checked for reasonableness; that is, the program should check to the extent possible that the values present in input data fields are values that could reasonably be expected to be there. Sometimes it is possible to test individual fields for reasonable values, and sometimes fields must be tested in combinations to see whether their contents are reasonable.

In programs that check numeric fields for all numbers, it is sometimes necessary to redefine the numeric fields as alphanumeric. The redefinition may be needed in case the numeric field is not found to contain all numbers, since then the program doesn't know what kinds of characters are in the field. An alphanumeric definition permits the field to be handled regardless of the characters in it.

The SIZE ERROR clause may be used with arithmetic verbs to check for arithmetic overflow. Overflow is considered to occur when a result of arithmetic has more places to the left of the decimal point than the programmer has allowed for in the result field. The COBOL statements following the words SIZE ERROR may include any number of steps but no conditional statements, such as IF, READ, or arithmetic with a SIZE ERROR clause. The SIZE ERROR processing is terminated with a period or an ELSE. If a size error occurs during execution of an arithmetic statement where the SIZE ERROR clause is specified, the result field remains unchanged from what it was before the operation. If a size error occurs in a statement where the SIZE ERROR clause is not specified, the results may be unpredictable.

Fill-in Exercises

1. A VALUE SPACES clause may appear in an 88-level entry only if the field is defined as ________.
2. To use a VALUE SPACES clause with a numeric field, the field must be ________.
3. The two classes of data that may be tested for with a class condition test are ________ and ________.
4. Testing the contents of fields to determine whether the fields contain valid codes or values within an expected range of values is called checking data for ________.
5. Arithmetic overflow can be tested for with the ________ clause.

6. Arithmetic overflow occurs if there are too many places in the answer to the ________ of the decimal point.
7. Overflow processing is terminated by a ________ or an ________.
8. ________ statements are prohibited in a SIZE ERROR clause.
9. If the ROUNDED option is specified in an arithmetic operation, rounding takes place ________ the field is tested for size error.
10. There may be no processing in the ________ path of a SIZE ERROR test.

Review Exercises

1. For each of the three fields defined below, write IF tests for the absence of data:
   a. without using any 88-level entries or REDEFINES clauses, if possible;
   b. using 88-level entries but no REDEFINES clause if possible;
   c. using 88-level entries and/or REDEFINES clauses.

```
  I. 05  OFFICE–NAME–IN          PIC X(25)
 II. 05  NO–OF–DAYS–IN           PIC 99.
III. 05  PURCHASE–IN             PIC 9(4)V99.
```

2. Write an IF test that will determine whether the field PURCHASE–IN, described in Review Exercise 1, contains all numbers, and execute the True path if it does not.
3. Given the following field:

```
05 SALE–AMOUNT–IN          PIC 999V99.
```

   write an 88-level entry and an IF test that will execute the True path if the contents of the field SALE–AMOUNT–IN is outside the reasonable range of $5 through $750.
4. Given the following field:

```
05 SKILL–CODE          PIC X.
```

   write 88-level entries giving suitable names to the following codes for the different skills: machinist, M; carpenter, C; press operator, P; riveter, R. Also write an 88-level entry giving a name to a missing code. Write IF tests to check for the absence of data in the field and for an invalid code.
5. Write a program to read data in the format shown in Figure 6.11. Have the program check the validity of the input as follows:

   a. Store number numeric
   b. Salesperson number numeric
   c. Customer number present

d. Sale amount numeric
e. Sale amount not less than $1 nor greater than $5,000.

List the contents of each error-free card. Accumulate the sale amounts from the error-free cards and print the total at the end of the report. Check for all types of errors in each card. If more than one error is found in a card, print an error line for each error.

*6. Write a program to read input data in the format shown in Figure 4.1. The program is to check that all Quantity and all Unit price fields contain only numbers. If a field is not found to contain only numbers, a suitable error line is to be printed. For each valid record, print the Customer number, Part number, Quantity, Unit price, and a Merchandise amount (the Quantity times the Unit price), except if the Merchandise amount is greater than 999,999.99. If the Merchandise amount is greater than 999,999.99, print only the Customer number, Part number, Quantity, and Unit price, and a message MERCHANDISE AMOUNT SUSPICIOUSLY LARGE. At the end of the report, print the total of the valid Merchandise amounts.

Subscripting and Indexing

9

KEY POINTS

Here are the key points you should learn from this chapter:

1. why subscripting and indexing are useful;
2. how to use subscripting and indexing on input data;
3. how to set up in COBOL one-dimensional tables and tables of more than one dimension;
4. how to use indexing to carry out a table search.

KEY WORDS

Key words to recognize and learn:

OCCURS	SET	ASCENDING KEY
subscript	relative indexing	key field
index	two-dimensional table	DESCENDING KEY
SEARCH	three-dimensional table	SEARCH ALL
INDEXED BY		
index name		

Subscripting and indexing may be used whenever a COBOL program has to refer to a group of related fields all having the same PICTURE. It has been mentioned earlier that programs can be reduced in size and made easier to read through the use of subscripting and indexing. For example, Program 15, one of the programs that can profit from the use of subscripting or indexing, contains several groups of related fields. One such group consists of SALE–1–IN, SALE–2–IN, and SALE–3–IN. Those fields are related in the sense that they must all be processed in an identical way. They all have the same PICTURE.

Another group of related fields in Program 15 consists of the three sale quotas JUNIOR–SALE–QUOTA, ASSOCIATE–SALE–QUOTA, and SENIOR–SALE–QUOTA. They too are all used in essentially the same way and have identical PICTUREs.

A Modification to Program 15 Using Subscripting

In Program 15 let us look at the repetitive coding that deals with SALE–1–IN, SALE–2–IN, and SALE–3–IN, in the routine WRITE–A–SALESPERSON–GROUP. There is separate coding in that routine to handle each sale field. The amount of coding could be reduced if there were some way to write the coding for one sale field and make that coding serve for the other two sale fields as well. In Program 18 we will use subscripting to accomplish that.

To enable a field to be subscripted you must define it with an **OCCURS** clause. The OCCURS clause tells COBOL exactly how many of the similar fields there are, so if we wanted to use subscripting on the three sale fields, we would replace the three individual descriptions of SALE–1–IN, SALE–2–IN, and SALE–3–IN with a single description that tells COBOL that there are three similar fields; that is, we would remove from Program 15 the descriptions:

```
05 SALE-1-IN        PIC 9(6)V99.
05 SALE-2-IN        PIC 9(6)V99.
05 SALE-3-IN        PIC 9(6)V99.
```

and insert instead the single entry:

```
05 SALE-IN OCCURS 3 TIMES   PIC 9(6)V99.
```

This tells COBOL that in the input record there are three fields, one right after the other, and that each has the PICTURE 9(6)V99. All three fields also have, in a sense, the same name too. They are all called SALE–IN now, and COBOL must have some way of distinguishing one from the other. This is accomplished by means of a **subscript.** In COBOL a subscript is a number (or an identifier defined as an integer) which is written in parentheses following the data name being subscripted. The three SALE–IN fields can be referred to as:

```
SALE-IN (1)     (pronounced "SALE-IN sub one")
SALE-IN (2)     (pronounced "SALE-IN sub two")
```

and

```
SALE-IN (3)     (pronounced "SALE-IN sub three")
```

anywhere in the Procedure Division. So the following would be a legal statement:

```
MOVE SALE-IN (1) TO SALE-OUT.
```

It would be read as "MOVE SALE–IN sub one TO SALE–OUT." Here are other legal statements:

```
IF SALE-IN (2) GREATER THAN 50.00
     PERFORM 50-DOLLAR-ROUTINE.
ADD SALE-IN (3) TO SALE-ACCUMULATOR ROUNDED.
```

Try reading the IF and ADD statements aloud before going on.

The subscript following the data name SALE–IN in each statement tells COBOL which of the three SALE–IN fields is being referred to in the statement. There must be a space between the data name and the left parenthesis and a space following the right parenthesis, unless it is followed by another right parenthesis.

Whenever a field is defined with an OCCURS clause, you must use a subscript (or an **index,** discussed later in this chapter) whenever you refer to that field in the Procedure Division (except with the **SEARCH** verb, also to be discussed later in this chapter). Also, whenever you want to use a subscript on a field, the field must be defined with an OCCURS clause.

A subscript doesn't have to be just a number; a subscript can be given in the form of an identifier, that is, a data name or a qualified data name. Any data name or qualified data name can be used as a subscript, provided it is defined as an integer. So we can make up the following field to serve as a subscript:

```
05 SALE-SUBSCRIPT        PIC S9.
```

The field SALE–SUBSCRIPT looks like an ordinary field and it is. It can be used in a Procedure Division according to all the rules of COBOL that we have studied so far. But, because it is defined as an integer, it can also be used as a subscript. So it is legal to refer to the three fields called SALE–IN this way:

```
SALE-IN (SALE-SUBSCRIPT)
```

(pronounced "SALE–IN subscripted by SALE–SUBSCRIPT").

The field SALE–SUBSCRIPT can be assigned the value 1 or 2 or 3, so that a statement like:

```
MOVE SALE-IN (SALE-SUBSCRIPT) TO SALE-OUT.
```

will execute in a way that depends on the value assigned to SALE–SUBSCRIPT at the time the statement is executed. If SALE–SUBSCRIPT happens to have the integer 1 assigned to it, the MOVE statement will execute as though it were written:

```
MOVE SALE-IN (1) TO SALE-OUT.
```

If SALE–SUBSCRIPT has a 2 assigned when the MOVE is executed, it will execute like:

```
MOVE SALE-IN (2) TO SALE-OUT.
```

It would be illegal for SALE–SUBSCRIPT to have a value larger than 3 when the MOVE statement executes, since the OCCURS clause in the definition of SALE–IN says that there are only three such fields. You can see that, by using an identifier as a subscript, it is possible to get a statement like:

```
MOVE SALE-IN (SALE-SUBSCRIPT) TO SALE-OUT.
```

to refer to all three of the fields SALE–IN merely by changing the value assigned to SALE–SUBSCRIPT. You will see exactly how such a situation is handled in Program 18.

Don't be fooled by the name of the field SALE–SUBSCRIPT. Names mean nothing to COBOL; COBOL doesn't understand English. COBOL doesn't care that the word SUBSCRIPT is part of the name SALE–SUBSCRIPT. The only reason that SALE–SUBSCRIPT can serve as a subscript is that it is defined as an integer.

COBOL also doesn't care that the word SALE is part of the name SALE–SUBSCRIPT. SALE–SUBSCRIPT doesn't have to be used to subscript only SALE–IN; it can be used to subscript other fields as well. A subscript in a COBOL program may be used on any field in the program that is defined with an OCCURS clause.

Exercise 1. Which of the following would be valid as a subscript:

a. 05 DOG PIC S99.
b. 05 SUBSCRIPT–1 PIC S99V99.
c. 05 SUBSCRIPT–2 PIC X(3).

Exercise 2. Given the following fields:

```
05 PURCHASES OCCURS 5 TIMES     PIC 9(4)V99.
05 RETURNS OCCURS 4 TIMES       PIC 9(3)V99.
05 SUBSC-1                      PIC S9.
05 SUBSC-2                      PIC S9.
```

a. Which of the following statements could be legal in a COBOL program?

1. ADD PURCHASES (SUBSC–1) TO ACCUMULATOR.
2. ADD PURCHASES (SUBSC–2) TO ACCUMULATOR.
3. ADD SUBSC–1 (PURCHASES) TO ACCUMULATOR.
4. ADD RETURNS (SUBSC–1) TO ACCUMULATOR.
5. ADD SUBSC–2 (RETURNS) TO ACCUMULATOR.

b. What is the highest value that may legally be assigned to SUBSC–1 at the time statement 1 above executes?
c. What is the highest value that may legally be assigned to SUBSC–2 at the time statement 2 above executes?

Another group of similar fields that contributes to the size of Program 15 consists of the sale-quota fields JUNIOR–SALE–QUOTA, ASSOCIATE–SALE–QUOTA, and SENIOR–SALE–QUOTA. The handling of those three fields in Program 15 is not quite so repetitive as the handling of the three sale fields, but Program 15 could nonetheless be shortened still further if we were able to refer to the sale-quota fields through subscripting. We might want to replace the description of the individual sale quota fields from Program 15:

```
05 JUNIOR-SALE-QUOTA         PIC S9(5) VALUE 0.
05 ASSOCIATE-SALE-QUOTA      PIC S9(5) VALUE 10000.
05 SENIOR-SALE-QUOTA         PIC S9(5) VALUE 50000.
```

with a single description like:

```
05 SALE-QUOTA OCCURS 3 TIMES PIC S9(5).
```

just as we replaced the descriptions of SALE–1–IN, SALE–2–IN, and SALE–3–IN with a single description for SALE–IN. But how can we get the values 0, 10000, and 50000 for the three classes of salespeople assigned to the three fields defined as SALE–QUOTA? COBOL does not permit a VALUE clause to appear in the same entry with an OCCURS clause. We had no need for a VALUE clause when we defined the three SALE–IN fields because the values of those fields were read as input. But in order to keep the VALUE clauses for the sale quotas, in Program 18 we will still be forced to use:

```
01 SALE-QUOTAS.
   05 JUNIOR-SALE-QUOTA        PIC S9(5) VALUE 0.
   05 ASSOCIATE-SALE-QUOTA     PIC S9(5) VALUE 10000.
   05 SENIOR-SALE-QUOTA        PIC S9(5) VALUE 50000.
```

We can use the REDEFINES clause to apply a new definition to this table; we can write:

```
01 SALE-QUOTA-TABLE REDEFINES SALE-QUOTAS.
   05 SALE-QUOTA OCCURS 3 TIMES PIC S9(5).
```

Then we can refer to JUNIOR–SALE–QUOTA by the name SALE–QUOTA (1), to ASSOCIATE–SALE–QUOTA by the name SALE–QUOTA (2), and to SENIOR–SALE–QUOTA by the name SALE–

QUOTA (3). More importantly, we can define a field to be used as a subscript, say:

```
05 CLASS-SUBSCRIPT        PIC S9.
```

Then we can refer to all three sale-quota fields as SALE–QUOTA (CLASS–SUBSCRIPT) by assigning the value 1 or 2 or 3 to the field CLASS–SUBSCRIPT. In Program 18 you will see exactly how the table of sale quotas is set up and how it is used.

Since it is so easy to set up tables in working storage, in Program 18 we will use a table that we didn't have in Program 15; it will be a table of the titles of the three classes of salespeople. The complete entry will look like this:

```
01 CLASS-TITLES.
   05 FILLER        PIC X(9) VALUE 'JUNIOR'.
   05 FILLER        PIC X(9) VALUE 'ASSOCIATE'.
   05 FILLER        PIC X(9) VALUE 'SENIOR'.
01 CLASS-TITLE-TABLE REDEFINES CLASS-TITLES.
   05 CLASS-TITLE OCCURS 3 TIMES PIC X(9).
```

Exercise 3. Referring to the CLASS–TITLE–TABLE above, tell whether the word JUNIOR or the word ASSOCIATE or the word SENIOR would be MOVEd to TITLE–OUT by the following pair of statements:

MOVE 3 TO CLASS–SUBSCRIPT.
MOVE CLASS–TITLE (CLASS–SUBSCRIPT) TO TITLE–OUT.

Exercise 4. Explain why all the FILLERs in CLASS–TITLES above have PICTURE X(9), even though the words JUNIOR and SENIOR are only six letters long.

Applying Subscripting

Program 18 is shown in Figure 9.1. Let us look at it to see how the use of subscripting makes it different from Program 15. In the Data Division you can find the fields that were discussed in the previous section: SALE–IN, with its OCCURS clause; the SALE–QUOTA–TABLE, with its REDEFINES clause; and the CLASS–TITLE–TABLE. In addition, the two fields that we need as subscripts, SALE–SUB-SCRIPT and CLASS–SUBSCRIPT, have been defined as integers; in Program 18 they have also both been made COMPUTATIONAL. Fields used as subscripts don't have to be COMPUTATIONAL, and the decision whether to make them COMPUTATIONAL is the same as with any other field: If a field is used only internally to the program and not read in or printed out, it should certainly be made COMPU-

Figure 9.1 Program 18, Part 1 of 4.

```
4-CB2 V4 RELEASE 1.5 10NOV77        IBM OS AMERICAN NATIONAL STANDARD COBOL



000010 IDENTIFICATION DIVISION.
000020 PROGRAM-ID. PROG18.
000030*AUTHOR. ELLEN M. LONGO.
000040*
000050*     THIS PROGRAM IS A MODIFICATION TO PROGRAM 15 USING
000060*     SUBSCRIPTING.
000070*
000080*****************************************************************
000090
000100 ENVIRONMENT DIVISION.
000110 CONFIGURATION SECTION.
000120 SPECIAL-NAMES.
000130      C01 IS NEW-PAGE.
000140 INPUT-OUTPUT SECTION.
000150 FILE-CONTROL.
000160      SELECT COMMISSION-REPORT ASSIGN TO UT-S-SYSPRINT.
000170      SELECT SALES-FILE-IN     ASSIGN TO UT-S-SYSIN.
000180
000190*****************************************************************
000200
000210 DATA DIVISION.
000220 FILE SECTION.
000230 FD   SALES-FILE-IN
000240      LABEL RECORDS ARE OMITTED.
000250
000260 01   SALES-RECORD-IN.
000270      05 SALESPERSON-NUMBER-IN     PIC 9(5).
000280      05 CLASS-IN                  PIC X.
000290         88 SALESPERSON-IS-JUNIOR     VALUES 'A' THRU 'F'.
000300         88 SALESPERSON-IS-ASSOCIATE VALUE 'G'.
000310         88 SALESPERSON-IS-SENIOR     VALUE 'H'.
000320         88 CLASS-CODE-IS-MISSING     VALUE SPACE.
000330      05 SALE-IN OCCURS 3 TIMES    PIC 9(6)V99.
000340      05 FILLER                    PIC X(50).
000350
000360 FD   COMMISSION-REPORT
000370      LABEL RECORDS ARE OMITTED.
000380
000390 01   REPORT-LINE                  PIC X(89).
000400
000410 WORKING-STORAGE SECTION.
000420 01   MORE-INPUT                   PIC X(3) VALUE 'YES'.
000430
000440 01   PAGE-HEAD-1.
000450      05 FILLER                    PIC X.
000460      05 FILLER                    PIC X(17) VALUE SPACES.
000470      05 FILLER                PIC X(19) VALUE 'COMMISSION REGISTER'.
000480
```

Figure 9.1 Part 1 (continued).

```
000490 01  PAGE-HEAD-2.
000500     05 FILLER                         PIC X.
000510     05 FILLER                         PIC X(43) VALUE SPACES.
000520     05 FILLER                         PIC X(5)  VALUE 'DATE'.
000530     05 RUN-MONTH                      PIC Z9.
000540     05 FILLER                         PIC X     VALUE '/'.
000550     05 RUN-DAY                        PIC 99.
000560     05 FILLER                         PIC X     VALUE '/'.
000570     05 RUN-YEAR                       PIC Z9.
000580
000590 01  PAGE-HEAD-3.
000600     05 FILLER                         PIC X.
000610     05 FILLER                         PIC X(10) VALUE SPACES.
000620     05 FILLER                         PIC X(12) VALUE 'SALES-'.
000630     05 FILLER                         PIC X(11) VALUE 'CLASS'.
000640     05 FILLER                         PIC X(10) VALUE 'SALE'.
000650     05 FILLER                         PIC X(10) VALUE 'COMMISSION'.
000660
000670 01  PAGE-HEAD-4.
000680     05 FILLER                         PIC X.
000690     05 FILLER                         PIC X(10) VALUE SPACES.
000700     05 FILLER                         PIC X(22) VALUE 'PERSON'.
000710     05 FILLER                         PIC X(6)  VALUE 'AMOUNT'.
000720
000730 01  PAGE-HEAD-5.
000740     05 FILLER                         PIC X.
000750     05 FILLER                         PIC X(10) VALUE SPACES.
000760     05 FILLER                         PIC X(6)  VALUE 'NUMBER'.
000770
000780 01  DETAIL-LINE.
000790     05 FILLER                         PIC X.
000800     05 SALESPERSON-NUMBER-OUT         PIC B(10)X(5).
000810     05 CLASS-TITLE-OUT                PIC B(5)X(9)B.
000820     05 SALE-AMOUNT-OUT                PIC ZZZ,ZZZ.99BBB.
000830     05 COMMISSION-OUT                 PIC ZZZ,ZZZ.99.
000840
000850 01  ERROR-LINE.
000860     05 FILLER                         PIC X.
000870     05 SALESPERSON-NUMBER-ERR         PIC B(10)X(5)B(5).
000880     05 FILLER                         PIC X(8)  VALUE 'ERROR -'.
000890     05 ERROR-MESSAGE                  PIC X(15).
000900     05 FILLER PIC X(29) VALUE 'CLASS SHOULD BE A, B, C, D, E'.
000910     05 FILLER PIC X(13) VALUE ', F, G, OR H.'.
000920
000930 01  ERROR-MESSAGES.
000940     05 CLASS-MISSING-MESSAGE          PIC X(15) VALUE 'CLASS MISSING.'.
000950     05 INVALID-CLASS-MESSAGE.
000960        10 FILLER                      PIC X(9) VALUE 'CLASS IS'.
000970        10 INVALID-CLASS               PIC X.
000980        10 FILLER                      PIC X    VALUE '.'.
000990
001000 01  AVERAGE-SALE-AMOUNT               PIC S9(6)V99.
001010 01  NUMBER-OF-SALES                   PIC S9     VALUE 3.
```

TATIONAL; if, on the other hand, a field is used as a subscript and also interacts with input or output, then a COMPUTATIONAL designation may improve or worsen program efficiency.

Let us see how the Procedure Division of Program 18 differs from the Procedure Division in Program 15. The MAIN–PROCESS paragraph is still largely the same as before, with only slight differences: It has now been written as a case and some of the IF tests have been simplified. The simplification was made possible by rearranging some of the program logic. The computation of AVERAGE–SALE–AMOUNT now refers to the three SALE–IN fields by their subscripts.

ASSIGNING A VALUE TO A SUBSCRIPT

The three paragraphs JUNIOR–SALESPERSON–COMMISSION, ASSOCIATE–SLSPSN–COMMISSION, and SENIOR–SALESPERSON–COMMISSION still do much of what was done by the equivalent paragraphs in Program 15 (JUNIOR–SALESPERSON–COMMISSION, LEVEL–2–COMMISSION, LEVEL–3–COMMISSION, LEVEL–4–COMMISSION, and LEVEL–5–COMMISSION) but with one important difference: Each of the three paragraphs assigns a value

Figure 9.1 Program 18, Part 2 of 4.

```
001020
001030 01  SALE-QUOTAS.
001040     05 JUNIOR-SALE-QUOTA          PIC S9(5) VALUE 0.
001050     05 ASSOCIATE-SALE-QUOTA       PIC S9(5) VALUE 10000.
001060     05 SENIOR-SALE-QUOTA          PIC S9(5) VALUE 50000.
001070 01  SALE-QUOTA-TABLE REDEFINES SALE-QUOTAS.
001080     05 SALE-QUOTA OCCURS 3 TIMES            PIC S9(5).
001090
001100 01  COMMISSION-RATES.
001110     05 LOW-RATE                   PIC SV99.
001120     05 HIGH-RATE                  PIC SV99.
001130     05 COMMISSION-RATE-1          PIC SV99  VALUE .10.
001140     05 COMMISSION-RATE-2          PIC SV99  VALUE .10.
001150     05 COMMISSION-RATE-3          PIC SV99  VALUE .05.
001160     05 COMMISSION-RATE-4          PIC SV99  VALUE .20.
001170     05 COMMISSION-RATE-5          PIC SV99  VALUE .20.
001180     05 COMMISSION-RATE-6          PIC SV99  VALUE .30.
001190     05 COMMISSION-RATE-7          PIC SV99  VALUE .30.
001200     05 COMMISSION-RATE-8          PIC SV99  VALUE .40.
001210
001220 01  TODAYS-DATE.
001230     05 RUN-YEAR                   PIC 99.
001240     05 RUN-MONTH                  PIC 99.
001250     05 RUN-DAY                    PIC 99.
001260
001270 01  CLASS-TITLES.
001280     05 FILLER                     PIC X(9)  VALUE 'JUNIOR'.
001290     05 FILLER                     PIC X(9)  VALUE 'ASSOCIATE'.
001300     05 FILLER                     PIC X(9)  VALUE 'SENIOR'.
001310 01  CLASS-TITLE-TABLE REDEFINES CLASS-TITLES.
001320     05 CLASS-TITLE OCCURS 3 TIMES           PIC X(9).
```

Figure 9.1 Part 2 (continued).

```
001330
001340 01  CLASS-SUBSCRIPTS.
001350     05 JUNIOR-SUBSCRIPT          PIC S9    VALUE 1 COMP.
001360     05 ASSOCIATE-SUBSCRIPT       PIC S9    VALUE 2 COMP.
001370     05 SENIOR-SUBSCRIPT          PIC S9    VALUE 3 COMP.
001380
001390 01  CLASS-SUBSCRIPT              PIC S9 COMP.
001400 01  SALE-SUBSCRIPT               PIC S9 COMP.
001410 01  LINE-SPACING                 PIC S9 COMP.
001420
001430************************************************************
001440
001450 PROCEDURE DIVISION.
001460     OPEN INPUT  SALES-FILE-IN,
001470          OUTPUT COMMISSION-REPORT.
001480     READ SALES-FILE-IN
001490         AT END
001500             MOVE 'NO' TO MORE-INPUT.
001510     IF MORE-INPUT = 'YES'
001520         PERFORM INITIALIZATION
001530         PERFORM MAIN-PROCESS UNTIL MORE-INPUT = 'NO'.
001540     CLOSE SALES-FILE-IN,
001550           COMMISSION-REPORT.
001560     STOP RUN.
001570
001580 INITIALIZATION.
001590     ACCEPT TODAYS-DATE FROM DATE.
001600     MOVE CORR TODAYS-DATE TO PAGE-HEAD-2.
001610     WRITE REPORT-LINE FROM PAGE-HEAD-1 AFTER NEW-PAGE.
001620     WRITE REPORT-LINE FROM PAGE-HEAD-2 AFTER 1.
001630     WRITE REPORT-LINE FROM PAGE-HEAD-3 AFTER 3.
001640     WRITE REPORT-LINE FROM PAGE-HEAD-4 AFTER 1.
001650     WRITE REPORT-LINE FROM PAGE-HEAD-5 AFTER 1.
001660
001670 MAIN-PROCESS.
001680     IF CLASS-CODE-IS-MISSING
001690         PERFORM CLASS-MISSING-ROUTINE
001700     ELSE
001710     IF SALESPERSON-IS-JUNIOR
001720         PERFORM JUNIOR-SALESPERSON-COMMISSION
001730     ELSE
001740         COMPUTE AVERAGE-SALE-AMOUNT ROUNDED =
001750             (SALE-IN (1) +
001760              SALE-IN (2) +
001770              SALE-IN (3)) / NUMBER-OF-SALES
001780     IF SALESPERSON-IS-ASSOCIATE
001790         PERFORM ASSOCIATE-SLSPSN-COMMISSION
001800     ELSE
001810     IF SALESPERSON-IS-SENIOR
001820         PERFORM SENIOR-SALESPERSON-COMMISSION
001830     ELSE
001840         PERFORM INVALID-CLASS-ROUTINE.
001850     READ SALES-FILE-IN
001860         AT END
001870             MOVE 'NO' TO MORE-INPUT.
001880
```

to the field CLASS–SUBSCRIPT. You will later see how CLASS–SUBSCRIPT is used. The paragraph JUNIOR–SALESPERSON–COMMISSION has the following:

```
MOVE JUNIOR–SUBSCRIPT TO CLASS–SUBSCRIPT.
```

JUNIOR–SUBSCRIPT is defined in working storage as being a 1, so the MOVE statement would have worked as well if it had been written:

```
MOVE 1 TO CLASS–SUBSCRIPT.
```

But, for purposes of program documentation and possible future modification, it is better to use the method given in the program. In any case, the paragraph ASSOCIATE–SLSPSN–COMMISSION MOVEs a 2 to CLASS–SUBSCRIPT, and the paragraph SENIOR–SALES-PERSON–COMMISSION MOVEs a 3.

Figure 9.1 Program 18, Part 3 of 4.

```
001890 JUNIOR-SALESPERSON-COMMISSION.
001900     MOVE JUNIOR-SUBSCRIPT  TO CLASS-SUBSCRIPT.
001910     MOVE COMMISSION-RATE-1 TO LOW-RATE.
001920     MOVE COMMISSION-RATE-2 TO HIGH-RATE.
001930     PERFORM PRODUCE-A-SALESPERSON-GROUP.
001940
001950 ASSOCIATE-SLSPSN-COMMISSION.
001960     MOVE ASSOCIATE-SUBSCRIPT TO CLASS-SUBSCRIPT.
001970     IF AVERAGE-SALE-AMOUNT LESS THAN ASSOCIATE-SALE-QUOTA
001980         MOVE COMMISSION-RATE-3 TO LOW-RATE
001990         MOVE COMMISSION-RATE-4 TO HIGH-RATE
002000     ELSE
002010         MOVE COMMISSION-RATE-5 TO LOW-RATE
002020         MOVE COMMISSION-RATE-6 TO HIGH-RATE.
002030     PERFORM PRODUCE-A-SALESPERSON-GROUP.
002040
002050 SENIOR-SALESPERSON-COMMISSION.
002060     MOVE SENIOR-SUBSCRIPT TO CLASS-SUBSCRIPT.
002070     IF AVERAGE-SALE-AMOUNT LESS THAN SENIOR-SALE-QUOTA
002080         MOVE COMMISSION-RATE-5 TO LOW-RATE
002090         MOVE COMMISSION-RATE-6 TO HIGH-RATE
002100     ELSE
002110         MOVE COMMISSION-RATE-7 TO LOW-RATE
002120         MOVE COMMISSION-RATE-8 TO HIGH-RATE.
002130     PERFORM PRODUCE-A-SALESPERSON-GROUP.
002140
002150 PRODUCE-A-SALESPERSON-GROUP.
002160     MOVE 2 TO LINE-SPACING.
002170     MOVE SALESPERSON-NUMBER-IN TO SALESPERSON-NUMBER-OUT.
002180     MOVE CLASS-TITLE (CLASS-SUBSCRIPT) TO CLASS-TITLE-OUT.
002190     PERFORM WRITE-A-SALESPERSON-GROUP
002200         VARYING SALE-SUBSCRIPT FROM 1 BY 1
002210         UNTIL SALE-SUBSCRIPT GREATER THAN NUMBER-OF-SALES.
002220
```

We come now to the paragraphs PRODUCE–A–SALESPERSON–GROUP and WRITE–A–SALESPERSON–GROUP, which replace the single old paragraph WRITE–A–SALESPERSON–GROUP from Program 15. In PRODUCE–A–SALESPERSON–GROUP, study the statement:

```
MOVE CLASS–TITLE (CLASS–SUBSCRIPT) TO CLASS–TITLE–OUT.
```

Remember that at this point in the program the field CLASS–SUBSCRIPT has already had a 1 assigned to it if the program is processing a junior salesperson, a 2 if an associate, and a 3 if a senior. Now find CLASS–TITLE in the Working Storage Section and do Exercise 5.

Exercise 5. In Program 18, refer to the statement:

MOVE CLASS–TITLE (CLASS–SUBSCRIPT) TO CLASS–TITLE–OUT.

a. If the program is processing a junior salesperson record, what value is assigned to CLASS–SUBSCRIPT and what value will be MOVEd to CLASS–TITLE–OUT?
b. If the program is processing an associate salesperson record, what value is assigned to CLASS–SUBSCRIPT and what value will be MOVEd to CLASS–TITLE–OUT?
c. If the program is processing a senior salesperson record, what value is assigned to CLASS–SUBSCRIPT and what value will be MOVEd to CLASS–TITLE–OUT?

Exercise 5 is a critical one in this book, and the concept it tests is critical for all COBOL programmers. In order to be able to program in COBOL in industry, you must understand and be able to use the ideas of subscripting. If you were not able to do Exercise 5 you will have to develop your own method for handling the subscripting concepts. Perhaps it would help to remember that when dealing with subscripts we are dealing with two different values at more or less the same time: the value of the subscript and the value of the subscripted field. In Exercise 5, the subscript is CLASS–SUBSCRIPT and the subscripted field is CLASS–TITLE. The following table might be a step toward a solution of Exercise 5:

Value of the subscript Value of CLASS–SUBSCRIPT	Value of the subscripted field Value of CLASS–TITLE (CLASS–SUBSCRIPT)
1	JUNIOR
2	ASSOCIATE
3	SENIOR

CONTROLLING A SUBSCRIPT WITH A PERFORM STATEMENT

We come now to a new form of the PERFORM verb, one containing the VARYING option. The VARYING option is often used when one or more subscripted fields are involved in the processing, but it can also be used in certain circumstances where there are no subscripted fields. The meaning of the PERFORM. . .VARYING statement in Program 18 is almost self-evident. The paragraph WRITE–A–SALES-PERSON–GROUP is to be PERFORMed over and over. The first time that WRITE–A–SALESPERSON–GROUP is PERFORMed, SALE–SUBSCRIPT will be set to 1 by the PERFORM statement. Then, each time WRITE–A–SALESPERSON–GROUP is PERFORMed again, the value of SALE–SUBSCRIPT will be increased by 1. This will go on until the value of SALE–SUBSCRIPT is greater than the value of NUMBER–OF–SALES. So the PERFORM statement causes WRITE–A–SALESPERSON–GROUP to be executed with SALE–SUBSCRIPT set to 1 and then 2 and then 3.

Now compare WRITE–A–SALESPERSON–GROUP in Program 18 with WRITE–A–SALESPERSON–GROUP in Program 15. In Program 15 you see three IF statements, each operating on a different sale field; in Program 18, the single IF statement operates on all three. In Program 15 we had three sale fields with the three different names SALE–1–IN, SALE–2–IN, and SALE–3–IN. In Program 18 the one name for all three sales fields, SALE–IN, is subscripted by SALE–SUBSCRIPT so that it can refer to all three. The value of SALE–SUBSCRIPT is controlled by the PERFORM. . .VARYING statement.

Figure 9.1 Program 18, Part 4 of 4.

```
002230 WRITE-A-SALESPERSON-GROUP.
002240     IF SALE-IN (SALE-SUBSCRIPT) LESS THAN
002250                                          SALE-QUOTA (CLASS-SUBSCRIPT)
002260         MULTIPLY SALE-IN (SALE-SUBSCRIPT) BY LOW-RATE
002270                                     GIVING COMMISSION-OUT ROUNDED
002280     ELSE
002290         MULTIPLY SALE-IN (SALE-SUBSCRIPT) BY HIGH-RATE
002300                                     GIVING COMMISSION-OUT ROUNDED.
002310     MOVE SALE-IN (SALE-SUBSCRIPT) TO SALE-AMOUNT-OUT.
002320     WRITE REPORT-LINE FROM DETAIL-LINE AFTER LINE-SPACING.
002330     MOVE 1 TO LINE-SPACING.
002340     MOVE SPACES TO DETAIL-LINE.
002350
002360 CLASS-MISSING-ROUTINE.
002370     MOVE CLASS-MISSING-MESSAGE TO ERROR-MESSAGE.
002380     PERFORM WRITE-ERROR-LINE.
002390
002400 INVALID-CLASS-ROUTINE.
002410     MOVE CLASS-IN              TO INVALID-CLASS.
002420     MOVE INVALID-CLASS-MESSAGE TO ERROR-MESSAGE.
002430     PERFORM WRITE-ERROR-LINE.
002440
002450 WRITE-ERROR-LINE.
002460     MOVE SALESPERSON-NUMBER-IN TO SALESPERSON-NUMBER-ERR.
002470     WRITE REPORT-LINE FROM ERROR-LINE AFTER 2.
```

If there had been ten sale fields in the input instead of three, the difference between WRITE–A–SALESPERSON–GROUP in Program 15 and the same paragraph in Program 18 would have been even more substantial.

There is one final use of subscripting in this program. The IF statement in WRITE–A–SALESPERSON–GROUP contains a reference to SALE–QUOTA (CLASS–SUBSCRIPT). At this point in the program, CLASS–SUBSCRIPT is still set to 1 if a junior salesperson is being processed and to 2 or to 3 for an associate or a senior, respectively. The subscripting on the name SALE–QUOTA permits the IF statement to get directly to the proper SALE–QUOTA amount without the intervening processing that was required in Program 15.

Program 18 was run with the same input as Program 15 and produced the output shown in Figure 9.2. As far as anyone can tell, it is the same as the output from Program 15 (except for the run date).

Figure 9.2 Output from Program 18.

```
          COMMISSION REGISTER
                                            DATE  2/19/80

SALES-        CLASS         SALE         COMMISSION
PERSON                     AMOUNT
NUMBER

09834        JUNIOR        1,985.00         198.50
                          15,550.00       1,555.00
                           8,575.80         857.58

02784        JUNIOR       14,753.00       1,475.30
                           9,850.00         985.00
                          20,050.50       2,005.05

26374        ASSOCIATE     1,985.00          99.25
                          15,550.00       3,110.00
                           8,575.80         428.79

27634        ASSOCIATE    14,753.00       4,425.90
                           9,850.00       1,970.00
                          20,050.50       6,015.15

36472        SENIOR       35,080.60       7,016.12
                          40,500.00       8,100.00
                          60,700.80      18,210.24

34782        SENIOR       55,689.00      22,275.60
                          46,098.00      13,829.40
                         567,000.00     226,800.00

23478        ERROR - CLASS MISSING.  CLASS SHOULD BE A, B, C, D, E, F, G, OR H.

84753        ERROR - CLASS IS U.     CLASS SHOULD BE A, B, C, D, E, F, G, OR H.
```

Exercise 6. What five Procedure Division statements were eliminated from Program 15 by using subscripting on the SALE–QUOTA field in Program 18?

Exercise 7. Modify Program 16 so that the three sale fields are subscripted. Take advantage of subscripting to rewrite the IF statements that test the sale fields for nonnumeric data and for being suspiciously large. You should be able to compress three case statements into one case.

Indexing

With indexing, a program can do many of the same things that can be done with subscripting, but more efficiently. In this discussion of indexing, we will first modify Program 18, replacing all of the subscripting functions with indexing. Then we will discuss the special features of indexing that make it more powerful than subscripting in certain contexts.

As you may remember, any field defined with an OCCURS clause must be subscripted or indexed whenever it is referred to in the Procedure Division (except with the SEARCH verb). A data name is indexed in much the same way that it is subscripted; that is, an index would appear in parentheses after the name of the field being indexed. For example, if the SALE–IN field had been set up to be indexed instead of subscripted, we could still refer to the three SALE–IN fields by the names SALE–IN (1), SALE–IN (2), and SALE–IN (3) or through the use of a variable index, as in SALE–IN (SALE–INDEX), pronounced "SALE–IN indexed by SALE–INDEX."

But the index field is set up differently from a subscript. You may remember that a subscript must be defined as an integer; an index field, however, need not be defined at all! All the programmer has to do is make up a name for the index, and COBOL provides the field automatically and in its most efficient form. The programmer tells COBOL the name of the index by using the **INDEXED BY** phrase. An INDEXED BY phrase must be included as part of an OCCURS clause whenever you wish to index a field. So to set up SALE–IN to be indexed instead of subscripted we could use:

```
05 SALE-IN
   OCCURS 3 TIMES INDEXED BY SALE-INDEX
                               PIC 9(6)V99.
```

That's all. COBOL provides the field SALE–INDEX without any further effort on the part of the programmer. When a name is given in an INDEXED BY phrase, it must not be otherwise defined.

A name given in an INDEXED BY phrase is called an **index name.** The rules for making up index names are the same as for making up any ordinary data names. But an index is not an ordinary field. While a subscript is an ordinary field and can be used in the same ways as any integer in a Procedure Division, an index is designed to be used only for indexing the one particular OCCURS clause that it is defined with.

In Program 18 we used the field CLASS–SUBSCRIPT to subscript both the CLASS–TITLE–TABLE and the SALE–QUOTA–TABLE. But in Program 19 each of those tables will need its own index. So we will need entries like:

```
05 CLASS–TITLE
    OCCURS 3 TIMES INDEXED BY TITLE–INDEX
                                    PIC X(9).
```

and

```
05 SALE–QUOTA
    OCCURS 3 TIMES INDEXED BY QUOTA–INDEX
                                    PIC 9(5).
```

In Program 18 the values of the subscripts were assigned by MOVE statements and assigned and controlled by a PERFORM statement with the VARYING option. An index too may have its value assigned and controlled by a PERFORM statement with the VARYING option, but it may not have its value assigned by a MOVE statement; instead, a **SET** statement must be used, as you will see in Program 19.

The SET Statement

Index names work essentially differently and more efficiently than subscripts. Each COBOL system uses its own method of assigning values to index names, each value corresponding to an element in the table being indexed. The SET statement can be used to assign to an index name the value that corresponds to some particular element of a table. So for example, the statement SET SALE–INDEX TO 1 will assign a value to SALE–INDEX that corresponds to the first SALE–IN field. The general format for this use of the SET verb is:

```
                                                  { index-name-3 }
SET index-name-1 [index-name-2] . . . TO { identifier-3 }
                                                  { literal-1    }
```

A Program Using Indexing

Program 19 is shown in Figure 9.3. It is nearly identical to Program 18 except in the ways already discussed. Find the SET statements in Program 19 and compare them to the MOVE statements in Program 18. For example, the statement in Program 18:

```
MOVE JUNIOR-SUBSCRIPT TO CLASS-SUBSCRIPT.
```

has been replaced in program 19 by the statement:

```
SET QUOTA-INDEX,
    TITLE-INDEX
        TO JUNIOR-INDEX.
```

You can see in the Working Storage Section that JUNIOR–INDEX has been given the VALUE 1, so the SET statement works as if it had been written:

```
SET QUOTA-INDEX,
    TITLE-INDEX
        TO 1.
```

The SET statement assigns to both TITLE–INDEX and QUOTA–INDEX values which correspond to the first items in their respective tables. That is, TITLE–INDEX is assigned a value which corresponds to the first CLASS–TITLE, and QUOTA–INDEX is assigned a value which corresponds to the first SALE–QUOTA. Notice that the SET statement works backwards: The receiving fields come before the word TO and the sending field after. SET is the only backwards statement in all of COBOL.

Figure 9.3 Program 19, Part 1 of 3.

```
4-CB2 V4 RELEASE 1.5 10NOV77        IBM OS AMERICAN NATIONAL STANDARD COBOL

  000010 IDENTIFICATION DIVISION.
  000020 PROGRAM-ID. PROG19.
  000030*AUTHOR. ELLEN M. LONGO.
  000040*
  000050*     THIS PROGRAM IS A MODIFICATION TO PROGRAM 18 USING
  000060*     INDEXING.
  000070*
  000080*****************************************************************
```

Figure 9.3 Part 1 (continued).

```
000090
000100 ENVIRONMENT DIVISION.
000110 CONFIGURATION SECTION.
000120 SPECIAL-NAMES.
000130     C01 IS NEW-PAGE.
000140 INPUT-OUTPUT SECTION.
000150 FILE-CONTROL.
000160     SELECT COMMISSION-REPORT ASSIGN TO UT-S-SYSPRINT.
000170     SELECT SALES-FILE-IN     ASSIGN TO UT-S-SYSIN.
000180
000190*****************************************************************
000200
000210 DATA DIVISION.
000220 FILE SECTION.
000230 FD  SALES-FILE-IN
000240     LABEL RECORDS ARE OMITTED.
000250
000260 01  SALES-RECORD-IN.
000270     05 SALESPERSON-NUMBER-IN      PIC 9(5).
000280     05 CLASS-IN                   PIC X.
000290        88 SALESPERSON-IS-JUNIOR      VALUES 'A' THRU 'F'.
000300        88 SALESPERSON-IS-ASSOCIATE VALUE 'G'.
000310        88 SALESPERSON-IS-SENIOR      VALUE 'H'.
000320        88 CLASS-CODE-IS-MISSING      VALUE SPACE.
000330     05 SALE-IN OCCURS 3 TIMES
000340        INDEXED BY SALE-INDEX      PIC 9(6)V99.
000350     05 FILLER                     PIC X(50).
000360
000370 FD  COMMISSION-REPORT
000380     LABEL RECORDS ARE OMITTED.
000390
000400 01  REPORT-LINE                   PIC X(89).
000410
000420 WORKING-STORAGE SECTION.
000430 01  MORE-INPUT                    PIC X(3) VALUE 'YES'.
000440
000450 01  PAGE-HEAD-1.
000460     05 FILLER                     PIC X.
000470     05 FILLER                     PIC X(17) VALUE SPACES.
000480     05 FILLER                PIC X(19) VALUE 'COMMISSION REGISTER'.
000490
000500 01  PAGE-HEAD-2.
000510     05 FILLER                     PIC X.
000520     05 FILLER                     PIC X(43) VALUE SPACES.
000530     05 FILLER                     PIC X(5)  VALUE 'DATE'.
000540     05 RUN-MONTH                  PIC Z9.
000550     05 FILLER                     PIC X     VALUE '/'.
000560     05 RUN-DAY                    PIC 99.
000570     05 FILLER                     PIC X     VALUE '/'.
000580     05 RUN-YEAR                   PIC Z9.
000590
000600 01  PAGE-HEAD-3.
000610     05 FILLER                     PIC X.
000620     05 FILLER                     PIC X(10) VALUE SPACES.
000630     05 FILLER                     PIC X(12) VALUE 'SALES-'.
000640     05 FILLER                     PIC X(11) VALUE 'CLASS'.
000650     05 FILLER                     PIC X(10) VALUE 'SALE'.
000660     05 FILLER                     PIC X(10) VALUE 'COMMISSION'.
```

Figure 9.3 Program 19, Part 2 of 3.

```
000670
000680 01   PAGE-HEAD-4.
000690      05 FILLER                   PIC X.
000700      05 FILLER                   PIC X(10) VALUE SPACES.
000710      05 FILLER                   PIC X(22) VALUE 'PERSON'.
000720      05 FILLER                   PIC X(6)  VALUE 'AMOUNT'.
000730
000740 01   PAGE-HEAD-5.
000750      05 FILLER                   PIC X.
000760      05 FILLER                   PIC X(10) VALUE SPACES.
000770      05 FILLER                   PIC X(6)  VALUE 'NUMBER'.
000780
000790 01   DETAIL-LINE.
000800      05 FILLER                   PIC X.
000810      05 SALESPERSON-NUMBER-OUT   PIC B(10)X(5).
000820      05 CLASS-TITLE-OUT          PIC B(5)X(9)B.
000830      05 SALE-AMOUNT-OUT          PIC ZZZ,ZZZ.99BBB.
000840      05 COMMISSION-OUT           PIC ZZZ,ZZZ.99.
000850
000860 01   ERROR-LINE.
000870      05 FILLER                   PIC X.
000880      05 SALESPERSON-NUMBER-ERR   PIC B(10)X(5)B(5).
000890      05 FILLER                   PIC X(8)  VALUE 'ERROR -'.
000900      05 ERROR-MESSAGE            PIC X(15).
000910      05 FILLER PIC X(29) VALUE 'CLASS SHOULD BE A, B, C, D, E'.
000920      05 FILLER PIC X(13) VALUE ', F, G, OR H.'.
000930
000940 01   ERROR-MESSAGES.
000950      05 CLASS-MISSING-MESSAGE    PIC X(15) VALUE 'CLASS MISSING.'.
000960      05 INVALID-CLASS-MESSAGE.
000970         10 FILLER                PIC X(9) VALUE 'CLASS IS'.
000980         10 INVALID-CLASS         PIC X.
000990         10 FILLER                PIC X    VALUE '.'.
001000
001010 01   AVERAGE-SALE-AMOUNT         PIC S9(6)V99.
001020 01   NUMBER-OF-SALES             PIC S9    VALUE 3.
001030
001040 01   SALE-QUOTAS.
001050      05 JUNIOR-SALE-QUOTA        PIC S9(5) VALUE 0.
001060      05 ASSOCIATE-SALE-QUOTA     PIC S9(5) VALUE 10000.
001070      05 SENIOR-SALE-QUOTA        PIC S9(5) VALUE 50000.
001080 01   SALE-QUOTA-TABLE REDEFINES SALE-QUOTAS.
001090      05 SALE-QUOTA OCCURS 3 TIMES
001100         INDEXED BY QUOTA-INDEX   PIC S9(5).
001110
001120 01   COMMISSION-RATES.
001130      05 LOW-RATE                 PIC SV99.
001140      05 HIGH-RATE                PIC SV99.
001150      05 COMMISSION-RATE-1        PIC SV99  VALUE .10.
001160      05 COMMISSION-RATE-2        PIC SV99  VALUE .10.
001170      05 COMMISSION-RATE-3        PIC SV99  VALUE .05.
001180      05 COMMISSION-RATE-4        PIC SV99  VALUE .20.
001190      05 COMMISSION-RATE-5        PIC SV99  VALUE .20.
001200      05 COMMISSION-RATE-6        PIC SV99  VALUE .30.
001210      05 COMMISSION-RATE-7        PIC SV99  VALUE .30.
001220      05 COMMISSION-RATE-8        PIC SV99  VALUE .40.
001230
```

Figure 9.3 Part 2 (continued).

```
001240 01  TODAYS-DATE.
001250     05 RUN-YEAR                    PIC 99.
001260     05 RUN-MONTH                   PIC 99.
001270     05 RUN-DAY                     PIC 99.
001280
001290 01  CLASS-TITLES.
001300     05 FILLER                      PIC X(9)  VALUE 'JUNIOR'.
001310     05 FILLER                      PIC X(9)  VALUE 'ASSOCIATE'.
001320     05 FILLER                      PIC X(9)  VALUE 'SENIOR'.
001330 01  CLASS-TITLE-TABLE REDEFINES CLASS-TITLES.
001340     05 CLASS-TITLE OCCURS 3 TIMES
001350        INDEXED BY TITLE-INDEX      PIC X(9).
001360
001370 01  CLASS-INDEXES.
001380     05 JUNIOR-INDEX                PIC S9    VALUE 1 COMP.
001390     05 ASSOCIATE-INDEX             PIC S9    VALUE 2 COMP.
001400     05 SENIOR-INDEX                PIC S9    VALUE 3 COMP.
001410
001420 01  LINE-SPACING                   PIC S9 COMP.
001430
001440*****************************************************************
001450
001460 PROCEDURE DIVISION.
001470     OPEN INPUT  SALES-FILE-IN,
001480          OUTPUT COMMISSION-REPORT.
001490     READ SALES-FILE-IN
001500         AT END
001510             MOVE 'NO' TO MORE-INPUT.
001520     IF MORE-INPUT = 'YES'
001530         PERFORM INITIALIZATION
001540         PERFORM MAIN-PROCESS UNTIL MORE-INPUT = 'NO'.
001550     CLOSE SALES-FILE-IN,
001560           COMMISSION-REPORT.
001570     STOP RUN.
001580
001590 INITIALIZATION.
001600     ACCEPT TODAYS-DATE FROM DATE.
001610     MOVE CORR TODAYS-DATE TO PAGE-HEAD-2.
001620     WRITE REPORT-LINE FROM PAGE-HEAD-1 AFTER NEW-PAGE.
001630     WRITE REPORT-LINE FROM PAGE-HEAD-2 AFTER 1.
001640     WRITE REPORT-LINE FROM PAGE-HEAD-3 AFTER 3.
001650     WRITE REPORT-LINE FROM PAGE-HEAD-4 AFTER 1.
001660     WRITE REPORT-LINE FROM PAGE-HEAD-5 AFTER 1.
001670
001680 MAIN-PROCESS.
001690     IF CLASS-CODE-IS-MISSING
001700         PERFORM CLASS-MISSING-ROUTINE
001710     ELSE
001720     IF SALESPERSON-IS-JUNIOR
001730         PERFORM JUNIOR-SALESPERSON-COMMISSION
001740     ELSE
001750         COMPUTE AVERAGE-SALE-AMOUNT ROUNDED =
001760             (SALE-IN (1) +
001770              SALE-IN (2) +
001780              SALE-IN (3)) / NUMBER-OF-SALES
001790     IF SALESPERSON-IS-ASSOCIATE
001800         PERFORM ASSOCIATE-SLSPSN-COMMISSION
```

Figure 9.3 Program 19, Part 3 of 3.

```
001810         ELSE
001820         IF SALESPERSON-IS-SENIOR
001830             PERFORM SENIOR-SALESPERSON-COMMISSION
001840         ELSE
001850             PERFORM INVALID-CLASS-ROUTINE.
001860         READ SALES-FILE-IN
001870             AT END
001880                 MOVE 'NO' TO MORE-INPUT.
001890
001900 JUNIOR-SALESPERSON-COMMISSION.
001910         SET QUOTA-INDEX,
001920             TITLE-INDEX
001930                 TO JUNIOR-INDEX.
001940         MOVE COMMISSION-RATE-1 TO LOW-RATE.
001950         MOVE COMMISSION-RATE-2 TO HIGH-RATE.
001960         PERFORM PRODUCE-A-SALESPERSON-GROUP.
001970
001980 ASSOCIATE-SLSPSN-COMMISSION.
001990         SET QUOTA-INDEX,
002000             TITLE-INDEX
002010                 TO ASSOCIATE-INDEX.
002020         IF AVERAGE-SALE-AMOUNT LESS THAN ASSOCIATE-SALE-QUOTA
002030             MOVE COMMISSION-RATE-3 TO LOW-RATE
002040             MOVE COMMISSION-RATE-4 TO HIGH-RATE
002050         ELSE
002060             MOVE COMMISSION-RATE-5 TO LOW-RATE
002070             MOVE COMMISSION-RATE-6 TO HIGH-RATE.
002080         PERFORM PRODUCE-A-SALESPERSON-GROUP.
002090
002100 SENIOR-SALESPERSON-COMMISSION.
002110         SET QUOTA-INDEX,
002120             TITLE-INDEX
002130                 TO SENIOR-INDEX.
002140         IF AVERAGE-SALE-AMOUNT LESS THAN SENIOR-SALE-QUOTA
002150             MOVE COMMISSION-RATE-5 TO LOW-RATE
002160             MOVE COMMISSION-RATE-6 TO HIGH-RATE
002170         ELSE
002180             MOVE COMMISSION-RATE-7 TO LOW-RATE
002190             MOVE COMMISSION-RATE-8 TO HIGH-RATE.
002200         PERFORM PRODUCE-A-SALESPERSON-GROUP.
002210
002220 PRODUCE-A-SALESPERSON-GROUP.
002230         MOVE 2                         TO LINE-SPACING.
002240         MOVE SALESPERSON-NUMBER-IN     TO SALESPERSON-NUMBER-OUT.
002250         MOVE CLASS-TITLE (TITLE-INDEX) TO CLASS-TITLE-OUT.
002260         PERFORM WRITE-A-SALESPERSON-GROUP
002270             VARYING SALE-INDEX FROM 1 BY 1
002280             UNTIL SALE-INDEX GREATER THAN NUMBER-OF-SALES.
002290
002300 WRITE-A-SALESPERSON-GROUP.
002310         IF SALE-IN (SALE-INDEX) LESS THAN SALE-QUOTA (QUOTA-INDEX)
002330             MULTIPLY SALE-IN (SALE-INDEX) BY LOW-RATE
002340                                     GIVING COMMISSION-OUT
002350         ELSE
002360             MULTIPLY SALE-IN (SALE-INDEX) BY HIGH-RATE
002370                                     GIVING COMMISSION-OUT.
002380         MOVE SALE-IN (SALE-INDEX) TO SALE-AMOUNT-OUT.
002390         WRITE REPORT-LINE FROM DETAIL-LINE AFTER LINE-SPACING.
002400         MOVE 1       TO LINE-SPACING.
002410         MOVE SPACES TO DETAIL-LINE.
```

Figure 9.3 Part 3 (continued).

```
002420
002430 CLASS-MISSING-ROUTINE.
002440      MOVE CLASS-MISSING-MESSAGE TO ERROR-MESSAGE.
002450      PERFORM WRITE-ERROR-LINE.
002460
002470 INVALID-CLASS-ROUTINE.
002480      MOVE CLASS-IN              TO INVALID-CLASS.
002490      MOVE INVALID-CLASS-MESSAGE TO ERROR-MESSAGE.
002500      PERFORM WRITE-ERROR-LINE.
002510
002520 WRITE-ERROR-LINE.
002530      MOVE SALESPERSON-NUMBER-IN TO SALESPERSON-NUMBER-ERR.
002540      WRITE REPORT-LINE FROM ERROR-LINE AFTER 2.
```

Program 19 was run using the same input as Program 18. The output from Program 19 is shown in Figure 9.4 and is evidently the same as from Program 18.

Figure 9.4 Output from Program 19.

```
        COMMISSION REGISTER
                                          DATE  2/13/80

SALES-         CLASS       SALE         COMMISSION
PERSON                    AMOUNT
NUMBER

09834       JUNIOR       1,985.00        198.50
                        15,550.00      1,555.00
                         8,575.80        857.58

02784       JUNIOR      14,753.00      1,475.30
                         9,850.00        985.00
                        20,050.50      2,005.05

26374       ASSOCIATE    1,985.00         99.25
                        15,550.00      3,110.00
                         8,575.80        428.79

27634       ASSOCIATE   14,753.00      4,425.90
                         9,850.00      1,970.00
                        20,050.50      6,015.15

36472       SENIOR      35,080.60      7,016.12
                        40,500.00      8,100.00
                        60,700.80     18,210.24

34782       SENIOR      55,689.00     22,275.60
                        46,098.00     13,829.40
                       567,000.00    226,800.00

23478       ERROR - CLASS MISSING. CLASS SHOULD BE A, B, C, D, E, F, G, OR H.

84753       ERROR - CLASS IS U.    CLASS SHOULD BE A, B, C, D, E, F, G, OR H.
```

Exercise 8. Why did the statement that assigned the value of the subscript in Program 18

```
MOVE JUNIOR-SUBSCRIPT TO CLASS-SUBSCRIPT
```

have only one receiving field, while in Program 19 the SET statement needed two?

```
SET QUOTA-INDEX,
    TITLE-INDEX
        TO JUNIOR-INDEX.
```

Exercise 9. Modify your solution to Exercise 7 so that the three sale fields are indexed instead of subscripted.

A Program with Table Searching

In Program 20, part of the processing requires the program to look up information in a table. Up to now, our programs always had some way of determining and using correct values for subscripts and indexes without having to look up anything in a table. In Program 20, the values of the indexes needed for processing can be found only by searching a table.

THE FEDERAL INCOME TAX TABLE

The tax brackets for computing Federal income tax are defined in the law in the form of a table. The table lists both a tax amount and a tax percentage for different amounts of taxable income. The table shows the tax on certain amounts of taxable income and the percentage rate that must be applied to any excess income over the table value. Naturally the law requires each taxpayer to use the table values that result in the highest possible tax. Here is part of the table for single taxpayers:

taxable income	tax amount	percentage to apply to excess taxable income
0	0	0
$2,300	0	14%
$3,400	$154	16%
$4,400	$314	18%
$6,500	$692	19%
$8,500	$1,072	21%
$10,800	$1,555	24%
$12,900	$2,059	26%
$15,000	$2,605	30%

taxable income	tax amount	percentage to apply to excess taxable income
$18,200	$3,565	34%
$23,500	$5,367	39%
$28,800	$7,434	44%
$34,100	$9,766	49%
$41,500	$13,392	55%
$55,300	$20,982	63%
$81,800	$37,677	68%

Suppose a taxpayer has taxable income of $3,500. The table entry for $3,400 would have to be used, and the tax would be $154 plus 16% of the income over $3,400. So the tax would be $154 plus 16% of $100, for a total of $170. Now suppose a taxpayer has income of $16,000. Then the entry for $15,000 would be used, and the tax amount would be $2,605 plus 30% of the excess income over $15,000, or a total tax of $2,905.

The tax rates in the table are rational only in the sense that you can never have an increase in taxable income and a decrease in after-tax net. Check and see that the tax on $12,900.00 is not larger than the tax on $12,899.99. The tax in both cases is $2,059.

A PROBLEM USING THE TAX TABLES

Program 20 uses input in the format shown in Figure 9.5. Each input record shows a taxpayer's Social Security number and a taxable income. The program is to compute the tax for each taxpayer and print the results in the format shown in Figure 9.6. For each input record, Program 20 must search the tax table to determine which tax amount and which percentage rate apply to the record. For programming convenience, each column from the tax table above is set up in Program 20 as a separate table, each with its own index.

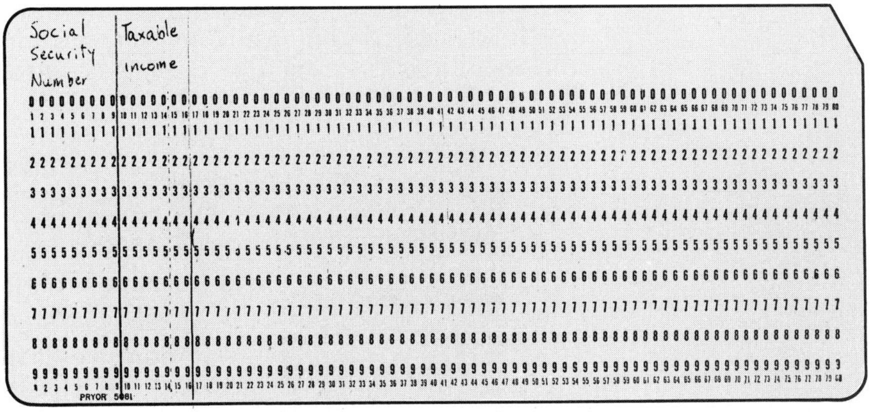

Figure 9.5
Data card format for Program 20.

150/10/6 PRINT CHART PROG. ID ________ PAGE ________
(SPACING: 150 POSITION SPAN, AT 10 CHARACTERS PER INCH, 6 LINES PER VERTICAL INCH) DATE ________
Fold back at dotted line.
PROGRAM TITLE ________
PROGRAMMER OR DOCUMENTALIST: ________
CHART TITLE Program 20
CARRIAGE CONTROL TAPE CHAN.

TAX REPORT

SOCIAL SECURITY NUMBER TAXABLE INCOME TAX AMOUNT

XXX-XX-XXXX ZZ,ZZ9.99 ZZ,ZZ9.99
XXX-XX-XXXX ZZ,ZZ9.99 ZZ,ZZ9.99
XXX-XX-XXXX ZZ,ZZ9.99 ZZ,ZZ9.99
XXX-XX-XXXX ZZ,ZZ9.99 ZZ,ZZ9.99

Figure 9.6 Output format for Program 20.

THE SEARCH VERB

The SEARCH statement permits a program to step through a table until one or more conditions specified by the programmer is satisfied. Before a SEARCH statement can be executed, the index for the table must be SET to a value corresponding to the position of the table element where the SEARCH is to begin. Usually the index is SET to 1 to start the SEARCH at the beginning of the table, but the index could be set to any value to start the SEARCH at any place in the table. For the given setting of the index, the SEARCH statement tests to see whether any of the specified conditions is satisfied. If any of the conditions is satisfied, the SEARCH terminates. If none of the conditions is satisfied, the SEARCH statement increases the value of the index so that it corresponds to the next table element and tests the conditions again. The testing and incrementing continue until one of the conditions is satisfied or the index runs off the end of the table. The general format of the SEARCH statement is:

```
SEARCH identifier-1 [ VARYING { identifier-2   } ]
                              { index-name-2   }
    [AT END imperative-statement-1]
    WHEN condition-1 { imperative-statement-2 }
                     { NEXT SENTENCE          }
    [ WHEN condition-2 { imperative-statement-3 } ] ...
                       { NEXT SENTENCE          }
```

Identifier–1 is the name of the table being searched; that is, it is a field that has been defined with an OCCURS clause and an INDEXED BY phrase. For each setting of the table index, the SEARCH statement first determines whether the index is off the end of the table. If the index is off the end of the table, and the AT END clause has been specified by the programmer, the program carries out the imperative statements given in the AT END clause. If the index is off the end of the table and the AT END clause has not been specified, the program goes to the next sentence. If the index is not off the end of the table, then the conditions listed in the WHEN clauses are tested in the order that they are given. As soon as one of the conditions is found to be true, the SEARCH terminates. The program then carries out the imperative statements listed in the WHEN clause that terminated the SEARCH or goes to the NEXT SENTENCE if that has been specified. The table index retains the value that it had when the SEARCH terminated until it is reset by either a SET statement, a PERFORM statement with the VARYING option, or another SEARCH statement.

THE TABLE SEARCH PROGRAM

Program 20 is shown in Figure 9.7. In the Data Division you can see how the Federal tax table has been coded. Each column of the tax table has been made a separate COBOL table. It would have been no problem to make the tax table appear as all one table in the program, but the method used provides better documentation. Since the input record has room for taxable income amounts of $99,999.99, income brackets up to $100,000 have been included in the program. In reality the Federal tax table goes higher than that but the tables, as they appear in the program, will properly process incomes up to and including $99,999.99.

Figure 9.7 Program 20, Part 1 of 3.

```
4-CB2 V4 RELEASE 1.5 10NOV77          IBM OS AMERICAN NATIONAL STANDARD COBOL


000010 IDENTIFICATION DIVISION.
000020 PROGRAM-ID. PROG20.
000030 AUTHOR. ELLEN M. LONGO.
000040*
000050*     THIS PROGRAM COMPUTES INCOME TAX BY FIRST
000060*     DOING A TABLE SEARCH TO FIND THE CORRECT
000070*     INCOME BRACKET.
000080*
000090*****************************************************
000100
000110 ENVIRONMENT DIVISION.
000120 CONFIGURATION SECTION.
000130 SPECIAL-NAMES.
000140      C01 IS NEW-PAGE.
```

Figure 9.7 Program 20, Part 2 of 3.

```
000150 INPUT-OUTPUT SECTION.
000160 FILE-CONTROL.
000170     SELECT TAXPAYER-FILE-IN ASSIGN TO UT-S-SYSIN.
000180     SELECT TAX-REPORT       ASSIGN TO UT-S-SYSPRINT.
000190
000200*****************************************************
000210
000220 DATA DIVISION.
000230 FILE SECTION.
000240 FD  TAXPAYER-FILE-IN
000250     LABEL RECORDS ARE OMITTED.
000260
000270 01  TAXPAYER-RECORD-IN.
000280     05  SOCIAL-SECURITY-NUMBER-IN.
000290         10  SSNO-1                 PIC X(3).
000300         10  SSNO-2                 PIC X(2).
000310         10  SSNO-3                 PIC X(4).
000320     05  TAXABLE-INCOME-IN          PIC 9(5)V99.
000330     05  FILLER                     PIC X(64).
000340
000350 FD  TAX-REPORT
000360     LABEL RECORDS ARE OMITTED.
000370
000380 01  REPORT-LINE                    PIC X(43).
000390
000400 WORKING-STORAGE SECTION.
000410 01  MORE-INPUT                     PIC X(3)  VALUE 'YES'.
000420
000430 01  PAGE-HEAD-1.
000440     05  FILLER                     PIC X.
000450     05  FILLER                     PIC X(20) VALUE SPACES.
000460     05  FILLER                     PIC X(10) VALUE 'TAX REPORT'.
000470
000480 01  PAGE-HEAD-2.
000490     05  FILLER                     PIC X.
000500     05  FILLER                     PIC X(10) VALUE SPACES.
000510     05  FILLER                     PIC X(13) VALUE 'SOCIAL'.
000520     05  FILLER                     PIC X(14) VALUE 'TAXABLE'.
000530     05  FILLER                     PIC X(3)  VALUE 'TAX'.
000540
000550 01  PAGE-HEAD-3.
000560     05  FILLER                     PIC X.
000570     05  FILLER                     PIC X(9)  VALUE SPACES.
000580     05  FILLER                     PIC X(14) VALUE 'SECURITY'.
000590     05  FILLER                     PIC X(13) VALUE 'INCOME'.
000600     05  FILLER                     PIC X(6)  VALUE 'AMOUNT'.
000610
000620 01  PAGE-HEAD-4.
000630     05  FILLER                     PIC X.
000640     05  FILLER                     PIC X(10) VALUE SPACES.
000650     05  FILLER                     PIC X(6)  VALUE 'NUMBER'.
000660
```

Figure 9.7 Part 2 (continued).

```
000670 01  BODY-LINE.
000680     05  FILLER                         PIC X.
000690     05  SOCIAL-SECURITY-NUMBER-OUT.
000700         10  SSNO-1                     PIC B(8)X(3).
000710         10  FILLER                     PIC X       VALUE '-'.
000720         10  SSNO-2                     PIC X(2).
000730         10  FILLER                     PIC X       VALUE '-'.
000740         10  SSNO-3                     PIC X(4)B(3).
000750     05  TAXABLE-INCOME-OUT             PIC ZZ,ZZ9.99.
000760     05  TAX-AMOUNT-OUT                 PIC B(2)ZZ,ZZ9.99.
000770
000780 01  TAX-BRACKETS.
000790     05  FILLER                         PIC S9(6) VALUE 0.
000800     05  FILLER                         PIC S9(6) VALUE 2300.
000810     05  FILLER                         PIC S9(6) VALUE 3400.
000820     05  FILLER                         PIC S9(6) VALUE 4400.
000830     05  FILLER                         PIC S9(6) VALUE 6500.
000840     05  FILLER                         PIC S9(6) VALUE 8500.
000850     05  FILLER                         PIC S9(6) VALUE 10800.
000860     05  FILLER                         PIC S9(6) VALUE 12900.
000870     05  FILLER                         PIC S9(6) VALUE 15000.
000880     05  FILLER                         PIC S9(6) VALUE 18200.
000890     05  FILLER                         PIC S9(6) VALUE 23500.
000900     05  FILLER                         PIC S9(6) VALUE 28800.
000910     05  FILLER                         PIC S9(6) VALUE 34100.
000920     05  FILLER                         PIC S9(6) VALUE 41500.
000930     05  FILLER                         PIC S9(6) VALUE 55300.
000940     05  FILLER                         PIC S9(6) VALUE 81800.
000950     05  FILLER                         PIC S9(6) VALUE 100000.
000960 01  TAX-BRACKET-TABLE REDEFINES TAX-BRACKETS.
000970     05  TAX-BRACKET
000980         OCCURS 17 TIMES INDEXED BY BRACKET-INDEX
000990                                        PIC S9(6).
001000
001010 01  TAX-AMOUNTS.
001020     05  FILLER                         PIC S9(5) VALUE 0.
001030     05  FILLER                         PIC S9(5) VALUE 0.
001040     05  FILLER                         PIC S9(5) VALUE 154.
001050     05  FILLER                         PIC S9(5) VALUE 314.
001060     05  FILLER                         PIC S9(5) VALUE 692.
001070     05  FILLER                         PIC S9(5) VALUE 1072.
001080     05  FILLER                         PIC S9(5) VALUE 1555.
001090     05  FILLER                         PIC S9(5) VALUE 2059.
001100     05  FILLER                         PIC S9(5) VALUE 2605.
001110     05  FILLER                         PIC S9(5) VALUE 3565.
001120     05  FILLER                         PIC S9(5) VALUE 5367.
001130     05  FILLER                         PIC S9(5) VALUE 7434.
001140     05  FILLER                         PIC S9(5) VALUE 9776.
001150     05  FILLER                         PIC S9(5) VALUE 13392.
001160     05  FILLER                         PIC S9(5) VALUE 20982.
001170     05  FILLER                         PIC S9(5) VALUE 37677.
001180 01  TAX-AMOUNT-TABLE REDEFINES TAX-AMOUNTS.
001190     05  TAX-AMOUNT
001200         OCCURS 16 TIMES INDEXED BY TAX-INDEX
001210                                        PIC S9(5).
```

Figure 9.7 Program 20, Part 3 of 3.

```
001220
001230 01  PERCENTAGES.
001240     05  FILLER                          PIC SV99  VALUE .0.
001250     05  FILLER                          PIC SV99  VALUE .14.
001260     05  FILLER                          PIC SV99  VALUE .16.
001270     05  FILLER                          PIC SV99  VALUE .18.
001280     05  FILLER                          PIC SV99  VALUE .19.
001290     05  FILLER                          PIC SV99  VALUE .21.
001300     05  FILLER                          PIC SV99  VALUE .24.
001310     05  FILLER                          PIC SV99  VALUE .26.
001320     05  FILLER                          PIC SV99  VALUE .30.
001330     05  FILLER                          PIC SV99  VALUE .34.
001340     05  FILLER                          PIC SV99  VALUE .39.
001350     05  FILLER                          PIC SV99  VALUE .44.
001360     05  FILLER                          PIC SV99  VALUE .49.
001370     05  FILLER                          PIC SV99  VALUE .55.
001380     05  FILLER                          PIC SV99  VALUE .63.
001390     05  FILLER                          PIC SV99  VALUE .68.
001400 01  PERCENTAGE-TABLE REDEFINES PERCENTAGES.
001410     05  PERCENTAGE
001420         OCCURS 16 TIMES INDEXED BY PERCENTAGE-INDEX
001430                                         PIC SV99.
001440
001450*****************************************************
001460
001470 PROCEDURE DIVISION.
001480     OPEN INPUT  TAXPAYER-FILE-IN,
001490          OUTPUT TAX-REPORT.
001500     READ TAXPAYER-FILE-IN
001510         AT END
001520             MOVE 'NO' TO MORE-INPUT.
001530     IF MORE-INPUT = 'YES'
001540         PERFORM INITIALIZATION
001550         PERFORM PRODUCE-DETAIL-LINES UNTIL MORE-INPUT = 'NO'.
001560     CLOSE TAXPAYER-FILE-IN,
001570           TAX-REPORT.
001580     STOP RUN.
001590
001600 INITIALIZATION.
001610     WRITE REPORT-LINE FROM PAGE-HEAD-1 AFTER NEW-PAGE.
001620     WRITE REPORT-LINE FROM PAGE-HEAD-2 AFTER 3.
001630     WRITE REPORT-LINE FROM PAGE-HEAD-3 AFTER 1.
001640     WRITE REPORT-LINE FROM PAGE-HEAD-4 AFTER 1.
001650     MOVE SPACES TO REPORT-LINE.
001660     WRITE REPORT-LINE AFTER 1.
001670
001680 PRODUCE-DETAIL-LINES.
001690     SET BRACKET-INDEX TO 1.
001700     SEARCH TAX-BRACKET
001710         WHEN TAXABLE-INCOME-IN
001720              EQUAL TO     TAX-BRACKET (BRACKET-INDEX) OR
001730              GREATER THAN TAX-BRACKET (BRACKET-INDEX) AND
001740              LESS THAN    TAX-BRACKET (BRACKET-INDEX + 1)
001750              NEXT SENTENCE.
001760     SET TAX-INDEX,
001770         PERCENTAGE-INDEX
001780             TO BRACKET-INDEX.
```

Figure 9.7 Part 3 (continued).

```
001790        COMPUTE TAX-AMOUNT-OUT ROUNDED =
001800            TAX-AMOUNT (TAX-INDEX) +
001810            PERCENTAGE (PERCENTAGE-INDEX) *
001820            (TAXABLE-INCOME-IN - TAX-BRACKET (BRACKET-INDEX)).
001830        MOVE CORR SOCIAL-SECURITY-NUMBER-IN TO
001840                                          SOCIAL-SECURITY-NUMBER-OUT.
001850        MOVE TAXABLE-INCOME-IN TO TAXABLE-INCOME-OUT.
001860        WRITE REPORT-LINE FROM BODY-LINE AFTER 1.
001870        READ TAXPAYER-FILE-IN
001880            AT END
001890                MOVE 'NO' TO MORE-INPUT.
```

The Procedure Division shows a typical use of the SEARCH verb. In this SEARCH, the TAXABLE–INCOME–IN is to be compared against the TAX–BRACKETS to find the proper income range. To start the SEARCH at the beginning of the table, the BRACKET–INDEX is first SET to 1. The SEARCH statement must name as its object a field defined with an OCCURS clause and an INDEXED BY phrase. In this case, we want to search all the 17 fields called TAX–BRACKET until the WHEN condition is met. Notice that the field name is written without an index when it appears in a SEARCH statement.

In the WHEN clause, the term TAX–BRACKET (BRACKET–INDEX + 1) shows the use of **relative indexing.** A relative index is formed from an index name followed by a space, then a plus sign or a minus sign and another space, and then a positive integer literal followed by a right parenthesis. Relative indexing enables the program to access elements of a table other than the element whose position in the table corresponds to the value of the index. Table 9.1 shows values of TAX–BRACKET (BRACKET–INDEX) and TAX–BRACKET (BRACKET–INDEX + 1) for different values of BRACKET–INDEX.

Using Table 9.1 as a guide, you can see that a SEARCH done with a TAXABLE–INCOME–IN of $3,500 would terminate with BRACKET–INDEX having a value that corresponds to the third element of the TAX–BRACKET–TABLE. The words NEXT SENTENCE appear in the WHEN clause, and so that is where the program goes when the SEARCH terminates. The TAX–INDEX and the PERCENTAGE–INDEX are then both SET so that all three indexes refer to corresponding positions in their respective tables. In this example, TAX–INDEX and PERCENTAGE–INDEX would both be set to the third positions in their tables, which are the TAX–AMOUNT and PERCENTAGE for a $3,500 income.

The COMPUTE statement which follows is a little involved, but if you study it you will see that it correctly COMPUTEs the tax as the

Position in TAX–BRACKET–TABLE corresponding to value of BRACKET–INDEX	Value of TAX–BRACKET (BRACKET–INDEX)	Value of TAX–BRACKET (BRACKET–INDEX + 1)
1	$ 0	$2,300
2	$2,300	$3,400
3	$3,400	$4,400

Table 9.1. Values of two indexed fields for different values of the index.

sum of the TAX–AMOUNT and the PERCENTAGE applied to the excess income. The input used for Program 20 is shown in Figure 9.8 and the output is shown in Figure 9.9.

```
0565060408000000
0895022501820000
0753197222000000
0123349520000000
0913247771000000
0125727411819900
```

Figure 9.8
Input data for Program 20.

```
                    TAX REPORT

  SOCIAL          TAXABLE           TAX
  SECURITY        INCOME           AMOUNT
  NUMBER

056-50-6040     80,000.00     36,543.00
089-50-2250     18,200.00      3,565.00
075-31-9722     20,000.00      4,177.00
012-35-4952          0.00          0.00
091-32-4777     10,000.00      1,387.00
012-57-2741     18,199.00      3,564.70
```

Figure 9.9
Output from Program 20.

Exercise 10. Write a program to read and process data cards in the format shown in Figure 9.E10. Each card contains a person's Social Security number and age. The program is to look up the age in a table and determine the letter code for the age bracket into which the person falls. For each input record, the program is to print a line showing the Social Security number, the age, and the letter code for the age bracket.

Use the following table of letter codes to make up the tables in your program:

Letter code	Range of ages
A	0–1
B	2–19
C	20–24
D	25–29
E	30–39
F	40–49
G	50–64
H	65–72
I	73–99

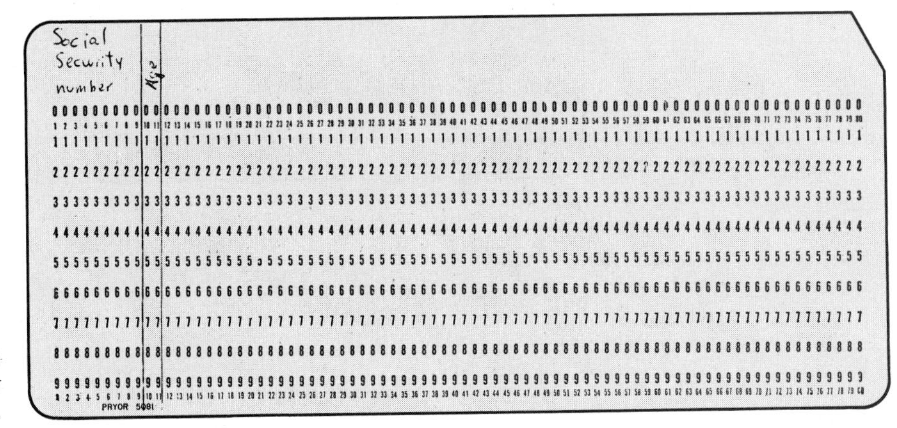

Figure 9.E10
Data card format for Exercise 10.

Tables of More Than One Dimension

COBOL is capable of handling **two-dimensional** and **three-dimensional tables.** A two-dimensional table needs two search values for finding any piece of information in the table. One example of a two-dimensional table is a table of insurance premiums where a person's age and sex are both needed to find a particular premium amount. Another example of a two-dimensional table is the one we will use in Program 21, a mileage table. There you need to know the city of departure and the destination city in order to find the mileage between them in the table.

An example of a three-dimensional table is a table of telephone charges. To find the charge for a particular call, you have to know the calling city, the called city, and the time of day. To handle real-life tables of more than three dimensions, the problem would have to be broken down into several smaller problems that could each be handled with a two- or three-dimensional table.

A Program Using a Two-Dimensional Table

Program 21 uses input in the format shown in Figure 9.10.

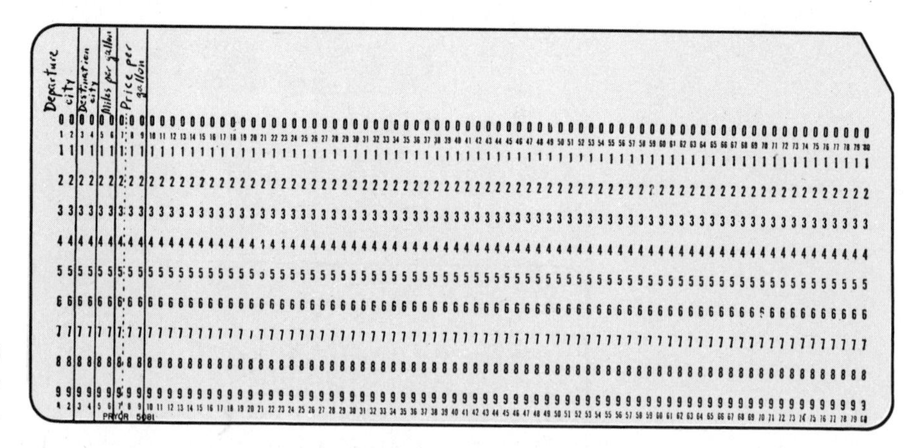

Figure 9.10
Data card format for Program 21.

Each record contains a two-digit code for a departure city, a two-digit code for a destination city, a field for miles per gallon, and a price per gallon. The two-digit city codes refer to the following cities in New York State:

Code	City
01	Albany
04	Binghamton
06	Buffalo
10	Glens Falls
12	Ithaca
14	Kingston
19	New York
20	Niagara Falls
36	Troy
38	Utica

The program is to find the road mileage between the departure and destination cities and estimate the cost of the gasoline needed to drive between them. For each record the program is to print the names and codes of the departure and destination cities, the mileage between them, and the estimated cost of the gasoline. The program is to allow for the possibility that either of the two city codes, or both, is invalid. The output should appear in the format shown in Figure 9.11. The mileage table used in Program 21 is given in Table 9.2.

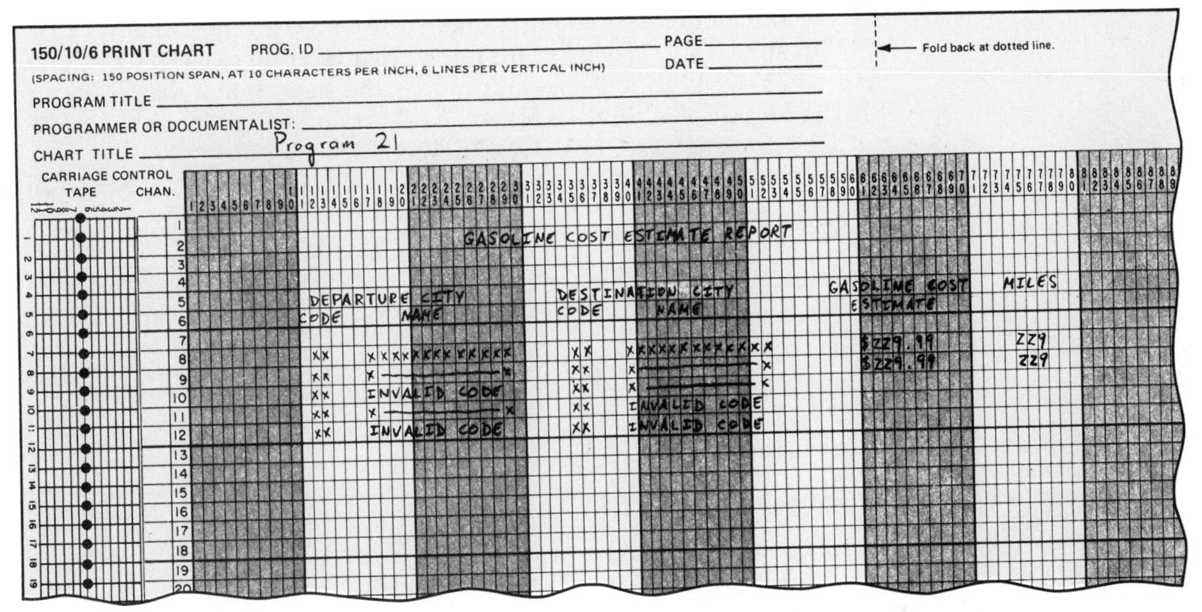

Figure 9.11 Output format for Program 21.

Destination city	01	04	06	10	12	14	19	20	36	38
Departure city										
01	000	140	290	54	166	51	148	302	8	91
04	140	000	205	169	50	132	184	217	143	93
06	290	205	000	305	155	341	438	21	290	199
10	54	169	305	000	199	105	202	317	47	106
12	166	50	155	199	000	182	234	167	169	93
14	51	132	341	105	182	000	97	353	59	142
19	148	184	438	202	234	97	000	450	156	239
20	302	217	21	317	167	353	450	000	302	211
36	8	143	290	47	169	59	156	302	000	91
38	91	93	199	106	93	142	239	211	91	000

Table 9.2. **The mileage table for Program 21.**

Program 21 is shown in Figure 9.12. The Data Division contains some new techniques of table construction. The CITY–CODE–AND–NAME–TABLE consists of ten entries, with each entry made up of a CITY–CODE and its corresponding CITY–NAME. Since CITY–CODE–AND–NAME is a group item, its OCCURS clause applies as well to the elementary items contained with the group, namely CITY–CODE and CITY–NAME. This means that references to CITY–CODE and CITY–NAME must be indexed by the same index that applies to the field CITY–CODE–AND–NAME; that index is given in the OCCURS clause as CITY–CODE–INDEX.

Figure 9.12 Program 21, Part 1 of 2.

```
4-CB2 V4 RELEASE 1.5 10NOV77          IBM OS AMERICAN NATIONAL STANDARD COBOL

000010 IDENTIFICATION DIVISION.
000020 PROGRAM-ID. PROG21.
000030*AUTHOR. PHIL RUBENSTEIN.
000040*
000050*      THIS PROGRAM FINDS THE MILEAGE BETWEEN TWO CITIES
000060*      AND ESTIMATES THE COST OF THE GASOLINE NEEDED
000070*      TO DRIVE BETWEEN THEM.
000080*
000090*****************************************************************
000100
000110 ENVIRONMENT DIVISION.
000120 CONFIGURATION SECTION.
000130 SPECIAL-NAMES.
000140      C01 IS NEW-PAGE.
000150 INPUT-OUTPUT SECTION.
000160 FILE-CONTROL.
000170      SELECT COST-ESTIMATE-REPORT ASSIGN TO UT-S-SYSPRINT.
000180      SELECT TRIP-FILE-IN          ASSIGN TO UT-S-SYSIN.
000190
000200*****************************************************************
000210
000220 DATA DIVISION.
000230 FILE SECTION.
000240 FD   TRIP-FILE-IN
000250      LABEL RECORDS ARE OMITTED.
000260
000270 01   TRIP-RECORD-IN.
000280      05 DEPARTURE-CITY-IN          PIC X(2).
000290      05 DESTINATION-CITY-IN        PIC X(2).
000300      05 MILES-PER-GALLON           PIC 99.
000310      05 PRICE-PER-GALLON           PIC 9V99.
000320      05 FILLER                     PIC X(71).
000330
000340 FD   COST-ESTIMATE-REPORT
000350      LABEL RECORDS ARE OMITTED.
```

Figure 9.12 Part 1 (continued)

```
000360
000370 01  REPORT-LINE                         PIC X(79).
000380
000390 WORKING-STORAGE SECTION.
000400 01  MORE-INPUT                          PIC X(3) VALUE 'YES'.
000410
000420 01  PAGE-HEAD-1.
000430     05 FILLER                           PIC X.
000440     05 FILLER                           PIC X(25) VALUE SPACES.
000450     05 FILLER  PIC X(29) VALUE 'GASOLINE COST ESTIMATE REPORT'.
000460
000470 01  PAGE-HEAD-2.
000480     05 FILLER                       PIC X.
000490     05 FILLER                       PIC X(11) VALUE SPACES.
000500     05 FILLER                       PIC X(22) VALUE 'DEPARTURE CITY'.
000510     05 FILLER                       PIC X(24) VALUE 'DESTINATION CITY'.
000520     05 FILLER                       PIC X(16) VALUE 'GASOLINE COST'.
000530     05 FILLER                       PIC X(5)  VALUE 'MILES'.
000540
000550 01  PAGE-HEAD-3.
000560     05 FILLER                           PIC X.
000570     05 FILLER                           PIC X(10) VALUE SPACES.
000580     05 FILLER                           PIC X(9)  VALUE 'CODE'.
000590     05 FILLER                           PIC X(14) VALUE 'NAME'.
000600     05 FILLER                           PIC X(9)  VALUE 'CODE'.
000610     05 FILLER                           PIC X(17) VALUE 'NAME'.
000620     05 FILLER                           PIC X(8)  VALUE 'ESTIMATE'.
000630
000640 01  BODY-LINE.
000650     05 FILLER                           PIC X.
000660     05 DEPARTURE-CITY-OUT               PIC B(11)X(2).
000670     05 DEPARTURE-CITY-NAME              PIC B(3)X(13).
000680     05 DESTINATION-CITY-OUT             PIC B(5)X(2).
000690     05 DESTINATION-CITY-NAME            PIC B(3)X(13)B(8).
000700     05 COST-ESTIMATE                    PIC $ZZ9.99.
000710     05 MILES-OUT                        PIC B(7)ZZ9.
000720
000730 01  CITY-CODES-AND-NAMES.
000740     05 FILLER    PIC X(15) VALUE '01ALBANY'.
000750     05 FILLER    PIC X(15) VALUE '04BINGHAMTON'.
000760     05 FILLER    PIC X(15) VALUE '06BUFFALO'.
000770     05 FILLER    PIC X(15) VALUE '10GLENS FALLS'.
000780     05 FILLER    PIC X(15) VALUE '12ITHACA'.
000790     05 FILLER    PIC X(15) VALUE '14KINGSTON'.
000800     05 FILLER    PIC X(15) VALUE '19NEW YORK'.
000810     05 FILLER    PIC X(15) VALUE '20NIAGARA FALLS'.
000820     05 FILLER    PIC X(15) VALUE '36TROY'.
000830     05 FILLER    PIC X(15) VALUE '38UTICA'.
000840 01  CITY-CODE-AND-NAME-TABLE
000850     REDEFINES CITY-CODES-AND-NAMES.
000860     05 CITY-CODE-AND-NAME
000870        OCCURS 10 TIMES
000880        ASCENDING KEY CITY-CODE
000890        INDEXED BY CITY-CODE-INDEX.
000900        10 CITY-CODE                     PIC X(2).
000910        10 CITY-NAME                     PIC X(13).
000920
```

The OCCURS clause for CITY–CODE–AND–NAME contains a new phrase, **ASCENDING KEY.** This phrase can be used in an OCCURS clause whenever the table items are in ascending order on some field. The field that is used for ordering the table is called the **key field.** If instead table items are in descending order on some key field, the **DESCENDING KEY** phrase may be used. The ASCENDING KEY and DESCENDING KEY phrases are useful in certain kinds of table SEARCHes, which will be discussed when we look at the Procedure Division.

The two-dimensional MILEAGE–TABLE is described with nested OCCURS clauses. There are ten departure cities and ten destination cities. The DEPARTURE–CITY and the DESTINATION–CITY must each have its own index, for it takes two indexes to find a particular MILEAGE value. The values for the MILEAGES are entered on ten lines of FILLERs, each line containing ten three-digit mileages. Each line of FILLER is one row from Table 9.2.

Here is a method you can use to construct a two-dimensional table in COBOL. First write the table on a piece of plain paper in the form shown in Table 9.2. Label the rows and columns of the table. Then write the rows of the table as FILLER entries in the program, as was done in MILEAGES. Then write the REDEFINES entry. The next entry

Figure 9.12 Program 21, Part 2 of 2.

```
000930 01  MILEAGES.
000940     05 FILLER PIC X(30) VALUE '000140290054166051148302008091'.
000950     05 FILLER PIC X(30) VALUE '140000205169050132184217143093'.
000960     05 FILLER PIC X(30) VALUE '290205000305155341438021290199'.
000970     05 FILLER PIC X(30) VALUE '054169305000199105202317047106'.
000980     05 FILLER PIC X(30) VALUE '166050155199000182234167169093'.
000990     05 FILLER PIC X(30) VALUE '051132341105182000097353059142'.
001000     05 FILLER PIC X(30) VALUE '148184438202234097000450156239'.
001010     05 FILLER PIC X(30) VALUE '302217021317167353450000302211'.
001020     05 FILLER PIC X(30) VALUE '008143290047169059156302000091'.
001030     05 FILLER PIC X(30) VALUE '091093199106093142239211091000'.
001040 01  MILEAGE-TABLE
001050     REDEFINES MILEAGES.
001060     05 DEPARTURE-CITY
001070        OCCURS 10 TIMES
001080        INDEXED BY DEPARTURE-CITY-INDEX.
001090        10 DESTINATION-CITY
001100           OCCURS 10 TIMES
001110           INDEXED BY DESTINATION-CITY-INDEX.
001120           15 MILEAGE                     PIC 9(3).
001130
001140 01  ANY-CITY-CODES-INVALID               PIC X(3).
001150 01  MILES-W                              PIC 9(3).
001160
001170*****************************************************************
```

Figure 9.12 Part 2 (continued).

```
001180
001190 PROCEDURE DIVISION.
001200     OPEN INPUT  TRIP-FILE-IN,
001210          OUTPUT COST-ESTIMATE-REPORT.
001220     READ TRIP-FILE-IN
001230         AT END
001240             MOVE 'NO' TO MORE-INPUT.
001250     IF MORE-INPUT = 'YES'
001260         PERFORM INITIALIZATION
001270         PERFORM MAIN-PROCESS UNTIL MORE-INPUT = 'NO'.
001280     CLOSE TRIP-FILE-IN,
001290           COST-ESTIMATE-REPORT.
001300     STOP RUN.
001310
001320 INITIALIZATION.
001330     WRITE REPORT-LINE FROM PAGE-HEAD-1 AFTER NEW-PAGE.
001340     WRITE REPORT-LINE FROM PAGE-HEAD-2 AFTER 3.
001350     WRITE REPORT-LINE FROM PAGE-HEAD-3 AFTER 1.
001360     MOVE SPACES TO REPORT-LINE.
001370     WRITE REPORT-LINE AFTER 1.
001380
001390 MAIN-PROCESS.
001400     MOVE SPACES              TO BODY-LINE.
001410     MOVE 'NO'                TO ANY-CITY-CODES-INVALID.
001420     MOVE DEPARTURE-CITY-IN TO DEPARTURE-CITY-OUT.
001430     SEARCH ALL CITY-CODE-AND-NAME
001440         AT END
001450             MOVE 'INVALID CODE' TO DEPARTURE-CITY-NAME
001460             MOVE 'YES'          TO ANY-CITY-CODES-INVALID
001470         WHEN CITY-CODE (CITY-CODE-INDEX) = DEPARTURE-CITY-IN
001480             MOVE CITY-NAME (CITY-CODE-INDEX) TO
001490                                         DEPARTURE-CITY-NAME
001500             SET DEPARTURE-CITY-INDEX TO CITY-CODE-INDEX.
001510     MOVE DESTINATION-CITY-IN TO DESTINATION-CITY-OUT.
001520     SEARCH ALL CITY-CODE-AND-NAME
001530         AT END
001540             MOVE 'INVALID CODE' TO DESTINATION-CITY-NAME
001550             MOVE 'YES'          TO ANY-CITY-CODES-INVALID
001560         WHEN CITY-CODE (CITY-CODE-INDEX) = DESTINATION-CITY-IN
001570             MOVE CITY-NAME (CITY-CODE-INDEX) TO
001580                                       DESTINATION-CITY-NAME
001590             SET DESTINATION-CITY-INDEX TO CITY-CODE-INDEX.
001600     IF ANY-CITY-CODES-INVALID = 'NO'
001610         MOVE MILEAGE (DEPARTURE-CITY-INDEX,
001620                                     DESTINATION-CITY-INDEX)
001630                                            TO MILES-OUT,
001640                                               MILES-W
001650         COMPUTE COST-ESTIMATE ROUNDED =
001660             (MILES-W / MILES-PER-GALLON) * PRICE-PER-GALLON.
001670     WRITE REPORT-LINE FROM BODY-LINE AFTER 1.
001680     READ TRIP-FILE-IN
001690         AT END
001700             MOVE 'NO' TO MORE-INPUT.
```

after the REDEFINES entry, at the 05 level, refers to the rows of the table, and the one after that, the level-10 entry, refers to the columns. The 05-level entry gives the caption of the rows, in this case DEPARTURE–CITY, and uses the OCCURS clause to say how many rows there are in the table. The level-10 entry gives the caption of the columns, in this case DESTINATION–CITY, and the number of columns. Finally, the level-15 entry gives the PICTURE of the individual table entry.

In the Procedure Division, the **SEARCH ALL** statement is used to SEARCH the CITY–CODE–AND–NAME–TABLE. You can see from the WHEN clause in the SEARCH ALL statement that the search will terminate WHEN the DEPARTURE–CITY–IN field matches one of the CITY–CODEs in the table. The SEARCH ALL statement will usually search a table faster than an ordinary SEARCH, and the improvement in speed becomes more noticeable for very large tables. The SEARCH ALL statement can be used, however, only if the ASCENDING KEY or DESCENDING KEY phrase is given in the OCCURS clause for the table and if the WHEN conditions in the SEARCH ALL specify tests for equal conditions only. There is no need to SET the table index before a SEARCH ALL statement because SEARCH ALL begins the SEARCH wherever it wants to for the fastest possible search.

The two SEARCH ALL statements in this program show the use of the AT END clause. If the SEARCH ALL examines the entire table without finding the equal condition specified in the WHEN clause, the two MOVE statements in the AT END clause are executed and then the program goes to the next sentence.

The statement

```
MOVE MILEAGE (DEPARTURE–CITY–INDEX, DESTINATION–CITY–INDEX)
                                    TO MILES–OUT, MILES–W
```

shows how a doubly-indexed table element is used. The name MILEAGE must always be accompanied by two indexes, since it is nested within two OCCURS clauses. The indexes must appear in the parentheses in the same order as they appear in the OCCURS clauses. The input for Program 21 is shown in Figure 9.13 and the output is shown in Figure 9.14.

```
123456789
090687324
000512398
121957443
193856734
010387645
012034523
361423453
101738383
```

Figure 9.13
Input data for Program 21.

Figure 9.14 Output from Program 21.

```
                GASOLINE COST ESTIMATE REPORT

 DEPARTURE CITY              DESTINATION CITY           GASOLINE COST   MILES
 CODE        NAME            CODE        NAME             ESTIMATE

  12    ITHACA                34    INVALID CODE
  09    INVALID CODE          06    BUFFALO
  00    INVALID CODE          05    INVALID CODE
  12    ITHACA                19    NEW YORK              $ 18.19         234
  19    NEW YORK              38    UTICA                 $ 31.32         239
  01    ALBANY                03    INVALID CODE
  01    ALBANY                20    NIAGARA FALLS         $ 46.45         302
  36    TROY                  14    KINGSTON              $ 11.62          59
  10    GLENS FALLS           17    INVALID CODE
```

Exercise 11. The insurance table below shows the amount of money that will be received by a male policyholder each year from a $100 policy. The amount received depends on the age at which he starts receiving money and his age at the time he bought the policy.

Age when the policy was purchased	Age when policyholder begins to receive payments		
	60	65	70
20	$17.13	$22.21	$29.17
21	$16.63	$21.56	$28.32
22	$16.15	$20.94	$27.49
23	$15.68	$20.33	$26.69
24	$15.22	$19.73	$25.91
25	$14.78	$19.16	$25.16
26	$14.35	$18.60	$24.42

Write a program to compute the amount of money a policyholder will receive each year. Use input in the format shown in Figure 9.E11. Each record shows the policy number, the age of the policyholder at the time the policy was purchased, the age at which the policyholder is to begin receiving money, and the amount of the policy in whole dollars. Remember that the money amounts in the table above are for a $100 policy.

For each input record, the program is to print a line showing the policy number and the annual money amount. If either of the ages in the input record is invalid, the program is to print an appropriate error message.

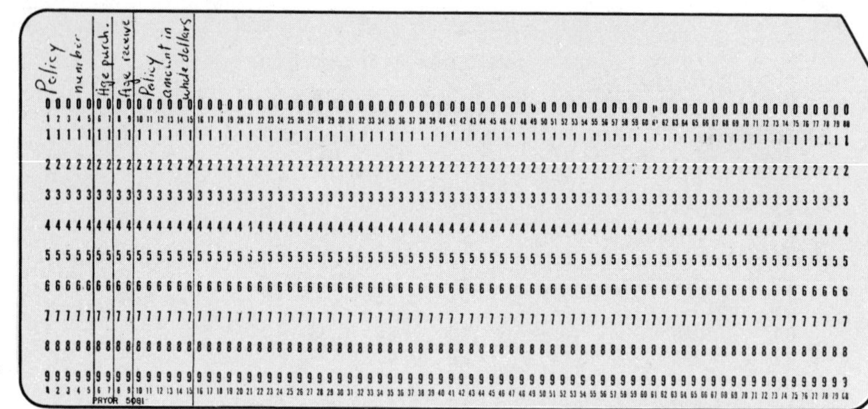

Figure 9.E11
Data card format for Exercise 11.

Summary

If a program has to process several related fields all having the same PICTURE, the fields may be defined with an OCCURS clause and then subscripted or indexed. A subscript may be an integer literal or an identifier defined as a integer. An index may be an integer literal or an index name.

If a field is to be indexed, its OCCURS clause must contain the INDEXED BY phrase. An index name, given in an INDEXED BY phrase, must not be otherwise defined in the program.

The SET statement and the PERFORM statement with the VARYING option may be used to assign a value to an index. The SEARCH statement may be used to step through a table until one or more specified conditions are met. The SEARCH ALL statement will usually give a faster search than an ordinary SEARCH statement. If SEARCH ALL is to be used, the OCCURS clause must contain either the ASCENDING KEY or DESCENDING KEY phrase, and the table items must be ordered on the key.

Fill-in Exercises

1. In order for a field to be subscripted or indexed, it must be defined with an ________ clause.
2. For a field to be indexed, its OCCURS clause must include an ________ phrase.
3. If a SEARCH ALL statement is to be used on a table, its OCCURS clause must include an ________ phrase or a ________ phrase.
4. The field "TAX–AMOUNT subscripted by TAX–SUBSCRIPT" would be written in COBOL as ________.

5. A subscript may be a positive integer numeric literal or an identifier defined as an ________.

6. The value of a subscript or index may be controlled by a PERFORM statement with the ________ option.

7. The value of an index may be changed only by a ________ statement, a PERFORM statement with the ________ option, a ________ statement, and a ________ ALL statement.

8. COBOL can process tables of up to ________ dimensions.

9. A name given in an INDEXED BY phrase is called an ________ ________.

10. The ________ clause can be used to specify the conditions under which a SEARCH or SEARCH ALL should terminate.

Review Exercises

1. Given the following fields:

```
05 AGE-RANGE OCCURS 12 TIMES            PIC 9(3).
05 OCCUPATION-CLASS OCCURS 9 TIMES      PIC X.
05 A-SUBSCRIPT                          PIC S99.
05 B-SUBSCRIPT                          PIC S99.
```

   a. Which of the following statements could be legal in a COBOL program?

      1. MOVE A-SUBSCRIPT (OCCUPATION-CLASS) TO LINE-OUT.
      2. MOVE OCCUPATION-CLASS (B-SUBSCRIPT) TO LINE-OUT.
      3. MOVE B-SUBSCRIPT (AGE-RANGE) TO LINE-OUT.

   b. What is the highest value that may legally be assigned to B-SUBSCRIPT at the time statement 2 executes?

2. Given the following table:

```
01 CITIES.
   05 FILLER    PIC X(12) VALUE 'PEEKSKILL'.
   05 FILLER    PIC X(12) VALUE 'PLATTSBURGH'.
   05 FILLER    PIC X(12) VALUE 'PORT JERVIS'.
   05 FILLER    PIC X(12) VALUE 'POUGHKEEPSIE'.
01 CITY-TABLE REDEFINES CITIES.
   05 CITY OCCURS 4 TIMES     PIC X(12).
```

   What will be MOVEd to LINE-OUT by the following pair of statements?

```
MOVE 2 TO CITY-SUBSCRIPT.
MOVE CITY (CITY-SUBSCRIPT) TO LINE-OUT.
```

3. Rewrite the table in Exercise 2 so that it may be indexed. You will have to change only the OCCURS clause. Then rewrite the statement MOVE 2 TO CITY–SUBSCRIPT as a SET statement to assign the value 2 to the table index. Finally, rewrite the statement MOVE CITY (CITY–SUBSCRIPT) TO LINE–OUT to show how your index would be used in a MOVE statement.
4. Given the following table:

```
01 PAYMENTS.
   05 FILLER PIC X(12) VALUE '171618192034'.
   05 FILLER PIC X(12) VALUE '145615471789'.
   05 FILLER PIC X(12) VALUE '135813981550'.
   05 FILLER PIC X(12) VALUE '121413751485'.
01 PAYMENTS–TABLE REDEFINES PAYMENTS.
   05 STARTING–AGE OCCURS 4 TIMES INDEXED BY
      STARTING–INDEX.
      10 ENDING–AGE OCCURS 3 TIMES INDEXED BY
         ENDING–INDEX.
         15 PAYMENT                PIC 99V99.
```

   What will be MOVEd to PAYMENT–OUT by the following group of statements?

```
SET STARTING–INDEX TO 4.
SET ENDING–INDEX TO 1.
MOVE PAYMENT (STARTING–INDEX, ENDING–INDEX)
  TO PAYMENT–OUT.
```

*5. Write a program to process cards in the format shown in Figure 9.RE5. Each record contains a branch office number and a sales

Figure 9.RE5 Data card format for Review Exercise 5.

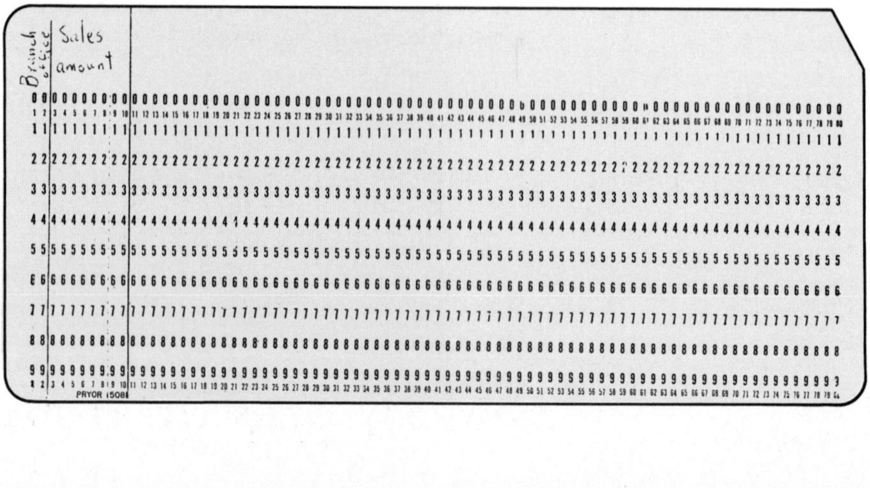

amount. For each input card, the program is to look up the name of the branch office and print on one line the branch office number, the branch office name, and the sales amount. At the end of the report print a total of all sales.

Use the following table of branch office names to set up the table(s) in your program:

Branch Office Name	Branch Office Number
Highbridge	04
Bayside	56
Brooklyn Heights	34
Borough Park	16
Bensonhurst	23
Flatlands	45

6. Write a program to process data cards in the format shown in Figure 9.RE6. Each record is for one employee and contains an Employee number, the number of years employed, a job classification, and the employee's current annual salary. Each employee is to receive a percentage salary increase based on the job classification and the number of years employed. The following

Figure 9.RE6 Data card format for Review Exercise 6.

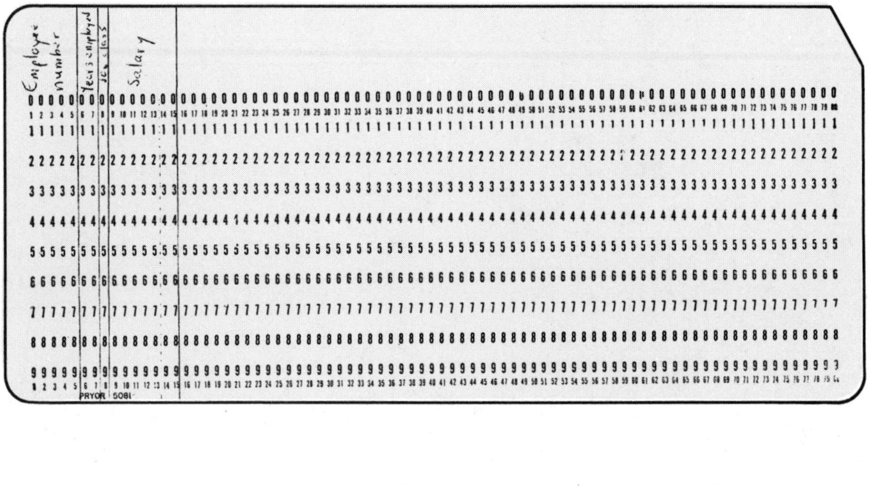

table shows the percentage increase for the different job classifications and years of employment:

Years employed	Job class 1	2	3	4	5
0–1	5%	5%	5%	5%	5%
2–5	10%	10%	15%	15%	15%
6–10	15%	15%	15%	15%	20%
11–20	20%	25%	25%	25%	25%
21–99	28%	28%	28%	28%	30%

Percentage salary increases

The program is to determine for each employee the dollar amount of the increase and print on one line the employee number, the old salary, the increase, and the new salary after the increase. At the end of the report, print the total of all the increases and the total of all the new salaries.

Report Writer 10

KEY POINTS

Here are the key points you should learn from this chapter:

1. the concept of automatic generation of coding for report output;
2. the features of the COBOL Report Writer;
3. how to use Report Writer to produce a variety of reports.

KEY WORDS

Key words to recognize and learn:

Report Section
REPORT IS
report name
report group
TYPE
DETAIL
REPORT HEADING
PAGE HEADING
CONTROL FOOTING
PAGE FOOTING
REPORT FOOTING
LINE
COLUMN
SOURCE
INITIATE
GENERATE
TERMINATE
CONTROL
FINAL
PAGE
FIRST DETAIL
SUM
USE BEFORE REPORTING
LAST DETAIL
FOOTING
PAGE–COUNTER
NEXT GROUP
nonprintable item
CONTROL HEADING
UPON
RESET ON

The Report Writer feature of COBOL makes programming for report output easier. The Report Writer provides automatically coding that would otherwise have to be written step by step by the programmer. Using the Report Writer, the programmer can describe certain characteristics that the output report is to have, and the Report Writer generates the coding needed to make the report look that way. For example, Report Writer can provide all the page overflow coding needed in a program. The programmer need only tell Report Writer how big the page is, and Report Writer provides coding that will count lines as they print, test for page overflow, and skip to a new page and print new headings when necessary. Report Writer can also provide coding that will take totals. If a total line is to print at the end of a report, the programmer need not code any of the totalling logic but just tell Report Writer which fields are to be totalled, and all the necessary coding will be provided automatically.

Best of all, Report Writer also contains control break logic. If control breaks are needed on a report, the programmer need only say what the control fields are, and Report Writer provides coding to test for control breaks and print the appropriate total lines.

Another feature of Report Writer that makes programming easier is the way that output line formats are specified in Report Writer. No more counting FILLER spaces and blanks. Just tell Report Writer in which print positions the fields are to print, and it provides all the necessary coding.

There have been many other schemes to automatically produce programs that print reports, but Report Writer is the best because it has all the vast logical power of COBOL behind it. For in using Report Writer, the programmer does not give up any of the capabilities of COBOL that we have studied; even when using Report Writer, the programmer still has command over all the output editing features, the nested IF, compound conditions, the PERFORM. . .VARYING, subscripting, indexing, and the SEARCH statements.

In this chapter we will not do any new reports. Instead, we will rewrite some of our earlier programs using Report Writer, and see how Report Writer handles such things as page overflow, totals, and control breaks, as well as ordinary detail line printing.

The Report Section and Report Groups

The first program to be redone using Report Writer is Program 4. This program reads cards and prints the contents of each one, reformatted, on one line. There are no page or column headings. You may find it useful to review Program 4 before going on (input format, Figure 2.10; input data, Figure 2.11; Program 4, Figure 2.12; output, Figure 2.13).

Program 22 is shown in Figure 10.1. The Data Division now has a new section, the **Report Section.** The Report Section must appear whenever Report Writer is to be used, and it must be the last section in the Data Division. In the File Section the output file FD entry no longer has an 01-level entry associated with it. Instead, the **REPORT IS** clause in the FD entry tells COBOL that Report Writer will be writing out a report on this file. When the REPORT IS clause appears, the programmer may not issue WRITE statements on this file; only Report Writer may write on it. The REPORT IS clause gives a name to the report that is being produced. In this program the **report name** is EMPLOYEE–REPORT. The rules for making up report names are the same as for making up file names.

Figure 10.1 Program 22, Part 1 of 2.

```
4-CB2 V4 RELEASE 1.5 10NOV77          IBM OS AMERICAN NATIONAL STANDARD COBOL



 000010 IDENTIFICATION DIVISION.
 000020 PROGRAM-ID.  PROG22.
 000030*AUTHOR. ELLEN M. LONGO.
 000040*
 000050*     THIS PROGRAM IS A MODIFICATION OF
 000060*     PROGRAM 4 USING REPORT WRITER.
 000070*
 000080*****************************************************************
 000090
 000100 ENVIRONMENT DIVISION.
 000110 INPUT-OUTPUT SECTION.
 000120 FILE-CONTROL.
 000130      SELECT EMPLOYEE-DATA-IN        ASSIGN TO UT-S-SYSIN.
 000140      SELECT EMPLOYEE-DATA-OUT       ASSIGN TO UT-S-SYSPRINT.
 000150
 000160*****************************************************************
 000170
 000180 DATA DIVISION.
 000190 FILE SECTION.
 000200 FD  EMPLOYEE-DATA-IN
 000210      LABEL RECORDS ARE OMITTED.
 000220
 000230 01   EMPLOYEE-RECORD-IN.
 000240      05  SS-NO-IN                                    PIC X(9).
 000250      05  IDENT-NO-IN                                 PIC X(5).
 000260      05  ANNUAL-SALARY-IN                            PIC X(7).
 000270      05  NAME-IN                                     PIC X(25).
 000280      05  FILLER                                      PIC X(34).
 000290
 000300 FD  EMPLOYEE-DATA-OUT
 000310      LABEL RECORDS ARE OMITTED
 000320      REPORT IS EMPLOYEE-REPORT.
 000330
 000340 WORKING-STORAGE SECTION.
 000350 01   MORE-INPUT                VALUE  'YES'          PIC X(3).
 000360
 000370 REPORT SECTION.
 000380 RD  EMPLOYEE-REPORT.
 000390
 000400 01   REPORT-LINE
 000410      TYPE DETAIL
 000420      LINE PLUS 1.
 000430      05  COLUMN  5             SOURCE IDENT-NO-IN      PIC X(5).
 000440      05  COLUMN 12             SOURCE SS-NO-IN         PIC X(9).
 000450      05  COLUMN 25             SOURCE NAME-IN          PIC X(25).
 000460      05  COLUMN 52             SOURCE ANNUAL-SALARY-IN PIC X(7).
 000470
 000480*****************************************************************
```

Every report name given in the File Section must appear in the Report Section as part of an **RD** entry (report description entry). Within the RD entry and the level numbers that follow it, many characteristics of the report are described. Report Writer uses these descriptions of the report to create the necessary coding.

Each 01-level entry following the RD entry describes a different type of line that may appear on the report or a group of related lines. The line or lines appearing under an 01-level entry is called a **report group.** In Program 22, the 01-level entry describes a report group called REPORT–LINE. In this case, the report group consists of only one line. The rules for making up report group names are the same as for making up data names.

The report group REPORT–LINE is shown as **TYPE DETAIL,** meaning that this is a detail line on the report. Some other TYPEs of report groups that may be described are:

a. **REPORT HEADING,** one or more lines to print only once on the report at the beginning;
b. **PAGE HEADING,** one or more lines to print at the top of every page;
c. **CONTROL FOOTING,** one or more lines to print after a control break has been detected;
d. **PAGE FOOTING,** one or more lines to print at the bottom of each page before skipping to a new page; and
e. **REPORT FOOTING,** one or more lines to print at the end of the report.

The clause **LINE** PLUS 1 tells Report Writer that the detail lines on this report are to be single-spaced. The clause LINE PLUS 1 means that each detail line is to be printed on whatever LINE the previous one was printed, PLUS 1. By one means or another, you must always explicitly tell Report Writer where to put every line it prints.

We come now to the 05-level entries, where we describe the individual fields that make up REPORT–LINE. Compare the description of the line here to the original description in Program 4 to see how much more convenient this is. Instead of having to count FILLERs, we tell Report Writer the print position where each field begins, using the **COLUMN** clause. The column numbers shown in these COLUMN clauses were taken directly from the original statement of the problem. Later we will be able to take column numbers right off the printer spacing chart, without having to count the number of blank spaces between fields.

The **SOURCE** clause tells Report Writer where the data comes from in the Data Division to fill each field. The SOURCE of a print field may be any identifier anywhere in the File Section or Working Storage Section or certain fields in the Report Section.

The Procedure Division introduces three new verbs. The **INITIATE**

Figure 10.1 Program 22, Part 2 of 2.

```
000490
000500 PROCEDURE DIVISION.
000510     OPEN INPUT  EMPLOYEE-DATA-IN,
000520          OUTPUT EMPLOYEE-DATA-OUT.
000530     READ EMPLOYEE-DATA-IN
000540         AT END
000550             MOVE 'NO' TO MORE-INPUT.
000560     IF MORE-INPUT = 'YES'
000570         INITIATE EMPLOYEE-REPORT
000580         PERFORM MAIN-PROCESS UNTIL MORE-INPUT = 'NO'
000590         TERMINATE EMPLOYEE-REPORT.
000600     CLOSE EMPLOYEE-DATA-IN,
000610           EMPLOYEE-DATA-OUT.
000620     STOP RUN.
000630
000640 MAIN-PROCESS.
000650     GENERATE REPORT-LINE.
000660     READ EMPLOYEE-DATA-IN
000670         AT END
000680             MOVE 'NO' TO MORE-INPUT.
```

statement must be issued to initialize the report file. It must be issued after the output file has been OPENed in the usual way and before a **GENERATE** verb is issued for that report. The INITIATE statement must include the report name as it appears in the RD entry. The GENERATE verb may be used to print a report group. In this case we have only the one report group REPORT–LINE. The statement GENERATE REPORT–LINE does everything: It blanks the output areas that should be blank, moves data from the input area to the print line, single-spaces the paper, and writes a line.

After the complete report is written a **TERMINATE** statement must be issued for the report name. This must be done before the output file is CLOSEd. Program 22 was run with the same input data as Program 4 and produced the output shown in Figure 10.2.

```
10503  100040002  MORALES, LUIS                 5000000
10890  101850005  JACOBSON, MRS. NELLIE         4651000
11277  201110008  GREENWOOD, JAMES              4302000
11664  209560011  COSTELLO, JOSEPH S.           3953000
12051  301810014  REITER, D.                    3604000
12438  304870017  MARRA, DITTA E.               3255000
12825  401710020  LIPKE, VINCENT R.             2906000
13212  407390023  KUGLER, CHARLES               2557000
13599  502070026  JAVIER, CARLOS                2208000
13986  505680029  GOODMAN, ISAAC                1859000
14373  604910032  FELDSOTT, MS. SALLY           1510000
14760  608250035  BUXBAUM, ROBERT               1161000
15147  703100038  DUMAY, MRS. MARY              0812000
15534  708020041  SMITH, R.                     0463000
15921  803220044  VINCENTE, MATTHEW J.          0114000
16308  901050047  THOMAS, THOMAS T.             4235000
```

Figure 10.2 Output from Program 22.

Exercise 1. Rewrite the program in Exercise 3, Chapter 2, page 26, using Report Writer.

A Final Total Using Report Writer

By redoing Program 10, we will show how Report Writer handles page and column headings, arithmetic manipulation of data before it is printed on a detail line, and totalling. Review Program 10 (input format for Programs 9 and 10, Figure 4.1; input data for Programs 9 and 10, Figure 4.4; output format, Figure 4.6; Program 10, Figure 4.9; output, Figure 4.10).

Program 23 is shown in Figure 10.3. The Working Storage Section

Figure 10.3 Program 23. Part 1 of 2.

```
4-CB2 V4 RELEASE 1.5 10NOV77          IBM OS AMERICAN NATIONAL STANDARD COBOL



000010 IDENTIFICATION DIVISION.
000020 PROGRAM-ID. PROG23.
000030*AUTHOR. ELLEN M. LONGO.
000040*
000050*     THIS PROGRAM IS A MODIFICATION TO
000060*     PROGRAM 10 USING REPORT WRITER.
000070*
000080*****************************************************************
000090
000100 ENVIRONMENT DIVISION.
000110 INPUT-OUTPUT SECTION.
000120 FILE-CONTROL.
000130     SELECT ORDER-FILE-IN    ASSIGN TO UT-S-SYSIN.
000140     SELECT ORDER-REPORT     ASSIGN TO UT-S-SYSPRINT.
000150
000160*****************************************************************
000170
000180 DATA DIVISION.
000190 FILE SECTION.
000200 FD  ORDER-FILE-IN
000210     LABEL RECORDS ARE OMITTED.
000220
000230 01  ORDER-RECORD-IN.
000240     05  CUSTOMER-NUMBER-IN                              PIC X(7).
000250     05  PART-NUMBER-IN                                  PIC X(8).
000260     05  FILLER                                          PIC X(7).
000270     05  QUANTITY-IN                                     PIC 9(3).
000280     05  UNIT-PRICE-IN                                   PIC 9(4)V99.
000290     05  HANDLING-IN                                     PIC 99V99.
000300     05  FILLER                                          PIC X(45).
000310
000320 FD  ORDER-REPORT
000330     LABEL RECORDS ARE OMITTED
000340     REPORT IS DAILY-ORDER-REPORT.
```

Figure 10.3 Part 1 (continued).

```
000350
000360 WORKING-STORAGE SECTION.
000370 01  MORE-INPUT                   VALUE  'YES'            PIC X(3).
000380 01  TAX-RATE                     VALUE  .07              PIC V99.
000390 01  MERCHANDISE-AMOUNT-W                                 PIC 9(6)V99.
000400 01  TAX-W                                                PIC 9(4)V99.
000410 01  ORDER-TOTAL-W                                        PIC 9(7)V99.
000420
000430 REPORT SECTION.
000440 RD  DAILY-ORDER-REPORT
000450     CONTROL FINAL
000460     PAGE 50 LINES, FIRST DETAIL 8.
000470
000480 01  TYPE PAGE HEADING.
000490     05  LINE 2.
000500         10 COLUMN 46    VALUE  'DAILY ORDER REPORT' PIC X(18).
000510     05  LINE 5.
000520         10 COLUMN 11    VALUE  'CUSTOMER'           PIC X(8).
000530         10 COLUMN 26    VALUE  'PART'               PIC X(4).
000540         10 COLUMN 38    VALUE  'QUANTITY'           PIC X(8).
000550         10 COLUMN 51    VALUE  'UNIT'               PIC X(4).
000560         10 COLUMN 64    VALUE  'MERCHANDISE'        PIC X(11).
000570         10 COLUMN 82    VALUE  'TAX'                PIC X(3).
000580         10 COLUMN 91    VALUE  'HANDLING'           PIC X(8).
000590         10 COLUMN 107   VALUE  'TOTAL'              PIC X(5).
000600     05  LINE 6.
000610         10 COLUMN 12    VALUE  'NUMBER'             PIC X(6).
000620         10 COLUMN 25    VALUE  'NUMBER'             PIC X(6).
000630         10 COLUMN 51    VALUE  'PRICE'              PIC X(5).
000640         10 COLUMN 66    VALUE  'AMOUNT'             PIC X(6).
000650
000660 01  DETAIL-LINE
000670     TYPE DETAIL
000680     LINE PLUS 1.
000690     05  COLUMN 12       SOURCE CUSTOMER-NUMBER-IN   PIC X(7).
000700     05  COLUMN 24       SOURCE PART-NUMBER-IN       PIC X(8).
000710     05  COLUMN 40       SOURCE QUANTITY-IN          PIC ZZ9.
000720     05  COLUMN 49       SOURCE UNIT-PRICE-IN        PIC Z,ZZZ.99.
000730     05  COLUMN 64       SOURCE MERCHANDISE-AMOUNT-W
000740                                           PIC ZZZ,ZZZ.99.
000750     05  COLUMN 80       SOURCE TAX-W                PIC Z,ZZZ.99.
000760     05  COLUMN 93       SOURCE HANDLING-IN          PIC ZZ.99.
000770     05  COLUMN 103      SOURCE ORDER-TOTAL-W
000780                                           PIC Z,ZZZ,ZZZ.99.
000790
000800 01  TYPE CONTROL FOOTING FINAL
000810     LINE PLUS 3.
000820     05  COLUMN 53       VALUE  'TOTALS'             PIC X(6).
000830     05  COLUMN 62       SUM    MERCHANDISE-AMOUNT-W
000840                                           PIC Z,ZZZ,ZZZ.99.
000850     05  COLUMN 79       SUM    TAX-W                PIC ZZ,ZZZ.99.
000860     05  COLUMN 92       SUM    HANDLING-IN          PIC ZZZ.99.
000870     05  COLUMN 102      SUM    ORDER-TOTAL-W
000880                                           PIC ZZ,ZZZ,ZZZ.99.
000890     05  COLUMN 116      VALUE  '**'                 PIC X(2).
000900
000910*****************************************************************
```

now contains only a few fields that we will need for intermediate arithmetic results. Everything else that was in the Working Storage Section in Program 10 will now be found somewhere in the Report Section.

The RD entry in Program 23 has a few clauses that we are seeing for the first time. The **CONTROL** clause tells Report Writer that **FINAL** totals are to be printed. In a later program we will see how totals for minor, intermediate, and major control breaks are indicated in the CONTROL clause. The **PAGE** clause is required if you want to control the vertical spacing of lines on the page. From the printer spacing chart in Figure 4.6, you can see that the first detail line of the report is to print on line 8 of the page, and we indicate this to the Report Writer by saying **FIRST DETAIL** 8. We are also required to tell Report Writer how many lines can fit on a page, and here we arbitrarily said 50.

For this report we have three report groups: a PAGE HEADING group that consists of three lines; a DETAIL group of one line; and a FINAL total line. So we need three 01-level entries, one for each report group. The first 01-level entry is for the PAGE HEADING group and is indicated by the clause TYPE PAGE HEADING. The printer spacing chart shows that the three lines that make up the PAGE HEADING are to print on lines 2, 5, and 6 of the page, and so those line numbers are indicated in the 05-level entries. The level-10 entries describe the individual fields that make up the three heading lines. Whenever you have more than one line in a group, the level number of the entry immediately following the 01-level entry can be any number in the range 02–48; here the level number is 05. Then the entries that describe the fields on the line can have level numbers in the range 03–49; here they have level number 10. Whenever there is only one line in a group, the entries that describe the fields on the line may have level numbers in the range 02–49.[1]

The DETAIL–LINE in this report is similar to the one in Program 22. But here we have some output editing. Note that the PICTUREs included in these descriptions are PICTUREs of the output fields as they are to print. Any PICTURE features may be used in the Report Section, including floating signs, check protection, and insertion, to obtain any kind of output editing.

The last 01-level entry describes the final total line. Since final total is considered a control break, the final total line must be described as CONTROL FOOTING FINAL.[2] The clause LINE PLUS 3 says that

---

1. An additional feature in some older COBOL systems provides that whenever there is only one field (on one line) in a report group, its description may be included at the 01 level.
2. Some COBOL systems take the presence of CONTROL FOOTING FINAL as sufficient indication that a final total line is to print and do not require the CONTROL FINAL clause in the RD entry.

the final total line is to print on whatever LINE the previous report group was printed, PLUS 3.

In the final total line we see the use of the **SUM** clause, which tells Report Writer that a particular field is to be printed as the SUM of the values of the field named. For example, the clause SUM MERCHANDISE–AMOUNT–W will cause Report Writer to set up an accumulator for the total of the MERCHANDISE–AMOUNT–W amounts. Whenever a GENERATE statement is executed for this report, Report Writer adds the value of MERCHANDISE–AMOUNT–W into the accumulator. The accumulator is defined by Report Writer as purely numeric. Editing the accumulator value takes place only just before the total is printed. The SUM clause may appear only in a CONTROL FOOTING report group.

The Procedure Division of this program has very little that is new. The MAIN–PROCESS routine shows that any regular COBOL processing may be done on an input record. In this case some arithmetic is done, and the results are assigned to working storage fields which are then used as SOURCE fields in the Report Section. Notice that nowhere in the Procedure Division do we tell Report Writer when to print a PAGE HEADING group or when to print the final total line,

Figure 10.3 Program 23. Part 2 of 2.

```
000920
000930 PROCEDURE DIVISION.
000940     OPEN INPUT  ORDER-FILE-IN,
000950          OUTPUT ORDER-REPORT.
000960     READ ORDER-FILE-IN
000970         AT END
000980             MOVE 'NO' TO MORE-INPUT.
000990     IF MORE-INPUT = 'YES'
001000         INITIATE DAILY-ORDER-REPORT
001010         PERFORM MAIN-PROCESS UNTIL MORE-INPUT = 'NO'
001020         TERMINATE DAILY-ORDER-REPORT.
001030     CLOSE ORDER-FILE-IN,
001040           ORDER-REPORT.
001050     STOP RUN.
001060
001070 MAIN-PROCESS.
001080     MULTIPLY QUANTITY-IN BY UNIT-PRICE-IN
001090                                  GIVING MERCHANDISE-AMOUNT-W.
001100     MULTIPLY MERCHANDISE-AMOUNT-W BY TAX-RATE
001110                                  GIVING TAX-W ROUNDED.
001120     ADD MERCHANDISE-AMOUNT-W,
001130         TAX-W,
001140         HANDLING-IN
001150             GIVING ORDER-TOTAL-W.
001160     GENERATE DETAIL-LINE.
001170     READ ORDER-FILE-IN
001180         AT END
001190             MOVE 'NO' TO MORE-INPUT.
```

nor do we code any of the logic for accumulating the sums or printing them. All of that is taken care of by Report Writer from information we provided in the Report Section.

Since we never had to refer to the PAGE HEADING group or the final total line group by name, their 01-level entries were written without names. The only group that needed a name was the DETAIL group, so that it could be referred to in the statement GENERATE DETAIL–LINE. An 01-level entry needs a name only if the name is referred to in a GENERATE statement, in a **USE BEFORE REPORTING** statement (which we will discuss briefly at the end of this chapter), or in an entry elsewhere in the Report Section.

Program 23 was run on the same input data as Program 10 and produced the output shown in Figure 10.4.

Figure 10.4 Output from Program 23.

DAILY ORDER REPORT

CUSTOMER NUMBER	PART NUMBER	QUANTITY	UNIT PRICE	MERCHANDISE AMOUNT	TAX	HANDLING	TOTAL
ABC1234	F2365-09	900	.10	90.00	6.30	.05	96.35
09G8239	836-7Y17	800	10.50	8,400.00	588.00	.50	8,988.50
ADGH784	091AN-07	50	250.00	12,500.00	875.00	5.00	13,380.00
9675473	23S-1287	6	7,000.29	42,001.74	2,940.12	50.00	44,991.86
			TOTALS	62,991.74	4,409.42	55.55	67,456.71

Exercise 2. Rewrite your solution for Exercise 9, Chapter 4, page 88, using Report Writer.

Multiple Control Breaks Using Report Writer

Program 14 has page and column headings, three levels of control breaks, the date, page overflow, page numbering, and group indication of the control fields. The materials for reviewing Program 14 are: input format, Figure 6.11; output, Figure 6.12; input data, Figure 6.13; output format, Figure 6.14; Program 14, Figure 6.17. Program 24, which has all the same features as Program 14, is shown in Figure 10.5.

In the Working Storage Section, the field TODAYS–DATE is set up to be used in the usual way. In the Procedure Division we will ACCEPT the DATE into TODAYS–DATE, and the fields TODAYS–YEAR, TODAYS–MONTH, and TODAYS–DAY will be used as SOURCE fields in the Report Section for printing the date.

Figure 10.5 Program 24, Part 1 of 4.

```
4-CB2 V4 RELEASE 1.5 10NOV77          IBM OS AMERICAN NATIONAL STANDARD COBOL

000010 IDENTIFICATION DIVISION.
000020 PROGRAM-ID. PROG24.
000030*AUTHOR. ELLEN M. LONGO.
000040*
000050*     THIS PROGRAM IS A MODIFICATION
000060*     TO PROGRAM 14 USING REPORT WRITER.
000070*
000080*****************************************************************
000090
000100 ENVIRONMENT DIVISION.
000110 INPUT-OUTPUT SECTION.
000120 FILE-CONTROL.
000130      SELECT SALES-FILE-IN ASSIGN TO UT-S-SYSIN.
000140      SELECT SALES-REPORT  ASSIGN TO UT-S-SYSPRINT.
000150
000160*****************************************************************
000170
000180 DATA DIVISION.
000190 FILE SECTION.
000200 FD   SALES-FILE-IN
000210      LABEL RECORDS ARE OMITTED.
000220
000230 01   SALES-RECORD-IN.
000240      05 STORE-NUMBER-IN                             PIC 9(3).
000250      05 SALESPERSON-NUMBER-IN                       PIC 9(3).
000260      05 CUSTOMER-NUMBER                             PIC 9(6).
000270      05 SALE-AMOUNT                                 PIC 9(4)V99.
000280      05 FILLER                                      PIC X(62).
000290
000300 FD   SALES-REPORT
000310      LABEL RECORDS ARE OMITTED
000320      REPORT IS SALES-REPORT-OUT.
000330
000340 WORKING-STORAGE SECTION.
000350 01   MORE-INPUT          VALUE   'YES'              PIC X(3).
000360
000370 01   TODAYS-DATE.
000380      05 TODAYS-YEAR                                 PIC 99.
000390      05 TODAYS-MONTH                                PIC 99.
000400      05 TODAYS-DAY                                  PIC 99.
000410
000420 REPORT SECTION.
000430 RD   SALES-REPORT-OUT
000440      CONTROLS FINAL, STORE-NUMBER-IN, SALESPERSON-NUMBER-IN
000450      PAGE 50 LINES, FIRST DETAIL 9, LAST DETAIL 42, FOOTING 48.
000460
000470 01   TYPE PAGE HEADING.
000480      05 LINE 2.
000490         10 COLUMN 36  VALUE  'SALES REPORT'     PIC X(12).
```

In the RD entry, we now must provide Report Writer with more information than in earlier programs. The CONTROL clause must now name all the levels of control breaks from the highest level to the most minor, in that order. Since a FINAL total is considered a control break, it must be named along with the two control fields STORE–NUMBER–IN and SALESPERSON–NUMBER–IN.[3]

Since we have page overflow in this program, the PAGE clause must be more elaborate than before. The size of the overall page is still arbitrarily given as 50. From the printer spacing chart, you can see that the FIRST DETAIL line is on line 9, so the PAGE clause includes FIRST DETAIL 9. We now include the **LAST DETAIL** clause to tell Report Writer on which line of the page it may print the last detail line. Here, LAST DETAIL 42 will give us approximately the same depth of page that we had in Program 14. Now remember from Program 14 that we didn't test for page overflow while printing total lines because we didn't want the total lines to cause page skipping. Report Writer recognizes that programmers will usually want to leave extra space at the bottom of the page for total lines; the **FOOTING** clause allows for this. With the FOOTING clause we can specify the last line of the page on which total lines (CONTROL FOOTING lines) may print. Since we can have a maximum of six lines of totals (the last set of totals: a salesperson total, a store total, and a final total, all double-spaced), we make the FOOTING limit 48. The depth of the full page, which we here made 50, could have been used to allow space beyond the limit of the total lines for PAGE FOOTINGs and a REPORT FOOTING. We have neither PAGE FOOTING nor REPORT FOOTING, so the depth of page could have been any number 48 or larger and the program would work the same way.

In the PAGE HEADING report group, LINE 3 is the most interesting. You can see how the fields TODAYS–MONTH, TODAYS–DAY, and TODAYS–YEAR are used to print the date. The SOURCE field used for printing the page number is a special register called **PAGE–COUNTER.** Report Writer sets up a PAGE–COUNTER for every RD entry. The INITIATE statement for a report sets the PAGE–COUNTER to 1, and then Report Writer adds 1 to PAGE–COUNTER every time page headings are printed. PAGE–COUNTER may be referred to and changed by ordinary Procedure Division statements. If a program contains more than one PAGE–COUNTER (because there is more than one RD entry in the program), then PAGE–COUNTER must be qualified by the RD name whenever it is referred to in the Procedure Division or in a report description other than the one it was set up for. If a PAGE–COUNTER is referred

---

3. In those systems where the presence of a CONTROL FOOTING FINAL line is sufficient indication that a FINAL total line is to be printed, the word FINAL may be omitted from the CONTROL clause of the RD entry.

Figure 10.5 Program 24, Part 2 of 4.

```
000500        05 LINE 3.
000510           10 COLUMN 20    VALUE   'DATE'                PIC X(4).
000520           10 COLUMN 25    SOURCE  TODAYS-MONTH          PIC Z9.
000530           10 COLUMN 27    VALUE   '/'                   PIC X.
000540           10 COLUMN 28    SOURCE  TODAYS-DAY            PIC 99.
000550           10 COLUMN 30    VALUE   '/'                   PIC X.
000560           10 COLUMN 31    SOURCE  TODAYS-YEAR           PIC 99.
000570           10 COLUMN 62    VALUE   'PAGE'                PIC X(4).
000580           10 COLUMN 67    SOURCE  PAGE-COUNTER          PIC Z9.
000590        05 LINE 6.
000600           10 COLUMN 25    VALUE   'STORE'               PIC X(5).
000610           10 COLUMN 34    VALUE   'SALES-'              PIC X(6).
000620           10 COLUMN 44    VALUE   'CUSTOMER'            PIC X(8).
000630           10 COLUMN 57    VALUE   'SALE'                PIC X(4).
000640        05 LINE 7.
000650           10 COLUMN 28    VALUE   'NO.'                 PIC X(3).
000660           10 COLUMN 34    VALUE   'PERSON'              PIC X(6).
000670           10 COLUMN 45    VALUE   'NUMBER'              PIC X(6).
000680           10 COLUMN 56    VALUE   'AMOUNT'              PIC X(6).
000690
000700 01     DETAIL-LINE
000710        TYPE DETAIL
000720        LINE PLUS 1.
000730        05 COLUMN 26         SOURCE STORE-NUMBER-IN      PIC 9(3)
000740                                                         GROUP INDICATE.
000750        05 COLUMN 35         SOURCE SALESPERSON-NUMBER-IN
000760                                                         PIC 9(3)
000770                                                         GROUP INDICATE.
000780        05 COLUMN 45         SOURCE CUSTOMER-NUMBER      PIC 9(6).
000790        05 COLUMN 55         SOURCE SALE-AMOUNT          PIC Z,ZZZ.99.
```

to only in its own report description, as ours is, it does not have to be qualified.

Report Writer also sets up a special register called LINE–COUNTER for every RD entry.[4] The LINE–COUNTER is used by Report Writer to position print lines on the page and to detect page overflow. LINE–COUNTER may be referred to in the Procedure Division but may not be changed. An INITIATE statement for a report sets LINE–COUNTER to 0. If there is more than one LINE–COUNTER in a program, the name LINE–COUNTER must be qualified by the RD name whenever it is referred to in the Procedure Division or outside of its own report description.

In this report description we have three different CONTROL FOOTING lines, one each for a control break on SALESPERSON–NUMBER–IN and STORE–NUMBER–IN and for a FINAL total. Each total line is defined with its own 01-level entry.

4. In some older COBOL systems, a LINE–COUNTER and a PAGE–COUNTER are set up only if the RD entry contains a PAGE clause.

Figure 10.5 Program 24, Part 3 of 4.

```
000800
000810 01   TYPE CONTROL FOOTING SALESPERSON-NUMBER-IN
000820      LINE PLUS 2
000830      NEXT GROUP PLUS 1.
000840      05 SALESPERSON-TOTAL
000850         COLUMN 54      SUM     SALE-AMOUNT          PIC ZZ,ZZZ.99.
000860      05 COLUMN 64      VALUE   '*'                  PIC X.
000870
000880 01   TYPE CONTROL FOOTING STORE-NUMBER-IN
000890      LINE PLUS 2
000900      NEXT GROUP PLUS 2.
000910      05 COLUMN 27      VALUE   'TOTAL FOR STORE NO.'
000920                                                     PIC X(19).
000930      05 COLUMN 47      SOURCE  STORE-NUMBER-IN      PIC 9(3).
000940      05 STORE-TOTAL
000950         COLUMN 52      SUM     SALESPERSON-TOTAL    PIC $ZZZ,ZZZ.99.
000960      05 COLUMN 64      VALUE   '**'                 PIC X(2).
000970
000980 01   TYPE CONTROL FOOTING FINAL
000990      LINE PLUS 2.
001000      05 COLUMN 35      VALUE   'GRAND TOTAL'        PIC X(11).
001010      05 COLUMN 50      SUM     STORE-TOTAL          PIC $Z,ZZZ,ZZZ.99.
001020      05 COLUMN 64      VALUE   '***'                PIC X(3).
001030
```

THE MINOR TOTAL LINE

The clause LINE PLUS 2 that appears in the description of the salesperson total line causes the total line to be double-spaced from the last detail line before it. In the description of the salesperson total line, we see for the first time the use of the **NEXT GROUP** clause. The NEXT GROUP clause is used for spacing or skipping the report output after printing a report group. Here we use NEXT GROUP PLUS 1 to give one blank line after a salesperson total line. When a NEXT GROUP clause appears in a CONTROL FOOTING group, as this one does, it is effective only for a control break at the same level as the CONTROL FOOTING line; that is, this NEXT GROUP PLUS 1 is effective only when a salesperson total line is being printed because of a salesperson break. The clause is ignored when a salesperson total line is printed as a result of a break on store number or at end-of-file. This is an extremely useful feature and allows the programmer great flexibility in spacing output lines of the report.

The total field itself is described as being the SUM of all the SALE–AMOUNT fields. The total field is given the name SALESPERSON–TOTAL. A SUMmed field has to be given a name only if it referred to somewhere in the Procedure Division or elsewhere in the Report Section. In earlier programs our SUMmed fields were not referred to, and so they were not given names.

THE INTERMEDIATE TOTAL LINE

The CONTROL FOOTING for the store total line follows. The clause LINE PLUS 2 will give double spacing for this total line. When a break on store number occurs, both a salesperson total line and a store total line print. First the salesperson total line prints, and the clause LINE PLUS 2 in the description of the salesperson total line causes it to be double-spaced from the last detail line. The clause NEXT GROUP PLUS 1 is ignored. The store total line is printed, and the clause LINE PLUS 2 in the description of the store total line causes it to be double-spaced from the salesperson total line.

In the description of the store total line is the clause NEXT GROUP PLUS 2. This will give two blank lines following the printing of a store total line, but only if the line is printed as a result of a break on store number. When the store total line is printed at end-of-file, the NEXT GROUP clause is ignored. This permits us to get exactly the spacing we want for the store total line. You can see from the printer spacing chart and the report output that after a store total line other than the last one, we want two blank lines before the detail lines for the next store begin. This line spacing sets off one store group from another. But at the end of the report, when a final total line is also printed, we want only one blank line following the store total line. By ignoring the NEXT GROUP clause at the right time, Report Writer enables the programmer to control this kind of variability in line spacing.

In the store total line, the total field is described as being the SUM of all the SALESPERSON–TOTAL values. As we have already seen, SALESPERSON–TOTAL is itself a SUM field and defined in the Report Section. Whenever a SUMmed field like SALESPERSON–TOTAL is itself used as the operand in some other SUM clause, it must either be defined at a level lower than where it is used in the control break hierarchy or at the same level. Our SUM fields meet that condition: SALESPERSON–TOTAL is defined in the salesperson line, which is at a lower level than the store total line, where it is used. The field which is the SUM of all the SALESPERSON–TOTAL amounts is given the name STORE–TOTAL. This name will be used elsewhere in the Report Section.

The field whose SOURCE is STORE–NUMBER–IN shows that when a control break occurs, the previous value of the control field is used in producing the total lines. That is, even though some new value of STORE–NUMBER–IN caused the break and is already assigned to the input area, Report Writer has saved and uses the correct, previous value.

THE FINAL TOTAL LINE

The total of all the totals is described in the description of CONTROL FOOTING FINAL as the SUM of all the STORE–TOTAL values. The grand total field itself needs no name, for it is not referred to anywhere.

Figure 10.5 Program 24, Part 4 of 4.

```
001040*****************************************************************
001050
001060 PROCEDURE DIVISION.
001070     OPEN INPUT  SALES-FILE-IN,
001080          OUTPUT SALES-REPORT.
001090     READ SALES-FILE-IN
001100         AT END
001110             MOVE 'NO' TO MORE-INPUT.
001120     IF MORE-INPUT = 'YES'
001130         INITIATE SALES-REPORT-OUT
001140         ACCEPT TODAYS-DATE FROM DATE
001150         PERFORM MAIN-PROCESS UNTIL MORE-INPUT = 'NO'
001160         TERMINATE SALES-REPORT-OUT.
001170     CLOSE SALES-FILE-IN,
001180           SALES-REPORT.
001190     STOP RUN.
001200
001210 MAIN-PROCESS.
001220     GENERATE DETAIL-LINE.
001230     READ SALES-FILE-IN
001240         AT END
001250             MOVE 'NO' TO MORE-INPUT.
```

THE PROCEDURE DIVISION

The single GENERATE statement in MAIN–PROCESS creates nearly the entire report. When the first GENERATE is issued, Report Writer prints the page and column headings and then the first detail line. After that, for every GENERATE statement that is issued, Report Writer checks for control breaks and page overflow, and takes the action indicated in the Report Section.

It is illegal to try to issue a GENERATE statement for a PAGE HEADING group or a CONTROL FOOTING group. Nobody tells Report Writer when to print page headings and total lines, for all the necessary logic is built in and tested. Program 24 was run with the same input as Program 14 and produced the output shown in Figure 10.6.

The GENERATE Statement

The general format of the GENERATE statement is:

```
GENERATE { data-name   }
         { report-name }
```

Figure 10.6 Output from Program 24, Part 1 of 3.

```
                    SALES REPORT
      DATE   2/13/80                                       PAGE  1

           STORE      SALES-      CUSTOMER       SALE
             NO.      PERSON       NUMBER       AMOUNT

           010        101         003001      1,234.56
                                  007002          2.24
                                  011003      6,665.70
                                  039004      8,439.20
                                  046005      8,446.48
                                  053006         12.34
                                  060006      9,494.93
                                  067007      4,000.03

                                             38,297.48 *

           010        102         074212      5,454.99
                                  081012           .33
                                  088013      5,849.58
                                  095015        393.90

                                             11,698.80 *

           010        103         003234        303.03

                                                303.03 *

            TOTAL FOR STORE NO. 010     $ 50,299.31 **

           020        011         007567      9,999.99
                                  011454        456.00
                                  015231      8,484.39
                                  019345      8,459.44
                                  023345      8,333.33

                                             35,733.15 *

           020        222         027345      4,343.43
```

If GENERATE data–name is used, data–name must be the name of a DETAIL group in the Report Section. These are the kinds of GENERATE statements we have been using. If GENERATE report–name is used, report–name must be the name given in the RD entry. In this form, Report Writer does everything except print detail lines; that is, it carries out all the summing, control break, and page overflow processing that it otherwise would. This is called group printing or summary reporting.

Figure 10.6 Output from Program 24, Part 2 of 3.

```
                         SALES REPORT
          DATE  2/13/80                                    PAGE  2

                 STORE      SALES-      CUSTOMER        SALE
                   NO.      PERSON      NUMBER        AMOUNT

                  020        222         031567      9,903.30
                                         035001         34.21

                                                    14,280.94 *

                  020        266         039903      4,539.87
                                         043854      5,858.30

                                                    10,398.17 *

                   TOTAL FOR STORE NO. 020       $ 60,412.26 **

                  030        193         047231      9,391.93
                                         051342      5,937.43
                                         055034      9,383.22
                                         059932      5,858.54
                                         063419      3,949.49

                                                    34,520.61 *

                   TOTAL FOR STORE NO. 030       $ 34,520.61 **

                  040        045         067333           .00
                                         071323      5,959.50

                                                     5,959.50 *

                  040        403         048399      3,921.47
                                         054392           .00

                                                     3,921.47 *
```

Figure 10.6 Output from Program 24, Part 3 of 3.

```
                    SALES REPORT
DATE  2/13/80                                    PAGE  3


          STORE      SALES-      CUSTOMER       SALE
            NO.      PERSON       NUMBER       AMOUNT

           040        412         060111     9,999.99

                                             9,999.99 *

             TOTAL FOR STORE NO. 040  $ 19,880.96 **

           046        012         013538          .00
                                  017521       690.78
                                  021504     1,381.56
                                  025487     2,072.34
                                  029470     2,763.12

                                             6,907.80 *

           046        028         033453     3,453.90
                                  037436     4,144.68
                                  041419     4,835.46

                                            12,434.04 *

             TOTAL FOR STORE NO. 046  $ 19,341.84 **

                  GRAND TOTAL      $  184,454.98 ***
```

Exercise 3. Rewrite the program in Exercise 2, Chapter 6, page 139, using Report Writer.

Exercise 4. Rewrite the program in Exercise 4, Chapter 6, page 152, using Report Writer.

Processing After End-of-File

In Program 25, a modification to Program 12, you will see different kinds of processing before and after end-of-file and other features of Report Writer. To review Program 12, see the input format for Programs 11 and 12, Figure 5.1; the input data for Programs 11 and

12, Figure 5.4; the output format, Figure 5.6; Program 12, Figure 5.10, and the output, Figure 5.11.

Program 25 is shown in Figure 10.7. The approach originally taken in Program 12 has been changed slightly for Program 25. When Program 12 was first done we didn't yet know how to use subscripting, but subscripting can be used to advantage in Program 25 and is. The benefits to be gained from subscripting (or indexing) with Report Writer are limited, though, because COBOL does not permit the use of the OCCURS clause in the Report Section.

Looking now at the Working Storage Section of Program 25, we see several numeric fields defined. The field ONE is defined so that Report Writer can count the number of students. The uses of the other fields will become clear when we look at the Procedure Division.

In the Report Section, the description of DETAIL–LINE shows that SOURCE fields may be subscripted. They may also be indexed and the subscript or index given as a literal, as it is here, or as an identifier.

DETAIL–LINE contains a **nonprintable item,** the 05-level entry whose SOURCE is ONE. The item is considered nonprintable by Report Writer because it lacks a COLUMN clause. It is included in this entry because of a peculiar requirement in some of the older COBOL systems: In order for an item to be added into a SUM accumulator, it must appear as a SOURCE item in a DETAIL or CONTROL FOOTING report group. Then, the item is added into its SUM accumulator whenever we print the DETAIL group or CONTROL FOOTING group where the item appears as a SOURCE. You will see shortly how the SUMming of ONE counts the number of students.

The CONTROL FOOTING FINAL group contains several nonprintable items, the ACCUMULATORS, all lacking their COLUMN clauses. The ACCUMULATORS use the SUM clause to total the grades on the four exams and to count the number of students. We never want to print the sums as part of the report output, though. We print the averages of the exam grades, not their sums; we need the sums only to compute the averages. We will refer to the individual ACCUMULATORS by name in the Procedure Division when we compute the class averages on the four exams.

You can see that one of the ACCUMULATORS, namely NUMBER–OF–STUDENTS, contains the clause SUM ONE. On the newer COBOL systems, that SUM accumulator would work properly wherever ONE was defined, but on the older systems ONE must appear as a SOURCE in order for the SUM to work (or else the **UPON** option, discussed in the next section, must be used). Even on the newer systems the appearance of ONE as a SOURCE field does no harm.

Most of the Procedure Division is self-explanatory. In MAIN–PROCESS, the only arithmetic we need to write explicitly is the computation of the student's average grade on the four exams. Report Writer doesn't do division. But it does addition, and, when the

Figure 10.7 Program 25, Part 1 of 2.

```
4-CB2 V4 RELEASE 1.5 10NOV77        IBM OS AMERICAN NATIONAL STANDARD COBOL

000010 IDENTIFICATION DIVISION.
000020 PROGRAM-ID, PROG25.
000030*AUTHOR. PHIL RUBENSTEIN.
000040*
000050*       THIS PROGRAM IS A MODIFICATION TO  PROGRAM 12
000060*       USING REPORT WRITER.
000070*
000080*****************************************************************
000090
000100 ENVIRONMENT DIVISION.
000110 CONFIGURATION SECTION.
000140 INPUT-OUTPUT SECTION.
000150 FILE-CONTROL.
000160     SELECT EXAM-GRADE-FILE-IN     ASSIGN TO UT-S-SYSIN.
000170     SELECT STUDENT-GRADE-REPORT  ASSIGN TO UT-S-SYSPRINT.
000180
000190*****************************************************************
000200
000210 DATA DIVISION.
000220 FILE SECTION.
000230 FD  EXAM-GRADE-FILE-IN
000240     LABEL RECORDS ARE OMITTED.
000250
000260 01  EXAM-GRADE-RECORD.
000270     05 STUDENT-NUMBER-IN  PIC X(9).
000280     05 FILLER             PIC X(6).
000290     05 GRADE-IN
000300        OCCURS 4 TIMES     PIC 9(3).
000310     05 FILLER             PIC X(51).
000320
000330 FD  STUDENT-GRADE-REPORT
000340     LABEL RECORDS ARE OMITTED
000350     REPORT IS GRADE-REPORT.
000360
000370 WORKING-STORAGE SECTION.
000380 01  MORE-INPUT            PIC X(3) VALUE 'YES'.
000390
000400 01  RUN-DATE.
000410     05 RUN-YEAR           PIC 99.
000420     05 RUN-MONTH          PIC 99.
000430     05 RUN-DAY            PIC 99.
000440
000450 01  NUMBER-OF-EXAMS       PIC S9    VALUE 4.
000460 01  ONE                   PIC S9    VALUE 1.
000470 01  STUDENT-AVERAGE-W     PIC S9(3)V9.
000480 01  CLASS-AVERAGE-W       PIC S9(3)V9.
000490 01  EXAM-AVERAGES.
000500     05 EXAM-AVERAGE
000510        OCCURS 4 TIMES     PIC S9(3)V9.
```

Figure 10.7 Program 25, Part 2 of 2.

```
000520
000530 REPORT SECTION.
000540 RD  GRADE-REPORT
000550     CONTROL FINAL
000560     PAGE 50 LINES, FIRST DETAIL 9.
000570
000580 01  TYPE PAGE HEADING.
000590     05 LINE 4.
000600        10 COLUMN 36 VALUE 'CLASS AVERAGE REPORT' PIC X(20).
000610        10 COLUMN 75 VALUE 'DATE'                 PIC X(4).
000620        10 COLUMN 80 SOURCE RUN-MONTH             PIC Z9.
000630        10 COLUMN 82 VALUE '/'                    PIC X.
000640        10 COLUMN 83 SOURCE RUN-DAY               PIC 99.
000650        10 COLUMN 85 VALUE '/'                    PIC X.
000660        10 COLUMN 86 SOURCE RUN-YEAR              PIC 99.
000670     05 LINE 6.
000680        10 COLUMN 21 VALUE 'STUDENT'              PIC X(7).
000690        10 COLUMN 45 VALUE 'G R A D E S'          PIC X(11).
000700     05 LINE 7.
000710        10 COLUMN 21 VALUE 'NUMBER'               PIC X(6).
000720        10 COLUMN 37 VALUE 'EXAM 1'               PIC X(6).
000730        10 COLUMN 45 VALUE 'EXAM 2'               PIC X(6).
000740        10 COLUMN 53 VALUE 'EXAM 3'               PIC X(6).
000750        10 COLUMN 61 VALUE 'EXAM 4'               PIC X(6).
000760        10 COLUMN 72 VALUE 'AVERAGE'              PIC X(7).
000770
000780 01  DETAIL-LINE
000790     TYPE DETAIL
000800     LINE PLUS 1.
000810     05 COLUMN 20 SOURCE STUDENT-NUMBER-IN        PIC X(9).
000820     05 COLUMN 38 SOURCE GRADE-IN (1)             PIC ZZ9.
000830     05 COLUMN 46 SOURCE GRADE-IN (2)             PIC ZZ9.
000840     05 COLUMN 54 SOURCE GRADE-IN (3)             PIC ZZ9.
000850     05 COLUMN 62 SOURCE GRADE-IN (4)             PIC ZZ9.
000860     05 COLUMN 72 SOURCE STUDENT-AVERAGE-W        PIC ZZZ.9.
000870     05           SOURCE ONE                      PIC S9.
000880
000890 01  TYPE CONTROL FOOTING FINAL.
000900     05 ACCUMALTORS.
000910        10 NUMBER-OF-STUDENTS SUM ONE             PIC S9(3).
000920        10 EXAM-1-SUM         SUM GRADE-IN (1)    PIC S9(5).
000930        10 EXAM-2-SUM         SUM GRADE-IN (2)    PIC S9(5).
000940        10 EXAM-3-SUM         SUM GRADE-IN (3)    PIC S9(5).
000950        10 EXAM-4-SUM         SUM GRADE-IN (4)    PIC S9(5).
000960     05 LINE PLUS 3.
000970        10 COLUMN 23 VALUE 'AVERAGES'             PIC X(8).
000980        10 COLUMN 38 SOURCE EXAM-AVERAGE (1)      PIC ZZZ.9.
000990        10 COLUMN 46 SOURCE EXAM-AVERAGE (2)      PIC ZZZ.9.
001000        10 COLUMN 54 SOURCE EXAM-AVERAGE (3)      PIC ZZZ.9.
001010        10 COLUMN 62 SOURCE EXAM-AVERAGE (4)      PIC ZZZ.9.
001020        10 COLUMN 72 SOURCE CLASS-AVERAGE-W       PIC ZZZ.9.
001030
```

Figure 10.8 Part 2 (continued).

```
001040*****************************************************************
001050
001060 PROCEDURE DIVISION.
001070     OPEN INPUT  EXAM-GRADE-FILE-IN,
001080          OUTPUT STUDENT-GRADE-REPORT.
001090     READ EXAM-GRADE-FILE-IN
001100         AT END
001110             MOVE 'NO' TO MORE-INPUT.
001120     IF MORE-INPUT = 'YES'
001130         PERFORM INITIALIZATION
001140         PERFORM MAIN-PROCESS UNTIL MORE-INPUT =  'NO'
001150         PERFORM FINAL-LINE-PROCESS.
001160     CLOSE EXAM-GRADE-FILE-IN,
001170           STUDENT-GRADE-REPORT.
001180     STOP RUN.
001190
001200 INITIALIZATION.
001210     INITIATE GRADE-REPORT.
001220     ACCEPT RUN-DATE FROM DATE.
001230
001240 MAIN-PROCESS.
001250     COMPUTE STUDENT-AVERAGE-W ROUNDED =
001260         (GRADE-IN (1) +
001270          GRADE-IN (2) +
001280          GRADE-IN (3) +
001290          GRADE-IN (4)) / NUMBER-OF-EXAMS.
001300     GENERATE DETAIL-LINE.
001310     READ EXAM-GRADE-FILE-IN
001320         AT END
001330             MOVE 'NO' TO MORE-INPUT.
001340
001350 FINAL-LINE-PROCESS.
001360     DIVIDE NUMBER-OF-STUDENTS INTO EXAM-1-SUM
001370                                     GIVING EXAM-AVERAGE (1).
001380     DIVIDE NUMBER-OF-STUDENTS INTO EXAM-2-SUM
001390                                     GIVING EXAM-AVERAGE (2).
001400     DIVIDE NUMBER-OF-STUDENTS INTO EXAM-3-SUM
001410                                     GIVING EXAM-AVERAGE (3).
001420     DIVIDE NUMBER-OF-STUDENTS INTO EXAM-4-SUM
001430                                     GIVING EXAM-AVERAGE (4).
001440     COMPUTE CLASS-AVERAGE-W ROUNDED =
001450         (EXAM-1-SUM +
001460          EXAM-2-SUM +
001470          EXAM-3-SUM +
001480          EXAM-4-SUM) / (NUMBER-OF-STUDENTS * NUMBER-OF-EXAMS).
001490     TERMINATE GRADE-REPORT.
```

GENERATE statement is executed, Report Writer accumulates the four grades and counts the student as indicated in the SUM clauses.

The paragraph FINAL–LINE–PROCESS shows that we can do some processing after end-of-file has been detected and before the TERMINATE statement is given. At that time, all the SUMs we need to compute the averages for the last line of the report are in the ACCUMULATORS. We first compute the class averages for the four exams with four DIVIDE statements. Then we COMPUTE the average of all the averages using the same computation method as in Program 12. The TERMINATE statement causes the final line to be printed. Program 25 was run with the same input as Program 12 and produced the output shown in Figure 10.8.

Figure 10.9 Output from Program 25.

```
                      CLASS AVERAGE REPORT                              DATE  2/13/80

   STUDENT                      G R A D E S
   NUMBER              EXAM 1   EXAM 2   EXAM 3   EXAM 4          AVERAGE

  070543242              100       78       98       84             90.0
  091020222               90       85       98        0             68.3
  075655343               22       67       76       57             55.5
  513467845               76       83       82       92             83.3


       AVERAGES           72.0     78.2     88.5     58.2           74.3
```

Exercise 5. Rewrite your solution to Exercise 2, Chapter 5, page 108, using Report Writer.

The UPON Option

Program 17 carries out validity checks on input and also checks for arithmetic overflow; it prints error messages mixed with regular output lines; it accumulates numeric data from error-free records and prints totals. Review Program 17 (input format, Figure 4.RE5.1; output format, Figure 8.6; Program 17, Figure 8.7; input data, Figure 8.8; output, Figure 8.9). Because it uses the OCCURS clause and the capabilities of the Report Writer, Program 26 is much more compact and readable than Program 17. Program 26 is shown in Figure 10.9.

Figure 10.9 Program 26, Part 1 of 3.

```
4-CB2 V4 RELEASE 1.5 10NOV77          IBM OS AMERICAN NATIONAL STANDARD COBOL

000010 IDENTIFICATION DIVISION.
000020 PROGRAM-ID. PROG26.
000030*AUTHOR.  E.M.L.
000040*
000050*     THIS PROGRAM IS A MODIFICATION
000060*     TO PROGRAM 17 USING REPORT WRITER.
000070*
000080*****************************************************************
000090
000100 ENVIRONMENT DIVISION.
000110 INPUT-OUTPUT SECTION.
000120 FILE-CONTROL.
000130     SELECT EMPLOYEE-LABOR-FILE-IN ASSIGN TO UT-S-SYSIN.
000140     SELECT LABOR-HOURS-REPORT     ASSIGN TO UT-S-SYSPRINT.
000150
000160*****************************************************************
000170
000180 DATA DIVISION.
000190 FILE SECTION.
000200 FD  EMPLOYEE-LABOR-FILE-IN
000210     LABEL RECORDS ARE OMITTED.
000220
000230 01  LABOR-RECORD-IN.
000240     05 SOCIAL-SECURITY-NUMBER                     PIC X(9).
000250     05 ALL-HOURS.
000260        10 HOURS OCCURS 6 TIMES                    PIC 99V9.
000270     05 ALL-HOURS-X REDEFINES ALL-HOURS.
000280        10 HOURS-X OCCURS 6 TIMES                  PIC X(3).
000290     05 FILLER                                     PIC X(53).
000300
000310 FD  LABOR-HOURS-REPORT
000320     LABEL RECORDS ARE OMITTED
000330     REPORT IS LABOR-REPORT.
000340
000350 WORKING-STORAGE SECTION.
000360 01  MORE-INPUT          VALUE  'YES'              PIC X(3).
000370 01  IS-THERE-AN-ERROR                             PIC X(3).
000380 01  TOTAL-FOR-WEEK-W                              PIC 99V9.
000390 01  ERROR-MESSAGE                                 PIC X(33).
000400
```

In the File Section, the six fields for the number of hours worked on Monday, Tuesday, Wednesday, Thursday, Friday, and Saturday have been defined with an OCCURS clause. For purposes of handling nonnumeric values in the HOURS fields, the six fields have also been redefined as alphanumeric. The following method of redefinition would have been illegal in COBOL:

05 HOURS OCCURS 6 TIMES PIC 99V9.
05 HOURS–X REDEFINES HOURS OCCURS 6 TIMES PIC X(3).

COBOL does not permit you to redefine a field whose entry contains an OCCURS clause.

In the Report Section we see for the first time more than one DETAIL report group defined. Two of them follow the approach taken in Program 17, where we use the REGULAR–BODY–LINE for records where the HOURS are all found to be numeric, and we use ALPHA–BODY–LINE for printing records containing nonnumeric HOURS fields. The third of the DETAIL groups, SUM–SIGNAL, is a nonprintable group; in fact, it's about as nonprintable as anything can get since it has no LINE clause, no COLUMN clause, and barely anything else. You will see shortly how we use SUM–SIGNAL to signal that we want Report Writer to do some accumulating. In the Procedure Division we will use IF statements to determine which type of DETAIL line should be printed and then issue either GENERATE REGULAR–BODY–LINE or GENERATE ALPHA–BODY–LINE. When the IF statements also determine that the HOURS should be added into the accumulators for the total line we will issue a GENERATE SUM–SIGNAL.

Figure 10.9 Program 26, Part 2 of 3.

```
000410 REPORT SECTION.
000420 RD   LABOR-REPORT
000430      CONTROL FINAL
000440      PAGE 50 LINES, FIRST DETAIL 9.
000450
000460 01   TYPE PAGE HEADING.
000470      05 LINE 3.
000480          10 COLUMN 21    VALUE  'WEEKLY HOURS REPORT'
000490                                                PIC X(19).
000500      05 LINE 6.
000510          10 COLUMN 11    VALUE  'SOC. SEC.'     PIC X(9).
000520          10 COLUMN 29    VALUE  'H O U R S   W O R K E D'
000530                                                PIC X(22).
000540          10 COLUMN 63    VALUE  'TOTAL'         PIC X(5).
000550      05 LINE 7.
000560          10 COLUMN 12    VALUE  'NUMBER'        PIC X(6).
000570          10 COLUMN 23    VALUE  'MON.'          PIC X(4).
000580          10 COLUMN 29    VALUE  'TUE.'          PIC X(4).
000590          10 COLUMN 35    VALUE  'WED.'          PIC X(4).
000600          10 COLUMN 41    VALUE  'THU.'          PIC X(4).
000610          10 COLUMN 48    VALUE  'FRI.'          PIC X(4).
000620          10 COLUMN 54    VALUE  'SAT.'          PIC X(4).
000630          10 COLUMN 61    VALUE  'FOR WEEK'      PIC X(8).
000640
000650 01   REGULAR-BODY-LINE
000660      TYPE DETAIL
000670      LINE PLUS 1.
000680      05 COLUMN 11          SOURCE SOCIAL-SECURITY-NUMBER
000690                                                PIC X(9).
000700      05 COLUMN 23          SOURCE HOURS (1)     PIC Z9.9.
```

Figure 10.9 Part 2 (continued).

```
000710      05 COLUMN 29          SOURCE HOURS (2)         PIC Z9.9.
000720      05 COLUMN 35          SOURCE HOURS (3)         PIC Z9.9.
000730      05 COLUMN 41          SOURCE HOURS (4)         PIC Z9.9.
000740      05 COLUMN 47          SOURCE HOURS (5)         PIC Z9.9.
000750      05 COLUMN 53          SOURCE HOURS (6)         PIC Z9.9.
000760      05 COLUMN 64          SOURCE TOTAL-FOR-WEEK-W  PIC ZZ.Z.
000770      05 COLUMN 70          SOURCE ERROR-MESSAGE     PIC X(33).
000780
000790 01   ALPHA-BODY-LINE
000800      TYPE DETAIL
000810      LINE PLUS 1.
000820      05 COLUMN 11          SOURCE SOCIAL-SECURITY-NUMBER
000830                                                     PIC X(9).
000840      05 COLUMN 24          SOURCE HOURS-X (1)       PIC X(3).
000850      05 COLUMN 30          SOURCE HOURS-X (2)       PIC X(3).
000860      05 COLUMN 36          SOURCE HOURS-X (3)       PIC X(3).
000870      05 COLUMN 42          SOURCE HOURS-X (4)       PIC X(3).
000880      05 COLUMN 48          SOURCE HOURS-X (5)       PIC X(3).
000890      05 COLUMN 54          SOURCE HOURS-X (6)       PIC X(3).
000900      05 COLUMN 70          VALUE  'HOURS WORKED NOT NUMERIC'
000910                                                     PIC X(24).
000920
000930 01   SUM-SIGNAL
000940      TYPE DETAIL.
000950
000960 01   TYPE CONTROL FOOTING FINAL
000970      LINE PLUS 3.
000980      05 COLUMN 15          VALUE  'TOTALS'          PIC X(6).
000990      05 COLUMN 22          SUM    HOURS (1)
001000                            UPON   SUM-SIGNAL        PIC ZZ9.9.
001010      05 COLUMN 28          SUM    HOURS (2)
001020                            UPON   SUM-SIGNAL        PIC ZZ9.9.
001030      05 COLUMN 34          SUM    HOURS (3)
001040                            UPON   SUM-SIGNAL        PIC ZZ9.9.
001050      05 COLUMN 40          SUM    HOURS (4)
001060                            UPON   SUM-SIGNAL        PIC ZZ9.9.
001070      05 COLUMN 46          SUM    HOURS (5)
001080                            UPON   SUM-SIGNAL        PIC ZZ9.9.
001090      05 COLUMN 52          SUM    HOURS (6)
001100                            UPON   SUM-SIGNAL        PIC ZZ9.9.
001110      05 COLUMN 61          SUM    TOTAL-FOR-WEEK-W
001120                            UPON   SUM-SIGNAL        PIC Z,ZZ9.9.
001130
```

In the CONTROL FOOTING FINAL line we see the use of the UPON option of the SUM clause. Remember that in the final line we want the totals of the HOURS in the error-free records only. Records containing errors are not to have their hours added into the totals. By using the UPON option, we here direct Report Writer to accumulate the indicated fields only when a SUM–SIGNAL is GENERATEd. The name given in an UPON phrase must be the name of a DETAIL report group in the same report as the CONTROL FOOTING line where the UPON phrase appears.

The Procedure Division in Program 25 is a mere fraction of the size it was in Program 17 mainly because many of the MOVE and ADD statements of Program 17 are now handled by Report Writer. The form of the two IF statements is still the same as in Program 17, and they carry out the same logic as before. Compare the two programs and see how Program 25 determines whether to GENERATE a REGULAR–BODY–LINE or an ALPHA–BODY–LINE and whether to GENERATE a SUM–SIGNAL. Program 26 was run with the same input data as Program 17 and produced the output shown in Figure 10.10.

Figure 10.9 Program 26, Part 3 of 3.

```
001140*****************************************************************
001150
001160 PROCEDURE DIVISION.
001170     OPEN INPUT  EMPLOYEE-LABOR-FILE-IN,
001180          OUTPUT LABOR-HOURS-REPORT.
001190     READ EMPLOYEE-LABOR-FILE-IN
001200         AT END
001210             MOVE 'NO' TO MORE-INPUT.
001220     IF MORE-INPUT = 'YES'
001230         INITIATE LABOR-REPORT
001240         PERFORM MAIN-PROCESS UNTIL MORE-INPUT = 'NO'
001250         TERMINATE LABOR-REPORT.
001260     CLOSE EMPLOYEE-LABOR-FILE-IN,
001270           LABOR-HOURS-REPORT.
001280     STOP RUN.
001290
001300 MAIN-PROCESS.
001310     MOVE 'NO' TO IS-THERE-AN-ERROR.
001320     IF ALL-HOURS NOT NUMERIC
001330         MOVE 'YES' TO IS-THERE-AN-ERROR
001340         GENERATE ALPHA-BODY-LINE
001350     ELSE
001360         ADD HOURS (1),
001370             HOURS (2),
001380             HOURS (3),
001390             HOURS (4),
001400             HOURS (5),
001410             HOURS (6)
001420                 GIVING TOTAL-FOR-WEEK-W
001430                     SIZE ERROR
001440                         MOVE 'YES' TO IS-THERE-AN-ERROR
001450                         MOVE 0 TO TOTAL-FOR-WEEK-W
001460                         MOVE 'TOTAL FOR WEEK EXCEEDS 99.9 HOURS'
001470                                                 TO ERROR-MESSAGE
001480                         GENERATE REGULAR-BODY-LINE.
001490     IF IS-THERE-AN-ERROR = 'NO'
001500         MOVE SPACES TO ERROR-MESSAGE
001510         GENERATE REGULAR-BODY-LINE
001520         GENERATE SUM-SIGNAL.
001530     READ EMPLOYEE-LABOR-FILE-IN
001540         AT END
001550             MOVE 'NO' TO MORE-INPUT.
```

Figure 10.10 Output from Program 26.

```
        WEEKLY HOURS REPORT

SOC. SEC.             H O U R S  W O R K E D               TOTAL
 NUMBER       MON.  TUE.  WED.   THU.    FRI.  SAT.      FOR WEEK

123344543      8.5   8.4   7.0    6.0    8.0   9.0          46.9
123321202      8.5   4.5   3.6   12.0    0.4   6.0          35.0
020556234      209   023   400    200    309   303                  HOURS WORKED NOT NUMERIC
123456789     80.5   8.5  85.0    3.2    7.3   3.3                  TOTAL FOR WEEK EXCEEDS 99.9 HOURS
987654321      9.5   8.5   7.5    6.5    5.5   4.5          42.0
098765432     12.5  11.5  10.6    9.3    8.4   7.2          59.5
345678901     80.3   2.1   3.2    4.3    5.4   7.6                  TOTAL FOR WEEK EXCEEDS 99.9 HOURS
456789012       85     6    45     65     77    77                  HOURS WORKED NOT NUMERIC
765432198      123   098   094    094    095   090                  HOURS WORKED NOT NUMERIC


     TOTALS   39.0  32.9  28.7   33.8   22.3  26.7         183.4
```

Exercise 6. Rewrite your solution to Exercise 6, Chapter 8, page 217, using Report Writer.

The CONTROL Clause

The general format of the CONTROL clause in the RD entry is:

$$\left\{\begin{array}{l}\underline{\textbf{CONTROL}}\ \textbf{IS}\\ \underline{\textbf{CONTROLS}}\ \textbf{ARE}\end{array}\right\}\left\{\begin{array}{l}\textbf{data-name-1 [data-name-2]} \ldots\\ \underline{\textbf{FINAL}}\ \textbf{[data-name-1 [data-name-2]} \ldots \textbf{]}\end{array}\right\}$$

Data–name–1, data–name–2, and so forth, are the control fields and may be any fields defined in the File Section or Working Storage Section. They may be qualified but not subscripted or indexed. The control fields must be listed in order from the highest level (major) to the lowest (minor). In some COBOL systems, the word FINAL may be omitted, and the presence of a CONTROL FOOTING FINAL line alone will give a FINAL break.

The PAGE Clause

The general format of the PAGE clause in the RD entry is:

$$\underline{\textbf{PAGE}}\left\{\begin{array}{l}\textbf{LIMIT IS}\\ \textbf{LIMITS ARE}\end{array}\right\}\textbf{integer-1}\left\{\begin{array}{l}\underline{\textbf{LINE}}\\ \underline{\textbf{LINES}}\end{array}\right\}$$

$$\textbf{[}\underline{\textbf{HEADING}}\ \textbf{integer-2]}\quad \textbf{[}\underline{\textbf{FIRST}}\ \underline{\textbf{DETAIL}}\ \textbf{integer-3]}$$

$$\textbf{[}\underline{\textbf{LAST}}\ \underline{\textbf{DETAIL}}\ \textbf{integer-4]}\quad \textbf{[}\underline{\textbf{FOOTING}}\ \textbf{integer-5]}$$

In some of the newest COBOL systems, the words LINE and LINES are optional.

It is illegal to provide conflicting or nonsensical information in the PAGE clause. For example, you cannot have a HEADING phrase here that conflicts with any LINE clauses that may appear in the definitions of any TYPE REPORT HEADING or TYPE PAGE HEADING report group. You cannot specify a line number for the FIRST DETAIL line that conflicts with the actual number of lines occupied by the REPORT HEADING or the PAGE HEADING; that is, if a REPORT HEADING and/or PAGE HEADING are to appear on the same page with DETAIL lines, then the headings must not run over into the line reserved for the FIRST DETAIL. If a REPORT HEADING or a REPORT FOOTING is to appear on a page by itself, then it may appear anywhere on the page.

The overall page depth (given by integer–1) must not be less than any of the other integers in the clause; integer–5 must not be less than integer–4; integer–4 must not be less than integer–3; and integer–3 must not be less than integer–2. All integers in the clause must be positive.

As mentioned earlier in this chapter, the line number given in the FOOTING phrase specifies the last line on which CONTROL FOOTING lines may print. Any additional space provided by the overall depth of the page can be used for PAGE FOOTING and REPORT FOOTING GROUPS.

The TYPE Clause

The TYPE clause is the only required clause in an 01-level entry in the Report Section. The general format of the TYPE clause is:

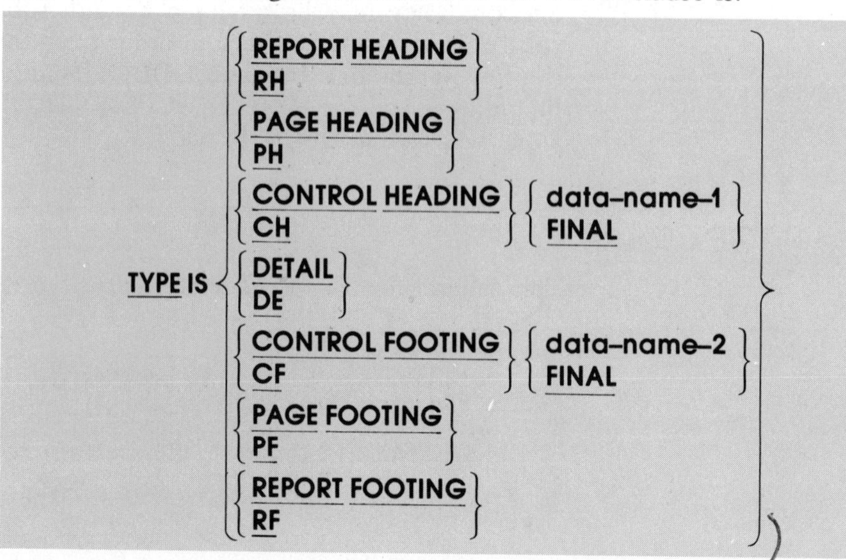

RH is the abbreviation for REPORT HEADING; PH is the abbreviation for PAGE HEADING; CH is the abbreviation for CONTROL HEADING; DE is the abbreviation for DETAIL; CF is the abbreviation for CONTROL FOOTING; PF is the abbreviation for PAGE FOOTING; and RF is the abbreviation for REPORT FOOTING. Data–name–1 and data–name–2 must be names which are given in the CONTROL clause of the RD entry; that is, data–name–1 and data–name–2 must be fields where control breaks are taken. We refrained from using abbreviations in the sample programs because their meanings are not at all obvious, as are PIC and COMP.

A **CONTROL HEADING** report group is one or more lines to be printed before the detail lines of a control group. It usually serves as a sort of title for the control group. For an example, consider the following entry:

```
RD  SAMPLE–REPORT
    CONTROL FINAL, YEAR, MONTH
    PAGE 50 LINES.
```

Then, if there were entries of:

```
01  TYPE CONTROL HEADING FINAL
01  TYPE CONTROL HEADING YEAR
01  TYPE CONTROL HEADING MONTH
01  TYPE CONTROL FOOTING MONTH
01  TYPE CONTROL FOOTING YEAR
01  TYPE CONTROL FOOTING FINAL
```

the report groups would print as follows:

```
final control heading (once on the report)
    year control heading
        month control heading
            first detail line on the report
            detail lines
                .
                .
        month control footing
        month control heading (for the next month)
            detail lines
                .
                .
        month control footing
    year control footing
    year control heading (for the next year)
        month control heading
            detail lines
```

```
                    .
                    .
            last detail line on the report
        month control footing
    year control footing
final control footing (once on the report)
```

CONTROL HEADING lines may appear in the same areas of a report page as DETAIL lines; that is, from the FIRST DETAIL line number through the LAST DETAIL line number as given in the PAGE clause.

The SUM Clause

The general format of the SUM clause is:

```
{SUM identifier-1 [identifier-2] . . .
 [UPON data-name-1 [data-name-2] . . . ]} . . .

 [ RESET ON { data-name-3 } ]
            { FINAL       }
```

The ellipsis after the brace shows that there may be as many SUM clauses as desired in a single entry. We have used only one SUM clause in an entry in our sample programs; if there is more than one SUM clause in an entry, all the identifiers named in all the SUM clauses are added into the SUM accumulator.

The data name(s) given in the UPON option must be the names of DETAIL report groups in the same report as the SUM clause.

The RESET ON phrase gives the programmer the ability to delay resetting SUM accumulators until after their normal time. Normally, a SUM accumulator is automatically reset to zero by Report Writer when a control break occurs and after the contents of the accumulator are printed. With the RESET phrase, the programmer can direct Report Writer not to reset the accumulator until some specified higher-level break occurs. This provides for printing ever growing lower-level subtotals. The RESET ON phrase should really be called the DON'T RESET UNTIL phrase. Using RESET ON FINAL delays resetting the accumulator for the entire report, and so the report output shows only ever growing subtotals for that SUM field.

USE BEFORE REPORTING

The USE BEFORE REPORTING statement allows the programmer to interrupt the normal flow of Report Writer operations and to execute any COBOL procedure immediately before printing a REPORT

HEADING, PAGE HEADING, CONTROL HEADING, CONTROL FOOTING, PAGE FOOTING, or REPORT FOOTING. The procedure may consist of as many ordinary Procedure Division statements as desired. The general format of the USE BEFORE REPORTING statement is:

USE BEFORE REPORTING identifier

Identifier may be the name of any report group defined as REPORT HEADING, PAGE HEADING, CONTROL HEADING, CONTROL FOOTING, PAGE FOOTING, or REPORT FOOTING.

If a USE BEFORE REPORTING procedure is named for a CONTROL FOOTING, the procedure will be executed during Report Writer's processing of a control break for that level of CONTROL FOOTING. During processing of the control break, Report Writer first carries out all SUMming that has been indicated for that level, then executes the USE BEFORE REPORTING procedure, and then edits and prints the CONTROL FOOTING report group.

If a USE BEFORE REPORTING procedure is named for a REPORT HEADING, PAGE HEADING, CONTROL HEADING, PAGE FOOTING, or REPORT FOOTING group, Report Writer first executes the procedure and then edits and prints the indicated report group. Any USE BEFORE REPORTING statements and their associated procedure statements must be at the very beginning of the Procedure Division, before the OPEN statement. Normal execution of the program would begin with the OPEN statement, and the USE BEFORE REPORTING procedures would be executed by Report Writer at the appropriate time.

Summary

Report Writer is a COBOL feature that permits the programmer to describe characteristics of a report rather than to write the step-by-step code needed to produce the report. Report Writer automatically provides coding needed to format output lines, sum into accumulators, print report headings, test for page overflow, skip to a new page, print page headings, test for control breaks, print total lines, reset accumulators, and print page footings and report footings. At the same time, the programmer has complete command over all the logical capabilities of COBOL. Before giving a GENERATE statement to print a DETAIL line, the program may carry out any desired processing on the input data. Before any other TYPEs of report groups are printed, the programmer may have a USE BEFORE REPORTING procedure.

The characteristics of all the reports produced by a program may be described in the Report Section. There, each report has an RD entry with a report name. Within the RD entry, report groups are described.

Report groups may be of TYPE REPORT HEADING, PAGE HEADING, CONTROL HEADING, DETAIL, CONTROL FOOTING, PAGE FOOTING, and REPORT FOOTING.

The LINE clause directs Report Writer to space or skip before printing a line. The NEXT GROUP clause directs Report Writer to space or skip after printing a line. When the NEXT GROUP clause appears in a CONTROL FOOTING group it is executed only when a control break occurs at the exact level of the CONTROL FOOTING.

Within each report group definition, printable lines and nonprintable items may be defined. A printable item is recognized by the appearance of a COLUMN clause and one of the clauses SOURCE, SUM, or VALUE. If an item contains a COLUMN clause, it must have a LINE clause properly related to it in order to tell Report Writer where it is to be printed.

The CONTROL clause in the RD entry indicates the fields where control breaks are to be taken. The PAGE clause in the RD entry indicates the areas of the report output page where different TYPEs of report groups may print.

The INITIATE, GENERATE, and TERMINATE statements are used in the Procedure Division to control the printing of the report. The INITIATE statement must be given for each report after the report output file is OPENed and before any GENERATE statements are given. The GENERATE statement directs Report Writer to produce a DETAIL report group and at the same time to test whether any other report groups should be printed. For each DETAIL group printed, Report Writer tests for page overflow and control breaks. If any conditions are found that require printing lines in addition to the DETAIL line, Report Writer automatically prints them with no further effort on the part of the programmer.

Fill-in Exercises

1. In the FD entry for a report file, the ___Report is___ clause must be used to name the report.

2. In the Report Section, each report name must have an ___RD___ entry.

3. The types of report groups that Report Writer can process are ___Report H___, ___Page H___, ___Control H___, ___Detail___, ___Control F___, ___Page F___, and ___Report F___.

4. The beginning print position of each printable item is given by a ___Column___ clause.

5. The vertical spacing of the paper is controlled by the ___LINE___ and ___NEXT GROUP___ clauses.

6. The statement that is used to begin processing a report is the INITIATE statement.
7. The statement that causes a DETAIL report group to print is the GENERATE statement.
8. The statement that is used to end processing a report is the TERMINATE statement.
9. The areas on the page where different TYPEs of report groups may print are described by the PAGE clause.
10. The control fields of a report are listed in the CONTROL clause in the RD entry.

Review Exercises

1. Rewrite your solution to Review Exercise 6, Chapter 2, page 32, using Report Writer.
2. Rewrite your solution to Review Exercise 6, Chapter 3, page 64, using Report Writer.
3. Rewrite your solution to Review Exercise 6, Chapter 4, page 92, using Report Writer.
4. Rewrite your solution to Review Exercise 2, Chapter 5, page 117, using Report Writer.
5. Rewrite your solution to Review Exercise 6, Chapter 6, page 155, using Report Writer.
6. Rewrite your solution to Review Exercise 6, Chapter 8, page 221, using Report Writer.
7. Rewrite your solution to Review Exercise 5, Chapter 9, page 264, using Report Writer.

Magnetic File Media

11

KEY POINTS

Here are the key points you should learn from this chapter:

1. the nature of files stored on magnetic media;
2. types of master files and their uses;
3. how to create, update, and list an indexed sequential file;
4. how to create, update, and list a standard sequential file.

KEY WORDS

Key words to recognize and learn:

magnetic storage media
magnetic tape
magnetic disk
magnetic drum
data cell
mass storage system
tape deck
tape transport
tape drive
read–write head
sequential
access
direct-access
random access
direct-access storage device
DASD
block
master file
transaction register
indexed sequential file
indexed file
file organization
relative
direct
ORGANIZATION INDEXED
RECORD KEY
BLOCK CONTAINS
DELETE
INVALID KEY
REWRITE
HIGH–VALUE
HIGH–VALUES
LOW–VALUE
LOW–VALUES
ACCESS IS RANDOM
NOMINAL KEY
continuation
I–O
sign condition
standard sequential
INTO

Up to now, all of our programs have written their output files onto paper, producing what is usually called a list, a listing, or a report, and all of our input files have been on punched cards or card equivalents. There are other kinds of materials, though, that computers can read input data from and write output onto. In this chapter we will discuss the COBOL handling of the **magnetic storage media.** There are a number of forms of magnetic media, such as **magnetic tape, magnetic**

disk, magnetic drum, data cell, and **mass storage system.**

Magnetic computer tape is very much like regular recording tape, except that computer tape is wider, thicker, and stronger. It is available on reels or cassettes just like recording tape, and just as music can be stored on recording tape as a series of magnetic impulses, computer tape can store characters as a series of magnetic impulses. The characters are stored one after another along the length of the tape, forming fields and records. Fields and records on tape can be thousands of characters long.

A computer installation that uses magnetic tape files will usually have one or more **tape decks** attached to the computer. The tape deck, or **tape transport** or **tape drive** as it is sometimes called, works just like a tape deck for recording tape: It moves the tape from one reel to the other and back, so that the computer can read data from the tape and write data onto it. Computer tapes can be removed from the transport when not in use and stored in a cabinet.

A magnetic disk looks like a large phonograph record coated on both sides with a thin layer of magnetizable material. Characters, fields, and records can be stored one after another in circular tracks on both sides of the disk. Data can be written onto and read from the disk by a device called a **read–write head.** The read–write head can be directed by the computer to any location on the disk surface to read any particular desired record or to write a record into any particular location.

You can now see the fundamental difference between files on tape and files on disk. Tape is a **sequential** medium; records on tape can be **accessed** only sequentially, which means that they can be read and written on tape only one after another, in order. Disk is a **direct-access** medium, which means that the computer can go to any location on the disk directly for reading or writing. So a disk file can enjoy **random access.** A disk file may also be accessed sequentially just as a phonograph record can be played through from beginning to end.

All other magnetic media, aside from magnetic tape, can be thought of as operating the same way as disk. Even though the actual principles of operation may be quite different, and in some cases the storage medium is brought to the read–write head instead of the read–write head being sent to the storage medium, for our purposes we need differentiate only between magnetic tape and magnetic disk. Disk and all the other media that work the same are called **direct-access storage devices (DASD).**[1]

On both sequential and direct-access media, records are often

1. Recent COBOL literature refers to direct-access storage devices as mass storage devices. We prefer the older term to avoid confusion with IBM's Model 3850 Mass Storage System.

grouped into larger units called **blocks.** Blocks have no logical meaning for a program; our tape and disk programs will still READ and WRITE one record at a time. The programmer has no responsibility regarding blocks of records except to decide how many records should be grouped into each block. There are many factors influencing such a decision. At this point in your study of COBOL it is best to ask your instructor to pick a number.

Advantages of Tape and Disk

Punched cards and printed output are good media for computers to use to receive data from the outside world and to communicate back results. They are easy for people to handle, and their contents can be read fairly easily by people. They have their role in data processing.

The big advantage of magnetic media is that the computer can write data onto them and then, at a later time, read the same data in again; the later time may be seconds later, weeks later, or, due to the extremely stable nature of magnetic media, years later. Other advantages of magnetic media over paper ones are:

a. obsolete data can be erased and new data written in their place, so magnetic media are reusable;
b. computers can read and write magnetic media much faster than paper media;
c. huge volumes of data can be stored more compactly on magnetic media than on paper.

Uses of Magnetic Media

The characteristics of magnetic media just given make them ideally suited for storing **master files** of data. Master files can contain quite different sorts of data in different kinds of applications and can serve quite different purposes.

One example of a master file is a payroll master file. In the file we might have one record for each employee. The record would contain some data which never changes, like the employee's Social Security number; some which changes infrequently, like the employee's name and address; some data that would be needed for reference, such as the rate of pay; and some that changes periodically, like the employee's current year-to-date FICA payments. Such a file could be used weekly by a program to produce paychecks. The program would also have to update the year-to-date FICA field; that is, it would have to change the year-to-date FICA amount as it appears on the file, so that the next time the file were used the FICA amount would be current. There would have to be some provision for updating the other data on the file, too. For example, if a person's address changes, that change must be reflected on the master file. Ordinarily, a computer installation will have a program to process such changes to a master.

Another sort of master file could be found in a sales application. The file could be a catalog master file with one record for every part number. Each record would contain a part number, a description of the part, a price, a shipping weight, and perhaps other information. In this application, the entire file is used for reference and is not changed until a new catalog is issued, perhaps every six months. Then the old master file could be thrown out (erased would be better) and a complete new catalog master created.

A catalog master could be used daily by a program that prints invoices. The program would read the day's orders and look in the master file to see that the part numbers in the orders were valid. Then it would get the description of the part for printing on the invoice and the price for computing the merchandise amount.

Exercise 1. Think of another application of a master file; tell the following about the file:

a. What does each record on the file represent? That is, is there one record for each employee or one record for each part number or one record for each what?
b. What are some of the fields that would be in the record?
c. Under what circumstances would each of the fields in the master have to be changed?
d. In what ways would the file be used?

Creating and Updating a Master File

Programs 27, 28, and 29, our next three programs, will all deal with the same application, but the three programs will do entirely different things. Program 27 will create a master file, and Programs 28 and 29 will do some processing with it. The overall problem to be solved by the three programs is: A bank wishes to create and maintain a master file of savings account information. There is to be one record on the file for each account. The record is to contain the account number, the depositor's name, the date of the last transaction to the account (either the date the account was opened or the date of the last deposit or withdrawal), and the current balance. When a new account is opened, it must be accompanied by an initial deposit; when an account is closed, the former balance is assumed to be the final withdrawal.

The bank wishes to update the account balances on the master file every day, to reflect all deposits and withdrawals made during the day, and produce a report. In addition, new accounts that were opened during the day should be added to the file, and accounts closed during the day should be deleted. If any depositors reported a change of name during the day, that, too, should be reflected on the master.

A report should be produced showing the individual transactions to the file and the total of all deposits and withdrawals made during the day. For control purposes, the report should also show the number of deposits and new accounts processed, the number of withdrawals and account closings, the number of name changes, the number of erroneous input transactions, and the total number of input transactions. The bank wants the daily report to have the format shown in Figure 11.1.

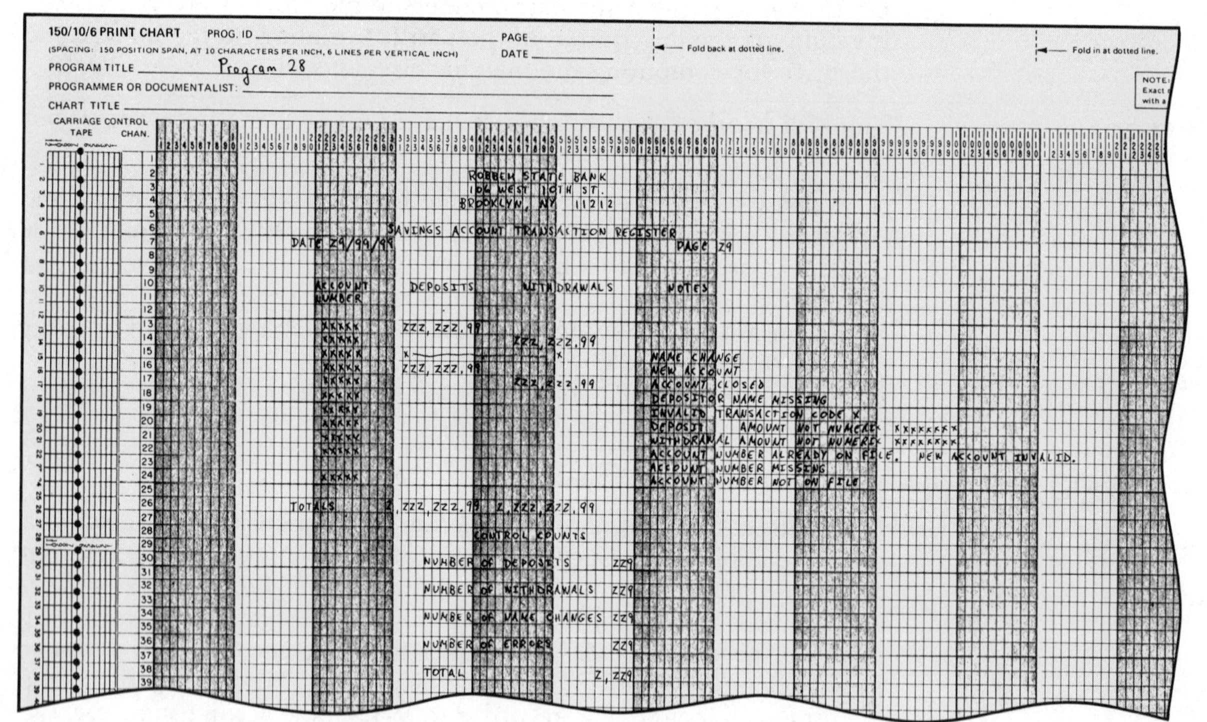
150/10/6 PRINT CHART PROG. ID PAGE
(SPACING: 150 POSITION SPAN, AT 10 CHARACTERS PER INCH, 6 LINES PER VERTICAL INCH) DATE
PROGRAM TITLE Program 28
PROGRAMMER OR DOCUMENTALIST:
CHART TITLE
CARRIAGE CONTROL TAPE CHAN.
Fold back at dotted line.
Fold in at dotted line.

ROBBEM STATE BANK
106 WEST 10TH ST.
BROOKLYN, NY 11212

SAVINGS ACCOUNT TRANSACTION REGISTER
DATE Z9/99/99 PAGE Z9

ACCOUNT NUMBER DEPOSITS WITHDRAWALS NOTES

XXXXX ZZZ,ZZZ.99
XXXXX ZZZ,ZZZ.99
XXXXX NAME CHANGE
XXXXX ZZZ,ZZZ.99 NEW ACCOUNT
XXXXX ZZZ,ZZZ.99 ACCOUNT CLOSED
XXXXX DEPOSITOR NAME MISSING
XXXXX INVALID TRANSACTION CODE X
XXXXX DEPOSIT AMOUNT NOT NUMERIC XXXXXXXX
XXXXX WITHDRAWAL AMOUNT NOT NUMERIC XXXXXXXX
XXXXX ACCOUNT NUMBER ALREADY ON FILE. NEW ACCOUNT INVALID.
ACCOUNT NUMBER MISSING
XXXXX ACCOUNT NUMBER NOT ON FILE

TOTALS Z,ZZZ,ZZZ.99 Z,ZZZ,ZZZ.99

CONTROL COUNTS

NUMBER OF DEPOSITS ZZ9
NUMBER OF WITHDRAWALS ZZ9
NUMBER OF NAME CHANGES ZZ9
NUMBER OF ERRORS ZZ9
TOTAL Z,ZZ9

Figure 11.1 Output format for Program 28.

Message formats are shown in Figure 11.1. The programs that process this master file will be expected to check for the kinds of errors indicated. It is particularly important to protect master files against garbage, for any garbage on the file may cause trouble in later runs when the file is used.

Notice that the **transaction register** in Figure 11.1 shows only the changes that were made to the master file in the daily run. Every now and then the bank may want to look at the contents of the entire file.

We will use Program 27 to create the master file from a deck of data cards; Program 28 to update the file with daily transactions, also punched in cards; and Program 29 to list the contents of the complete file whenever it may be required.

From the statement of the problem, a programmer couldn't know whether the master file ought to be created on tape or on disk. That kind of decision would have to come from a programmer's supervisor or other experienced computer professional. In this case it comes from the author. We will create the file on disk, and we will make it an **indexed sequential file.**

Indexed Sequential Files on Disk

An indexed sequential file, also called an **indexed file,** has most of its records ordered sequentially on some key field. The file also has indexes which are used by the COBOL system to find records in the file and to find a place to slip in new records when we wish to add new accounts to the file. We use the indexed sequential form of **file organization** in this first problem because indexed sequential files are easiest to update. Other kinds of organization that you may use for COBOL files are sequential, **relative,** and **direct.** Indexed sequential files can be established only on direct-access devices, such as magnetic disk. Relative files and direct files must also be on DASD; sequential files may be created on tape or on DASD.

The input to Program 27 will be a batch of new accounts to create the account master file. The new account information will be punched in cards. The format for new account information is shown in Figure 11.2.

The 1 in column 1 of the card identifies this as a new account. The amount field is the initial deposit. To create an indexed sequential file, the input records must be in order on the key field, so that the system

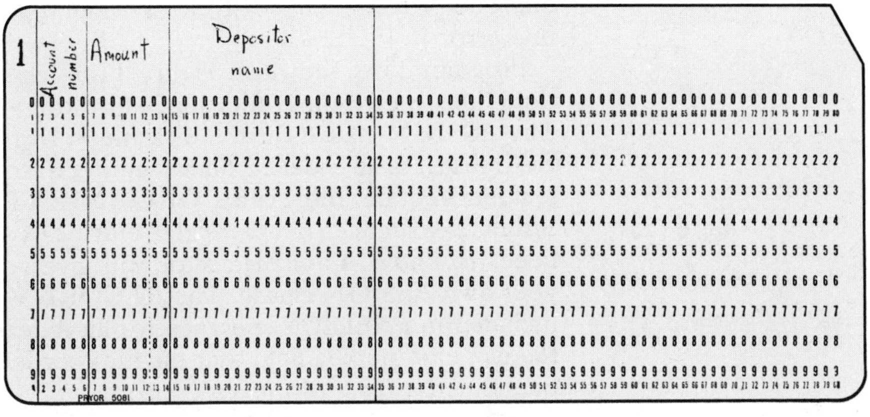

Figure 11.2
New account input format for Programs 27 and 28.

can create the file and the indexes properly. In this application, the key field is the account number. As the master file is created, COBOL will automatically check that the master records are in ascending order on account number and that there are no duplicate account numbers in the master file.

Program 27 will READ the card input file, do some necessary validity checks, create an indexed sequential file out of the good input records, and produce a report telling us what it did. Program 27 can use the same output format as Program 28 (Figure 11.1), except that, when we create the file with Program 27, we will have all new accounts with their initial deposits (and possibly some errors). We won't have any withdrawals, account closings, or name changes.

Creating an Indexed Sequential File

Program 27 will make five validity checks on the input data in this order:

a. new account code 1 in column 1
b. account number present
c. money amount numeric
d. depositor name present
e. account number in sequence and not a duplicate.

If the account number is missing or not in sequence or a duplicate of an already existing account number, the account cannot be added to the file. If the money amount is not numeric, we would be in terrible trouble later if we put the account on the file. For even though there is nothing to stop us in Program 27 from putting a nonnumeric amount into the current balance field on the master, later in Program 28 when we tried to use that amount in addition or subtraction we would get one or more of the unpleasant results discussed in Chapter 8.

If the depositor name is missing, we will not put the account on the file. The program will cease checking an input record as soon as it finds one error in it.

Program 27 is shown in Figure 11.3. This is our first program with three files: a card input file, a printer output file, and a disk output file. The ASSIGN clause for the disk file in this program indicates to COBOL that it is an indexed file on a direct-access device. Some COBOL systems require the clause **ORGANIZATION INDEXED** in the SELECT sentence, but here the entry DA–I tells the system what it needs to know. Your instructor will give you the entry you need for your own ASSIGN clause. The **RECORD KEY** clause must always be included in a SELECT sentence for an indexed file. It tells the system the name of the key field that the file is ordered on. The key field must be defined as part of the disk record.

Figure 11.3 Program 27, Part 1 of 3.

```
4-CB2 V4 RELEASE 1.5 10NOV77          IBM OS AMERICAN NATIONAL STANDARD COBOL



  000010 IDENTIFICATION DIVISION.
  000020 PROGRAM-ID. PROG27.
  000030*AUTHOR. ELLEN M. LONGO.
  000040*
  000050*     THIS PROGRAM CREATES AN INDEXED SEQUENTIAL
  000060*     MASTER FILE OF SAVINGS ACCOUNT RECORDS.
  000070*
  000080*****************************************************************
  000090
  000100 ENVIRONMENT DIVISION.
  000110 INPUT-OUTPUT SECTION.
  000120 FILE-CONTROL.
  000130     SELECT TRANSACTION-REGISTER-FILE ASSIGN TO UT-S-SYSPRINT.
  000140     SELECT TRANSACTION-FILE-IN       ASSIGN TO UT-S-SYSIN.
  000150     SELECT ACCOUNT-MASTER-FILE       ASSIGN TO DA-I-DISK
  000160         RECORD KEY IS ACCOUNT-NUMBER-M.
  000170
  000180*****************************************************************
  000190
  000200 DATA DIVISION.
  000210 FILE SECTION.
  000220 FD  TRANSACTION-FILE-IN
  000230     LABEL RECORDS ARE OMITTED.
  000240
  000250 01  TRANSACTION-RECORD-IN.
  000260     05 TRANSACTION-CODE-IN                       PIC X.
  000270        88 TRANSACTION-IS-NEW-ACCOUNT                 VALUE '1'.
  000280     05 ACCOUNT-NUMBER-IN                         PIC X(5).
  000290        88 ACCOUNT-NUMBER-IS-MISSING                  VALUE SPACES.
  000300     05 INITIAL-DEPOSIT-IN                        PIC 9(6)V99.
  000310     05 INITIAL-DEPOSIT-IN-X
  000320        REDEFINES INITIAL-DEPOSIT-IN              PIC X(8).
  000330     05 DEPOSITOR-NAME-IN                         PIC X(20).
  000340        88 DEPOSITOR-NAME-IS-MISSING                  VALUE SPACES.
  000350     05 FILLER                                    PIC X(46).
  000360
  000370 FD  ACCOUNT-MASTER-FILE
  000380     LABEL RECORDS ARE STANDARD
  000390     BLOCK CONTAINS 10 RECORDS.
  000400
  000410 01  ACCOUNT-MASTER-RECORD.
  000420     05 DELETE-CODE                               PIC X.
  000430     05 DATE-OF-LAST-TRANSACTION-M                PIC 9(6).
  000440     05 ACCOUNT-NUMBER-M                          PIC X(5).
  000450     05 CURRENT-BALANCE-M                         PIC S9(7)V99.
  000460     05 DEPOSITOR-NAME-M                          PIC X(20).
  000470
  000480 FD  TRANSACTION-REGISTER-FILE
  000490     LABEL RECORDS ARE OMITTED
  000500     REPORT IS TRANSACTION-REGISTER.
  000510
```

Figure 11.3 Program 27, Part 2 of 3.

```
000520 WORKING-STORAGE SECTICN.
000530 01  MORE-INPUT          VALUE  'YES'                PIC X(3).
000540
000550 01  RUN-DATE.
000560     05 RUN-YEAR                                     PIC 99.
000570     05 RUN-MONTH                                    PIC 99.
000580     05 RUN-DAY                                      PIC 99.
000590
000600 01  ONE                 VALUE  1                    PIC S9.
000610 01  MESSAGE-W                                       PIC X(53).
000620 01  HAS-INPUT-BEEN-CHECKED                          PIC X(3).
000630
000640 01  ERROR-MESSAGES.
000650     05 INVALID-TRANSACTICN-CODE.
000660        10 FILLER        VALUE  'INVALID TRANSACTION CODE'
000670                                                     PIC X(25).
000680        10 TRANSACTICN-CODE-E                        PIC X.
000690     05 DEPOSIT-INVALID.
000700        10 FILLER        VALUE  'DEPCSIT AMCUNT NCT NUMERIC'
000710                                                     PIC X(30).
000720        10 INITIAL-DEPOSIT-E                         PIC X(8).
000730     05 OUT-OF-SEQUENCE
000740        VALUE        'DUPLICATE OR CUT-OF-SECUENCE ACCOUNT NUMBER.'
000750                                                     PIC X(44).
000760 REPORT SECTION.
000770 RD  TRANSACTICN-REGISTER
000780     CONTROL FINAL
000790     PAGE 52 LINES,
000800     FIRST DETAIL 13,
000810     LAST DETAIL 38,
000820     FOOTING 52.
000830
000840 01  TYPE REPORT HEADING.
000850     05 LINE 2.
000860        10 COLUMN 40  VALUE  'ROEBEM STATE BANK'
000870                                                     PIC X(17).
000880     05 LINE 3.
000890        10 COLUMN 40  VALUE  '106 WEST 10TH ST.'
000900                                                     PIC X(17).
000910     05 LINE 4.
000920        10 COLUMN 39  VALUE  'BRCCKLYN, NY  11212'
000930                                                     PIC X(19).
000940
000950 01  TYPE PAGE HEADING.
000960     05 LINE 6.
000970        10 COLUMN 30  VALUE
000980                      'SAVINGS ACCCUNT TRANSACTION REGISTER'
000990                                                     PIC X(36).
001000     05 LINE 7.
001010        10 COLUMN 18  VALUE  'DATE'                  PIC X(4).
001020        10 COLUMN 23  SOURCE RUN-MONTH               PIC Z9.
001030        10 COLUMN 25  VALUE  '/'                     PIC X.
001040        10 COLUMN 26  SOURCE RUN-DAY                 PIC 99.
001050        10 COLUMN 28  VALUE  '/'                     PIC X.
001060        10 COLUMN 29  SOURCE RUN-YEAR                PIC 99.
001070        10 COLUMN 66  VALUE  'PAGE'                  PIC X(4).
001080        10 COLUMN 71  SOURCE PAGE-COUNTER            PIC Z9.
```

Figure 11.3 Part 2 (continued).

```
001090      05 LINE 10.
001100         10 COLUMN 21  VALUE  'ACCOUNT'          PIC X(7).
001110         10 COLUMN 33  VALUE  'DEPOSITS'         PIC X(8).
001120         10 COLUMN 47  VALUE  'WITHDRAWALS'      PIC X(11).
001130         10 COLUMN 65  VALUE  'NOTES'            PIC X(5).
001140      05 LINE 11.
001150         10 COLUMN 21  VALUE  'NUMBER'           PIC X(6).
001160
001170 01   DEPOSIT-LINE
001180      TYPE DETAIL
001190      LINE PLUS 1.
001200      05 COLUMN 22     SOURCE ACCOUNT-NUMBER-IN  PIC X(5).
001210      05 COLUMN 32     SOURCE INITIAL-DEPOSIT-IN PIC ZZZ,ZZZ.99.
001220      05 COLUMN 63     VALUE  'NEW ACCOUNT'      PIC X(11).
001230
001240 01   ERROR-LINE
001250      TYPE DETAIL
001260      LINE PLUS 1.
001270      05 COLUMN 22     SOURCE ACCOUNT-NUMBER-IN  PIC X(5).
001280      05 COLUMN 63     SOURCE MESSAGE-W          PIC X(53).
001290
001300 01   TYPE CONTROL FOOTING FINAL.
001310      05 LINE PLUS 2.
001320         10 COLUMN 18  VALUE  'TOTALS'           PIC X(6).
001330         10 COLUMN 30  SUM    INITIAL-DEPOSIT-IN
001340                       UPON   DEPOSIT-LINE       PIC Z,ZZZ,ZZZ.99.
001350      05 LINE PLUS 2
001360            COLUMN 41  VALUE  'CONTROL COUNTS'   PIC X(14).
001370      05 LINE PLUS 2.
001380         10 COLUMN 35  VALUE  'NUMBER OF DEPOSITS'
001390                                                 PIC X(18).
001400         10 NUMBER-OF-DEPOSITS
001410            COLUMN 58  SUM    ONE
001420                       UPON   DEPOSIT-LINE       PIC ZZ9.
001430      05 LINE PLUS 2.
001440         10 COLUMN 35  VALUE  'NUMBER OF WITHDRAWALS'
001450                                                 PIC X(21).
001460         10 COLUMN 60  VALUE  0                  PIC 9.
001470      05 LINE PLUS 2.
001480         10 COLUMN 35  VALUE  'NUMBER OF NAME CHANGES'
001490                                                 PIC X(22).
001500         10 COLUMN 60  VALUE  0                  PIC 9.
001510      05 LINE PLUS 2.
001520         10 COLUMN 35  VALUE  'NUMBER OF ERRORS'
001530                                                 PIC X(16).
001540         10 NUMBER-OF-ERRORS
001550            COLUMN 58  SUM    ONE
001560                       UPON   ERROR-LINE         PIC ZZ9.
001570      05 LINE PLUS 2.
001580         10 COLUMN 35  VALUE  'TOTAL'            PIC X(5).
001590         10 COLUMN 56  SUM    NUMBER-OF-DEPOSITS,
001600                              NUMBER-OF-ERRORS   PIC Z,ZZ9.
001610
001620*****************************************************************
```

Figure 11.3 Program 27, Part 3 of 3.

```
001630
001640 PROCEDURE DIVISION.
001650     OPEN INPUT  TRANSACTION-FILE-IN,
001660          OUTPUT TRANSACTION-REGISTER-FILE,
001670                 ACCOUNT-MASTER-FILE.
001680     READ TRANSACTION-FILE-IN
001690         AT END
001700             MOVE 'NO' TO MORE-INPUT.
001710     IF MORE-INPUT = 'YES'
001720         ACCEPT RUN-DATE FROM DATE
001730         INITIATE TRANSACTION-REGISTER
001740         PERFORM MAIN-PROCESS UNTIL MORE-INPUT = 'NO'
001750         TERMINATE TRANSACTION-REGISTER.
001760     CLOSE TRANSACTION-FILE-IN,
001770           TRANSACTION-REGISTER-FILE,
001780           ACCOUNT-MASTER-FILE.
001790     STOP RUN.
001800
001810 MAIN-PROCESS.
001820     MOVE 'NO' TO HAS-INPUT-BEEN-CHECKED.
001830     IF NOT TRANSACTION-IS-NEW-ACCOUNT
001840         MOVE TRANSACTION-CODE-IN      TO TRANSACTION-CODE-E
001850         MOVE INVALID-TRANSACTION-CODE TO MESSAGE-W
001860         GENERATE ERROR-LINE
001870     ELSE
001880     IF ACCOUNT-NUMBER-IS-MISSING
001890         MOVE 'ACCOUNT NUMBER MISSING' TO MESSAGE-W
001900         GENERATE ERROR-LINE
001910     ELSE
001920     IF INITIAL-DEPOSIT-IN NOT NUMERIC
001930         MOVE INITIAL-DEPOSIT-IN-X TO INITIAL-DEPOSIT-E
001940         MOVE DEPOSIT-INVALID      TO MESSAGE-W
001950         GENERATE ERROR-LINE
001960     ELSE
001970     IF DEPOSITOR-NAME-IS-MISSING
001980         MOVE 'DEPOSITOR NAME MISSING' TO MESSAGE-W
001990         GENERATE ERROR-LINE
002000     ELSE
002010         MOVE 'YES'                 TO HAS-INPUT-BEEN-CHECKED
002020         MOVE RUN-DATE              TO DATE-OF-LAST-TRANSACTION-M
002030         MOVE ACCOUNT-NUMBER-IN     TO ACCOUNT-NUMBER-M
002040         MOVE INITIAL-DEPOSIT-IN    TO CURRENT-BALANCE-M
002050         MOVE DEPOSITOR-NAME-IN     TO DEPOSITOR-NAME-M
002060         MOVE SPACE                 TO DELETE-CODE
002070         WRITE ACCOUNT-MASTER-RECORD
002080             INVALID KEY
002090                 MOVE 'NO' TO HAS-INPUT-BEEN-CHECKED
002100                 MOVE OUT-OF-SEQUENCE TO MESSAGE-W
002110                 GENERATE ERROR-LINE.
002120     IF HAS-INPUT-BEEN-CHECKED = 'YES'
002130         GENERATE DEPOSIT-LINE.
002140     READ TRANSACTION-FILE-IN
002150         AT END
002160             MOVE 'NO' TO MORE-INPUT.
```

In the FD entry for the disk file, ACCOUNT–MASTER–FILE, we see LABEL RECORDS ARE STANDARD. Label records are special records written onto tape and disk files by the COBOL system and used by the system to identify the file. You should always use standard labels on all your tape and disk files unless there is some compelling reason to do otherwise.

In the **BLOCK CONTAINS** clause we have indicated that the system should process ten records as a single block. In many cases a block can be much larger than we have here, usually thousands of characters long, but with indexed sequential files there are good reasons to keep the blocks shorter. Later when we use ordinary sequential files we will make the blocks larger.

In the disk master record, ACCOUNT–MASTER–RECORD, the first character is reserved for a DELETE–CODE. In the COBOL system used to run the programs in this book, deletion of records from an indexed sequential file is handled through the first character of the record. For now we will just reserve it, and, when we have to delete records from the file in Program 28, we will see how it is used. Some COBOL systems have a **DELETE** verb which can be used instead to remove records.

The RECORD KEY, in this case ACCOUNT–NUMBER–M, may be anywhere in the record, except in the first character position if that position is needed for processing deletions.

There is nothing new in the Working Storage Section. When we look at the Procedure Division you will see how the working storage fields are used.

In the Procedure Division, the nested IF statement at the beginning of MAIN–PROCESS makes the first four of the five validity checks on the input. If an error is found, the program GENERATEs an ERROR–LINE and goes on to the next input record. If no error is found, the program constructs the disk record by using five MOVE statements and attempts to WRITE it. The **INVALID KEY** clause in the WRITE statement permits the system to check for three kinds of errors:

a. the account number being written is out of sequence;
b. the account number being written duplicates an account number already on the file;
c. there is no more room on the disk to write more records.

If any of these three conditions arises, the program executes the imperative statements given in the INVALID KEY clause. In this program we are assuming that there will always be plenty of space on the disk, so the error message that is printed after an INVALID KEY condition has to do only with duplicate and out-of-sequence account numbers.

Program 27 was run with the input shown in Figure 11.4. The output is shown in Figure 11.5.

Figure 11.4 Input for Program 27.

```
10000700100784ROSEBUCCI
10001400001000ROBERT DAVIS M.D.
10002100012500LORICE MONTI
10002800700159MICHAEL SMITH
10003500015000JOHN J. LEHMAN
10003200002500JOSEPH CAMILLO
10004900015000JAY GREENE
10005600000100EVELYN SLATER
10006300007500LORRAINE SMALL
10007000050000PATRICK J. LEE
10007700001037LESLIE MINSKY
10008400150000JOHN DAPRINO
10009100010000JOE'S DELI
10009800050000GEORGE CULHANE
10010500005000LENORE MILLER
10011200025000ROSEMARY LANE
10011900050000MICHELE CAPUANO
10012600075000JAMES BUDD
10013300100000PAUL LERNER, D.D.S.
10014000002575BETH FALLON
100    00002000JANE HALEY
10015400005000ONE DAY CLEANERS
10016100002450ROBERT RYAN
10016800012550KELLY HEDERMAN
10017500001000MARY KEATING
20018200007500BOB LANIGAN
10018900003500J. & L. CAIN
10019600015000IMPERIAL FLORIST
10020300000500JOYCE MITCHELL
10021000025000JERRY PARKS
10021700005000CARL CALDERON
10022400017550JOHN WILLIAMS
10023100055500BILL WILLIAMS
10023800001000KEVIN PARKER
10024500003500FRANK CAPUTO
10025200001500
10025900002937
10026600009957MARTIN LANG
10027300027500VITO CACACI
10028000002000COMMUNITY DRUGS
10028700001500SOLOMON CHAPELS
10029400150000JOHN BURKE
10030100015750PAT P. POWERS
10030800200000JOE GARCIA
10031500025000GRACE MICELI
10032200002000MARIA FASANO
10032900001000GUY VOLPONE
100336)))) %))SALVATORE CALI
10034300100000JOE & MARY SESSA
10035000250000ROGER SHAW
```

Figure 11.5 Output from Program 27, Part 1 of 2.

```
                    ROBBEM STATE BANK
                     106 WEST 10TH ST.
                   BROOKLYN, NY  11212

          SAVINGS ACCOUNT TRANSACTION REGISTER
E  2/26/80                                     PAGE  1

ACCOUNT     DEPOSITS        WITHDRAWALS       NOTES
NUMBER

 00007       1,C07.84                       NEW ACCOUNT
 00014          1C.00                       NEW ACCOUNT
 00021         125.00                       NEW ACCOUNT
 00028       7,C01.59                       NEW ACCOUNT
 00035         150.00                       NEW ACCOUNT
 00032                                      DUPLICATE OR OUT-OF-SEQUENCE ACCOUNT NUMBER.
 00049         150.00                       NEW ACCOUNT
 00056           1.00                       NEW ACCOUNT
 00063          75.C0                       NEW ACCOUNT
 00070         500.00                       NEW ACCOUNT
 00077          10.37                       NEW ACCOUNT
 00084       1,500.C0                       NEW ACCOUNT
 00091         1CC.C0                       NEW ACCOUNT
 00098         500.00                       NEW ACCOUNT
 00105          50.00                       NEW ACCOUNT
 00112         250.00                       NEW ACCOUNT
 00119         5C0.00                       NEW ACCOUNT
 00126         750.00                       NEW ACCOUNT
 00133       1,CC0.C0                       NEW ACCOUNT
 00140          25.75                       NEW ACCOUNT
 00                                         DUPLICATE OR OUT-OF-SEQUENCE ACCOUNT NUMBER.
 00154          50.00                       NEW ACCOUNT
 00161          24.50                       NEW ACCOUNT
 00168         125.50                       NEW ACCOUNT
 00175          10.00                       NEW ACCOUNT
 00182                                      INVALID TRANSACTICN CODE 2
```

Figure 11.5 Output from Program 27, Part 2 of 2.

```
                SAVINGS ACCOUNT TRANSACTION REGISTER
  DATE   2/26/80                                           PAGE   2

      ACCOUNT        DEPOSITS        WITHDRAWALS         NOTES
      NUMBER

       00189           35.00                           NEW ACCOUNT
       00196          150.00                           NEW ACCOUNT
       00203            5.00                           NEW ACCOUNT
       00210          250.00                           NEW ACCOUNT
       00217           50.00                           NEW ACCOUNT
       00224          175.50                           NEW ACCOUNT
       00231          555.00                           NEW ACCOUNT
       00238           10.00                           NEW ACCOUNT
       00245           35.00                           NEW ACCOUNT
       00252                                           DEPOSITOR NAME MISSING
       00259                                           DEPOSITOR NAME MISSING
       00266           99.57                           NEW ACCOUNT
       00273          275.00                           NEW ACCOUNT
       00280           20.00                           NEW ACCOUNT
       00287           15.00                           NEW ACCOUNT
       00294        1,500.00                           NEW ACCOUNT
       00301          157.50                           NEW ACCOUNT
       00308        2,000.00                           NEW ACCOUNT
       00315          250.00                           NEW ACCOUNT
       00322           20.00                           NEW ACCOUNT
       00329           10.00                           NEW ACCOUNT
       00336                                           DEPOSIT AMOUNT NOT NUMERIC     )))) %)
       00343        1,000.00                           NEW ACCOUNT
       00350        2,500.00                           NEW ACCOUNT

  TOTALS           23,029.12

                          CONTROL COUNTS

                    NUMBER OF DEPOSITS        44

                    NUMBER OF WITHDRAWALS      0

                    NUMBER OF NAME CHANGES     0

                    NUMBER OF ERRORS           6

                    TOTAL                     50
```

*Exercise 2.

Write a program to create an accounts receivable master file. Each master file record should contain the following fields:

a. Customer account number
b. Customer name
c. Customer street address
d. Customer city, state, and ZIP code
e. Credit limit
f. Current balance

Use the input format shown in Figure 11.E2 to create the file. Make all suitable validity checks on input records. For now, make the current balance field in each master record 0.

Design an appropriate report for your program to show the error messages and the good records that were placed on the file. Save the master file so that you may update it in Exercise 3.

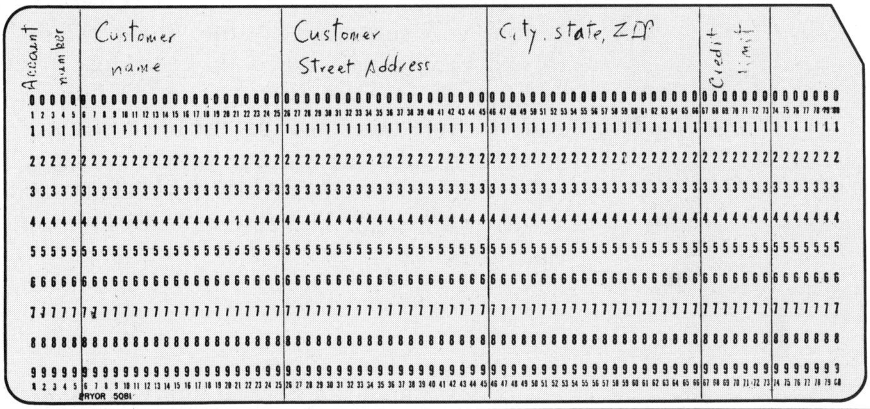

Figure 11.E2
Data card format for Exercise 2.

Updating an Indexed Sequential File

Program 28 reads a batch of transactions and makes changes to our savings account master file. Each transaction record will have a code in column 1 so that the program can recognize what type of transaction it is. We already know that code 1 is for a new account. The complete list of transaction types and their codes is:

Transaction Code	Transaction Type
1	New account
2	Deposit
3	Withdrawal
4	Depositor name change
5	Account closed

For each different type of transaction, the program makes different validity checks on the input and then processes the record in different ways.

ADDING A RECORD TO AN INDEXED SEQUENTIAL FILE

For a code-1 transaction, a new account, the program will have to build a new master record from the information it finds in the transaction record and then try to WRITE the new record onto the master file.

First, the program will make the validity checks appropriate to a new account. It will check that the account number and depositor name are present and that the initial deposit is numeric. If an error is found during those checks, an error message is written and the program goes on to the next record.

If the input data passes the validity checks, the program makes up a new master record and tries to WRITE it onto the master file. The system will automatically find the place on the disk where the new record should be slipped in, rearrange a few records if necessary to make room for the new one, and WRITE the new master record. If the system finds that there is already a record on the file with the same account number as the one we are trying to WRITE, it will signal the error through the INVALID KEY clause so that the program can print an error message. The system will in that case not put the new record onto the disk.

UPDATING A RECORD ON AN INDEXED SEQUENTIAL FILE

For codes 2 and 3, a deposit and withdrawal, respectively, and code 4, a depositor name change, the program must change the contents of an existing master record. For codes 2 and 3, the program checks that the money amount is numeric. For code 4, a depositor's name must be present.

If the input passes the validity checks, the program tries to find the master record that has the same account number as the transaction record. COBOL uses its random-access capability to find the desired master record and READ it from the disk. If the record being sought is not in the file, the READ statement signals through the INVALID KEY clause.

If the desired record is found, the READ statement assigns it to the ACCOUNT–MASTER–RECORD area in the File Section. There it may be changed: Deposits can be added and withdrawals subtracted from the CURRENT–BALANCE–M field; the DEPOSITOR–NAME–M field can be replaced. Then the changed record can be rewritten back onto the disk. The **REWRITE** statement finds the place on the disk where the old record is and puts the new changed record right in over it, erasing the old one.

DELETING A RECORD FROM AN INDEXED SEQUENTIAL FILE

A code 5 transaction, to close an account, treat the last CURRENT–BALANCE–M as a final withdrawal, and delete the record from the file, should contain only the transaction code and the account number.

A READ statement first tries to find the master record whose account number matches the account number in the transaction. If the record is found and assigned to the input area, it can be processed as a withdrawal and deleted from the file.

Some COBOL systems have a DELETE verb which can be used to delete a record from a file after the record has been read. In the system used to run the programs in this book, two steps are required instead in order to delete a record after it has been read. First, a special character called **HIGH–VALUE** must be MOVEd into the first character position of the record and then the record must be rewritten.

HIGH–VALUE (or **HIGH–VALUES,** which has the identical meaning) is a COBOL reserved word. It is a figurative constant standing for the highest value that can be assigned to a single character. On the computer used to run the programs in this book, HIGH–VALUE is higher than the letter *Z* and higher than 9. Another figurative constant **LOW–VALUE** (or **LOW–VALUES**) stands for the lowest value that can be assigned to a character. On this computer, LOW–VALUE is lower than 0 and lower than blank. HIGH–VALUE and LOW–VALUE are alphanumeric constants and can be used only where alphanumeric constants are legal.

A Program to Update an Indexed Sequential File

The processing that Program 28 carries out is summarized in the hierarchy diagram in Figure 11.6. Here we have a hierarchy diagram that is too big and complicated to fit onto one piece of paper. So in part a of the diagram we show only the first three levels and in parts b through e the remaining details. Each of the boxes from the third level of part a appears at the top level in one of the parts b through e. So, for example, the box "Process a new account," which is one of the boxes at the third level in part a, is fully detailed out in part b. In the diagram taken as a whole, all the steps required to process each type of transaction are shown left to right. The diagram also shows the construction of all the different types of error messages.

The input format for Program 28 is shown in Figure 11.7. It is the same as the input format for Program 27, except that column 1 can now contain any transaction code 1 through 5 instead of just the new account code 1. Program 28 is shown in Figure 11.8.

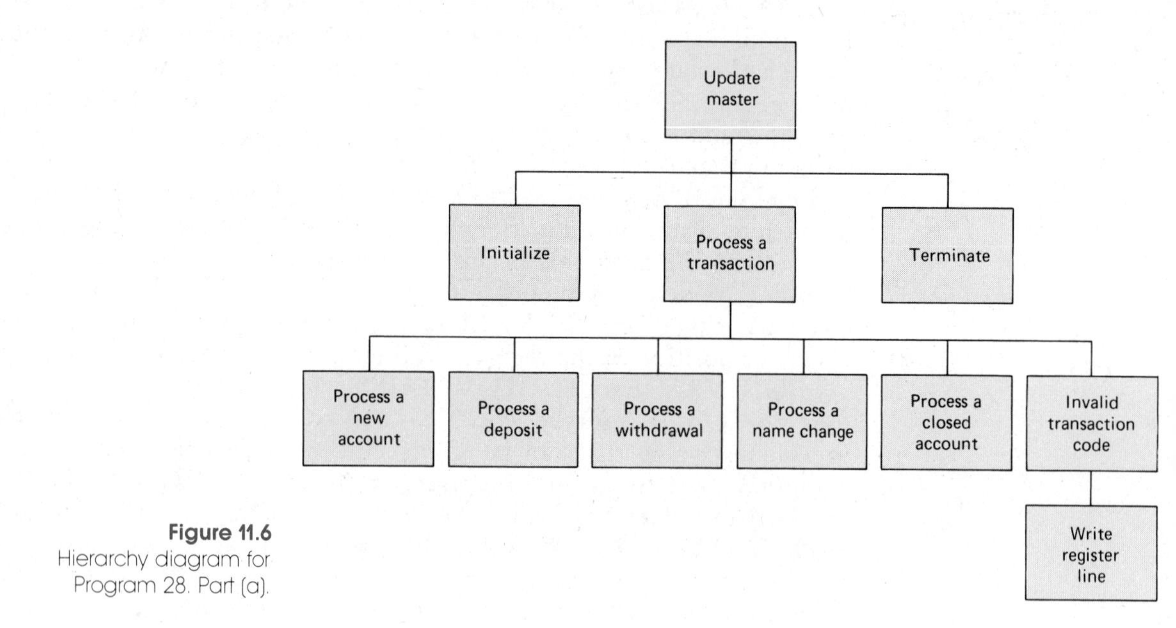

Figure 11.6
Hierarchy diagram for Program 28. Part (a).

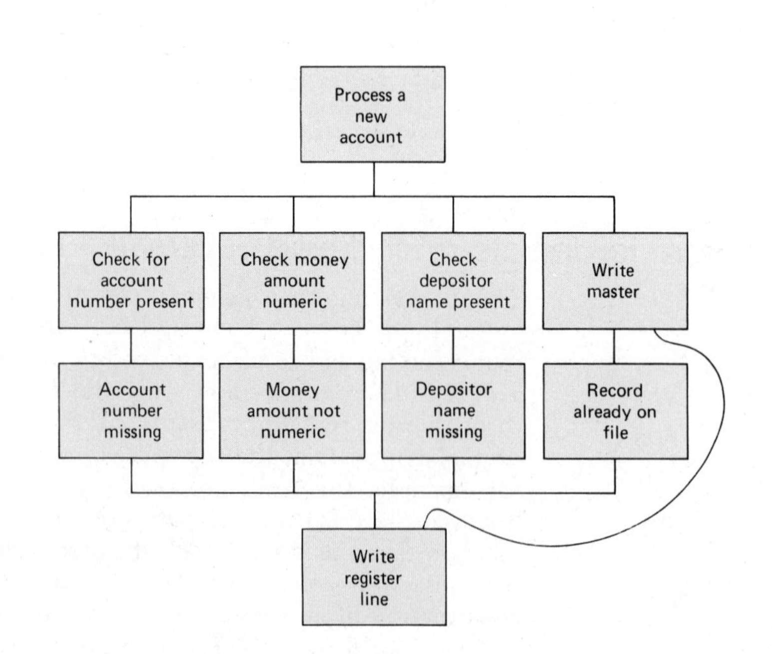

Figure 11.6
Part (b).

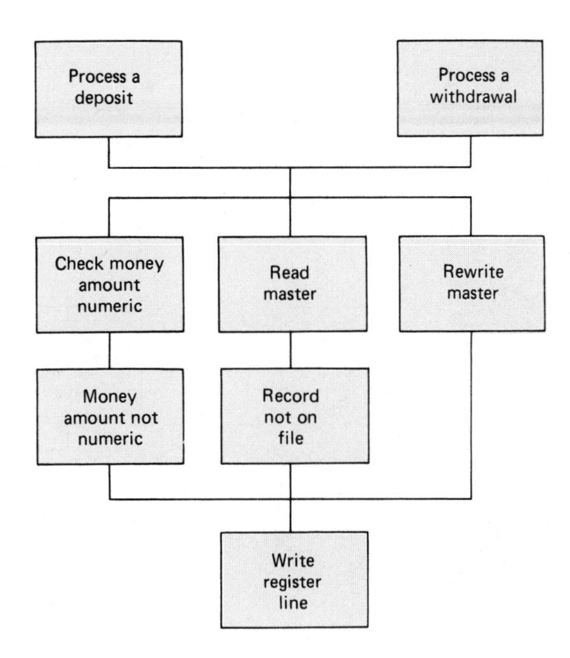

Figure 11.6 Part (c).

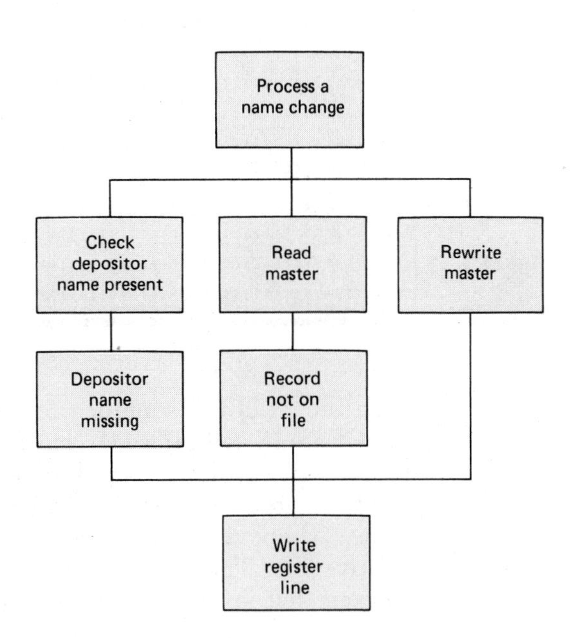

Figure 11.6 Part (d).

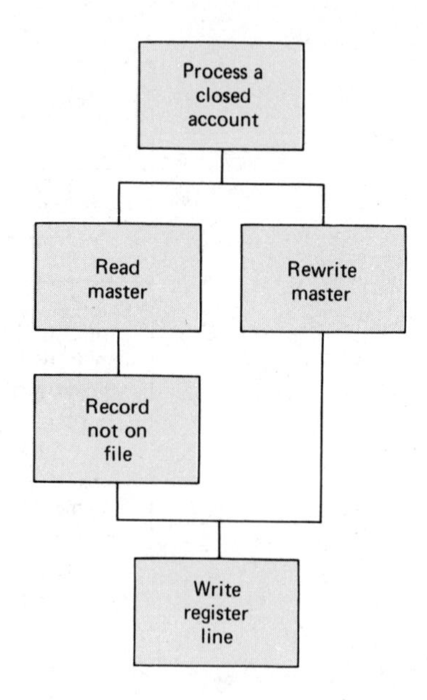

Figure 11.6
Part (e).

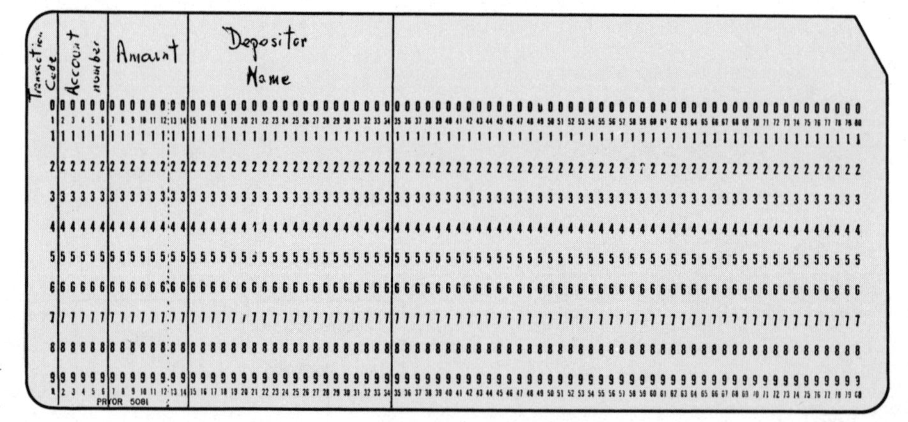

Figure 11.7
Transaction input format for Program 28.

In the SELECT sentence for the ACCOUNT–MASTER–FILE, the **ACCESS IS RANDOM** clause tells the COBOL system that in this program we will need random access on this file. The COBOL system used to run the programs in this book requires the **NOMINAL KEY** clause whenever random access is to be carried out on an indexed or relative file. The NOMINAL KEY is the name of a field in working storage that is used to find a desired record on the file. You will see how it is used when we look at the Procedure Division. Some COBOL systems do not require the NOMINAL KEY clause in order to carry out random access; they use the RECORD KEY for the purpose instead.

Figure 11.8 Program 28, Part 1 of 4.

```
4-CB2 V4 RELEASE 1.5 10NOV77          IBM OS AMERICAN NATIONAL STANDARD COBOL


  000010 IDENTIFICATION DIVISION.
  000020 PROGRAM-ID. PROG28.
  000030*AUTHOR. ELLEN M. LONGO.
  000040*
  000050*    THIS PROGRAM UPDATES AN INDEXED SEQUENTIAL FILE.
  000060*
  000070*****************************************************************
  000080
  000090 ENVIRONMENT DIVISION.
  000100 INPUT-OUTPUT SECTION.
  000110 FILE-CONTROL.
  000120     SELECT TRANSACTION-FILE-IN       ASSIGN TO UT-S-SYSIN.
  000130     SELECT TRANSACTION-REGISTER-FILE ASSIGN TO UT-S-SYSPRINT.
  000140     SELECT ACCOUNT-MASTER-FILE       ASSIGN TO DA-I-DISK
  000150         RECORD KEY IS ACCOUNT-NUMBER-M
  000160         ACCESS IS RANDOM
  000170         NOMINAL KEY IS ACCOUNT-NUMBER-W.
  000180
  000190*****************************************************************
  000200
  000210 DATA DIVISION.
  000220 FILE SECTION.
  000230 FD  TRANSACTION-FILE-IN
  000240     LABEL RECORDS ARE OMITTED.
  000250 01  TRANSACTION-RECORD-IN.
  000260     05 TRANSACTION-CODE-IN                     PIC X.
  000270        88 TRANSACTION-IS-NEW-ACCOUNT              VALUE '1'.
  000280        88 TRANSACTION-IS-DEPOSIT                  VALUE '2'.
  000290        88 TRANSACTION-IS-WITHDRAWAL               VALUE '3'.
  000300        88 TRANSACTION-IS-CHANGE-NAME              VALUE '4'.
  000310        88 TRANSACTION-IS-CLOSE-ACCOUNT            VALUE '5'.
  000320     05 ACCOUNT-NUMBER-IN                       PIC X(5).
  000330        88 ACCOUNT-NUMBER-IS-MISSING               VALUE SPACES.
  000340     05 MONEY-AMOUNT-IN                         PIC 9(6)V99.
  000350     05 MONEY-AMOUNT-IN-X
  000360        REDEFINES MONEY-AMOUNT-IN               PIC X(8).
  000370     05 DEPOSITOR-NAME-IN                       PIC X(20).
  000380        88 DEPOSITOR-NAME-IS-MISSING               VALUE SPACES.
  000390     05 FILLER                                  PIC X(46).
  000400
  000410 FD  ACCOUNT-MASTER-FILE
  000420     LABEL RECORDS ARE STANDARD
  000430     BLOCK CONTAINS 10 RECORDS.
  000440
  000450 01  ACCOUNT-MASTER-RECORD.
  000460     05 DELETE-CODE                             PIC X.
  000470     05 DATE-OF-LAST-TRANSACTION-M              PIC 9(6).
  000480     05 ACCOUNT-NUMBER-M                        PIC X(5).
  000490     05 CURRENT-BALANCE-M                       PIC S9(7)V99.
  000500     05 DEPOSITOR-NAME-M                        PIC X(20).
  000510
```

Figure 11.8 Program 28, Part 2 of 4.

```
C00520 FD   TRANSACTICN-REGISTER-FILE
000530      LABEL RECCRDS ARE CMITTED
000540      REPORT IS TRANSACTICN-REGISTER.
C00550
C00560 WORKING-STORAGE SECTION.
CC0570 01   MORE-INPUT           VALUE  'YES'                PIC X(3).
CC0580
CC0590 01   RUN-DATE.
C00600      05 RUN-YEAR                                      PIC 99.
CC0610      05 RUN-MONTH                                     PIC 99.
000620      05 RUN-DAY                                       PIC 99.
CC0630
000640 01   CNE                  VALLE  1                    PIC S9.
CC0650 01   MESSAGE-W                                        PIC X(53).
000660 01   HAVE-ANY-ERRORS-BEEN-FOUND                       PIC X(3).
CC0670
CC0680 01   ERROR-MESSAGES.
000690      05 INVALID-TRANSACTICN-CCDE.
000700         10 FILLER         VALUE  'INVALID TRANSACTION CODE'
000710                                                       PIC X(25).
CC0720         10 TRANSACTICN-CCDE-E                         PIC X.
CC0730      05 MCNEY-AMOUNT-INVALID.
000740         10 TRANSACTICN-TYPE                           PIC X(11).
000750         10 FILLER         VALUE  'AMCLNT NCT NUMERIC'
000760                                                       PIC X(19).
CC0770         10 INVALID-MCNEY-AMCLNT                       PIC X(8).
CCC780      05 ALREADY-ON-FILE
C00790         VALUE 'ACCOUNT NUMBER ALREADY CN FILE.  NEW ACCOUNT INVALI
CC0800-              'D.'                                    PIC X(53).
CC0810
0CC820 01   ACCOUNT-NUMBER-W                                 PIC X(5).
000830
000840 REPORT SECTION.
CC0850 RD   TRANSACTICN-REGISTER
0C0860      CCNTROL FINAL
C00870      PAGE 52 LINES,
CCC880      FIRST DETAIL 13,
CCC890      LAST DETAIL 38,
CCC900      FOOTING 52.
C00910
CCC920 01   TYPE REPORT HEACING.
CCC930      05 LINE 2.
CC0940         10 COLUMN 40  VALUE  'ROBBEM STATE BANK'
C00950                                                       PIC X(17).
C00960      05 LINE 3.
CCC970         10 CCLUMN 40  VALUE  '106 WEST 10TH ST.'
C00980                                                       PIC X(17).
CCC990      05 LINE 4.
C01000         10 CCLUMN 39  VALUE  'BRCCKLYN, NY  11212'
C01010                                                       PIC X(19).
CC1020
C01030 01   TYPE PAGE HEADING.
001040      05 LINE 6.
001050         10 CCLUMN 30  VALUE
001060                       'SAVINGS ACCCUNT TRANSACTICN REGISTER'
001070                                                       PIC X(36).
```

Figure 11.8 Part 2 (continued)

```
001080        05 LINE 7.
001090            10 COLUMN 18   VALUE   'DATE'                PIC X(4).
001100            10 COLUMN 23   SOURCE  RUN-MONTH             PIC Z9.
001110            10 COLUMN 25   VALUE   '/'                   PIC X.
001120            10 COLUMN 26   SOURCE  RUN-DAY               PIC 99.
001130            10 COLUMN 28   VALUE   '/'                   PIC X.
001140            10 COLUMN 29   SOURCE  RUN-YEAR              PIC 99.
001150            10 COLUMN 66   VALUE   'PAGE'                PIC X(4).
001160            10 COLUMN 71   SOURCE  PAGE-COUNTER          PIC Z9.
001170        05 LINE 10.
001180            10 COLUMN 21   VALUE   'ACCOUNT'             PIC X(7).
001190            10 COLUMN 33   VALUE   'DEPOSITS'            PIC X(8).
001200            10 COLUMN 47   VALUE   'WITHDRAWALS'         PIC X(11).
001210            10 COLUMN 65   VALUE   'NOTES'               PIC X(5).
001220        05 LINE 11.
001230            10 COLUMN 21   VALUE   'NUMBER'              PIC X(6).
001240
001250 01     DEPOSIT-LINE
001260        TYPE DETAIL
001270        LINE PLUS 1.
001280        05 COLUMN 22       SOURCE ACCOUNT-NUMBER-IN     PIC X(5).
001290        05 COLUMN 32       SOURCE MONEY-AMOUNT-IN       PIC ZZZ,ZZZ.99.
001300        05 COLUMN 63       SOURCE MESSAGE-W             PIC X(53).
001310
001320 01     WITHDRAWAL-LINE
001330        TYPE DETAIL
001340        LINE PLUS 1.
001350        05 COLUMN 22       SOURCE ACCOUNT-NUMBER-IN     PIC X(5).
001360        05 COLUMN 46       SOURCE MONEY-AMOUNT-IN       PIC ZZZ,ZZZ.99.
001370        05 COLUMN 63       SOURCE MESSAGE-W             PIC X(53).
001380
```

In the Data Division we see for the first time a nonnumeric literal that is too long to fit on one line. It is the literal in the field called ALREADY–ON–FILE. The complete literal is supposed to be:

ACCOUNT NUMBER ALREADY ON FILE. NEW ACCOUNT INVALID.

The final D and its period do not fit. Whenever a nonnumeric literal cannot fit on a line, you must write as much of it as you can all the way up through column 72 of the card, even if it means breaking a word in the middle. All spaces and punctuation that you write right up through column 72 also count as part of the literal. Then, in column 7 of the next card, you must use a hyphen to indicate that this card is a continuation. Then a single quote mark must go in area B to indicate that a nonnumeric literal is being continued and then the rest of the literal followed by its closing quote mark. Please remember that we use the **continuation** mark in column 7 only for continuing a literal from one card to the next and not to continue ordinary statements and entries from one card to the next. Remember that we have gotten this far without previously using the continuation indicator.

Figure 11.8 Program 28, Part 3 of 4.

```
001390 01   NAME-CHANGE-LINE
001400      TYPE DETAIL
001410      LINE PLUS 1.
001420      05 COLUMN 22       SOURCE ACCOUNT-NUMBER-IN   PIC X(5).
001430      05 COLUMN 32       SOURCE DEPOSITOR-NAME-IN   PIC X(20).
001440      05 COLUMN 63       VALUE  'NAME CHANGE'       PIC X(11).
001450
001460 01   ERROR-LINE
001470      TYPE DETAIL
001480      LINE PLUS 1.
001490      05 COLUMN 22       SOURCE ACCOUNT-NUMBER-IN   PIC X(5).
001500      05 COLUMN 63       SOURCE MESSAGE-W           PIC X(53).
001510
001520 01   TYPE CONTROL FOOTING FINAL.
001530      05 LINE PLUS 2.
001540          10 COLUMN 18  VALUE   'TOTALS'           PIC X(6).
001550          10 COLUMN 30  SUM     MONEY-AMOUNT-IN
001560                        UPON    DEPOSIT-LINE       PIC Z,ZZZ,ZZZ.99.
001570          10 COLUMN 44  SUM     MONEY-AMOUNT-IN
001580                        UPON    WITHDRAWAL-LINE    PIC Z,ZZZ,ZZZ.99.
001590      05 LINE PLUS 2
001600              COLUMN 41  VALUE  'CONTROL COUNTS'   PIC X(14).
001610      05 LINE PLUS 2.
001620          10 COLUMN 35  VALUE   'NUMBER OF DEPOSITS'
001630                                                   PIC X(18).
001640          10 NUMBER-OF-DEPOSITS
001650              COLUMN 58  SUM    ONE
001660                        UPON    DEPOSIT-LINE       PIC ZZ9.
001670      05 LINE PLUS 2.
001680          10 COLUMN 35  VALUE   'NUMBER OF WITHDRAWALS'
001690                                                   PIC X(21).
001700          10 NUMBER-OF-WITHDRAWALS
001710              COLUMN 58  SUM    ONE
001720                        UPON    WITHDRAWAL-LINE    PIC ZZ9.
001730      05 LINE PLUS 2.
001740          10 COLUMN 35  VALUE   'NUMBER OF NAME CHANGES'
001750                                                   PIC X(22).
001760          10 NUMBER-OF-NAME-CHANGES
001770              COLUMN 58  SUM    ONE
001780                        UPON    NAME-CHANGE-LINE   PIC ZZ9.
001790      05 LINE PLUS 2.
001800          10 COLUMN 35  VALUE   'NUMBER OF ERRORS'
001810                                                   PIC X(16).
001820          10 NUMBER-OF-ERRORS
001830              COLUMN 58  SUM    ONE
001840                        UPON    ERROR-LINE         PIC ZZ9.
001850      05 LINE PLUS 2.
001860          10 COLUMN 35  VALUE   'TOTAL'            PIC X(5).
001870          10 COLUMN 56  SUM     NUMBER-OF-DEPOSITS,
001880                                NUMBER-OF-WITHDRAWALS,
001890                                NUMBER-OF-NAME-CHANGES,
001900                                NUMBER-OF-ERRORS   PIC Z,ZZ9.
001910
001920*****************************************************************
001930
001940 PROCEDURE DIVISION.
001950     OPEN INPUT  TRANSACTION-FILE-IN,
001960          OUTPUT TRANSACTION-REGISTER-FILE,
```

Figure 11.8 Part 3 (continued).

```
001970          I-O     ACCOUNT-MASTER-FILE.
001980      READ TRANSACTION-FILE-IN
001990          AT END
002000              MOVE 'NO' TO MORE-INPUT.
002010      IF MORE-INPUT = 'YES'
002020          ACCEPT RUN-DATE FROM DATE
002030          INITIATE TRANSACTION-REGISTER
002040          PERFORM MAIN-PROCESS UNTIL MORE-INPUT = 'NO'
002050          TERMINATE TRANSACTION-REGISTER.
002060      CLOSE TRANSACTION-FILE-IN,
002070            TRANSACTION-REGISTER-FILE,
002080            ACCOUNT-MASTER-FILE.
002090      STOP RUN.
```

In the Procedure Division the OPEN statement shows a new way to OPEN a file. A file must be OPENed as **I–O** if it is to be used for both input and output. Our ACCOUNT–MASTER–FILE will be read from, written onto, and rewritten onto. The REWRITE verb can be used on a file only if the file has been OPENed as I–O.

The MAIN–PROCESS paragraph tests the transaction code in the input record and PERFORMs the appropriate procedure. If the transaction code is not one of the valid ones, an error message is printed.

The NEW–ACCOUNT–ROUTINE is similar to the new account processing that was done in Program 27, but while in Program 27 the ACCOUNT–MASTER–FILE had to be created sequentially, here we can slip in new accounts anywhere we like. To do this kind of random insertion of records, we must MOVE the account number of the new record to the NOMINAL KEY field before writing the new record. You may remember that the NOMINAL KEY field was given as ACCOUNT–NUMBER–W, and the statement:

```
MOVE ACCOUNT-NUMBER-IN TO ACCOUNT-NUMBER-M,
                          ACCOUNT-NUMBER-W
```

MOVEs the new account number both to its place in the new master record and to the NOMINAL KEY.

The WRITE statement in the NEW–ACCOUNT–ROUTINE tries to insert a new record into the file. If the account number of the new record is the same as the account number of a record already on the file, the steps given in the INVALID KEY clause will be executed.

In the paragraph READ–MASTER, an attempt is made to READ randomly the record whose account number is in ACCOUNT–NUMBER–W. If the READ statement cannot find such a record on the file, the steps in the INVALID KEY clause are executed. If the record is found, it is assigned to the ACCOUNT–MASTER–RECORD area in the File Section.

Figure 11.8 Program 28, Part 4 of 4.

```
002100
002110 MAIN-PROCESS.
002120     MOVE 'NO'    TO HAVE-ANY-ERRORS-BEEN-FOUND.
002130     MOVE SPACES TO MESSAGE-W.
002140     IF TRANSACTION-IS-NEW-ACCOUNT
002150         PERFORM NEW-ACCOUNT-ROUTINE
002160     ELSE
002170     IF TRANSACTION-IS-DEPOSIT
002180         PERFORM DEPOSIT-ROUTINE
002190     ELSE
002200     IF TRANSACTION-IS-WITHDRAWAL
002210         PERFORM WITHDRAWAL-ROUTINE
002220     ELSE
002230     IF TRANSACTION-IS-CLOSE-ACCOUNT
002240         PERFORM CLOSE-ACCOUNT-ROUTINE
002250     ELSE
002260     IF TRANSACTION-IS-CHANGE-NAME
002270         PERFORM NAME-CHANGE-ROUTINE
002280     ELSE
002290         MOVE TRANSACTION-CODE-IN       TO TRANSACTION-CODE-E
002300         MOVE INVALID-TRANSACTION-CODE TO MESSAGE-W
002310         GENERATE ERROR-LINE.
002320     READ TRANSACTION-FILE-IN
002330         AT END
002340             MOVE 'NO' TO MORE-INPUT.
002350
002360 NEW-ACCOUNT-ROUTINE.
002370     IF ACCOUNT-NUMBER-IS-MISSING
002380         MOVE 'YES' TO HAVE-ANY-ERRORS-BEEN-FOUND
002390         MOVE 'ACCOUNT NUMBER MISSING' TO MESSAGE-W
002400         GENERATE ERROR-LINE
002410     ELSE
002420         MOVE 'DEPOSIT' TO TRANSACTION-TYPE
002430         PERFORM CHECK-MONEY-AMOUNT.
002440     IF HAVE-ANY-ERRORS-BEEN-FOUND = 'NO'
002450         PERFORM CHECK-FOR-DEPOSITOR-NAME
002460         IF HAVE-ANY-ERRORS-BEEN-FOUND = 'NO'
002470             MOVE RUN-DATE            TO DATE-OF-LAST-TRANSACTION-M
002480             MOVE ACCOUNT-NUMBER-IN TO ACCOUNT-NUMBER-M,
002490                                                    ACCOUNT-NUMBER-W
002500             MOVE MONEY-AMOUNT-IN     TO CURRENT-BALANCE-M
002510             MOVE DEPOSITOR-NAME-IN TO DEPOSITOR-NAME-M
002520             MOVE SPACE               TO DELETE-CODE
002530             WRITE ACCOUNT-MASTER-RECORD
002540                 INVALID KEY
002550                     MOVE 'YES' TO HAVE-ANY-ERRORS-BEEN-FOUND
002560                     MOVE ALREADY-ON-FILE TO MESSAGE-W
002570                     GENERATE ERROR-LINE.
002580     IF HAVE-ANY-ERRORS-BEEN-FOUND = 'NO'
002590         MOVE 'NEW ACCOUNT' TO MESSAGE-W
002600         GENERATE DEPOSIT-LINE.
002610
002620 DEPOSIT-ROUTINE.
002630     MOVE 'DEPOSIT' TO TRANSACTION-TYPE.
002640     PERFORM CHECK-MONEY-AMOUNT.
002650     IF HAVE-ANY-ERRORS-BEEN-FOUND = 'NO'
002660         PERFORM READ-MASTER
```

Figure 11.8 Part 4 (continued).

```
002670             IF HAVE-ANY-ERRORS-BEEN-FOUND = 'NO'
002680                 ADD MONEY-AMOUNT-IN TO CURRENT-BALANCE-M
002690                 PERFORM REWRITE-MASTER
002700                 GENERATE DEPOSIT-LINE.
002710
002720 WITHDRAWAL-ROUTINE.
002730     MOVE 'WITHDRAWAL' TO TRANSACTION-TYPE.
002740     PERFORM CHECK-MONEY-AMOUNT.
002750     IF HAVE-ANY-ERRORS-BEEN-FOUND = 'NO'
002760         PERFORM READ-MASTER
002770         IF HAVE-ANY-ERRORS-BEEN-FOUND = 'NO'
002780             SUBTRACT MONEY-AMOUNT-IN FROM CURRENT-BALANCE-M
002790             PERFORM REWRITE-MASTER
002800             GENERATE WITHDRAWAL-LINE.
002810
002820 CLOSE-ACCOUNT-ROUTINE.
002830     PERFORM READ-MASTER.
002840     IF HAVE-ANY-ERRORS-BEEN-FOUND = 'NO'
002850         MOVE HIGH-VALUE          TO DELETE-CODE
002860         MOVE CURRENT-BALANCE-M  TO MONEY-AMOUNT-IN
002870         MOVE 'ACCOUNT CLOSED'   TO MESSAGE-W
002880         GENERATE WITHDRAWAL-LINE
002890         PERFORM REWRITE-MASTER.
002900
002910 NAME-CHANGE-ROUTINE.
002920     PERFORM CHECK-FOR-DEPOSITOR-NAME.
002930     IF HAVE-ANY-ERRORS-BEEN-FOUND = 'NO'
002940         PERFORM READ-MASTER
002950         IF HAVE-ANY-ERRORS-BEEN-FOUND = 'NO'
002960             MOVE DEPOSITOR-NAME-IN TO DEPOSITOR-NAME-M
002970             PERFORM REWRITE-MASTER
002980             GENERATE NAME-CHANGE-LINE.
002990
003000 CHECK-MONEY-AMOUNT.
003010     IF MONEY-AMOUNT-IN NOT NUMERIC
003020         MOVE MONEY-AMOUNT-IN-X      TO INVALID-MONEY-AMOUNT
003030         MOVE MONEY-AMOUNT-INVALID TO MESSAGE-W
003040         MOVE 'YES'                  TO HAVE-ANY-ERRORS-BEEN-FOUND
003050         GENERATE ERROR-LINE.
003060
003070 CHECK-FOR-DEPOSITOR-NAME.
003080     IF DEPOSITOR-NAME-IS-MISSING
003090         MOVE 'DEPOSITOR NAME MISSING' TO MESSAGE-W
003100         MOVE 'YES' TO HAVE-ANY-ERRORS-BEEN-FOUND
003110         GENERATE ERROR-LINE.
003120
003130 READ-MASTER.
003140     MOVE ACCOUNT-NUMBER-IN TO ACCOUNT-NUMBER-W.
003150     READ ACCOUNT-MASTER-FILE
003160         INVALID KEY
003170             MOVE 'ACCOUNT NUMBER NOT ON FILE' TO MESSAGE-W
003180             MOVE 'YES' TO HAVE-ANY-ERRORS-BEEN-FOUND
003190             GENERATE ERROR-LINE.
003200
003210 REWRITE-MASTER.
003220     MOVE RUN-DATE TO DATE-OF-LAST-TRANSACTION-M.
003230     REWRITE ACCOUNT-MASTER-RECORD.
```

Program 28 was run using the transaction input data shown in Figure 11.9. The output is shown in Figure 11.10.

Figure 11.9 Transaction input for Program 28.

```
300014001CC000
200091001500000
300168000011C00
200308000005000
400266
10035700001C00ROBIN RATANSKI
400140            BETH DENNY
200210000025000
200063000007500
200077000015000
200280002317O0
200161001256З4
100364001500000JOSE TORRES
100308000017000AL MARRELLA
700427000025000GREG PRUITT
200133002563000
5
200182
80031500013798
10037100015000AL ISE MARKOVITZ
10037100001000THOMAS HERR
20002800015327
20009100120000
20021000025000
30011900002735
400084            JOHN & SALLY DUPRINO
20008400357429
20027300172500
70038500007500JAMES WASHINGTCN
10039200007500INEZ WASHINGTON
10039900014Z00GARY NASTI
10039800001000JUAN ALVAREZ
20002100012750
500232
30002800150050
400098            GEORGE & ANN CULFANE
500182
300259293S0000
20030100005763
20037100015000
10040600120000
20012600035000
20017500019202
500168
10041300010000BILL HAYES
10042000150000JOHN RICE
20021000025000
200007)))$%)))
20032200006875
30009100050000
```

Figure 11.10 Transaction register output from Program 28, Part 1 of 2.

```
                          ROBBEM STATE BANK
                          106 WEST 10TH ST.
                          BROOKLYN, NY  11212

             SAVINGS ACCOUNT TRANSACTION REGISTER
DATE  2/27/80                                         PAGE  1


   ACCOUNT       DEPOSITS          WITHDRAWALS          NOTES
   NUMBER

    00014                            1,000.00
    00091         1,500.00
    00168                              110.00
    00308            50.00
    00266                                              DEPOSITOR NAME MISSING
    00357            10.00                             NEW ACCOUNT
    00140       BETH DENNY                             NAME CHANGE
    00210           250.00
    00063            75.00
    00077           150.00
    00280         2,317.00
    00161         1,256.34
    00364         1,500.00                             NEW ACCOUNT
    00308                                              ACCOUNT NUMBER ALREADY ON FILE.  NEW ACCOUNT INVALID.
    00427                                              INVALID TRANSACTION CODE 7
    00133         2,563.00
                                                       ACCOUNT NUMBER NOT ON FILE
    00182                                              DEPOSIT     AMOUNT NOT NUMERIC
    00315                                              INVALID TRANSACTION CODE 8
    00371           150.00                             NEW ACCOUNT
    00371                                              ACCOUNT NUMBER ALREADY ON FILE.  NEW ACCOUNT INVALID.
    00028           153.27
    00091         1,200.00
    00210           250.00
    00119                               27.35
    00084       JOHN & SALLY DUPRINC                   NAME CHANGE
```

Figure 11.10 Transaction register output from Program 28, Part 2 of 2.

```
                SAVINGS ACCOUNT TRANSACTION REGISTER
DATE   2/27/80                                                PAGE  2

     ACCOUNT       DEPOSITS        WITHDRAWALS         NOTES
     NUMBER

      00084       3,574.29
      00273       1,725.00
      00385                                          INVALID TRANSACTION CODE 7
      00392          75.00                           NEW ACCOUNT
      00399                                          DEPOSIT     AMOUNT NOT NUMERIC 00014Z00
      00398          10.00                           NEW ACCOUNT
      00021         127.50
      00232                                          ACCOUNT NUMBER NOT ON FILE
      00028                         1,500.50
      00098      GEORGE & ANN OULHANE                NAME CHANGE
      00182                                          ACCOUNT NUMBER NOT ON FILE
      00259                                          ACCOUNT NUMBER NOT ON FILE
      00301          57.63
      00371         150.00
      00406                                          DEPOSITOR NAME MISSING
      00126         350.00
      00175         192.02
      00168                            15.50         ACCOUNT CLOSED
      00413         100.00                           NEW ACCOUNT
      00420       1,500.00                           NEW ACCOUNT
      00210         250.00
      00007                                          DEPOSIT     AMOUNT NOT NUMERIC )))$%)))
      00322          68.75
      00091                           500.00

TOTALS           19,604.80         3,153.35

                         CONTROL COUNTS

                  NUMBER OF DEPOSITS          27

                  NUMBER OF WITHDRAWALS        6

                  NUMBER OF NAME CHANGES       3

                  NUMBER OF ERRORS            14

                  TOTAL                       50
```

***Exercise 3.**

Write a program to update the accounts receivable master file that you created in Exercise 2. Use the input transaction format shown in Figure 11.E3. Use the following transaction codes:

Transaction Code	Transaction Type
1	New account
2	Purchase
3	Payment
4	Credit limit change
5	Customer name change
6	Customer street address change
7	City, state, ZIP code change
8	Account closed

For a purchase, check that the purchase amount is numeric and add it to the correct balance in the master. For a payment, check for numeric and subtract it from the balance. For a credit limit change, check that the new limit is numeric and use it to replace the old limit in the master. For a new account, the fields that must be present are the account number, the credit limit, and the customer name and full address. For a closed account, delete the record from the master file.

Design a transaction register to show the disposition of every transaction, whether the transaction is in error or whether it is successfully applied to the master. If any transaction for any record causes the current balance to exceed the credit limit, print the words CREDIT LIMIT EXCEEDED on the register next to the transaction.

At the end of the transaction register, show the total of the purchase amounts and the total of the payment amounts. Also show the number of error-free transactions of each type, the number of erroneous transactions, and the total number of transactions.

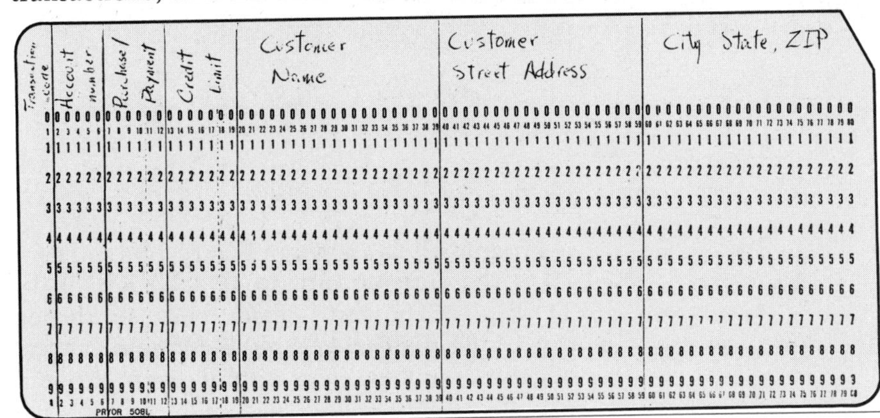

Figure 11.E3
Transaction input format for Exercise 3.

Listing the Contents of an Indexed Sequential File

An indexed sequential file can easily be processed sequentially, and Program 29 READs our savings account file from beginning to end and lists its contents. You will see in the listing that all the updates we did in Program 28 took effect: the deleted record is gone, the new accounts are in, the current balances are correct, and all records have their correct last transaction date. The output format for the list is shown in Figure 11.11.

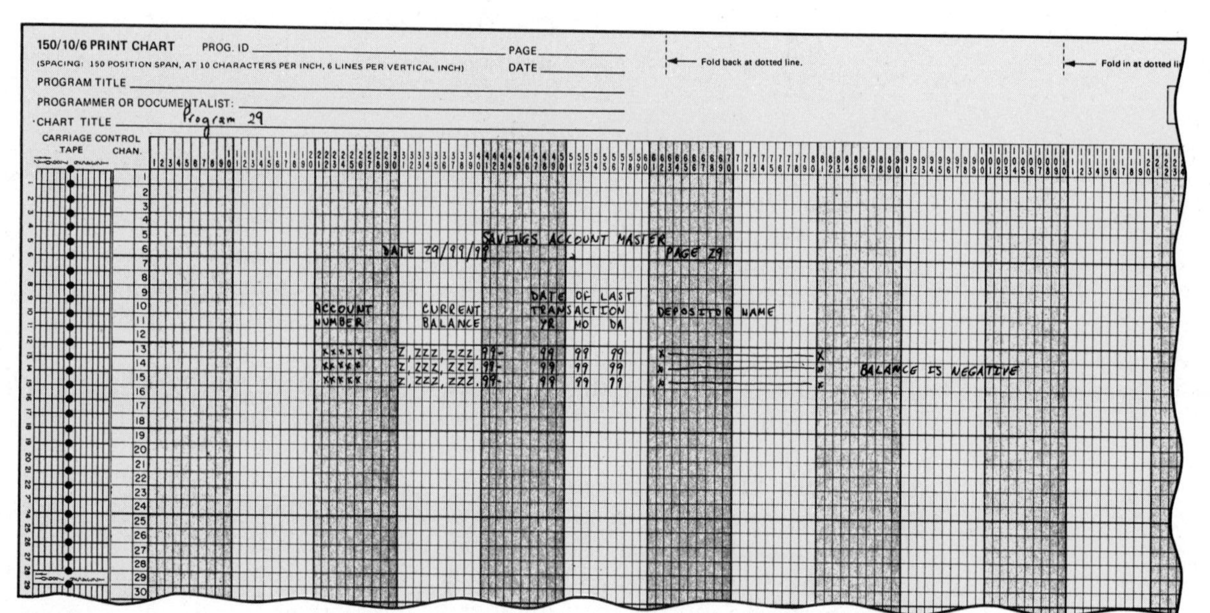

Figure 11.11 Output format for Program 29.

Program 29 is shown in Figure 11.12. The only thing new in Program 29 is the **sign condition** IF CURRENT–BALANCE–M IS NEGATIVE . . . . The test is made because, through an error of some sort, the CURRENT–BALANCE–M field on the master might contain a negative number. The output from Program 29 is shown in Figure 11.13.

Figure 11.12 Program 29, Part 1 of 2.

```
4-CB2 V4 RELEASE 1.5 10NOV77          IBM OS AMERICAN NATIONAL STANDARD COBOL



000010 IDENTIFICATION DIVISION.
000020 PROGRAM-ID. PROG29.
000030*AUTHOR. ELLEN M. LONGO.
000040*
000050*    THIS PROGRAM LISTS AN INDEXED SEQUENTIAL FILE.
000060*
000070*****************************************************************
000080
000090 ENVIRONMENT DIVISION.
000100 INPUT-OUTPUT SECTION.
000110 FILE-CONTROL.
000120     SELECT MASTER-LIST-FILE     ASSIGN TO UT-S-SYSPRINT.
000130     SELECT ACCOUNT-MASTER-FILE ASSIGN TO DA-I-DISK
000140        RECORD KEY IS ACCOUNT-NUMBER-M.
000150
000160*****************************************************************
000170
000180 DATA DIVISION.
000190 FILE SECTION.
000200 FD  ACCOUNT-MASTER-FILE
000210     LABEL RECORDS ARE STANDARD
000220     BLOCK CONTAINS 10 RECORDS.
000230
000240 01  ACCOUNT-MASTER-RECORD.
000250     05 DELETE-CODE                                    PIC X.
000260     05 DATE-OF-LAST-TRANSACTION-M.
000270        10 TRANSACTION-YEAR-M                          PIC 99.
000280        10 TRANSACTION-MONTH-M                         PIC 99.
000290        10 TRANSACTION-DAY-M                           PIC 99.
000300     05 ACCOUNT-NUMBER-M                               PIC X(5).
000310     05 CURRENT-BALANCE-M                              PIC S9(7)V99.
000320     05 DEPOSITOR-NAME-M                               PIC X(20).
000330
000340 FD  MASTER-LIST-FILE
000350     LABEL RECORDS ARE OMITTED
000360     REPORT IS MASTER-LIST.
000370
000380 WORKING-STORAGE SECTION.
000390 01  MORE-INPUT          VALUE   'YES'                 PIC X(3).
000400 01  MESSAGE-W                                         PIC X(19).
000410
000420 01  RUN-DATE.
000430     05 RUN-YEAR                                       PIC 99.
000440     05 RUN-MONTH                                      PIC 99.
000450     05 RUN-DAY                                        PIC 99.
000460
000470 REPORT SECTION.
000480 RD  MASTER-LIST
000490     PAGE 38 LINES, FIRST DETAIL 13.
```

Figure 11.12 Program 29, Part 2 of 2.

```
CCC500
C00510 01   TYPE PAGE HEADING.
CC0520      05 LINE 5.
000530          10 COLUMN 41  VALUE  'SAVINGS ACCCUNT MASTER'
000540                                                  PIC X(22).
C00550      05 LINE 6.
CC0560          10 COLUMN 29  VALUE  'DATE'             PIC X(4).
000570          10 COLUMN 34  SOURCE RUN-MONTH          PIC Z9.
C00580          10 COLUMN 36  VALUE  '/'                PIC X.
000590          10 COLUMN 37  SOURCE RUN-DAY            PIC 99.
C00600          10 COLUMN 39  VALUE  '/'                PIC X.
0C0610          10 COLUMN 40  SOURCE RUN-YEAR           PIC 99.
CC0620          10 COLUMN 63  VALUE  'PAGE'             PIC X(4).
000630          10 COLUMN 68  SOURCE PAGE-CCUNTER       PIC Z9.
CC0640      05 LINE 9
C00650             COLUMN 47  VALUE  'DATE OF LAST'     PIC X(12).
000660      05 LINE 10.
C00670          10 COLUMN 21  VALUE  'ACCCUNT'          PIC X(7).
000680          10 COLUMN 34  VALUE  'CURRENT'          PIC X(7).
000690          10 COLUMN 47  VALUE  'TRANSACTION'      PIC X(11).
CC0700          10 COLUMN 62  VALUE  'DEPCSITCR NAME'   PIC X(14).
CC0710      05 LINE 11.
0C0720          10 COLUMN 21  VALUE  'NUMBER'           PIC X(6).
000730          10 COLUMN 34  VALUE  'BALANCE'          PIC X(7).
000740          10 COLUMN 48  VALUE  'YR  MO  DA'       PIC X(10).
CC0750
CC0760 01   DETAIL-LINE
CC0770      TYPE DETAIL
000780      LINE PLUS 1.
CC0790      05 COLUMN 22        SOURCE ACCCUNT-NUMBER-M    PIC X(5).
CC0800      05 COLUMN 31        SOURCE CURRENT-BALANCE-M
00C810                          PIC Z,ZZZ,ZZZ.99-.
000820      05 COLUMN 48        SOURCE TRANSACTION-YEAR-M  PIC 99.
000830      05 COLUMN 52        SOURCE TRANSACTION-MONTH-M PIC 99.
0C0840      05 COLUMN 56        SOURCE TRANSACTION-DAY-M   PIC 99.
CC0850      05 COLUMN 62        SOURCE DEPCSITOR-NAME-M    PIC X(20).
CC0860      05 COLUMN 86        SOURCE MESSAGE-M           PIC X(19).
CC0870
CC0880****************************************************************************
CCC890
CCC900 PROCEDURE DIVISION.
000910      OPEN INPUT  ACCOUNT-MASTER-FILE,
CC0920           OUTPUT MASTER-LIST-FILE.
CCC930      READ ACCOUNT-MASTER-FILE
0C0940          AT END
CC0950              MOVE 'NO' TO MORE-INPUT.
000960      IF MORE-INPUT = 'YES'
CC0970          ACCEPT RUN-DATE FROM DATE
000980          INITIATE MASTER-LIST
CCC990          PERFORM MAIN-PROCESS UNTIL MORE-INPUT = 'NO'
001000          TERMINATE MASTER-LIST.
CC1010      CLOSE ACCCUNT-MASTER-FILE,
001020            MASTER-LIST-FILE.
001030      STOP RUN.
```

Figure 11.12 Part 2 (continued).

```
001040
001050 MAIN-PROCESS.
001060     IF CURRENT-BALANCE-W IS NEGATIVE
001070         MOVE 'BALANCE IS NEGATIVE' TO MESSAGE-W
001080     ELSE
001090         MOVE SPACES TO MESSAGE-W.
001100     GENERATE DETAIL-LINE.
001110     READ ACCOUNT-MASTER-FILE
001120         AT END
001130             MOVE 'NO' TO MORE-INPUT.
```

Figure 11.13 Output from Program 29, Part 1 of 2.

```
                        SAVINGS ACCOUNT MASTER
         DATE  2/27/80                                    PAGE  1


                              DATE OF LAST
ACCOUNT        CURRENT        TRANSACTION     DEPOSITOR NAME
NUMBER         BALANCE         YR  MO  DA

 00007        1,007.84         80  02  26     ROSEBUCCI
 00014          990.00-        80  02  27     ROBERT DAVIS M.D.           BALANCE IS NEGATIVE
 00021          252.50         80  02  27     LORICE MONTI
 00028        5,654.36         80  02  27     MICHAEL SMITH
 00035          150.00         80  02  26     JOHN J. LEHMAN
 00049          150.00         80  02  26     JAY GREENE
 00056            1.00         80  02  26     EVELYN SLATER
 00063          150.00         80  02  27     LORRAINE SMALL
 00070          500.00         80  02  26     PATRICK J. LEE
 00077          160.37         80  02  27     LESLIE MINSKY
 00084        5,074.29         80  02  27     JOHN & SALLY DUPRINO
 00091        2,300.00         80  02  27     JOE'S DELI
 00098          500.00         80  02  27     GEORGE & ANN CULHANE
 00105           50.00         80  02  26     LENORE MILLER
 00112          250.00         80  02  26     ROSEMARY LANE
 00119          472.65         80  02  27     MICHELE CAPUANO
 00126        1,100.00         80  02  27     JAMES BUDD
 00133        3,563.00         80  02  27     PAUL LERNER, D.D.S.
 00140           25.75         80  02  27     BETH DENNY
 00154           50.00         80  02  26     ONE DAY CLEANERS
 00161        1,280.84         80  02  27     ROBERT RYAN
 00175          202.02         80  02  27     MARY KEATING
 00189           35.00         80  02  26     J. & L. CAIN
 00196          150.00         80  02  26     IMPERIAL FLORIST
 00203            5.00         80  02  26     JOYCE MITCHELL
 00210        1,000.00         80  02  27     JERRY PARKS
```

Figure 11.13 Output from Program 29, Part 2 of 2.

SAVINGS ACCOUNT MASTER

DATE 2/27/80 PAGE 2

ACCOUNT NUMBER	CURRENT BALANCE	DATE OF LAST TRANSACTION YR	MO	DA	DEPOSITOR NAME
00217	50.00	80	02	26	CARL CALDERON
00224	175.50	80	02	26	JOHN WILLIAMS
00231	555.00	80	02	26	BILL WILLIAMS
00238	10.00	80	02	26	KEVIN PARKER
00245	35.00	80	02	26	FRANK CAPUTO
00266	99.57	80	02	26	MARTIN LANG
00273	2,000.00	80	02	27	VITO CACACI
00280	2,337.00	80	02	27	COMMUNITY DRUGS
00287	15.00	80	02	26	SOLOMON CHAPELS
00294	1,500.00	80	02	26	JOHN BURKE
00301	215.13	80	02	27	PAT P. POWERS
00308	2,050.00	80	02	27	JOE GARCIA
00315	250.00	80	02	26	GRACE MICELI
00322	88.75	80	02	27	MARIA FASANO
00329	10.00	80	02	26	GUY VOLPONE
00343	1,000.00	80	02	26	JOE & MARY SESSA
00350	2,500.00	80	02	26	ROGER SHAW
00357	10.00	80	02	27	ROBIN RATANSKI
00364	1.500.00	80	02	27	JOSE TORRES
00371	300.00	80	02	27	ALISE MARKOVITZ
00392	75.00	80	02	27	INEZ WASHINGTON
00398	10.00	80	02	27	JUAN ALVAREZ
00413	100.00	80	02	27	BILL HAYES
00420	1,500.00	80	02	27	JOHN RICE

The Sign Condition

The general format of the sign condition is:

arithmetic-expression IS [NOT] { POSITIVE | NEGATIVE | ZERO }

Zero values are considered neither positive nor negative; unsigned fields are considered positive unless the value assigned is zero.

Exercise 4. Write a program to list the contents of the accounts-receivable master file that you updated in Exercise 3. Print the words CREDIT LIMIT EXCEEDED next to any record whose current balance exceeds its credit limit.

Creating a Sequential File on Tape

We will now do the entire savings account application over again, this time using as our master file a **standard sequential** file on tape. We will create the master file by running Program 30 with the same input transactions that we used for Program 27. Of course the transactions must be in order on account number and so will the master be. The format of the transaction register is the same as in Program 27 when we created an indexed sequential master.

Program 30 is shown in Figure 11.14. There are very few differences between this and Program 27. The SELECT sentence for the ACCOUNT–MASTER–FILE no longer has the RECORD KEY clause, for that clause is not permitted on sequential files. The DELETE–CODE field is no longer needed in the ACCOUNT–MASTER–RECORD, for COBOL has no built-in deletion procedure for sequential files. You will see how to delete a record from a sequential file when we look at the update program later. COBOL also does not check for out-of-sequence and duplicate key fields when it is processing sequential files, so we need the ACCOUNT–NUMBER–SAVE field in working storage in order to make the sequence check ourselves.

Figure 11.14 Program 30, Part 1 of 3.

```
4-CB2 V4 RELEASE 1.5 10NOV77          IBM OS AMERICAN NATIONAL STANDARD COBOL


  000010 IDENTIFICATION DIVISION.
  000020 PROGRAM-ID. PROG30.
  000030*AUTHOR. ELLEN M. LONGO.
  000040*
  000050*     THIS PROGRAM CREATES A SEQUENTIAL MASTER
  000060*     FILE OF SAVINGS ACCOUNT RECORDS.
  000070*
  000080*****************************************************************
```

Figure 11.14 Program 30, Part 2 of 3.

```
000090
000100 ENVIRONMENT DIVISION.
000110 INPUT-OUTPUT SECTION.
000120 FILE-CONTROL.
000130     SELECT TRANSACTION-REGISTER-FILE ASSIGN TO UT-S-SYSPRINT.
000140     SELECT TRANSACTION-FILE-IN       ASSIGN TO UT-S-SYSIN.
000150     SELECT ACCOUNT-MASTER-FILE       ASSIGN TO UT-S-TAPE.
000160
000170*****************************************************************
000180
000190 DATA DIVISION.
000200 FILE SECTION.
000210 FD  TRANSACTION-FILE-IN
000220     LABEL RECORDS ARE OMITTED.
000230
000240 01  TRANSACTION-RECORD-IN.
000250     05 TRANSACTION-CODE-IN                          PIC X.
000260        88 TRANSACTION-IS-NEW-ACCOUNT                    VALUE '1'.
000270     05 ACCOUNT-NUMBER-IN                            PIC X(5).
000280        88 ACCOUNT-NUMBER-IS-MISSING                     VALUE SPACES.
000290     05 INITIAL-DEPOSIT-IN                           PIC 9(6)V99.
000300     05 INITIAL-DEPOSIT-IN-X
000310        REDEFINES INITIAL-DEPOSIT-IN                 PIC X(8).
000320     05 DEPOSITOR-NAME-IN                            PIC X(20).
000330        88 DEPOSITOR-NAME-IS-MISSING                     VALUE SPACES.
000340     05 FILLER                                       PIC X(46).
000350
000360 FD  ACCOUNT-MASTER-FILE
000370     LABEL RECORDS ARE STANDARD
000380     BLOCK CONTAINS 50 RECORDS.
000390
000400 01  ACCOUNT-MASTER-RECORD.
000410     05 DATE-OF-LAST-TRANSACTION-M                   PIC 9(6).
000420     05 ACCOUNT-NUMBER-M                             PIC X(5).
000430     05 CURRENT-BALANCE-M                            PIC S9(7)V99.
000440     05 DEPOSITOR-NAME-M                             PIC X(20).
000450
000460 FD  TRANSACTION-REGISTER-FILE
000470     LABEL RECORDS ARE OMITTED
000480     REPORT IS TRANSACTION-REGISTER.
000490
000500 WORKING-STORAGE SECTION.
000510 01  ACCOUNT-NUMBER-SAVE
000520                          VALUE  LOW-VALUES         PIC X(5).
000530 01  MORE-INPUT           VALUE  'YES'              PIC X(3).
000540
000550 01  RUN-DATE.
000560     05 RUN-YEAR                                    PIC 99.
000570     05 RUN-MONTH                                   PIC 99.
000580     05 RUN-DAY                                     PIC 99.
000590
000600 01  ONE                  VALUE  1                  PIC S9.
000610 01  MESSAGE-W                                      PIC X(53).
000620
000630 01  ERROR-MESSAGES.
000640     05 INVALID-TRANSACTION-CODE.
```

Figure 11.14 Part 2 (continued).

```
000650         10 FILLER      VALUE   'INVALID TRANSACTICN CODE'
000660                                                        PIC X(25).
000670         10 TRANSACTICN-CODE-E                          PIC X.
000680      05 DEPOSIT-INVALID.
000690         10 FILLER      VALUE   'DEPCSIT AMCLNT NOT NUMERIC'
000700                                                        PIC X(30).
000710         10 INITIAL-DEPOSIT-E                           PIC X(8).
000720      05 OUT-OF-SEQUENCE
000730         VALUE         'DUPLICATE OR CUT-OF-SEQUENCE ACCOUNT NUMBER.'
000740                                                        PIC X(44).
000750 REPORT SECTION.
000760 RD   TRANSACTICN-REGISTER
000770      CONTROL FINAL
000780      PAGE 52 LINES,
000790      FIRST DETAIL 13,
000800      LAST DETAIL 38,
000810      FOOTING 52.
000820
000830 01   TYPE REPORT HEADING.
000840      05 LINE 2.
000850         10 COLUMN 40   VALUE   'ROEBEM STATE BANK'
000860                                                        PIC X(17).
000870      05 LINE 3.
000880         10 COLUMN 40   VALUE   '106 WEST 10TH ST.'
000890                                                        PIC X(17).
000900      05 LINE 4.
000910         10 COLUMN 39   VALUE   'BRCCKLYN, NY  11212'
000920                                                        PIC X(19).
000930
000940 01   TYPE PAGE HEADING.
000950      05 LINE 6.
000960         10 COLUMN 30   VALUE
000970                        'SAVINGS ACCCUNT TRANSACTION REGISTER'
000980                                                        PIC X(36).
000990      05 LINE 7.
001000         10 COLUMN 18   VALUE   'DATE'                  PIC X(4).
001010         10 COLUMN 23   SOURCE RUN-MONTH                PIC Z9.
001020         10 COLUMN 25   VALUE   '/'                     PIC X.
001030         10 COLUMN 26   SOURCE RUN-DAY                  PIC 99.
001040         10 COLUMN 28   VALUE   '/'                     PIC X.
001050         10 COLUMN 29   SOURCE RUN-YEAR                 PIC 99.
001060         10 COLUMN 66   VALUE   'PAGE'                  PIC X(4).
001070         10 COLUMN 71   SOURCE PAGE-CCLNTER             PIC Z9.
001080      05 LINE 10.
001090         10 COLUMN 21   VALUE   'ACCCUNT'               PIC X(7).
001100         10 COLUMN 33   VALUE   'DEPOSITS'              PIC X(8).
001110         10 COLUMN 47   VALUE   'WITHDRAWALS'           PIC X(11).
001120         10 COLUMN 65   VALUE   'NOTES'                 PIC X(5).
001130      05 LINE 11.
001140         10 COLUMN 21   VALUE   'NUMBER'                PIC X(6).
001150
001160 01   DEPOSIT-LINE
001170      TYPE DETAIL
001180      LINE PLUS 1.
001190      05 COLUMN 22      SOURCE ACCCUNT-NUMBER-IN  PIC X(5).
001200      05 COLUMN 32      SOURCE INITIAL-DEPCSIT-IN PIC ZZZ,ZZZ.99.
001210      05 COLUMN 63      VALUE 'NEW ACCCUNT'       PIC X(11).
001220
```

In the Procedure Division, the structure of the IF case has been changed slightly, since we no longer have the INVALID KEY clause in the WRITE statement. (The INVALID KEY clause was used in Program 27 to detect duplicate and out-of-sequence account numbers, and that feature is not available in this program.) Instead, whenever we WRITE an ACCOUNT–MASTER–RECORD, we save its account number in ACCOUNT–NUMER–SAVE. Then, before we WRITE another ACCOUNT–MASTER–RECORD, we check that the account number we are about to WRITE is greater than the previous one written. If the account number we are about to WRITE is NOT GREATER than the previous, we have an out-of-sequence or duplicate number.

Figure 11.14 Program 30, Part 3 of 3.

```
001230 01   ERROR-LINE
001240      TYPE DETAIL
001250      LINE PLUS 1.
001260      05 COLUMN 22      SOURCE ACCOUNT-NUMBER-IN    PIC X(5).
001270      05 COLUMN 63      SOURCE MESSAGE-W            PIC X(53).
001280
001290 01   TYPE CONTROL FOOTING FINAL.
001300      05 LINE PLUS 2.
001310         10 COLUMN 18   VALUE   'TOTALS'            PIC X(6).
001320         10 COLUMN 30   SUM     INITIAL-DEPOSIT-IN
001330                        UPON    DEPOSIT-LINE        PIC Z,ZZZ,ZZZ.99.
001340      05 LINE PLUS 2
001350            COLUMN 41   VALUE   'CONTROL COUNTS'    PIC X(14).
001360      05 LINE PLUS 2.
001370         10 COLUMN 35   VALUE   'NUMBER OF DEPOSITS'
001380                                                    PIC X(18).
001390         10 NUMBER-OF-DEPOSITS
001400            COLUMN 58   SUM     ONE
001410                        UPON    DEPOSIT-LINE        PIC ZZ9.
001420      05 LINE PLUS 2.
001430         10 COLUMN 35   VALUE   'NUMBER OF WITHDRAWALS'
001440                                                    PIC X(21).
001450         10 COLUMN 60   VALUE   0                   PIC 9.
001460      05 LINE PLUS 2.
001470         10 COLUMN 35   VALUE   'NUMBER OF NAME CHANGES'
001480                                                    PIC X(22).
001490         10 COLUMN 60   VALUE   0                   PIC 9.
001500      05 LINE PLUS 2.
001510         10 COLUMN 35   VALUE   'NUMBER OF ERRORS'
001520                                                    PIC X(16).
001530         10 NUMBER-OF-ERRORS
001540            COLUMN 58   SUM     ONE
001550                        UPON    ERROR-LINE          PIC ZZ9.
001560      05 LINE PLUS 2.
001570         10 COLUMN 35   VALUE   'TOTAL'             PIC X(5).
001580         10 COLUMN 56   SUM     NUMBER-OF-DEPOSITS,
001590                                NUMBER-OF-ERRORS    PIC Z,ZZ9.
```

Figure 11.14 Part 3 (continued).

```
001600
001610*****************************************************************************
001620
001630 PROCEDURE DIVISION.
001640     OPEN INPUT  TRANSACTION-FILE-IN,
001650          OUTPUT TRANSACTION-REGISTER-FILE,
001660                 ACCOUNT-MASTER-FILE.
001670     READ TRANSACTION-FILE-IN
001680         AT END
001690             MOVE 'NO' TO MORE-INPUT.
001700     IF MORE-INPUT = 'YES'
001710         ACCEPT RUN-DATE FROM DATE
001720         INITIATE TRANSACTION-REGISTER
001730         PERFORM MAIN-PROCESS UNTIL MORE-INPUT = 'NO'
001740         TERMINATE TRANSACTION-REGISTER.
001750     CLOSE TRANSACTION-FILE-IN,
001760           TRANSACTION-REGISTER-FILE,
001770           ACCOUNT-MASTER-FILE.
001780     STOP RUN.
001790
001800 MAIN-PROCESS.
001810     IF NOT TRANSACTION-IS-NEW-ACCOUNT
001820         MOVE TRANSACTION-CODE-IN      TO TRANSACTION-CODE-E
001830         MOVE INVALID-TRANSACTION-CODE TO MESSAGE-W
001840         GENERATE ERROR-LINE
001850     ELSE
001860     IF ACCOUNT-NUMBER-IS-MISSING
001870         MOVE 'ACCOUNT NUMBER MISSING' TO MESSAGE-W
001880         GENERATE ERROR-LINE
001890     ELSE
001900     IF INITIAL-DEPOSIT-IN NOT NUMERIC
001910         MOVE INITIAL-DEPOSIT-IN-X TO INITIAL-DEPOSIT-E
001920         MOVE DEPOSIT-INVALID      TO MESSAGE-W
001930         GENERATE ERROR-LINE
001940     ELSE
001950     IF DEPOSITOR-NAME-IS-MISSING
001960         MOVE 'DEPOSITOR NAME MISSING' TO MESSAGE-W
001970         GENERATE ERROR-LINE
001980     ELSE
001990     IF ACCOUNT-NUMBER-IN NOT GREATER THAN ACCOUNT-NUMBER-SAVE
002000         MOVE OUT-OF-SEQUENCE TO MESSAGE-W
002010         GENERATE ERROR-LINE
002020     ELSE
002030         MOVE RUN-DATE             TO  DATE-OF-LAST-TRANSACTION-M
002040         MOVE ACCOUNT-NUMBER-IN   TO ACCOUNT-NUMBER-M
002050         MOVE INITIAL-DEPOSIT-IN  TO CURRENT-BALANCE-M
002060         MOVE DEPOSITOR-NAME-IN   TO DEPOSITOR-NAME-M
002070         MOVE ACCOUNT-NUMBER-IN   TO ACCOUNT-NUMBER-SAVE
002080         WRITE ACCOUNT-MASTER-RECORD
002090         GENERATE DEPOSIT-LINE.
002100     READ TRANSACTION-FILE-IN
002110         AT END
002120             MOVE 'NO' TO MORE-INPUT.
```

Program 30 produced the transaction register shown in Figure 11.15.

Figure 11.15 Output from Program 30, Part 1 of 2.

```
                         ROBBEM STATE BANK
                         106 WEST 10TH ST.
                        BROCKLYN, NY  11212

                SAVINGS ACCCUNT TRANSACTICN REGISTER
  DATE  2/26/8C                                          PAGE  1

     ACCOUNT     DEPCSITS        WITHDRAWALS          NOTES
     NUMBER

      C000C7     1,CC7.84                           NEW ACCOUNT
      00014         1C.C0                           NEW ACCCUNT
      C0021        125.00                           NEW ACCCUNT
      00028      7,C01.59                           NEW ACCCUNT
      00035        15C.00                           NEW ACCCUNT
      00032                                         DUPLICATE OR CUT-CF-SEQUENCE ACCOUNT NUMBER.
      00049        15C.00                           NEW ACCOUNT
      00056          1.C0                           NEW ACCOUNT
      00063         75.00                           NEW ACCOUNT
      00070        5CC.00                           NEW ACCOUNT
      00077         10.37                           NEW ACCOUNT
      00084      1,5CC.C0                           NEW ACCOUNT
      00091        1CC.C0                           NEW ACCOUNT
      00098        500.C0                           NEW ACCOUNT
      00105         50.C0                           NEW ACCCUNT
      00112        250.00                           NEW ACCCUNT
      00119        5CC.C0                           NEW ACCCUNT
      00126        750.00                           NEW ACCOUNT
      00133      1,CCC.00                           NEW ACCOUNT
      00140         25.75                           NEW ACCOUNT
      00                                            DUPLICATE OR CUT-CF-SEQUENCE ACCOUNT NUMBER.
      00154         5C.00                           NEW ACCOUNT
      00161         24.50                           NEW ACCOUNT
      00168        125.50                           NEW ACCOUNT
      00175         10.C0                           NEW ACCOUNT
      00182                                         INVALID TRANSACTICN CODE 2
```

Figure 11.15 Output from Program 30, Part 2 of 2.

```
                SAVINGS ACCOUNT TRANSACTION REGISTER
   DATE   2/26/80                                          PAGE   2

      ACCOUNT       DEPOSITS        WITHDRAWALS        NOTES
      NUMBER

       00189           35.00                         NEW ACCOUNT
       00196          150.00                         NEW ACCOUNT
       00203            5.00                         NEW ACCOUNT
       00210          250.00                         NEW ACCOUNT
       00217           50.00                         NEW ACCOUNT
       00224          175.50                         NEW ACCOUNT
       00231          555.00                         NEW ACCOUNT
       00238           10.00                         NEW ACCOUNT
       00245           35.00                         NEW ACCOUNT
       00252                                         DEPOSITOR NAME MISSING
       00259                                         DEPOSITOR NAME MISSING
       00266           99.57                         NEW ACCOUNT
       00273          275.00                         NEW ACCOUNT
       00280           20.00                         NEW ACCOUNT
       00287           15.00                         NEW ACCOUNT
       00294        1,500.00                         NEW ACCOUNT
       00301          157.50                         NEW ACCOUNT
       00308        2,000.00                         NEW ACCOUNT
       00315          250.00                         NEW ACCOUNT
       00322           20.00                         NEW ACCOUNT
       00329           10.00                         NEW ACCOUNT
       00336                                         DEPOSIT AMOUNT NOT NUMERIC      )))) %))
       00343        1,000.00                         NEW ACCOUNT
       00350        2,500.00                         NEW ACCOUNT

   TOTALS          23,029.12

                          CONTROL COUNTS

                  NUMBER OF DEPOSITS          44

                  NUMBER OF WITHDRAWALS        0

                  NUMBER OF NAME CHANGES       0

                  NUMBER OF ERRORS             6

                  TOTAL                       50
```

***Exercise 5.** Write a program to create a sequential accounts receivable file. Use the same input that you used in Exercise 2.

Create your sequential file on magnetic tape if your computer has facilities for doing so. Otherwise, create the sequential file on disk.

Updating a Sequential File on Tape

Two problems arise when we try to update a master file on tape that we did not have when updating one on disk. First, we don't have random-access capability on tape. This means that, if we want to fetch a master record off the file, we can't rely on the COBOL system to just go and READ it. The program will have to find the master record it wants by READing record after record on the master file and looking at each until it finds the right one.

Another difficulty is that once a master record has been updated in computer storage and is ready to be rewritten back onto the file, tape does not provide a way for an existing record to be erased and overlaid with a new one. If our sequential file were stored on disk, this second difficulty would be eliminated: Once a proper master record were found, read, updated in storage, and ready to be rewritten, the new record could be placed directly into its proper location in the sequential file on disk. Of course, even if our file were on disk, the first problem, that of finding the correct master record in the first place, would still be with us, for in a sequential file on disk the program must still READ every record and look at it to see if it's the one it wants.

All of these matters can be resolved by having the update transactions come into the program in account number sequence. The program READs the first transaction and then READs through the master trying to find the matching account number. As it is READing through unwanted master records, it WRITEs the unchanged master records onto a new tape that will become our new updated master file. When the program finds the proper master record, it updates it in the input area and then WRITEs the updated master record onto the new tape. If a record is to be added to the master file, the program constructs the new master record and then WRITEs it onto the new master tape in its proper sequence. If a master record is to be deleted, the program simply refrains from writing it onto the new master tape. During all this, the account numbers in the transaction records must be sequence checked, for in a sequential update an out-of-sequence transaction can play havoc with the program logic.

Program 31 allows for the possibility that there may be more than one transaction against a single master record; that is, an account may have a deposit and a withdrawal in the same day. Program 31 even handles the situation where an account is created and has some other transaction against it in the same day. Similarly, an account may have some transaction and then be closed in the same day. In such a case, the transaction that opens the account must come before any other transactions for the account and a transaction that closes an account must be the last one of all the transactions against the account.

A hierarchy diagram for Program 31 is shown in Figure 11.16. Each numbered box corresponds to a numbered paragraph in the Procedure

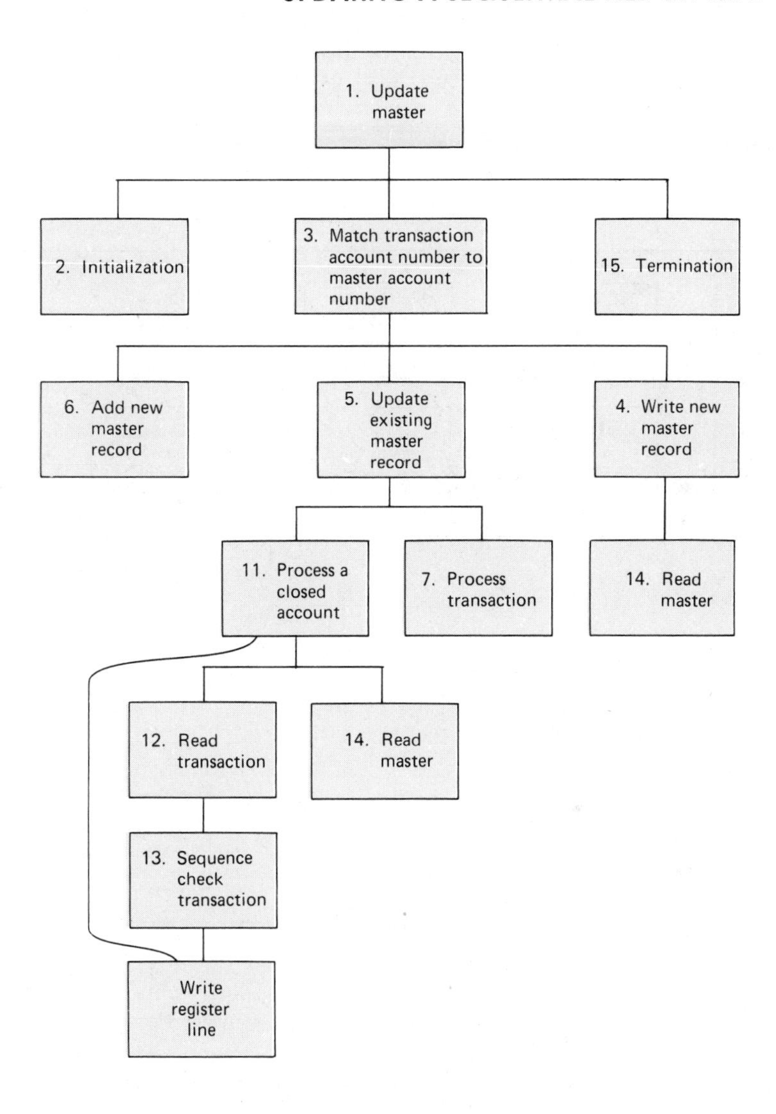

Figure 11.16
Hierarchy diagram for Program 31. Part (a).

Division of the program. Boxes without numbers represent functions so small that they were pushed up into the paragraphs that PERFORM them.

The box "Match transaction account number to master account number" shows that the program has to READ through the master file looking for a record to match a given transaction. In its search through the master file, it may sometimes find that there is no master record to match the transaction, which will occur when a new account is to be added to the file or when the transaction is in error. Program 31 allows for both possibilities.

Figure 11.16 Part (b).

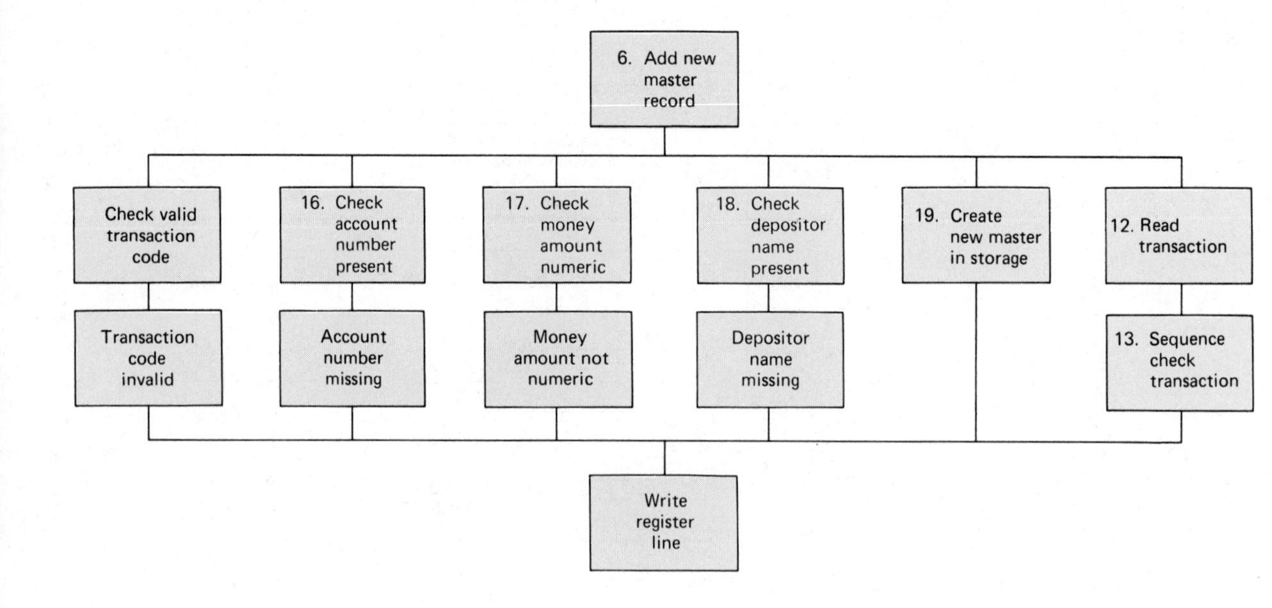

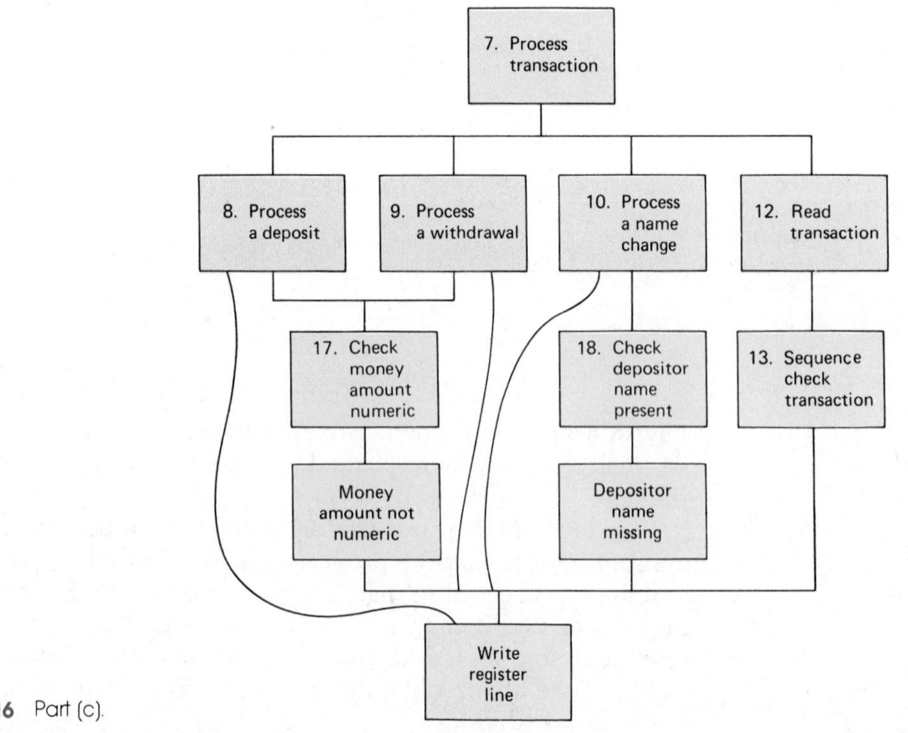

Figure 11.16 Part (c).

Program 31 is shown in Figure 11.17. We now need four files since the master file is represented by both an old input master tape and a new updated master tape.

In this program all of our field descriptions are in working storage. All the entries in the File Section are merely dummy entries required by COBOL. In the Working Storage Section there are two areas for master records, ACCOUNT–MASTER–W and ACCOUNT–MAS-

Figure 11.17 Program 31, Part 1 of 5.

```
4-CB2 V4 RELEASE 1.5 10NOV77          IBM OS AMERICAN NATIONAL STANDARD COBOL

000010 IDENTIFICATION DIVISION.
000020 PROGRAM-ID. PROG31.
000030*AUTHOR. ELLEN M. LONGO.
000040*
000050*     THIS PROGRAM UPDATES A
000060*     SEQUENTIAL MASTER FILE ON TAPE.
000070*
000080*****************************************************************
000090
000100 ENVIRONMENT DIVISION.
000110 INPUT-OUTPUT SECTION.
000120 FILE-CONTROL.
000130     SELECT ACCOUNT-MASTER-IN           ASSIGN TO UT-S-TAPE1.
000140     SELECT ACCOUNT-MASTER-OUT          ASSIGN TO UT-S-TAPE2.
000150     SELECT TRANSACTION-FILE-IN         ASSIGN TO UT-S-SYSIN.
000160     SELECT TRANSACTION-REGISTER-FILE   ASSIGN TO UT-S-SYSPRINT.
000170
000180*****************************************************************
000190
000200 DATA DIVISION.
000210 FILE SECTION.
000220 FD  ACCOUNT-MASTER-IN
000230     LABEL RECORDS ARE STANDARD
000240     BLOCK CONTAINS 50 RECORDS.
000250
000260 01  ACCOUNT-RECORD-IN                          PIC X(40).
000270
000280 FD  ACCOUNT-MASTER-OUT
000290     LABEL RECORDS ARE STANDARD
000300     BLOCK CONTAINS 50 RECORDS.
000310
000320 01  ACCOUNT-RECORD-OUT                         PIC X(40).
000330
000340 FD  TRANSACTION-FILE-IN
000350     LABEL RECORDS ARE OMITTED.
000360
000370 01  TRANSACTION-RECORD-IN                      PIC X(80).
000380
000390 FD  TRANSACTION-REGISTER-FILE
000400     LABEL RECORDS ARE OMITTED
000410     REPORT IS TRANSACTION-REGISTER.
```

TER–N. The field ACCOUNT–MASTER–N is used to build new master records when new accounts are to be added to the file; it is also used as the update area when fields are to be changed in existing master records. Records will always be written onto the new master tape from ACCOUNT–MASTER–N. The other area, ACCOUNT–MASTER–W, is used as a temporary storage area for each master record read in from the old master tape. You will see how both areas are used when we look at the Procedure Division.

In the Working Storage Section the field ACCOUNT–NUMBER–IN–SAVE is used to check that the transactions all come in in sequence. As each transaction is processed, its account number is saved. When the next transaction is read, the program checks that its account number is equal to or higher than the account number in the previous transaction. Remember that, since this program allows for multiple transactions to a single master record, we could get several transactions, all with the same account number. When all the transactions with the same account number have been processed and a new higher transaction account number is read, then we know that the master record is completely updated and can be written onto the new master tape. The field TRANSACTION–SEQUENCE–CHECKED is used to signal the program that the current transaction is in proper sequence and may be processed.

In the Procedure Division the paragraphs have numbers as well as names to facilitate finding them. The main loop in this program, MATCH–TRANSACTION–TO–MASTER, continues to execute until both the TRANSACTION–FILE–IN and the ACCOUNT–MASTER–IN have reached end-of-file. You will see shortly how the figurative constant HIGH–VALUES can be used as a convenient end-of-file indicator.

We must process both input files to end-of-file in order to be sure that we have processed all the input transactions and have also copied onto the new tape all of the unchanged master records.

When MATCH–TRANSACTION–TO–MASTER is entered the first time, the account number of the just-read transaction is in ACCOUNT–NUMBER–IN and the account number of the just-read master record is in ACCOUNT–NUMBER–N and ACCOUNT–NUMBER–W, while the IF case tries to figure out what to do with this master record. You can see how the IF statement works if you imagine the very first transaction record and the very first master record. If the

Figure 11.17 Program 31, Part 2 of 5.

```
000420
000430 WORKING-STORAGE SECTION.
000440 01  TRANSACTION-RECORD-W.
000450     05 TRANSACTION-CODE-IN                          PIC X.
000460        88 TRANSACTION-IS-NEW-ACCOUNT                   VALUE '1'.
```

Figure 11.17 Part 2 (continued).

```
000470           88 TRANSACTION-IS-DEPOSIT                         VALUE '2'.
000480           88 TRANSACTION-IS-WITHDRAWAL                      VALUE '3'.
000490           88 TRANSACTION-IS-NAME-CHANGE                     VALUE '4'.
000500           88 TRANSACTION-IS-CLOSE-ACCOUNT                   VALUE '5'.
000510        05 ACCOUNT-NUMBER-IN                              PIC X(5).
000520           88 ACCOUNT-NUMBER-IS-MISSING                      VALUE SPACES.
000530        05 MONEY-AMOUNT-IN                                PIC 9(6)V99.
000540        05 MONEY-AMOUNT-IN-X
000550           REDEFINES MONEY-AMOUNT-IN                      PIC X(8).
000560        05 DEPOSITOR-NAME-IN                              PIC X(20).
000570           88 DEPOSITOR-NAME-IS-MISSING                      VALUE SPACES.
000580        05 FILLER                                         PIC X(26).
000590
000600 01     ACCOUNT-MASTER-W.
000610        05 DATE-OF-LAST-TRANSACTION-W                     PIC 9(6).
000620        05 ACCOUNT-NUMBER-W         VALUE LOW-VALUES      PIC X(5).
000630        05 CURRENT-BALANCE-W                              PIC S9(7)V99.
000640        05 DEPOSITOR-NAME-W                               PIC X(20).
000650
000660 01     ACCOUNT-MASTER-N.
000670        05 DATE-OF-LAST-TRANSACTION-N                     PIC 9(6).
000680        05 ACCOUNT-NUMBER-N         VALUE LOW-VALUES      PIC X(5).
000690        05 CURRENT-BALANCE-N                              PIC S9(7)V99.
000700        05 DEPOSITOR-NAME-N                               PIC X(20).
000710
000720 01     MORE-INPUT           VALUE   'YES'                PIC X(3).
000730
000740 01     RUN-DATE.
000750        05 RUN-YEAR                                       PIC 99.
000760        05 RUN-MONTH                                      PIC 99.
000770        05 RUN-DAY                                        PIC 99.
000780
000790 01     ONE                  VALUE   1                    PIC S9.
000800 01     MESSAGE-W                                         PIC X(53).
000810 01     HAVE-ANY-ERRORS-BEEN-FOUND                        PIC X(3).
000820 01     ACCOUNT-NUMBER-IN-SAVE
000830                             VALUE  LOW-VALUES            PIC X(5).
000840 01     TRANSACTION-SEQUENCE-CHECKED                      PIC X(3).
000850
000860 01     ERROR-MESSAGES.
000870        05 INVALID-TRANSACTION-CODE.
000880           10 FILLER         VALUE   'INVALID TRANSACTION CODE'
000890                                                          PIC X(25).
000900           10 TRANSACTION-CODE-E                          PIC X.
000910        05 MONEY-AMOUNT-INVALID.
000920           10 TRANSACTION-TYPE                            PIC X(11).
000930           10 FILLER         VALUE   'AMOUNT NOT NUMERIC'
000940                                                          PIC X(19).
000950           10 INVALID-MONEY-AMOUNT                        PIC X(8).
000960        05 ALREADY-ON-FILE
000970           VALUE 'ACCOUNT NUMBER ALREADY ON FILE.  NEW ACCOUNT INVALI
000980-                'D.'                                     PIC X(53).
000990        05 OUT-OF-SEQUENCE
001000                             VALUE   'TRANSACTION OUT OF SEQUENCE'
001010                                                          PIC X(27).
```

Figure 11.17 Program 31, Part 3 of 5.

```
001020
001030 REPORT SECTION.
001040 RD   TRANSACTION-REGISTER
001050      CONTROL FINAL
001060      PAGE 52 LINES,
001070      FIRST DETAIL 13,
001080      LAST DETAIL 38,
001090      FOOTING 52.
001100
001110 01   TYPE REPORT HEADING.
001120      05 LINE 2.
001130         10 COLUMN 40  VALUE  'ROBBEM STATE BANK'
001140                                                    PIC X(17).
001150      05 LINE 3.
001160         10 COLUMN 40  VALUE  '106 WEST 10TH ST.'
001170                                                    PIC X(17).
001180      05 LINE 4.
001190         10 COLUMN 39  VALUE  'BROOKLYN, NY  11212'
001200                                                    PIC X(19).
001210
001220 01   TYPE PAGE HEADING.
001230      05 LINE 6.
001240         10 COLUMN 30  VALUE
001250                       'SAVINGS ACCOUNT TRANSACTION REGISTER'
001260                                                    PIC X(36).
001270      05 LINE 7.
001280         10 COLUMN 18  VALUE  'DATE'                PIC X(4).
001290         10 COLUMN 23  SOURCE RUN-MONTH             PIC Z9.
001300         10 COLUMN 25  VALUE  '/'                   PIC X.
001310         10 COLUMN 26  SOURCE RUN-DAY               PIC 99.
001320         10 COLUMN 28  VALUE  '/'                   PIC X.
001330         10 COLUMN 29  SOURCE RUN-YEAR              PIC 99.
001340         10 COLUMN 66  VALUE  'PAGE'                PIC X(4).
001350         10 COLUMN 71  SOURCE PAGE-COUNTER          PIC Z9.
001360      05 LINE 10.
001370         10 COLUMN 21  VALUE  'ACCOUNT'             PIC X(7).
001380         10 COLUMN 33  VALUE  'DEPOSITS'            PIC X(8).
001390         10 COLUMN 47  VALUE  'WITHDRAWALS'         PIC X(11).
001400         10 COLUMN 65  VALUE  'NOTES'               PIC X(5).
001410      05 LINE 11.
001420         10 COLUMN 21  VALUE  'NUMBER'              PIC X(6).
001430
001440 01   DEPOSIT-LINE
001450      TYPE DETAIL
001460      LINE PLUS 1.
001470      05 COLUMN 22     SOURCE ACCOUNT-NUMBER-IN     PIC X(5).
001480      05 COLUMN 32     SOURCE MONEY-AMOUNT-IN       PIC ZZZ,ZZZ.99.
001490      05 COLUMN 63     SOURCE MESSAGE-W             PIC X(53).
001500
001510 01   WITHDRAWAL-LINE
001520      TYPE DETAIL
001530      LINE PLUS 1.
001540      05 COLUMN 22     SOURCE ACCOUNT-NUMBER-IN     PIC X(5).
001550      05 COLUMN 46     SOURCE MONEY-AMOUNT-IN       PIC ZZZ,ZZZ.99.
001560      05 COLUMN 63     SOURCE MESSAGE-W             PIC X(53).
001570
```

Figure 11.17 Part 3 (continued).

```
001580 01   NAME-CHANGE-LINE
001590      TYPE DETAIL
001600      LINE PLUS 1.
001610      05 COLUMN 22        SCLRCE ACCCLNT-NUMBER-IN   PIC X(5).
001620      05 COLUMN 32        SOURCE DEPCSITCR-NAME-IN   PIC X(20).
001630      05 COLUMN 63        VALUE  'NAME CHANGE'       PIC X(11).
001640
001650 01   ERROR-LINE
001660      TYPE DETAIL
001670      LINE PLUS 1.
001680      05 COLUMN 22        SCLRCE ACCCLNT-NUMBER-IN   PIC X(5).
001690      05 COLUMN 63        SOURCE MESSAGE-W           PIC X(53).
001700
001710 01   TYPE CONTRCL FOOTING FINAL.
001720      05 LINE PLUS 2.
001730          10 COLUMN 18    VALUE  'TOTALS'            PIC X(6).
001740          10 COLUMN 30    SUM    MONEY-AMCUNT-IN
001750                          UPCN   DEPCSIT-LINE        PIC Z,ZZZ,ZZZ.99.
001760          10 COLUMN 44    SUM    MONEY-AMCUNT-IN
001770                          UPCN   WITHDRAWAL-LINE     PIC Z,ZZZ,ZZZ.99.
001780      05 LINE PLUS 2
001790              COLUMN 41   VALUE  'CONTRCL CCUNTS'    PIC X(14).
001800      05 LINE PLUS 2.
001810          10 COLUMN 35    VALUE  'NUMBER CF DEPCSITS'
001820                                                     PIC X(18).
001830          10 NUMBER-OF-DEPOSITS
001840              COLUMN 58   SUM    ONE
001850                          UPCN   DEPCSIT-LINE        PIC ZZ9.
001860      05 LINE PLUS 2.
001870          10 COLUMN 35    VALUE  'NUMBER CF WITHDRAWALS'
001880                                                     PIC X(21).
001890          10 NUMBER-OF-WITHDRAWALS
001900              COLUMN 58   SUM    ONE
001910                          UPCN   WITHDRAWAL-LINE     PIC ZZ9.
001920      05 LINE PLUS 2.
001930          10 COLUMN 35    VALUE  'NUMBER CF NAME CHANGES'
001940                                                     PIC X(22).
001950          10 NUMBER-OF-NAME-CHANGES
001960              COLUMN 58   SUM    ONE
001970                          UPCN   NAME-CHANGE-LINE    PIC ZZ9.
001980      05 LINE PLUS 2.
001990          10 COLUMN 35    VALUE  'NUMBER CF ERRCRS'
002000                                                     PIC X(16).
002010          10 NUMBER-OF-ERRCRS
002020              COLUMN 58   SUM    ONE
002030                          UPCN   ERRCR-LINE          PIC ZZ9.
002040      05 LINE PLUS 2.
002050          10 COLUMN 35    VALUE  'TOTAL'             PIC X(5).
002060          10 COLUMN 56    SUM    NUMBER-CF-DEPCSITS,
002070                                 NUMBER-OF-WITHCRAWALS,
002080                                 NUMBER-OF-NAME-CHANGES,
002090                                 NUMBER-CF-ERRORS    PIC Z,ZZ9.
002100
002110*****************************************************************
```

Figure 11.17 Program 31, Part 4 of 5.

```
002120
002130 PROCEDURE DIVISION.
002140 1-UPDATE-MASTER.
002150     PERFORM 2-INITIALIZATION.
002160     PERFORM 3-MATCH-TRANSACTION-TO-MASTER UNTIL
002170         ACCOUNT-NUMBER-W  = HIGH-VALUES AND
002180         ACCOUNT-NUMBER-IN = HIGH-VALUES.
002190     PERFORM 15-TERMINATION.
002200     STOP RUN.
002210
002220 2-INITIALIZATION.
002230     OPEN INPUT  TRANSACTION-FILE-IN,
002240                 ACCOUNT-MASTER-IN,
002250          OUTPUT TRANSACTION-REGISTER-FILE,
002260                 ACCOUNT-MASTER-OUT.
002270     ACCEPT RUN-DATE FROM DATE.
002280     INITIATE TRANSACTION-REGISTER.
002290     PERFORM 12-READ-TRANSACTION.
002300     PERFORM 14-READ-MASTER.
002310
002320 3-MATCH-TRANSACTION-TO-MASTER.
002330     IF ACCOUNT-NUMBER-IN LESS THAN ACCOUNT-NUMBER-N
002340         PERFORM 6-ADD-NEW-MASTER-RECORD
002350     ELSE
002360     IF ACCOUNT-NUMBER-IN EQUAL TO ACCOUNT-NUMBER-N
002370         PERFORM 5-UPDATE-EXISTING-MSTR-RECORD
002380     ELSE
002390     IF ACCOUNT-NUMBER-IN GREATER THAN ACCOUNT-NUMBER-N
002400         PERFORM 4-WRITE-NEW-MASTER-RECORD.
002410
002420 4-WRITE-NEW-MASTER-RECORD.
002430     WRITE ACCOUNT-RECORD-OUT FROM ACCOUNT-MASTER-N.
002440     PERFORM 14-READ-MASTER.
002450
002460 5-UPDATE-EXISTING-MSTR-RECORD.
002470     IF TRANSACTION-IS-CLOSE-ACCOUNT
002480         PERFORM 11-PROCESS-A-CLOSED-ACCOUNT
002490     ELSE
002500         PERFORM 7-PROCESS-TRANSACTION.
002505
002510 6-ADD-NEW-MASTER-RECORD.
002520     IF NOT TRANSACTION-IS-NEW-ACCOUNT
002530         MOVE 'ACCOUNT NUMBER NOT ON FILE' TO MESSAGE-W
002540         GENERATE ERROR-LINE
002550     ELSE
002560         MOVE 'NO' TO HAVE-ANY-ERRORS-BEEN-FOUND
002570         PERFORM 16-CHECK-ACCOUNT-NUMB-PRESENT
002580         IF HAVE-ANY-ERRORS-BEEN-FOUND = 'NO'
002590             MOVE 'DEPOSIT' TO TRANSACTION-TYPE
002600             PERFORM 17-CHECK-MONEY-AMOUNT-NUMERIC
002610             IF HAVE-ANY-ERRORS-BEEN-FOUND = 'NO'
002620                 PERFORM 18-CHECK-DEPOSTOR-NAME-PRESENT
002630                 IF HAVE-ANY-ERRORS-BEEN-FOUND = 'NO'
002640                     PERFORM 19-CREATE-NEW-MSTR-IN-STORAGE.
002650     PERFORM 12-READ-TRANSACTION.
002655
```

Figure 11.17 Part 4 (continued).

```
002660 7-PROCESS-TRANSACTION.
002670     MOVE SPACES TO MESSAGE-W.
002680     MOVE 'NO' TO HAVE-ANY-ERRORS-BEEN-FOUND.
002690     IF TRANSACTION-IS-DEPOSIT
002700        PERFORM 8-PROCESS-A-DEPOSIT
002710     ELSE
002720     IF TRANSACTION-IS-WITHDRAWAL
002730        PERFORM 9-PROCESS-A-WITHDRAWAL
002740     ELSE
002750     IF TRANSACTION-IS-NAME-CHANGE
002760        PERFORM 10-PROCESS-A-NAME-CHANGE
002770     ELSE
002780        MOVE TRANSACTION-CODE-IN      TO TRANSACTION-CODE-E
002790        MOVE INVALID-TRANSACTION-CODE TO MESSAGE-W
002800        GENERATE ERROR-LINE.
002810     PERFORM 12-READ-TRANSACTION.
002820
002830 8-PROCESS-A-DEPOSIT.
002840     MOVE 'DEPOSIT' TO TRANSACTION-TYPE.
002850     PERFORM 17-CHECK-MONEY-AMOUNT-NUMERIC.
002860     IF HAVE-ANY-ERRORS-BEEN-FOUND = 'NO'
002870         GENERATE DEPOSIT-LINE
002880         MOVE RUN-DATE TO DATE-OF-LAST-TRANSACTION-N
002890         ADD MONEY-AMOUNT-IN TO CURRENT-BALANCE-N.
002900
002910 9-PROCESS-A-WITHDRAWAL.
002920     MOVE 'WITHDRAWAL' TO TRANSACTION-TYPE.
002930     PERFORM 17-CHECK-MONEY-AMOUNT-NUMERIC.
002940     IF HAVE-ANY-ERRORS-BEEN-FOUND = 'NO'
002950         GENERATE WITHDRAWAL-LINE
002960         MOVE RUN-DATE TO DATE-OF-LAST-TRANSACTION-N
002970         SUBTRACT MONEY-AMOUNT-IN FROM CURRENT-BALANCE-N.
002980
002990 10-PROCESS-A-NAME-CHANGE.
003000     PERFORM 18-CHECK-DEPOSTOR-NAME-PRESENT.
003010     IF HAVE-ANY-ERRORS-BEEN-FOUND = 'NO'
003020         MOVE DEPOSITOR-NAME-IN TO DEPOSITOR-NAME-N
003030         MOVE RUN-DATE TO DATE-OF-LAST-TRANSACTION-N
003040         GENERATE NAME-CHANGE-LINE.
003050
003060 11-PROCESS-A-CLOSED-ACCOUNT.
003070     MOVE 'ACCOUNT CLOSED' TO MESSAGE-W.
003080     MOVE CURRENT-BALANCE-N TO MONEY-AMOUNT-IN.
003090     GENERATE WITHDRAWAL-LINE.
003100     PERFORM 12-READ-TRANSACTION.
003110     PERFORM 14-READ-MASTER.
003120
003130 12-READ-TRANSACTION.
003140     MOVE 'NO' TO TRANSACTION-SEQUENCE-CHECKED.
003150     PERFORM 13-SEQUENCE-CHECK-TRANSACTION
003160         UNTIL
003170               TRANSACTION-SEQUENCE-CHECKED = 'YES'
003180     MOVE ACCOUNT-NUMBER-IN TO ACCOUNT-NUMBER-IN-SAVE.
003210
```

Figure 11.17 Program 31, Part 5 of 5.

```
003220 13-SEQUENCE-CHECK-TRANSACTION.
003230     READ TRANSACTION-FILE-IN INTO TRANSACTION-RECORD-W
003240         AT END
003250             MOVE HIGH-VALUES TO ACCOUNT-NUMBER-IN.
003260     IF ACCOUNT-NUMBER-IN LESS THAN ACCOUNT-NUMBER-IN-SAVE
003270         MOVE OUT-OF-SEQUENCE TO MESSAGE-W
003280         GENERATE ERROR-LINE
003290     ELSE
003300         MOVE 'YES' TO TRANSACTION-SEQUENCE-CHECKED.
003310
003320
003330 14-READ-MASTER.
003340     IF ACCOUNT-NUMBER-W EQUAL TO ACCOUNT-NUMBER-N
003350         READ ACCOUNT-MASTER-IN INTO ACCOUNT-MASTER-W
003360             AT END
003370                 MOVE HIGH-VALUES TO ACCOUNT-NUMBER-W.
003390     MOVE ACCOUNT-MASTER-W TO ACCOUNT-MASTER-N.
003400
003410 15-TERMINATION.
003420     TERMINATE TRANSACTION-REGISTER.
003430     CLOSE     ACCOUNT-MASTER-IN,
003440               ACCOUNT-MASTER-OUT,
003450               TRANSACTION-FILE-IN,
003460               TRANSACTION-REGISTER-FILE.
003470
003480 16-CHECK-ACCOUNT-NUMB-PRESENT.
003490     IF ACCOUNT-NUMBER-IS-MISSING
003500         MOVE 'ACCOUNT NUMBER MISSING' TO MESSAGE-W
003510         MOVE 'YES' TO HAVE-ANY-ERRORS-BEEN-FOUND
003520         GENERATE ERROR-LINE.
003530
003540 17-CHECK-MONEY-AMOUNT-NUMERIC.
003550     IF MONEY-AMOUNT-IN NOT NUMERIC
003560         MOVE MONEY-AMOUNT-IN-X TO INVALID-MONEY-AMOUNT
003570         MOVE MONEY-AMOUNT-INVALID TO MESSAGE-W
003580         MOVE 'YES' TO HAVE-ANY-ERRORS-BEEN-FOUND
003590         GENERATE ERROR-LINE.
003600
003610 18-CHECK-DEPOSITOR-NAME-PRESENT.
003620     IF DEPOSITOR-NAME-IS-MISSING
003630         MOVE 'DEPOSITOR NAME MISSING' TO MESSAGE-W
003640         MOVE 'YES' TO HAVE-ANY-ERRORS-BEEN-FOUND
003650         GENERATE ERROR-LINE.
003660
003670 19-CREATE-NEW-MSTR-IN-STORAGE.
003680     MOVE 'NEW ACCOUNT' TO MESSAGE-W.
003690     GENERATE DEPOSIT-LINE.
003700     MOVE ACCOUNT-NUMBER-IN      TO ACCOUNT-NUMBER-N.
003710     MOVE MONEY-AMOUNT-IN        TO CURRENT-BALANCE-N.
003720     MOVE DEPOSITOR-NAME-IN      TO DEPOSITOR-NAME-N.
003730     MOVE RUN-DATE               TO DATE-OF-LAST-TRANSACTION-N.
003740
```

transaction ACCOUNT–NUMBER–IN is LESS THAN the master ACCOUNT–NUMBER–N, then the transaction is probably trying to create a new account to slip in in front of the existing master record. If the transaction ACCOUNT–NUMBER–IN is EQUAL TO the master ACCOUNT–NUMBER–N, then the transaction is probably trying to update an existing master record. If the ACCOUNT–NUMBER–IN is GREATER THAN ACCOUNT–NUMBER–N, this means that the transaction applies to some master record further along in the file and the current master record can be written onto the new tape.

The paragraph READ–MASTER provides for the possibility that a new account may be created and updated in a single day or created and closed in a single day. If ACCOUNT–NUMBER–W is not equal to ACCOUNT–NUMBER–N, this means that a new record was just created in the ACCOUNT–NUMBER–N area and that the next master record is already in the ACCOUNT–NUMBER–W area.

In the two routines SEQUENCE–CHECK–TRANSACTION and READ–MASTER, we find a new way to handle an end-of-file condition. If either routine finds an end-of-file, it MOVEs HIGH–VALUES into the account number field in the input area. This is a method that is used to cause the routine MATCH–TRANSACTION–TO–MASTER to produce the desired results when either the transaction file or the input master file has reached end-of-file. For example, if the transaction file reaches end-of-file before the input master file does, we have the situation shown in Figure 11.18: The transaction records are shown as matching some input master records, but the transactions run out before the master does. The remaining records on the input master tape have to be copied, unchanged, onto the output master. By moving HIGH–VALUES into the transaction ACCOUNT–NUMBER–IN

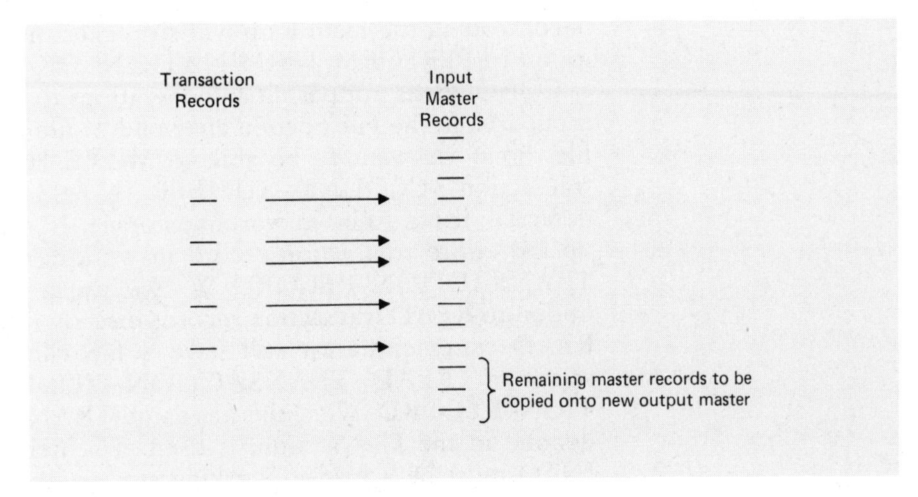

Figure 11.18 Transaction end-of-file occurs before input master end-of-file does.

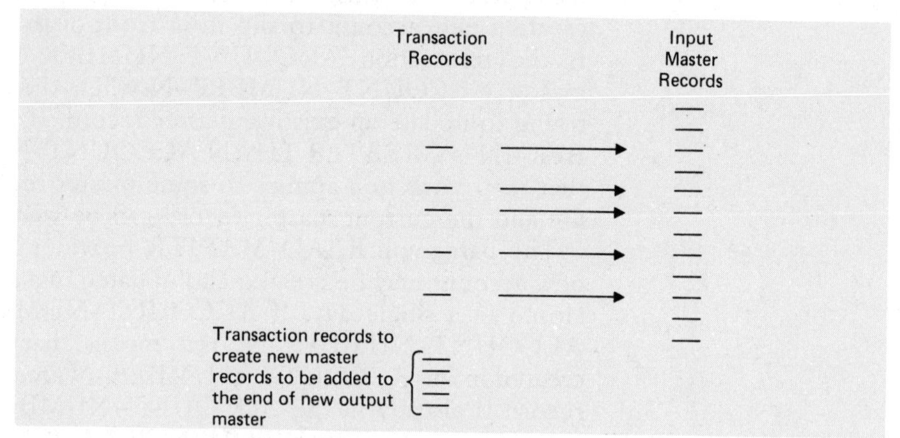

Figure 11.19
Input master end-of-file occurs before transaction end-of-file does.

field, we can be sure that it will always be GREATER THAN the ACCOUNT–NUMBER–N field of the master input file. In this way, MATCH–TRANSACTION–TO–MASTER will always PERFORM the routine WRITE–NEW–MASTER–RECORD, which is what we want. On the other hand, if the input master file reaches end-of-file before the transaction file does, we have the situation shown in Figure 11.19. There, the extra transaction records are probably trying to create new account records to be added to the end of the output master. By moving HIGH–VALUES into the master ACCOUNT–NUMBER–W field, we can be sure that the transaction ACCOUNT–NUMBER–IN will always be LESS. In that way, MATCH–TRANSACTION–TO–MASTER will PERFORM ADD–NEW–MASTER–RECORD, which is what we would want in this case.

When both input files are AT END, the PERFORM statement that is controlling the main loop will stop executing, and the program will go on to PERFORM TERMINATION.

COBOL does not permit any operations to be carried out on any file input area in the File Section after end-of-file has been reached on that file. So if we want to be able to MOVE HIGH–VALUES into the transaction ACCOUNT–NUMBER–IN field after end-of-file has been detected, it has to be in working storage. In this program we have set up the entire transaction record in working storage under the name TRANSACTION–RECORD–W. We use a READ with the **INTO** option to READ transaction records directly into working storage. The READ statement must still have a file name as its object. In the statement READ TRANSACTION–FILE–IN INTO TRANSACTION–RECORD–W. . . the file name TRANSACTION–FILE–IN, as defined in the File Section, is used. The field named after the word INTO must be a field in working storage or the record definition of an output file in the File Section.

Program 31 was run using the file created in Program 30 as its input master, and input transactions shown in Figure 11.20. The transaction register output is shown in Figure 11.21.

Figure 11.20 Transaction input for Program 31.

```
200007)))$%)))
300014001CCC00
20002100012750
20002800015327
300028C0150050
20006300C07500
20007700015000
400084         JOHN & SALLY DUPRINO
20008400357429
20009100150000
20009100120000
300091C0050000
400098         GEORGE & ANN CULHANE
30011900002735
20012600035000
20013300256300
400140         BETH DENNY
20016100125634
30016800011000
500168
20017500019202
200182
500182
20021000025000
20021000025000
20021000025000
500232
30025929390C00
400266
20027300172500
20028000231700
20030100005763
5
10030800017000AL MARRELLA
20030800C05000
80031500013798
20032200006875
10035700001C00ROBIN RATANSKI
10036400150000JOSE TORRES
10037100015000ALISE MARKOVITZ
20037100015000
100371000C1C00THOMAS HERR
70038500007500JAMES WASHINGTON
10039200007500INEZ WASHINGTON
100398C0001000JUAN ALVAREZ
10039900014Z00GARY NASTI
1004060012C000
1004130001CC00BILL HAYES
1004200015C000JOHN RICE
700427000C2500GREG PRUITT
```

Figure 11.21 Transaction register output from Program 31.

```
                              ROBBEM STATE BANK
                              106 WEST 10TH ST.
                             BROOKLYN, NY  11212

                   SAVINGS ACCOUNT TRANSACTION REGISTER
DATE  2/27/80
                                                          PAGE  1


    ACCOUNT       DEPOSITS        WITHDRAWALS         NOTES
    NUMBER


     00007                                           DEPOSIT     AMOUNT NOT NUMERIC )))$%)))
     00014                          1,000.00
     00021         127.50
     00028         153.27
     00028                          1,500.50
     00063          75.00
     00077         150.00
     00084       JOHN & SALLY DUPRINO                NAME CHANGE
     00084       3,574.29
     00091       1,500.00
     00091       1,200.00
     00091                            500.00
     00098       GEORGE & ANN CULHANE                NAME CHANGE
     00119                             27.35
     00126         350.00
     00133       2,563.00
     00140       BETH DENNY                          NAME CHANGE
     00161       1,256.34
     00168                            110.00
     00168                             15.50         ACCOUNT CLOSED
     00175         192.02
     00182                                           ACCOUNT NUMBER NOT ON FILE
     00182                                           ACCOUNT NUMBER NOT ON FILE
     00210         250.00
     00210         250.00
     00210         250.00
```

Figure 11.21 Continued.

```
            SAVINGS ACCOUNT TRANSACTION REGISTER
DATE   2/27/80                                         PAGE  2

   ACCOUNT       DEPOSITS        WITHDRAWALS          NOTES
   NUMBER

                                                    ACCOUNT NUMBER NOT ON FILE
    00232                                           ACCOUNT NUMBER NOT ON FILE
    00259                                           DEPOSITOR NAME MISSING
    00266
    00273        1,725.00
    00280        2,317.00
    00301           57.63
                                                    TRANSACTION OUT OF SEQUENCE
    00308                                           INVALID TRANSACTION CODE 1
    00308           50.00
    00315                                           INVALID TRANSACTION CODE 8
    00322           68.75
    00357           10.00                           NEW ACCOUNT
    00364        1,500.00                           NEW ACCOUNT
    00371          150.00                           NEW ACCOUNT
    00371          150.00
    00371                                           INVALID TRANSACTION CODE 1
    00385                                           ACCOUNT NUMBER NOT ON FILE
    00392           75.00                           NEW ACCOUNT
    00398           10.00                           NEW ACCOUNT
    00399                                           DEPOSIT    AMOUNT NOT NUMERIC 00014Z00
    00406                                           DEPOSITOR NAME MISSING
    00413          100.00                           NEW ACCOUNT
    00420        1,500.00                           NEW ACCOUNT
    00427                                           ACCOUNT NUMBER NOT ON FILE

TOTALS          19,604.80       3,153.35

                      CONTROL COUNTS

                NUMBER OF DEPOSITS          27

                NUMBER OF WITHDRAWALS        6

                NUMBER OF NAME CHANGES       3

                NUMBER OF ERRORS            14

                TOTAL                       50
```

***Exercise 6.** Write a program to update the file you created in Exercise 5. Allow for the possibility that there may be more than one transaction against a single master record. Run your program at least twice with different transactions to check that the program works:

a. if the transaction end-of-file occurs before the input master end-of-file does; and
b. if the input master end-of-file occurs before the transaction end-of-file does.

Have your program check for invalid fields in the transactions. Also have your program check for invalid combinations of transactions, such as a transaction to create a new record followed by another transaction to create a record with the same customer number.

Listing a Sequential File on Tape

In Chapter 2 we had a program to list a sequential card file. Listing a sequential tape file presents no new concepts. Program 32, to list the file that was created by Program 31, is shown in Figure 11.22. The output produced by Program 32 is shown in Figure 11.23.

Figure 11.22 Program 32, Part 1 of 2.

```
4-CB2 V4 RELEASE 1.5 10NOV77          IBM OS AMERICAN NATIONAL STANDARD COBOL


  000010 IDENTIFICATION DIVISION.
  000020 PROGRAM-ID. PROG32.
  000030*AUTHOR. ELLEN M. LONGO.
  000040*
  000050*     THIS PROGRAM LISTS A SEQUENTIAL TAPE FILE.
  000060*
  000070*****************************************************************
  000080
  000090 ENVIRONMENT DIVISION.
  000100 INPUT-OUTPUT SECTION.
  000110 FILE-CONTROL.
  000120     SELECT MASTER-LIST-FILE      ASSIGN TO UT-S-SYSPRINT.
  000130     SELECT ACCOUNT-MASTER-FILE ASSIGN TO UT-S-TAPE.
  000150
  000160*****************************************************************
```

Figure 11.22 Part 1 (continued).

```
000170
000180 DATA DIVISION.
000190 FILE SECTION.
000200 FD  ACCOUNT-MASTER-FILE
000210     LABEL RECORDS ARE STANDARD
000220     BLOCK CONTAINS 50 RECORDS.
000230
000240 01  ACCOUNT-MASTER-RECORD.
000260     05 DATE-OF-LAST-TRANSACTION-M.
000270        10 TRANSACTION-YEAR-M                       PIC 99.
000280        10 TRANSACTION-MONTH-M                      PIC 99.
000290        10 TRANSACTION-DAY-M                        PIC 99.
000300     05 ACCOUNT-NUMBER-M                            PIC X(5).
000310     05 CURRENT-BALANCE-M                           PIC S9(7)V99.
000320     05 DEPOSITOR-NAME-M                            PIC X(20).
000330
000340 FD  MASTER-LIST-FILE
000350     LABEL RECORDS ARE OMITTED
000360     REPORT IS MASTER-LIST.
000370
000380 WORKING-STORAGE SECTION.
000390 01  MORE-INPUT          VALUE  'YES'               PIC X(3).
000400 01  MESSAGE-W                                      PIC X(19).
000410
000420 01  RUN-DATE.
000430     05 RUN-YEAR                                    PIC 99.
000440     05 RUN-MONTH                                   PIC 99.
000450     05 RUN-DAY                                     PIC 99.
000460
000470 REPORT SECTION.
000480 RD  MASTER-LIST
000490     PAGE 38 LINES, FIRST DETAIL 13.
000500
000510 01  TYPE PAGE HEADING.
000520     05 LINE 5.
000530        10 COLUMN 41  VALUE  'SAVINGS ACCOUNT MASTER'
000540                                                    PIC X(22).
000550     05 LINE 6.
000560        10 COLUMN 29  VALUE  'DATE'                 PIC X(4).
000570        10 COLUMN 34  SOURCE RUN-MONTH              PIC Z9.
000580        10 COLUMN 36  VALUE  '/'                    PIC X.
000590        10 COLUMN 37  SOURCE RUN-DAY                PIC 99.
000600        10 COLUMN 39  VALUE  '/'                    PIC X.
000610        10 COLUMN 40  SOURCE RUN-YEAR               PIC 99.
000620        10 COLUMN 63  VALUE  'PAGE'                 PIC X(4).
000630        10 COLUMN 68  SOURCE PAGE-COUNTER           PIC Z9.
000640     05 LINE 9
000650           COLUMN 47  VALUE  'DATE OF LAST'         PIC X(12).
000660     05 LINE 10.
000670        10 COLUMN 21  VALUE  'ACCOUNT'              PIC X(7).
000680        10 COLUMN 34  VALUE  'CURRENT'              PIC X(7).
000690        10 COLUMN 47  VALUE  'TRANSACTION'          PIC X(11).
000700        10 COLUMN 62  VALUE  'DEPOSITOR NAME'       PIC X(14).
000710     05 LINE 11.
000720        10 COLUMN 21  VALUE  'NUMBER'               PIC X(6).
000730        10 COLUMN 34  VALUE  'BALANCE'              PIC X(7).
000740        10 COLUMN 48  VALUE  'YR  MO  DA'              PIC X(10).
```

Figure 11.22 Program 32, Part 2 of 2.

```
000750
000760 01   DETAIL-LINE
000770      TYPE DETAIL
000780      LINE PLUS 1.
000790      05 COLUMN 22      SOURCE ACCOUNT-NUMBER-M    PIC X(5).
000800      05 COLUMN 31      SOURCE CURRENT-BALANCE-M   PIC Z,ZZZ,ZZZ.99-.
000810      05 COLUMN 48      SOURCE TRANSACTION-YEAR-M    PIC 99.
000820      05 COLUMN 52      SOURCE TRANSACTION-MONTH-M   PIC 99.
000830      05 COLUMN 56      SOURCE TRANSACTION-DAY-M     PIC 99.
000840      05 COLUMN 62      SOURCE DEPOSITOR-NAME-M    PIC X(20).
000850      05 COLUMN 86      SOURCE MESSAGE-W           PIC X(19).
000860
000870*****************************************************************
000880
000890 PROCEDURE DIVISION.
000900      OPEN INPUT ACCOUNT-MASTER-FILE, OUTPUT MASTER-LIST-FILE.
000910      READ ACCOUNT-MASTER-FILE AT END MOVE 'NO' TO MORE-INPUT.
000920      IF MORE-INPUT = 'YES'
000930          ACCEPT RUN-DATE FROM DATE
000940          INITIATE MASTER-LIST
000950          PERFORM MAIN-PROCESS UNTIL MORE-INPUT = 'NO'
000960          TERMINATE MASTER-LIST.
000970      CLOSE ACCOUNT-MASTER-FILE, MASTER-LIST-FILE.
000980      STOP RUN.
000990
001000 MAIN-PROCESS.
001010      IF CURRENT-BALANCE-M IS NEGATIVE
001020          MOVE 'BALANCE IS NEGATIVE' TO MESSAGE-W
001030      ELSE
001040          MOVE SPACES TO MESSAGE-W.
001050      GENERATE DETAIL-LINE.
001060      READ ACCOUNT-MASTER-FILE AT END MOVE 'NO' TO MORE-INPUT.
```

Figure 11.23 Output from Program 32, Part 1 of 2.

SAVINGS ACCOUNT MASTER

DATE 2/27/80 PAGE 1

ACCOUNT NUMBER	CURRENT BALANCE	DATE OF LAST TRANSACTION YR	MO	DA	DEPOSITOR NAME	
00007	1,007.84	80	02	26	ROSEBUCCI	
00014	990.00-	80	02	27	ROBERT DAVIS M.D.	BALANCE IS NEGATIVE
00021	252.50	80	02	27	LORICE MONTI	
00028	5,654.36	80	02	27	MICHAEL SMITH	
00035	150.00	80	02	26	JOHN J. LEHMAN	
00049	150.00	80	02	26	JAY GREENE	
00056	1.00	80	02	26	EVELYN SLATER	
00063	150.00	80	02	27	LORRAINE SMALL	
00070	500.00	80	02	26	PATRICK J. LEE	
00077	160.37	80	02	27	LESLIE MINSKY	
00084	5,074.29	80	02	27	JOHN & SALLY DUPRINO	
00091	2,300.00	80	02	27	JOE'S DELI	
00098	500.00	80	02	27	GEORGE & ANN CULHANE	
00105	50.00	80	02	26	LENORE MILLER	
00112	250.00	80	02	26	ROSEMARY LANE	
00119	472.65	80	02	27	MICHELE CAPUANO	
00126	1,100.00	80	02	27	JAMES BUDD	
00133	3,563.00	80	02	27	PAUL LERNER, D.D.S.	
00140	25.75	80	02	27	BETH DENNY	
00154	50.00	80	02	26	ONE DAY CLEANERS	
00161	1,280.84	80	02	27	ROBERT RYAN	
00175	202.02	80	02	27	MARY KEATING	
00189	35.00	80	02	26	J. & L. CAIN	
00196	150.00	80	02	26	IMPERIAL FLORIST	
00203	5.00	80	02	26	JOYCE MITCHELL	
00210	1,000.00	80	02	27	JERRY PARKS	

Figure 11.23 Output from Program 32, Part 2 of 2.

```
                         SAVINGS ACCCUNT MASTER
           CATE  2/27/8C                                  PAGE  2

                                       DATE CF LAST
  ACCOUNT          CURRENT             TRANSACTION        DEPOSITOR NAME
  NUMBER           BALANCE              YR   MO   DA

   00217               50.00            8C   02   26      CARL CALDERCN
   00224              175.50            8C   02   26      JOHN WILLIAMS
   00231              555.00            8C   02   26      BILL WILLIAMS
   00238               1C.00            8C   02   26      KEVIN PARKER
   00245               35.00            8C   C2   26      FRANK CAPUTC
   00266               99.57            8C   02   26      MARTIN LANG
   00273            2,CCC.00            8C   C2   27      VITO CACACI
   C0280            2,337.00            8C   02   27      COMMUNITY DRUGS
   00287               15.00            80   02   26      SOLOMON CHAPELS
   00294            1,5CC.00            8C   02   26      JOHN BURKE
   00301              215.13            80   02   27      PAT P. POWERS
   00308            2,050.00            8C   02   27      JOE GARCIA
   00315              250.00            80   02   26      GRACE MICELI
   00322               88.75            80   02   27      MARIA FASANO
   00329               1C.00            8C   02   26      GUY VOLPONE
   00343            1,CCC.00            80   02   26      JOE & MARY SESSA
   00350            2,500.00            8C   02   26      ROGER SHAW
   00357               1C.00            80   C2   27      ROBIN RATANSKI
   00364            1,5CC.00            80   02   27      JOSE TORRES
   00371              3CC.00            80   02   27      ALISE MARKOVITZ
   00392               75.00            80   02   27      INEZ WASHINGTON
   00398               10.00            8C   02   27      JUAN ALVAREZ
   00413              1CC.00            80   02   27      BILL HAYES
   00420            1,500.00            80   02   27      JCHN RICE
```

Exercise 7. Write a program to list the file you created in Exercise 6.

Summary

Master files may be stored on tape or on direct-access storage devices. Files may be stored using sequential organization or using one of the other forms of organization indexed sequential, relative, and direct.

An indexed sequential file must be created sequentially. During file creation, COBOL also creates indexes for the file which are later used by the system. An indexed sequential file may be accessed randomly or sequentially. During random access, COBOL uses the file indexes to find any desired record or to find a place to insert a new record. An indexed sequential file can be created only on DASD, as can relative and direct files.

A sequential file may be created on tape or on DASD. It must be created sequentially. A sequential file can be accessed only sequentially, whether it is on tape or on disk. For this reason, transactions against a sequential master file must be in key sequence.

A record on disk, whether in an indexed sequential or standard sequential file, can be replaced by having a new record written over it; on tape, however, it is not possible to replace a record. For this reason, whenever a master file on tape is updated, a completely new master tape must be written, containing changed records, new records, and unchanged master records that were just copied from the old master.

Fill-in Exercises

1. Three kinds of magnetic storage media that computers can read input data from and write output onto are ________, ________, and ________.
2. The only access method available for tape files is ________ access.
3. Three advantages of magnetic media over paper media are
   a. ________________________________
   b. ________________________________
   c. ________________________________
4. The four kinds of file organization methods that can be handled by COBOL programs are ________, ________, ________, and ________.
5. The only kind of files that can be stored on magnetic tape are files that are organized ________.
6. In creating an indexed sequential file, the ________ clause tells the system the name of the field on which the records are ordered.
7. The ________ clause tells the system how many records are in a block.
8. The INVALID KEY clause can be used in ________ and ________ statements.
9. The ________ verb is used to replace an existing record on a direct-access storage device.
10. The figurative constant ________ stands for the lowest value that can be assigned to a character.

Review Exercises

*1. A school wishes to maintain a file of alumni names and addresses. Write a program to create an indexed sequential file on disk using transaction input in the format shown in Figure 11.RE1.

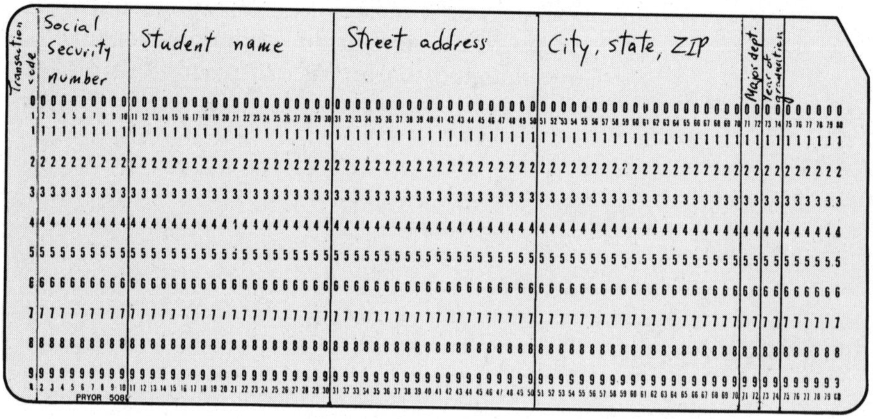

Figure 11.RE1
Data card format for Review Exercise 1.

Each master record should contain:

a. the student's Social Security number (the key field)
b. the student's name
c. the student's street address
d. city, state, and ZIP code
e. the student's major department
f. the student's year of graduation (two digits).

Use transaction code 1 for all transactions in this program. Check each transaction for the presence of all fields.

*2. Write a program to randomly update the file you created in Review Exercise 1. Use the following transaction codes for the update transactions:

Transaction Code	Transaction Type
1	New student record
2	Name change
3	Street address change
4	City, state, ZIP code change
5	Major department change
6	Year of graduation change
7	Delete record

For each type of transaction, check for the presence of the appropriate field(s). Provide suitable error messages for INVALID KEY conditions.

3. Write a program to list the indexed sequential file you updated in Exercise 2.

*4. Write a program to create a sequential file from the same input data you used in Review Exercise 1. Create the file on tape or disk as your facilities allow.

*5. Write a program to update the sequential file you created in Review Exercise 4. Provide for all suitable error checking. Test your program with more than one set of transactions.

6. Write a program to list the file you updated in Review Exercise 5.

Sorting and String Processing

12

KEY POINTS

Here are the key points you should learn from this chapter:

1. how to use the SORT verb;
2. how to write procedures that process data before or after they are sorted;
3. how to use the STRING verb;
4. how to use the UNSTRING verb;
5. the use of pointers with STRING and UNSTRING.

KEY WORDS

Key words to recognize and learn:

SORT	sort file description	UNSTRING
INPUT PROCEDURE	USING	DELIMITED BY
OUTPUT PROCEDURE	RETURN	POINTER
SD entry	GO	OVERFLOW
	RELEASE	delimiter
	STRING	COUNT

The **SORT** feature of COBOL permits a program to rearrange the order of input or output records. The order of the records may be rearranged immediately after the records have been read in and before any processing is done, or the records may be reordered after all the processing is done and immediately before the records are written out, or at any time in between.

Using the SORT Verb

Our first program using the SORT verb is a rewrite of Program 22, which was a rewrite of Program 4. Review Program 22: input format,

Figure 2.10; input data, Figure 2.11; Program 22, Figure 10.1; output, Figure 10.2.

Program 33 uses a SORT statement to SORT the input data into alphabetical order before listing the records using Report Writer. Once the data are in alphabetical order, we can have Report Writer reformat the records into print lines and print them.

Any processing that is done on records before or after SORTing must be carried out by an **INPUT PROCEDURE** or an **OUTPUT PROCEDURE.** An INPUT PROCEDURE consists of ordinary COBOL statements that are executed before any SORTing is done, and an OUTPUT PROCEDURE consists of statements executed after SORTing. In this program we need only an OUTPUT PROCEDURE. Program 33 is shown in Figure 12.1.

The Environment Division shows a SELECT sentence for a SORT–WORK–FILE. The work file used by the SORT verb must be provided by the programmer, as we have done here with SORT–WORK–FILE. SORT–WORK–FILE is a programmer-supplied name. The name of a work file for the SORT operation can be any legal name made up in accordance with the rules for making up file names.

In the Data Division, an **SD entry** (**sort file description** entry) must contain the file name made up in the SELECT sentence. The SD entry must have associated with it at least one 01-level entry. Ordinarily, the 01-level entry is broken down into fields, and at least one of the fields is a key field that the records are to be sorted on.

Figure 12.1 Program 33, Part 1 of 2.

```
4-CB2 V4 RELEASE 1.5 10NOV77        IBM OS AMERICAN NATIONAL STANDARD COBOL


 000040 IDENTIFICATION DIVISION.
 000050 PROGRAM-ID.  PROG33.
 000060*AUTHOR. PHIL RUBENSTEIN.
 000070*
 000080*     THIS PROGRAM SORTS THE INPUT RECORDS OF PROGRAM
 000090*     22 INTO ALHPABETICAL ORDER AND THEN PRINTS
 000095*     THEM IN THE FORMAT USED IN PROGRAM 22.
 000100*
 000110*****************************************************************
 000120
 000130 ENVIRONMENT DIVISION.
 000140 INPUT-OUTPUT SECTION.
 000150 FILE-CONTROL.
 000160      SELECT EMPLOYEE-DATA-IN       ASSIGN TO UT-S-SORTIN.
 000170      SELECT EMPLOYEE-DATA-OUT      ASSIGN TO UT-S-SYSPRINT.
 000180      SELECT SORT-WORK-FILE         ASSIGN TO UT-S-SORTWK.
 000190
 000200*****************************************************************
```

The input record, EMPLOYEE–RECORD–IN, though, did not have to be broken down into its component fields. Since we are doing no processing on the input records, we don't need names for their fields. As a general rule, if a program contains a SORT statement and no INPUT PROCEDURE, then the input file records do not have to be broken down into fields.

In the Procedure Division, a SORT statement always has four required clauses. First, we must name the work file that the SORT statement is to use, in this case SORT–WORK–FILE. Second, we must say whether the records are to be SORTed in ASCENDING or DESCENDING order on their key field, and we must say what the KEY field is. In this case, we want to SORT the records into alphabetical order, so we say ASCENDING KEY NAME. More than one key field may be given in this clause. You will see how when we look at the general format of the SORT statement.

The third entry in a SORT statement tells whether there is an INPUT PROCEDURE for the SORT. Since we need no INPUT PROCEDURE in this program, we indicate instead that the entire input file is simply to be SORTed by saying **USING** EMPLOYEE–DATA–IN. The fourth entry in the SORT statement tells whether there is an OUTPUT PROCEDURE. In our case there is one, and its name is OUTPUT–PROCEDURE.

Figure 12.1 Program 33, Part 2 of 2.

```
000210
000220 DATA DIVISION.
000230 FILE SECTION.
000240 SD   SORT-WORK-FILE.
000250
000260 01   SORT-RECORD.
000270      05 SS-NO                                             PIC X(9).
000280      05 IDENT-NO                                          PIC X(5).
000290      05 ANNUAL-SALARY                                     PIC X(7).
000300      05 NAME                                              PIC X(25).
000310      05 FILLER                                            PIC X(34).
000320
000330 FD   EMPLOYEE-DATA-IN
000340      LABEL RECORDS ARE OMITTED.
000350
000360 01   EMPLOYEE-RECORD-IN                                   PIC X(80).
000370
000380 FD   EMPLOYEE-DATA-OUT
000390      LABEL RECORDS ARE OMITTED
000400      REPORT IS EMPLOYEE-REPORT.
000410
000420 WORKING-STORAGE SECTION.
000430 01   MORE-INPUT                VALUE  'YES'               PIC X(3).
000440
000450 REPORT SECTION.
000460 RD   EMPLOYEE-REPORT.
000470
```

Figure 12.1 Part 2 (continued).

```
000480 01  REPORT-LINE
000490     TYPE DETAIL
000500     LINE PLUS 1.
000510     05  COLUMN  5             SOURCE IDENT-NO          PIC X(5).
000520     05  COLUMN 12             SOURCE SS-NO             PIC X(9).
000530     05  COLUMN 25             SOURCE NAME              PIC X(25).
000540     05  COLUMN 52             SOURCE ANNUAL-SALARY     PIC X(7).
000550
000560*****************************************************************
000570
000580 PROCEDURE DIVISION.
000590     SORT SORT-WORK-FILE
000600            ASCENDING KEY NAME
000610            USING EMPLOYEE-DATA-IN
000620            OUTPUT PROCEDURE OUTPUT-PROCEDURE.
000630     STOP RUN.
000640
000650 OUTPUT-PROCEDURE SECTION.
000660     OPEN OUTPUT EMPLOYEE-DATA-OUT.
000670     RETURN SORT-WORK-FILE
000680          AT END
000690               MOVE 'NO' TO MORE-INPUT.
000700     IF MORE-INPUT = 'YES'
000710          INITIATE EMPLOYEE-REPORT
000720          PERFORM MAIN-PROCESS UNTIL MORE-INPUT = 'NO'
000730          TERMINATE EMPLOYEE-REPORT.
000740     GO TO OUTPUT-PROCEDURE-EXIT.
000750
000760 MAIN-PROCESS.
000770     GENERATE REPORT-LINE.
000780     RETURN SORT-WORK-FILE
000790          AT END
000800               MOVE 'NO' TO MORE-INPUT.
000810
000820 OUTPUT-PROCEDURE-EXIT.
000830     CLOSE EMPLOYEE-DATA-OUT.
```

Any INPUT or OUTPUT PROCEDURE used by the SORT must be defined as a SECTION in the Procedure Division. This is the first time we have used a SECTION in the Procedure Division. We have previously used required SECTIONs in the Environment and Data Divisions, but program logic never dictates a need for a SECTION in the Procedure Division. We use one here only because the rules of COBOL require that any INPUT or OUTPUT PROCEDURE used with a SORT be named as a SECTION; perhaps some future COBOL system will relax this unproductive requirement.

The word SECTION appears only in the section header OUTPUT–PROCEDURE SECTION. It does not appear in the SORT statement. SECTION names are made up in accordance with the rules for making up paragraph names. A SECTION in the Procedure Division may contain as many paragraphs as desired.

The OUTPUT–PROCEDURE SECTION is made up of regular

Procedure Division statements, except that a **RETURN** verb is used where we would ordinarily expect a READ.)

The RETURN statement is used in place of a READ when already SORTed records are to be processed. A RETURN statement may be used only in an OUTPUT PROCEDURE, and every OUTPUT PROCEDURE must contain at least one RETURN statement. Notice that the file named in the RETURN statement is the work file used for the SORT. You must not give an OPEN, CLOSE, READ, or WRITE statement for a SORT work file. The AT END clause in the RETURN statement works the same as the AT END clause in a READ. The RETURN statement may use the INTO option in the same circumstances that a READ would.

(Additional restrictions in some COBOL systems effectively require the last executable statement in an INPUT or OUTPUT PROCEDURE to also physically be the last statement in the PROCEDURE. We have seen that there is no logical need for this restriction, for in almost all the programs in this book our last executable statement was not the physically last statement. This additional requirement leads us to need a new verb, **GO.** The GO verb transfers control to any paragraph or section name in the Procedure Division.)

The least disruptive way to handle this useless restriction is as shown in the OUTPUT–PROCEDURE SECTION of Program 33. Make up a paragraph at the end of the program containing only a CLOSE statement. Give the paragraph any legal name; here we call it OUTPUT–PROCEDURE–EXIT. Then, in the OUTPUT–PROCEDURE SECTION, in the place after the IF statement where we would usually expect to find a CLOSE statement, put a GO statement instead to transfer control to the end. Program 33 was run with the same input data as Program 22 and produced the output shown in Figure 12.2.

Figure 12.2 Output from Program 33.

```
14760  608250035    BUXBAUM, ROBERT              1161000
11664  209560011    COSTELLO, JOSEPH S.          3953000
15147  703100038    DUMAY, MRS. MARY             0812000
14373  604910032    FELDSOTT, MS. SALLY          1510000
13986  505680029    GOODMAN, ISAAC               1859000
11277  201110008    GREENWOOD, JAMES             4302000
10890  101850005    JACOBSON, MRS. NELLIE        4651000
13599  502070026    JAVIER, CARLOS               2208000
13212  407390023    KUGLER, CHARLES              2557000
12825  401710020    LIPKE, VINCENT R.            2906000
12438  304870017    MARRA, DITTA E.              3255000
10503  100040002    MORALES, LUIS                5000000
12051  301810014    REITER, D.                   3604000
15534  708020041    SMITH, R.                    0463000
16308  901050047    THOMAS, THOMAS T.            4235000
15921  803220044    VINCENTE, MATTHEW J.         0114000
```

The SORT Statement

One widely used general format of the SORT statement is:

```
SORT file-name-1 ON {DESCENDING} KEY {data-name-1} . . .
                    {ASCENDING }

   [ON {DESCENDING} KEY {data-name-2} . . .] . . .
       {ASCENDING }

   {INPUT PROCEDURE IS section-name-1 [THRU section-name-2]}
   {USING file-name-2                                      }

   {OUTPUT PROCEDURE IS section-name-3 [THRU section-name-4]}
   {GIVING file-name-3                                      }
```

The ellipses show that the programmer may specify as many key fields as desired to control the SORT. The fields must be specified in order from major to minor. ASCENDING KEYs and DESCENDING KEYs may be mixed in any combination. As an example, a list of employees may be SORTed with DESCENDING date of hire as its major field and ASCENDING name as minor field. Then the list would come out with those most recently hired first, and, within each date of hire, the employees' names would be listed alphabetically.

If no OUTPUT PROCEDURE is used, the word GIVING tells the SORT statement to simply place the SORTed records onto the output file given as file–name–3.

Exercise 1. Rewrite your solution to Exercise 3, Chapter 2, page 26. Use Report Writer to print the output. Use a SORT statement to SORT the input data by employee name within company name before printing.

A Program with an INPUT PROCEDURE

In Program 34 we will use an INPUT PROCEDURE to select certain input records for SORTing. An INPUT PROCEDURE may also be used to modify input records before they are SORTed. Program 34 is a modification of Program 7. Review Program 7: input format for Programs 6 and 7, Figure 3.3; input data for Programs 6 and 7, Figure 3.6; main loop for Program 7, Figure 3.11; output 3.12.

In Program 34 we print only the parts having a quantity on hand of less than 400, as we did in Program 7. But in Program 34, shown in Figure 12.3, the output is listed in part number order. In Program 34 we use an INPUT PROCEDURE to select the records to be SORTed. We want to SORT only the records that are going to be printed, namely

Figure 12.3 Program 34.

```
4-CB2 V4 RELEASE 1.5 10NOV77        IBM OS AMERICAN NATIONAL STANDARD COBOL



000010 IDENTIFICATION DIVISION.
000020 PROGRAM-ID. PROG34.
000030*AUTHOR. PHIL RUBENSTEIN.
000040*
000050*     THIS PROGRAM IS A MODIFICATION TO PROGRAM 7.
000060*     INPUT RECORDS ARE SORTED ON PART NUMBER
000070*     AFTER BEING SELECTED.
000080*
000090*****************************************************************
000100
000110 ENVIRONMENT DIVISION.
000120 INPUT-OUTPUT SECTION.
000130 FILE-CONTROL.
000140      SELECT INVENTORY-LIST     ASSIGN TO UT-S-SYSPRINT.
000150      SELECT INVENTORY-FILE-IN ASSIGN TO UT-S-SYSIN.
000160      SELECT SORT-FILE          ASSIGN TO UT-S-SORTWK.
000170
000180*****************************************************************
000190
000200 DATA DIVISION.
000210 FILE SECTION.
000220 FD   INVENTORY-LIST
000230      LABEL RECORDS ARE OMITTED
000240      REPORT IS INVENTORY-REPORT.
000250
000260 SD   SORT-FILE.
000270
000280 01   SORT-RECORD.
000290      05 PART-NUMBER-S              PIC X(8).
000300      05 PART-DESCRIPTION-S         PIC X(20).
000310      05 QUANTITY-S                 PIC 9(3).
000320      05 FILLER                     PIC X(49).
000330
000340 FD   INVENTORY-FILE-IN
000350      LABEL RECORDS ARE OMITTED.
000360
000370 01   INVENTORY-RECORD-IN.
000380      05 PART-NUMBER-IN             PIC X(8).
000390      05 PART-DESCRIPTION-IN        PIC X(20).
000400      05 QUANTITY-IN                PIC 9(3).
000410      05 FILLER                     PIC X(49).
000420
000430 WORKING-STORAGE SECTION.
000440 01   MORE-INPUT                    PIC X(3) VALUE 'YES'.
000450
000460 REPORT SECTION.
000470 RD   INVENTORY-REPORT.
000480
000490 01   DETAIL-LINE
```

Figure 12.3 Program 34 (continued).

```
000500      TYPE DETAIL
000510      LINE PLUS 1.
000515      05 COLUMN 9   SOURCE PART-NUMBER-S        PIC X(8).
000520      05 COLUMN 20  SOURCE QUANTITY-S           PIC 9(3).
000530      05 COLUMN 41  SOURCE PART-DESCRIPTION-S   PIC X(20).
000540
000550************************************************************
000560
000570 PROCEDURE DIVISION.
000580      SORT SORT-FILE
000590           ASCENDING KEY PART-NUMBER-S
000600           INPUT  PROCEDURE INPUT-PROCEDURE
000610           OUTPUT PROCEDURE OUTPUT-PROCEDURE.
000620      STOP RUN.
000630
000640 INPUT-PROCEDURE SECTION.
000650      OPEN INPUT INVENTORY-FILE-IN.
000660      READ INVENTORY-FILE-IN
000670          AT END
000680              MOVE 'NO' TO MORE-INPUT.
000690      IF MORE-INPUT = 'YES'
000700          PERFORM MAIN-PROCESS-1 UNTIL MORE-INPUT = 'NO'.
000710      GO TO INPUT-PROCEDURE-EXIT.
000720
000730 MAIN-PROCESS-1.
000740      IF QUANTITY-IN LESS THAN 400
000750          RELEASE SORT-RECORD FROM INVENTORY-RECORD-IN.
000760      READ INVENTORY-FILE-IN
000770          AT END
000780              MOVE 'NO' TO MORE-INPUT.
000790
000800 INPUT-PROCEDURE-EXIT.
000810      CLOSE INVENTORY-FILE-IN.
000820
000830 OUTPUT-PROCEDURE SECTION.
000840      MOVE 'YES' TO MORE-INPUT.
000850      OPEN OUTPUT INVENTORY-LIST.
000860      RETURN SORT-FILE
000870          AT END
000880              MOVE 'NO' TO MORE-INPUT.
000890      IF MORE-INPUT = 'YES'
000900          INITIATE INVENTORY-REPORT
000910          PERFORM MAIN-PROCESS-2 UNTIL MORE-INPUT = 'NO'
000920          TERMINATE INVENTORY-REPORT.
000930      GO TO OUTPUT-PROCEDURE-EXIT.
000940
000950 MAIN-PROCESS-2.
000960      GENERATE DETAIL-LINE.
000970      RETURN SORT-FILE
000980          AT END
000990              MOVE 'NO' TO MORE-INPUT.
001000
001010 OUTPUT-PROCEDURE-EXIT.
001020      CLOSE INVENTORY-LIST.
```

the ones whose quantity on hand is less than 400. You can see in the Data Division that the input record, INVENTORY–RECORD–IN, is this time broken down into its fields. The field names are needed so that the INPUT PROCEDURE can refer to them.

We also need in Program 34 an OUTPUT PROCEDURE to format and print the records after they have been SORTed. You can see in the SORT statement in the Procedure Division that both an INPUT PROCEDURE and an OUTPUT PROCEDURE have been named. The two PROCEDUREs, INPUT–PROCEDURE and OUTPUT–PROCEDURE, have both been defined as SECTIONs. Each SECTION must have its own last executable statement at the physical end of the SECTION. So you can see where the paragraphs INPUT–PROCEDURE–EXIT and OUTPUT–PROCEDURE–EXIT have been placed.

Notice that each PROCEDURE deals with only its associated input or output file. INPUT–PROCEDURE OPENs and CLOSEs only the input file, INVENTORY–FILE–IN. OUTPUT–PROCEDURE OPENs and CLOSEs only the output file, INVENTORY–LIST. Of course, neither the INPUT nor the OUTPUT PROCEDURE OPENs or CLOSEs the sort work file, SORT–WORK, for it is illegal to do so.

This program uses the **RELEASE** verb. A RELEASE statement may be used only in an INPUT PROCEDURE, and every INPUT PROCEDURE must have at least one RELEASE statement. The RELEASE statement is issued when the INPUT processing is complete and the record is ready for SORTing. In Program 34 there is very little INPUT processing. Each input record is examined to see whether its QUANTITY–IN is LESS THAN 400. If the QUANTITY–IN is equal to 400 or greater, the program goes on to the next input record.

Notice that the RELEASE statement names the 01-level record name associated with the sort work file. The RELEASE statement may use the FROM option in the same way that a WRITE statement would. Program 34 was run with the same input data as Program 7 and produced the results shown in Figure 12.4.

Figure 12.4 Output from Program 34.

```
C0123435    207             AXE
C1256749    399             LEVEL
02023406    367             PAINT REMOVER
20033547    301             TWELVE INCH WRENCH
32109886    019             PAINT BRUSH
34554311    259             SEARCH LIGHT
```

Exercise 2. Modify your solution to Exercise 4, Chapter 3, page 46, so that your program selects for printing only those employees who have fewer than 25 hours worked. Have your program SORT the employee records by employee number. Use Report Writer to print the SORTed records.

String Manipulation

Two COBOL verbs, **STRING** and **UNSTRING,** permit COBOL programs to attach two or more fields end to end in a single field or to break down a single field into two or more smaller fields.

USING THE STRING VERB

Two applications of the STRING verb will be shown. The first uses input data in the format shown in Figure 12.5. Each input card contains a date in the usual American form MMDDYY.

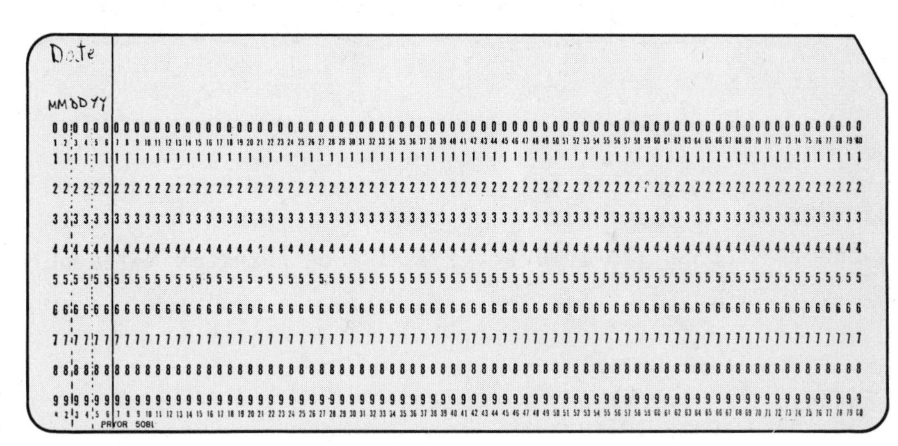

Figure 12.5
Data card format for Program 35.

For each card read, Program 35 expands the date and prints the input and the expanded date on one line, as shown in Figure 12.6. Notice that there is only one space between the name of the month and the day, regardless of the length of the name or whether the day is a one-digit number or two. The STRING verb allows us to place the output components exactly where we want them in the output field.

Program 35 is shown in Figure 12.7. The names of the 12 months are arranged in a MONTHS–TABLE, so that the month number in the input can be used directly as a subscript to obtain the name of the month.

The field EXPANDED–DATE is used for assigning the complete expanded date in English ready to be printed. The field COMMA–SPACE–CENTURY is a constant that is inserted into every date processed.

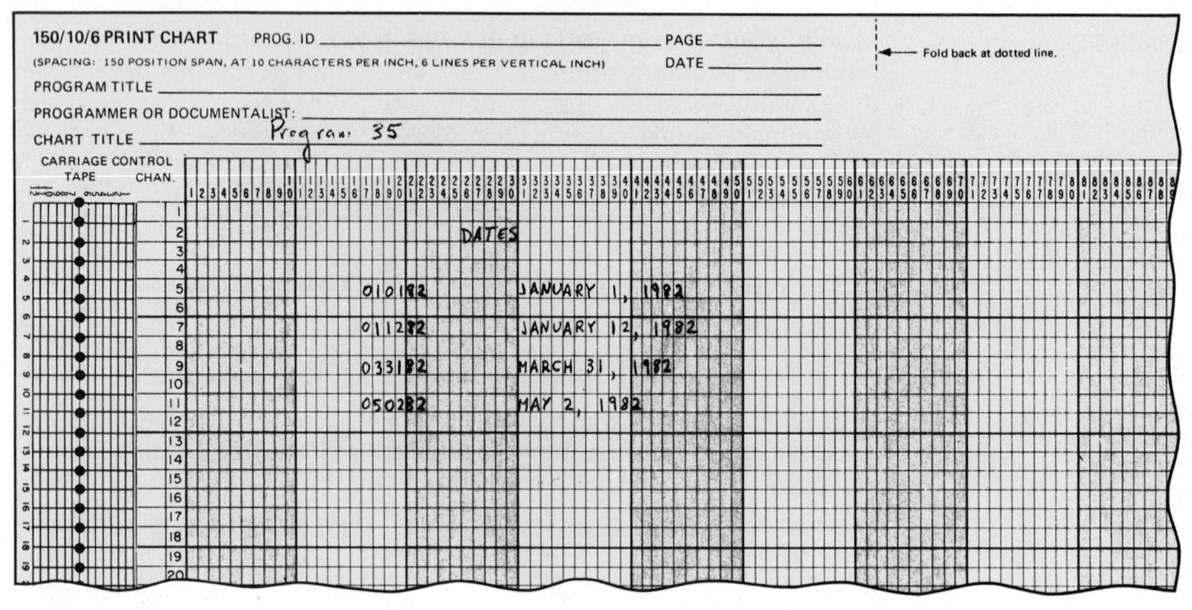

Figure 12.6 Output format for Program 35.

Figure 12.7 Program 35, Part 1 of 2.

```
4-CB2 V4 RELEASE 1.5 10NOV77        IBM OS AMERICAN NATIONAL STANDARD COBOL



000010 IDENTIFICATION DIVISION.
000020 PROGRAM-ID. PROG35.
000030*AUTHOR. PHIL RUBENSTEIN.
000040*
000050*     THIS PROGRAM READS DATES IN MMDDYY FORM
000060*     AND EXPANDS THEM.
000070*
000080*****************************************************************
000090
000100 ENVIRONMENT DIVISION.
000110 INPUT-OUTPUT SECTION.
000120 FILE-CONTROL.
000130     SELECT DATE-LIST-FILE  ASSIGN TO UT-S-SYSPRINT.
000140     SELECT DATE-FILE-IN ASSIGN TO UT-S-SYSIN.
000150
000160*****************************************************************
000170
000180 DATA DIVISION.
000190 FILE SECTION.
000200 FD  DATE-FILE-IN
000210     LABEL RECORDS ARE OMITTED.
000220
```

Figure 12.7 Part 1 (continued).

```
000230 01   DATE-RECORD-IN.
000240      05 DATE-IN.
000250         10 MONTH-IN                               PIC 99.
000260         10 DAY-IN.
000270            15 TENS-PLACE                          PIC 9.
000280            15 UNITS-PLACE                         PIC 9.
000290         10 YEAR-IN                                PIC 99.
000300      05 FILLER                                    PIC X(74).
000310
000320 FD   DATE-LIST-FILE
000330      LABEL RECORDS ARE OMITTED
000340      REPORT IS DATE-LIST.
000350
000360 WORKING-STORAGE SECTION.
000370 01   MORE-INPUT                                   PIC X(3) VALUE 'YES'.
000380
000390 01   MONTHS.
000400      05 FILLER VALUE 'JANUARY .'                  PIC X(10).
000410      05 FILLER VALUE 'FEBRUARY .'                 PIC X(10).
000420      05 FILLER VALUE 'MARCH .'                    PIC X(10).
000430      05 FILLER VALUE 'APRIL .'                    PIC X(10).
000440      05 FILLER VALUE 'MAY .'                      PIC X(10).
000450      05 FILLER VALUE 'JUNE .'                     PIC X(10).
000460      05 FILLER VALUE 'JULY .'                     PIC X(10).
000470      05 FILLER VALUE 'AUGUST .'                   PIC X(10).
000480      05 FILLER VALUE 'SEPTEMBER '                 PIC X(10).
000490      05 FILLER VALUE 'OCTOBER .'                  PIC X(10).
000500      05 FILLER VALUE 'NOVEMBER .'                 PIC X(10).
000510      05 FILLER VALUE 'DECEMBER .'                 PIC X(10).
000520 01   MONTHS-TABLE
000530      REDEFINES MONTHS.
000540      05 MONTH              OCCURS 12 TIMES        PIC X(10).
000550
000560 01   EXPANDED-DATE                                PIC X(18).
000570 01   COMMA-SPACE-CENTURY
000580                            VALUE ', 19'           PIC X(4).
000590
000600 REPORT SECTION.
000610 RD   DATE-LIST
000620      PAGE 30 LINES, FIRST DETAIL 5.
000630
000640 01   TYPE PAGE HEADING
000650      LINE 2.
000660      05 COLUMN 26          VALUE 'DATES'          PIC X(5).
000670
000680 01   DETAIL-LINE
000690      TYPE DETAIL
000700      LINE PLUS 2.
000710      05 COLUMN 17          SOURCE DATE-IN         PIC 9(6).
000720      05 COLUMN 31          SOURCE EXPANDED-DATE PIC X(18).
000730
000740*****************************************************************
000750
000760 PROCEDURE DIVISION.
000770      OPEN INPUT  DATE-FILE-IN,
000780           OUTPUT DATE-LIST-FILE.
000790      READ DATE-FILE-IN
```

In the Procedure Division, one STRING statement or another is executed, depending on whether the day of the month has a nonzero tens' place. In either case, the statement simply STRINGs the sending fields, one after another, end to end INTO the receiving field EXPANDED–DATE and assigns the complete string to it.

The **DELIMITED BY** clause tells the STRING statement which characters from each of the sending fields should be assigned to the receiving field. The clause DELIMITED BY '.' causes all characters in the sending field MONTH (MONTH–IN), up to but not including the dot, to be assigned to the receiving field EXPANDED–DATE. If no dot is found in the sending field, as in SEPTEMBER, the entire sending field is assigned to the receiving field.

The clause DELIMITED BY SIZE causes the whole lengths of the other sending fields to be added to the string of fields assigned to EXPANDED–DATE. You can see that there is no need to write a DELIMITED BY clause for every sending field. A whole group of sending fields can share a DELIMITED BY clause, as in the two STRING statements in Program 35. You may use as many DELIMITED BY clauses as you like in a STRING statement, and they may be mixed in any desired way.

Figure 12.7 Program 35, Part 2 of 2.

```
000800          AT END
000810              MOVE 'NO' TO MORE-INPUT.
000820      IF MORE-INPUT = 'YES'
000830          INITIATE DATE-LIST
000840          PERFORM MAIN-PROCESS UNTIL MORE-INPUT = 'NO'
000850          TERMINATE DATE-LIST.
000860      CLOSE DATE-FILE-IN,
000870              DATE-LIST-FILE.
000880      STOP RUN.
000890
000900 MAIN-PROCESS.
000910      MOVE SPACES TO EXPANDED-DATE.
000920      IF TENS-PLACE = 0
000930          STRING MONTH (MONTH-IN)        DELIMITED BY '.',
000940                 UNITS-PLACE,
000950                 COMMA-SPACE-CENTURY,
000960                 YEAR-IN                 DELIMITED BY SIZE
000970                 INTO EXPANDED-DATE
000980      ELSE
000990          STRING MONTH (MONTH-IN)        DELIMITED BY '.',
001000                 DAY-IN,
001010                 COMMA-SPACE-CENTURY,
001020                 YEAR-IN                 DELIMITED BY SIZE
001030                 INTO EXPANDED-DATE.
001040      GENERATE DETAIL-LINE.
001050      READ DATE-FILE-IN
001060          AT END
001070              MOVE 'NO' TO MORE-INPUT.
```

Any alphanumeric data may be specified as a delimiter. If no delimiter is found in the sending field, the entire sending field is added to the string. The input data for Program 35 are shown in Figure 12.8. The output is in Figure 12.9.

```
010283
010182
011282
033182
050282
121483
073183
112482
021483
```

Figure 12.8
Input for Program 35.

```
                    DATES

010283          JANUARY 2, 1983
010182          JANUARY 1, 1982
011282          JANUARY 12, 1982
033182          MARCH 31, 1982
050282          MAY 2, 1982
121483          DECEMBER 14, 1983
073183          JULY 31, 1983
112482          NOVEMBER 24, 1982
021483          FEBRUARY 14, 1983
```

Figure 12.9
Output from Program 35.

GENERAL FORMAT OF THE STRING STATEMENT

The general format of the STRING statement is:

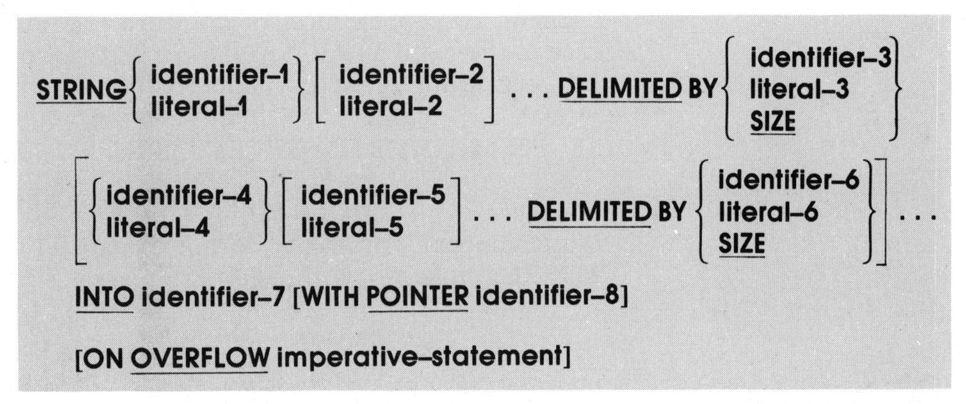

The **POINTER** option is used when more than one STRING statement is needed to construct a complete receiving field. The POINTER field is then used to record where each STRING statement leaves off in the receiving field, so that the next STRING statement can continue in the right place.

The **OVERFLOW** option is used if there is the possibility that the completed string will be longer than the receiving field.

ANOTHER APPLICATION OF THE STRING VERB

We now develop an elaborate application of the STRING verb which deals with selective use of fields. Program 36 reads input data in the format shown in Figure 12.10. Each input record contains an eight-digit dollars-and-cents amount, from $.00 up to $999,999.99. The program is to read each card and print the money amount, edited, and also print the six-digit dollar amount in words, as shown in Figure 12.11.

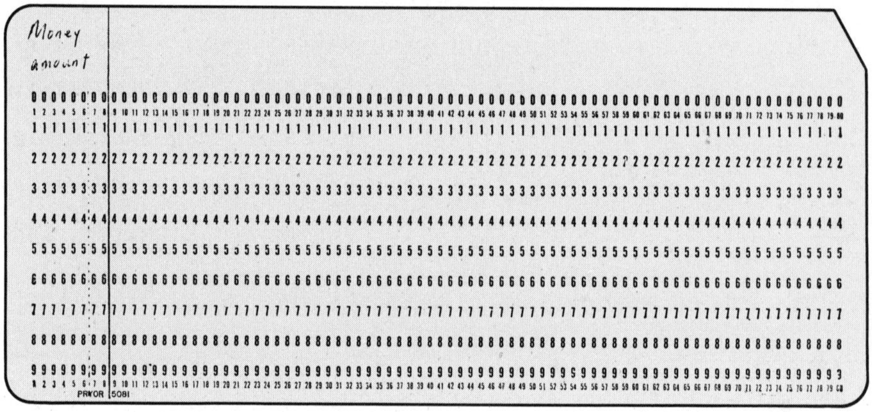

Figure 12.10 Data card format for Program 36.

A number can sometimes be expressed in few words, like THREE or TWO HUNDRED or TWENTY-SIX or sometimes in many words. In fact, there can be as many as 11 different pieces of data needed to write a six-digit number in English; six words for the digits, two hyphens, two appearances of the word HUNDRED, and one appearance of the word THOUSAND. An example of a number that uses all 11 pieces, with its pieces numbered is:

① SEVEN ② HUNDRED ③ SEVENTY ④ - ⑤ SEVEN ⑥ THOUSAND ⑦ SEVEN ⑧ HUNDRED ⑨ SEVENTY ⑩ - ⑪ SEVEN

Most numbers written out do not contain all 11 pieces. The number THREE contains only the eleventh piece,

⑪ THREE

The other two short numbers are shown with their components numbered:

⑦ ⑧
TWO HUNDRED

⑨ ⑩ ⑪
TWENTY-SIX

Program 36 assumes that for each input number none of the 11 pieces is needed to write the number out. Then, with IF statements, it determines piece by piece which of the 11 are needed. After all the IF tests are complete, the program STRINGs together all the needed pieces.

Peculiarities in the way that numbers are written complicate the program. For example, component 6, the word THOUSAND, appears in the number if either component 1, component 3, or component 5 does. Every word used in the expression of a number is always followed by a space, except for components 3 and 9 when they are larger than TEN. Components 3 and 9 may be followed by a space or by a hyphen, depending on the word following. Values between 11

```
150/10/6 PRINT CHART   PROG. ID ________   PAGE ____   Fold back at dotted line.   Fold in at dotted l
(SPACING: 150 POSITION SPAN, AT 10 CHARACTERS PER INCH, 6 LINES PER VERTICAL INCH)   DATE ____
PROGRAM TITLE ________
PROGRAMMER OR DOCUMENTALIST: ________
CHART TITLE ________
CARRIAGE CONTROL TAPE   CHAN.

 2              MONEY
 4 $900,000.00  NINE HUNDRED THOUSAND DOLLARS AND 00/XX
 6 $ 87,017.06  EIGHTY-SEVEN THOUSAND SEVENTEEN DOLLARS AND 06/XX
 8 $777,777.77  SEVEN HUNDRED SEVENTY-SEVEN THOUSAND SEVEN HUNDRED SEVENTY-SEVEN DOLLARS AND 77/XX
10 $       .05  ONLY 05/XX
12 $  1,119.19  ONE THOUSAND ONE HUNDRED NINETEEN DOLLARS AND 19/XX
14 $ 20,000.00  TWENTY THOUSAND DOLLARS AND 00/XX
16 $       .00  ONLY 00/XX
18 $      1.10  ONE DOLLAR AND 10/XX
```

Figure 12.11 Output format for Program 36.

and 19 (and between 11,000 and 19,000) have to be handled in a special way because they are irregular forms.[1]

A hierarchy diagram for Program 36 is shown in Figure 12.12. In the step "Determine values of all places," the program breaks down the input number into separate fields for each of the place values. "Set all word fields to assume none is present" clears all 11 of the fields that could make up the dollar amount. Then in "Fill all required word fields," the program determines which words are needed and assigns each to a field reserved for the purpose. For example, the box "Hundreds of thousands" assigns a word to the component-1 field to say how many hundreds of thousands there are in the number. It also assigns the word HUNDRED to the component-2 field and the word THOUSAND to the component-6 field, since the presence of a hundreds of thousands amount means that the words HUNDRED and THOUSAND must also appear in the number. The right end of the diagram shows that the words DOLLARS AND might be assigned to a field to be eventually strung into the output. If the money amount is less than $1, the word ONLY would appear instead. If the money amount is greater than $.99 and less than $2, the words DOLLAR AND would appear. Program 36 is shown in Figure 12.13.

Figure 12.12
Hierarchy diagram for Program 36.

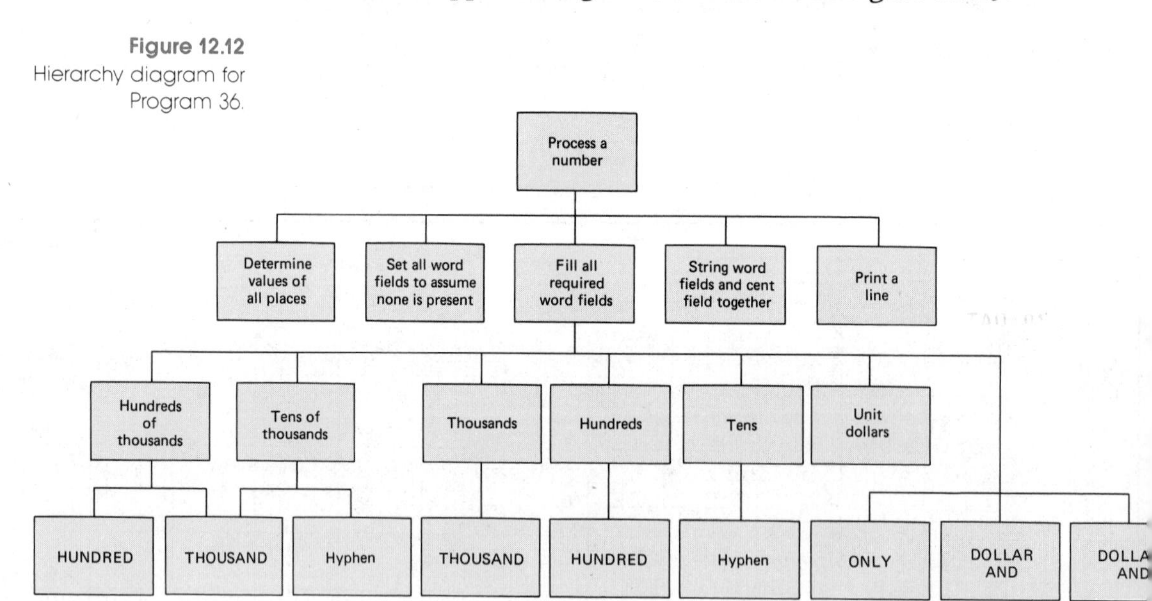

1. In French and Spanish, the numbers 17–19 are regular.

In the Working Storage Section, an OCCURS clause has been used in the definition of the 11 fields that may have assigned to them any or all of the 11 components needed to express a dollar amount. The 11 fields are called MONEY–WORD, and the uses of the 11 fields are:

MONEY–WORD (1)	**If there are any hundreds of thousands in the amount, the number of hundreds of thousands is assigned to this field.**

Figure 12.13 Program 36, Part 1 of 4.

```
4-CB2 V4 RELEASE 1.5 10NOV77          IBM OS AMERICAN NATIONAL STANDARD COBOL



000010 IDENTIFICATION DIVISION.
000020 PROGRAM-ID. PROG36.
000030
000040*     THIS PROGRAM READS IN MONEY AMOUNTS AND PRINTS
000050*     THEM IN WORDS.
000060*
000070*****************************************************************
000080
000090 ENVIRONMENT DIVISION.
000100 INPUT-OUTPUT SECTION.
000110 FILE-CONTROL.
000120     SELECT MONEY-FILE-IN  ASSIGN TO UT-S-SYSIN.
000130     SELECT MONEY-FILE-OUT ASSIGN TO UT-S-SYSPRINT.
000140
000150*****************************************************************
000160
000170 DATA DIVISION.
000180 FILE SECTION.
000190 FD  MONEY-FILE-IN
000200     LABEL RECORDS ARE OMITTED.
000210
000220 01  MASTER-RECORD-IN.
000230     05 MONEY-IN                                  PIC 9(6)V99.
000240     05 FILLER                                    PIC X(72).
000250
000260 FD  MONEY-FILE-OUT
000270     LABEL RECORDS ARE OMITTED
000280     REPORT IS MONEY-REPORT.
000290
000300 WORKING-STORAGE SECTION.
000310 01  MORE-INPUT      VALUE 'YES'                  PIC X(3).
000320 01  MONEY-IN-SAVE                                PIC 9(6)V99.
000330
000340 01  ENGLISH-WORDS.
000350     05 MONEY-WORD
000360        OCCURS 11 TIMES INDEXED BY WORD-INDEX     PIC X(10).
000370     05 DOLLARS-OR-ONLY                           PIC X(12).
000380     05 EXPANDED-MONEY                            PIC X(82).
000390
```

MONEY–WORD (2)	If there are any hundreds of thousands in the amount, the word HUNDRED is assigned to this field.
MONEY–WORD (3)	This field is for the number of tens of thousands.
MONEY–WORD (4)	If there are any tens of thousands in the amount, then either a space or a hyphen is assigned to this field—a hyphen if there are thousands and a space if not.
MONEY–WORD (5)	The number of thousands.
MONEY–WORD (6)	If there are any hundreds of thousands, or any tens of thousands, or any thousands, the word THOUSAND is assigned to this field.
MONEY–WORD (7)	The number of hundreds.
MONEY–WORD (8)	The word HUNDRED, if there are any hundreds.
MONEY–WORD (9)	The number of tens.
MONEY–WORD (10)	A space or a hyphen, if there are any tens.
MONEY–WORD (11)	The number of unit dollars.

The field DOLLARS–OR–ONLY is set aside to have the words DOLLARS AND, DOLLAR AND, or ONLY assigned to it, depending on the magnitude of the money amount.

A table of UNITS–PLACES and a table of TENS–PLACES are set up to provide words to be used in the output. We need no table for higher place values because all place values higher than tens use the same words as either the units' place or the tens' place. All the numbers up through 19 are included in the UNITS–PLACES–TABLE to facilitate programming for the irregular forms of the numbers between 11 and 19. In this program, all numbers below 20 are treated as units, and programming for the tens' place handles numbers of 20 or larger. A space is included after each word in the UNITS–PLACES–TABLE because the space is part of the field. The units' words are always followed by a space whenever they appear in the output. There is no such space in the entries in the TENS–PLACES–TABLE, because the tens' words are not always followed by a space; they are sometimes followed by a hyphen. An IF test in the program determines whether any particular tens' word is to be followed by a space or by a hyphen, and inserts the appropriate character. In the Procedure Division, the paragraph PROCESS–A–NUMBER follows the hierarchy diagram.

Figure 12.13 Program 36, Part 2 of 4.

```
000400 01   UNITS-PLACES.
000410      05 FILLER         VALUE 'ONE .'               PIC X(10).
000420      05 FILLER         VALUE 'TWO .'               PIC X(10).
000430      05 FILLER         VALUE 'THREE .'             PIC X(10).
000440      05 FILLER         VALUE 'FOUR .'              PIC X(10).
000450      05 FILLER         VALUE 'FIVE .'              PIC X(10).
000460      05 FILLER         VALUE 'SIX .'               PIC X(10).
000470      05 FILLER         VALUE 'SEVEN .'             PIC X(10).
000480      05 FILLER         VALUE 'EIGHT .'             PIC X(10).
000490      05 FILLER         VALUE 'NINE .'              PIC X(10).
000500      05 FILLER         VALUE 'TEN .'               PIC X(10).
000510      05 FILLER         VALUE 'ELEVEN .'            PIC X(10).
000520      05 FILLER         VALUE 'TWELVE .'            PIC X(10).
000530      05 FILLER         VALUE 'THIRTEEN .'          PIC X(10).
000540      05 FILLER         VALUE 'FOURTEEN .'          PIC X(10).
000550      05 FILLER         VALUE 'FIFTEEN .'           PIC X(10).
000560      05 FILLER         VALUE 'SIXTEEN .'           PIC X(10).
000570      05 FILLER         VALUE 'SEVENTEEN '          PIC X(10).
000580      05 FILLER         VALUE 'EIGHTEEN .'          PIC X(10).
000590      05 FILLER         VALUE 'NINETEEN .'          PIC X(10).
000600 01   UNITS-PLACES-TABLE
000610      REDEFINES UNITS-PLACES.
000620      05 UNITS-PLACE  OCCURS 19 TIMES               PIC X(10).
000630
000640 01   TENS-PLACES.
000650      05 FILLER         VALUE 'TEN.'                PIC X(8).
000660      05 FILLER         VALUE 'TWENTY.'             PIC X(8).
000670      05 FILLER         VALUE 'THIRTY.'             PIC X(8).
000680      05 FILLER         VALUE 'FORTY.'              PIC X(8).
000690      05 FILLER         VALUE 'FIFTY.'              PIC X(8).
000700      05 FILLER         VALUE 'SIXTY.'              PIC X(8).
000710      05 FILLER         VALUE 'SEVENTY.'            PIC X(8).
000720      05 FILLER         VALUE 'EIGHTY.'             PIC X(8).
000730      05 FILLER         VALUE 'NINETY.'             PIC X(8).
000740 01   TENS-PLACES-TABLE
000750      REDEFINES TENS-PLACES.
000760      05 TENS-PLACE   OCCURS 9 TIMES                PIC X(8).
000770
000780 01   PLACE-VALUES.
000790      05 HUNDREDS-OF-THOUSANDS                      PIC S9.
000800      05 TENS-OF-THOUSANDS                          PIC S9.
000810      05 THOUSANDS                                  PIC S99.
000820      05 HUNDREDS                                   PIC S9.
000830      05 TENS                                       PIC S9.
000840      05 UNIT-DOLLARS                               PIC S99.
000850      05 CENTS-UNEDITED                             PIC V99.
000860      05 WHOLE-CENTS
000870         REDEFINES CENTS-UNEDITED                   PIC 99.
000880
000890 REPORT SECTION.
000900 RD   MONEY-REPORT
000910      PAGE 50 LINES, FIRST DETAIL 4.
000920
000930 01   TYPE PAGE HEADING
000940      LINE 2.
000950      05 COLUMN 16    VALUE 'MONEY'                 PIC X(5).
000960
```

In SET–WORD–FIELDS–TO–NONE, a dot is assigned to each of the MONEY–WORDs 1 through 11 to indicate that none of the 11 words is needed yet. The paragraph FILL–REQUIRED–WORD–FIELDS determines which of the 11 are needed and assigns to each its appropriate word. For example, IF the field HUNDREDS–OF–THOUSANDS is not found to equal 0, then the number of hundreds thousands is MOVEd to MONEY–WORD (1) to serve as component 1. Once the program sees that there are some hundreds of thousands in the number, it knows also that the words HUNDRED and THOUSAND must appear in the output as components 2 and 6, respectively, and so it MOVEs those words to MONEY–WORD (2) and MONEY–WORD (6). The last IF statement in FILL–REQUIRED–WORD–FIELDS fills not a MONEY–WORD field but the DOLLARS–OR–ONLY field.

Figure 12.13 Program 36, Part 3 of 4.

```
000970 01   DETAIL-LINE
000980      TYPE DETAIL
000990      LINE PLUS 2.
001000      05 COLUMN 3     SOURCE MONEY-IN-SAVE          PIC $ZZZ,ZZZ.99.
001010      05 COLUMN 16    SOURCE EXPANDED-MONEY         PIC X(82).
001020
001030****************************************************************
001040
001050 PROCEDURE DIVISION.
001060      OPEN INPUT  MONEY-FILE-IN,
001070           OUTPUT MONEY-FILE-OUT.
001080      READ MONEY-FILE-IN
001090          AT END
001100              MOVE 'NO' TO MORE-INPUT.
001110      IF MORE-INPUT = 'YES'
001120          INITIATE MONEY-REPORT
001130          PERFORM PROCESS-A-NUMBER UNTIL MORE-INPUT = 'NO'
001140          TERMINATE MONEY-REPORT.
001150      CLOSE MONEY-FILE-IN,
001160            MONEY-FILE-OUT.
001170      STOP RUN.
001180
001190 PROCESS-A-NUMBER.
001200      PERFORM DETERMINE-VALUES-OF-ALL-PLACES.
001210      PERFORM SET-WORD-FIELDS-TO-NONE.
001220      PERFORM FILL-REQUIRED-WORD-FIELDS.
001225      PERFORM STRING-WORDS-AND-CENT-FIELDS.
001230      GENERATE DETAIL-LINE.
001240      READ MONEY-FILE-IN
001250          AT END
001260              MOVE 'NO' TO MORE-INPUT.
```

Figure 12.13 Part 3 (continued).

```
001270
001280 DETERMINE-VALUES-OF-ALL-PLACES.
001290     MOVE MONEY-IN TO MONEY-IN-SAVE.
001300     DIVIDE MONEY-IN BY 100000 GIVING HUNDREDS-OF-THOUSANDS
001310                                       REMAINDER MONEY-IN.
001320     IF MONEY-IN GREATER THAN 19999.99
001330         DIVIDE MONEY-IN BY 10000 GIVING TENS-OF-THOUSANDS
001340                                       REMAINDER MONEY-IN
001350     ELSE
001360         MOVE 0 TO TENS-OF-THOUSANDS.
001370     DIVIDE MONEY-IN BY 1000 GIVING THOUSANDS  REMAINDER MONEY-IN.
001380     DIVIDE MONEY-IN BY 100  GIVING HUNDREDS   REMAINDER MONEY-IN.
001390     IF MONEY-IN GREATER THAN 19.99
001400         DIVIDE MONEY-IN BY 10 GIVING TENS     REMAINDER MONEY-IN
001410     ELSE
001420         MOVE 0 TO TENS.
001430     MOVE MONEY-IN TO UNIT-DOLLARS.
001440     MOVE MONEY-IN TO CENTS-UNEDITED.
001450
001460 SET-WORD-FIELDS-TO-NONE.
001470     PERFORM DOT-FILL
001480         VARYING WORD-INDEX FROM 1 BY 1
001490         UNTIL
001500             WORD-INDEX GREATER THAN 11.
001510
001520 FILL-REQUIRED-WORD-FIELDS.
001530     IF HUNDREDS-OF-THOUSANDS NOT = 0
001540         MOVE UNITS-PLACE (HUNDREDS-OF-THOUSANDS)
001550                                           TO MONEY-WORD (1)
001560         MOVE 'HUNDRED .'                  TO MONEY-WORD (2)
001570         MOVE 'THOUSAND .'                 TO MONEY-WORD (6).
001580     IF TENS-OF-THOUSANDS NOT = 0
001590         MOVE TENS-PLACE (TENS-OF-THOUSANDS)   TO MONEY-WORD (3)
001600         MOVE 'THOUSAND .'                 TO MONEY-WORD (6)
001610         IF THOUSANDS NOT = 0
001620             MOVE '-.'                     TO MONEY-WORD (4)
001630         ELSE
001640             MOVE ' .'                     TO MONEY-WORD (4).
001650     IF THOUSANDS NOT = 0
001660         MOVE UNITS-PLACE (THOUSANDS)      TO MONEY-WORD (5)
001670         MOVE 'THOUSAND .'                 TO MONEY-WORD (6).
001680     IF HUNDREDS NOT = 0
001690         MOVE UNITS-PLACE (HUNDREDS)       TO MONEY-WORD (7)
001700         MOVE 'HUNDRED .'                  TO MONEY-WORD (8).
001710     IF TENS NOT = 0
001720         MOVE TENS-PLACE (TENS)            TO MONEY-WORD (9)
001730         IF UNIT-DOLLARS NOT = 0
               MOVE '-.'                     TO MONEY-WORD (10)
001750         ELSE
001760             MOVE ' .'                     TO MONEY-WORD (10).
001770     IF UNIT-DOLLARS NOT = 0
001780         MOVE UNITS-PLACE (UNIT-DOLLARS)   TO MONEY-WORD (11).
001790     IF MONEY-IN-SAVE LESS THAN 1.00
001800         MOVE 'ONLY .'          TO DOLLARS-OR-ONLY
001810     ELSE
001815     IF MONEY-IN-SAVE LESS THAN 2.00
001820         MOVE 'DOLLAR AND .'  TO DOLLARS-OR-ONLY
```

Figure 12.13 Program 36. Part 4 of 4.

```
001830         ELSE
001840              MOVE 'DOLLARS AND ' TO DOLLARS-OR-ONLY.
001850
001860 STRING-WORDS-AND-CENT-FIELDS.
001865      MOVE SPACES TO EXPANDED-MONEY.
001870      STRING MONEY-WORD (1),
001880              MONEY-WORD (2),
001890              MONEY-WORD (3),
001900              MONEY-WORD (4),
001910              MONEY-WORD (5),
001920              MONEY-WORD (6),
001930              MONEY-WORD (7),
001940              MONEY-WORD (8),
001950              MONEY-WORD (9),
001960              MONEY-WORD (10),
001970              MONEY-WORD (11),
001980              DOLLARS-OR-ONLY DELIMITED BY '.',
001990              WHOLE-CENTS,
002000              '/XX'              DELIMITED BY SIZE
002010              INTO EXPANDED-MONEY.
002020
002030 DOT-FILL.
002040      MOVE '.' TO MONEY-WORD (WORD-INDEX).
```

In the STRING statement in STRING–WORDS–AND–CENT–FIELDS, those MONEY–WORD fields still having only a dot assigned are ignored. Program 36 was run with the input data shown in Figure 12.14 and produced the output shown in Figure 12.15.

```
90000000
08701706
77777777
00000005
00111919
02000000
00000000
00000110
00000099
00000199
00000200
00000100
```

Figure 12.14
Input for Program 36.

Figure 12.15 Output from Program 36.

```
              MONEY
$900,000.00   NINE HUNDRED THOUSAND DOLLARS AND 00/XX
$ 87,017.06   EIGHTY-SEVEN THOUSAND SEVENTEEN DOLLARS AND 06/XX
$777,777.77   SEVEN HUNDRED SEVENTY-SEVEN THOUSAND SEVEN HUNDRED SEVENTY-SEVEN DOLLARS AND 77/XX
$        .05  ONLY 05/XX
$   1,119.19  ONE THOUSAND ONE HUNDRED NINETEEN DOLLARS AND 19/XX
$ 20,000.00   TWENTY THOUSAND DOLLARS AND 00/XX
$        .00  ONLY 00/XX
$       1.10  ONE DOLLAR AND 10/XX
$        .99  ONLY 99/XX
$       1.99  ONE DOLLAR AND 99/XX
$       2.00  TWO DOLLARS AND 00/XX
$       1.00  ONE DOLLAR AND 00/XX
```

Exercise 3. Modify Program 36 so that it can handle nine-digit dollars-and-cents amounts, up through $9,999,999.99.

A PROGRAM USING THE UNSTRING VERB

The UNSTRING verb can be used to select out certain parts of fields for processing. The UNSTRING statement examines a field from left to right looking for characters or combinations of characters specified by the programmer. The characters or combinations of characters specified for the search are called delimiters. When any delimiter is found in the field being examined, the UNSTRING statement assigns all the examined characters, up to but not including the delimiter, to some other field specified by the programmer. On request, the UNSTRING statement will also assign the delimiter itself to another field specified by the programmer and assign to still another field a count of the number of characters that were examined before the delimiter was found.

A single UNSTRING statement can be made to scan an entire field in one operation, looking for all occurrences of delimiters. In that case, the programmer can provide as many fields as desired to contain the pieces of the unstrung subject field, any or all of the delimiters that were found, and any or all of the counts.

Program 37 examines English words, phrases, and sentences and replaces all occurrences of the word MAN with the word PERSON and all occurrences of the word MEN with PERSONS. Although the goal of such a program may seem admirable, the results are perhaps not exactly what one would expect or want. Typical input, like:

MANY DISTINCTLY HUMAN ACTIVITIES EMANATED FROM ERAS AS DISTANT AS CAVEMAN DAYS.

MEN AND WOMEN ENGAGED IN MENDING, IN MANUFACTURING, IN MENTAL MANIPULATIONS,

and

THE MAN–EATING TIGER MANAGED ANYWAY TO MANGLE THE WOMAN'S MANDIBLE.

could produce output of the sort shown in Figure 12.16. The input text is punched in cards starting in column 1. If the edited text occupies more than 120 print positions, Program 37 inserts the words CHARACTERS MISSING at the right end of the edited print line.

Figure 12.16 Output format for Program 37.

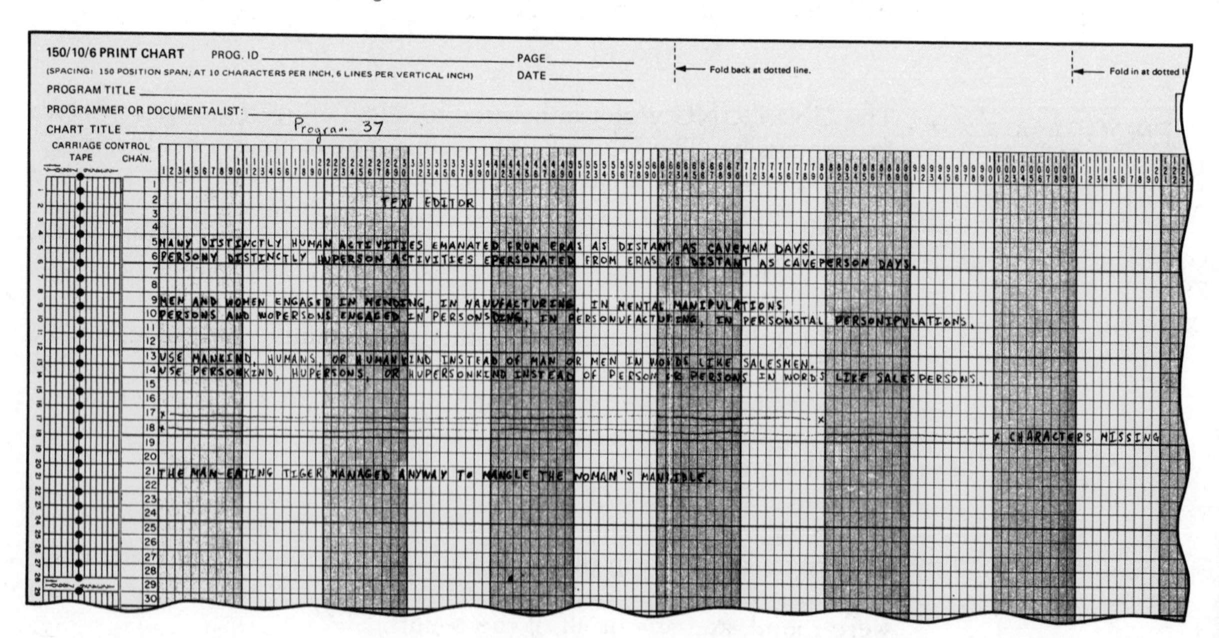

Program 37 is shown in Figure 12.17. The field SINGULAR–OR–PLURAL is used to contain each delimiter as it is found. So SINGULAR–OR–PLURAL might contain the word MAN or MEN. If

Figure 12.17 Program 37, Part 1 of 3.

```
4-CB2 V4 RELEASE 1.5 10NOV77          IBM OS AMERICAN NATIONAL STANDARD COBOL

 000040 IDENTIFICATION DIVISION.
 000050 PROGRAM-ID. PROG37.
 000060*AUTHOR. PHIL RUBENSTEIN.
 000070*
 000080*     THIS PROGRAM EDITS ENGLISH TEXT, REPLACING ALL
 000090*     APPEARANCES OF MAN WITH PERSON AND ALL APPEARANCES
 000100*     OF MEN WITH PERSONS.
 000110*
 000120*****************************************************************
 000130
 000140 ENVIRONMENT DIVISION.
 000150 INPUT-OUTPUT SECTION.
 000160 FILE-CONTROL.
 000170      SELECT TEXT-OUT  ASSIGN TO UT-S-SYSPRINT.
 000180      SELECT TEXT-IN   ASSIGN TO UT-S-SYSIN.
 000190
 000200*****************************************************************
 000210
 000220 DATA DIVISION.
 000230 FILE SECTION.
 000240 FD   TEXT-IN
 000250      LABEL RECORDS ARE OMITTED.
 000260
 000270 01   SENDING-FIELD                           PIC X(80).
 000280
 000290 FD   TEXT-OUT
 000300      LABEL RECORDS ARE OMITTED
 000310      REPORT IS EDITED-TEXT.
 000320
 000330 WORKING-STORAGE SECTION.
 000340 01   MORE-INPUT                              PIC X(3) VALUE 'YES'.
 000350
 000360 01   SINGULAR-OR-PLURAL                      PIC X(3).
 000370 01   RECEIVING-FIELD                         PIC X(120 .
 000380 01   INTERMEDIATE-FIELD                      PIC X(81).
 000390
 000400 01   SIZES                        COMPUTATIONAL.
 000410      05 SIZE-OF-SENDING-FIELD     VALUE 80    PIC S99.
 000420      05 SIZE-OF-RECEIVING-FIELD   VALUE 120   PIC S999.
 000430
 000440 01   POINTERS                     COMPUTATIONAL.
 000450      05 SENDING-FIELD-POINTER                 PIC S99.
 000460      05 RECEIVING-FIELD-POINTER               PIC S999.
 000470      05 INTERMEDIATE-POINTER                  PIC S99.
 000480
```

neither MAN nor MEN is found in the portion of input text being scanned, then SINGULAR–OR–PLURAL will contain blanks.

The RECEIVING–FIELD is used to build up the line of edited text piece by piece as the words MAN and MEN are found in the SENDING–FIELD.

The INTERMEDIATE–FIELD is used as temporary storage. When a delimiter is found in the SENDING–FIELD or the end of the SENDING–FIELD is reached, the characters up to the end or up to but not including the delimiter are assigned to the INTERMEDIATE–FIELD. The characters wait there while the program determines whether they will be strung INTO the RECEIVING–FIELD with the word PERSON attached, or whether they will be strung INTO the RECEIVING–FIELD with the word PERSONS attached, or whether they will just be strung INTO the RECEIVING–FIELD.

Since it will usually take several separate examinations of different parts of the SENDING–FIELD to find all the occurrences of MAN and MEN, we use a SENDING–FIELD–POINTER. Each execution of the UNSTRING statement sets the SENDING–FIELD–POINTER to say where it left off in the scan of the SENDING–FIELD. Then the next execution of the UNSTRING statement can pick up where the previous one left off. Similarly, the RECEIVING–FIELD is usually built up by several executions of the STRING statement. Each execution of the STRING statement sets the RECEIVING–FIELD–POINTER to record where it left off STRINGing fields INTO the RECEIVING–FIELD. Then the next execution of the STRING statement can pick up where the previous one left off. The operation of these two pointers is largely automatic. The programmer merely names the pointer fields in the POINTER clauses of the STRING and UNSTRING statements. You will see exactly how the POINTERs work when we look at the Procedure Division.

The INTERMEDIATE–POINTER is used to record how many characters are assigned to the INTERMEDIATE–FIELD by the UNSTRING statement. You will see how it is used when we look at the Procedure Division.

The REPLACEMENTS–TABLE contains the three possible VALUEs that can be strung INTO the RECEIVING–FIELD under different conditions: PERSON, PERSONS, or nothing at all.

In the Procedure Division, the MAIN–PROCESS paragraph MOVEs 1 to the SENDING–FIELD–POINTER and to the RECEIVING–FIELD–POINTER to tell the STRING and UNSTRING statements to start at the beginning (left end) of the SENDING–FIELD for the scanning, and at the beginning (left end) of the RECEIVING–FIELD in building up the edited text. After clearing the RECEIVING–FIELD, we PERFORM the paragraph DESEX–TEXT until either the SENDING–FIELD has been completely scanned or the RECEIVING–FIELD is completely full, whichever comes first. At that time both the

Figure 12.17 Program 37, Part 2 of 3.

```
000490 01  REPLACEMENTS.
000500     05 FILLER         VALUE 'PERSON.'          PIC X(7).
000510     05 FILLER         VALUE 'PERSONS'          PIC X(7).
000520     05 FILLER         VALUE '.'                PIC X(7).
000530 01  REPLACEMENTS-TABLE
000540     REDEFINES REPLACEMENTS.
000550     05 REPLACEMENT
000560        OCCURS 3 TIMES INDEXED BY REPLACEMENT-INDEX
000570                                                PIC X(7).
000580 REPORT SECTION.
000590 RD  EDITED-TEXT
000600     PAGE 30 LINES, FIRST DETAIL 5.
000610
000620 01  TYPE PAGE HEADING
000630     LINE 2.
000640     05 COLUMN 28      VALUE 'TEXT EDITOR'      PIC X(11).
000650
000660 01  DETAIL-LINES
000670     TYPE DETAIL.
000680     05 LINE PLUS 3
000690        COLUMN 1       SOURCE SENDING-FIELD     PIC X(80).
000700     05 LINE PLUS 1
000710        COLUMN 1       SOURCE RECEIVING-FIELD PIC X(120).
000720
000730*****************************************************************
000740
000750 PROCEDURE DIVISION.
000760     OPEN INPUT  TEXT-IN,
000770          OUTPUT TEXT-OUT.
000780     READ TEXT-IN
000790         AT END
000800             MOVE 'NO' TO MORE-INPUT.
000810     IF MORE-INPUT = 'YES'
000820         INITIATE EDITED-TEXT
000830         PERFORM MAIN-PROCESS UNTIL MORE-INPUT = 'NO'
000840         TERMINATE EDITED-TEXT.
000850     CLOSE TEXT-IN,
000860           TEXT-OUT.
000870     STOP RUN.
000880
000890 MAIN-PROCESS.
000900     MOVE 1       TO SENDING-FIELD-POINTER,
000905                     RECEIVING-FIELD-POINTER.
000910     MOVE SPACES TO RECEIVING-FIELD.
000920     PERFORM DESEX-TEXT
000925         UNTIL
000930             SENDING-FIELD-POINTER GREATER THAN
000940                                       SIZE-OF-SENDING-FIELD OR
000950             RECEIVING-FIELD-POINTER GREATER THAN
000960                                       SIZE-OF-RECEIVING-FIELD.
000970     GENERATE DETAIL-LINES.
000980     READ TEXT-IN
000990         AT END
001000             MOVE 'NO' TO MORE-INPUT.
001010
```

SENDING–FIELD and the RECEIVING–FIELD are ready to be printed, and we GENERATE the pair of DETAIL–LINES.

The paragraph DESEX–TEXT executes over and over UNTIL either the SENDING–FIELD is completely scanned or the RECEIVING–FIELD is full. In each execution of SCAN–FOR–MAN–OR–MEN, only one occurrence of a delimiter is looked for. If one is found or the end of the SENDING–FIELD is reached, all the characters that were examined up to, but not including the delimiter, if any, are assigned to the INTERMEDIATE–FIELD.

The IF statement tests to see whether anything was found in the SCAN–FOR–MAN–OR–MEN. IF INTERMEDIATE–FIELD is NOT = SPACES, then there was some text found in the SENDING–FIELD. If SINGULAR–OR–PLURAL and INTERMEDIATE–FIELD are both equal to SPACES, that means that only spaces were found in the SENDING–FIELD, and the end of the SENDING–FIELD was reached without a delimiter being found. In such a case there is no need to process the record further.

If either SINGULAR–OR–PLURAL or INTERMEDIATE–FIELD is NOT = SPACES, we then PERFORM DELIMIT–TEXT. The operation DELIMIT–TEXT assigns the character HIGH–VALUE to the right of the characters that were just assigned to INTERMEDATE–FIELD by the scan. You will see why we need to delimit the INTERMEDIATE–FIELD in this way when we look at the paragraph STRING–TEXT–WITH–REPLACEMENT.

STRING–TEXT–WITH–REPLACEMENT takes the characters assigned to the INTERMEDIATE–FIELD and takes the suitable entry from the REPLACEMENTS–TABLE and STRINGs them INTO the RECEIVING–FIELD, adding them to the right end of any characters that might have been assigned to the RECEIVING–FIELD by an earlier execution of STRING–TEXT–WITH–REPLACEMENT.

The paragraph SCAN–FOR–MAN–OR–MEN contains just the UNSTRING statement. In an UNSTRING statement the programmer first names the field to be scanned, in this case SENDING–FIELD. Then the delimiters are named. There may be as many delimiters as desired connected by OR. The delimiters must be alphanumeric; in this case our delimiters are MAN and MEN. The INTO clause tells the UNSTRING statement where to assign the characters it has examined while looking for the delimiter, in this case INTERMEDIATE–FIELD.

The optional clauses which follow tell the UNSTRING statement where to assign the delimiter if any is found. In this program the delimiter is to be assigned to the field SINGULAR–OR–PLURAL. If the end of a scanned field is encountered before a delimiter is found, the UNSTRING statement assigns either zeros or spaces to the DELIMITER field, depending on how the DELIMITER field is defined. The **COUNT** clause tells the UNSTRING statement where to assign the COUNT of the number of characters that were examined before

Figure 12.17 Program 37, Part 3 of 3.

```
001020 DESEX-TEXT.
001030     PERFORM SCAN-FOR-MAN-OR-MEN.
001040     IF INTERMEDIATE-FIELD NOT = SPACES OR
001050        SINGULAR-OR-PLURAL NOT = SPACES
001060         PERFORM DELIMIT-TEXT
001070         PERFORM STRING-TEXT-WITH-REPLACEMENT.
001080
001090 SCAN-FOR-MAN-OR-MEN.
001100     UNSTRING SENDING-FIELD
001110                 DELIMITED BY 'MAN' OR 'MEN'
001120                 INTO INTERMEDIATE-FIELD
001130                      DELIMITER IN SINGULAR-OR-PLURAL
001140                      COUNT     IN INTERMEDIATE-POINTER
001150                      POINTER      SENDING-FIELD-POINTER.
001160
001170 DELIMIT-TEXT.
001180     ADD 1 TO INTERMEDIATE-POINTER.
001190     STRING HIGH-VALUE DELIMITED BY SIZE
001200            INTO INTERMEDIATE-FIELD
001210                 POINTER INTERMEDIATE-POINTER.
001220
001230 STRING-TEXT-WITH-REPLACEMENT.
001240     IF SINGULAR-OR-PLURAL = 'MAN'
001250         SET REPLACEMENT-INDEX TO 1
001260     ELSE
001270     IF SINGULAR-OR-PLURAL = 'MEN'
001280         SET REPLACEMENT-INDEX TO 2
001290     ELSE
001300         SET REPLACEMENT-INDEX TO 3.
001310     STRING INTERMEDIATE-FIELD DELIMITED BY HIGH-VALUE,
001320            REPLACEMENT (REPLACEMENT-INDEX) DELIMITED BY '.'
001330            INTO RECEIVING-FIELD
001340                 POINTER RECEIVING-FIELD-POINTER
001350                 OVERFLOW
001360                     PERFORM LINE-TOO-LONG.
001370
001380 LINE-TOO-LONG.
001390     MOVE 102 TO RECEIVING-FIELD-POINTER.
001400     STRING ' CHARACTERS MISSING' DELIMITED BY SIZE
001410            INTO RECEIVING-FIELD
001420                 POINTER RECEIVING-FIELD-POINTER.
```

the delimiter was found. In this case, the count is assigned to INTERMEDIATE–POINTER and is the same as the number of characters assigned to INTERMEDIATE–FIELD. The POINTER clause is included so that the next time this UNSTRING statement is executed it will pick up in the right place. The POINTER is handled automatically by the UNSTRING statement so that each scan starts to the right of the delimiter that stopped the previous scan.

We enter DELIMIT–TEXT with some characters assigned to INTERMEDIATE–FIELD and with a COUNT of the number of these characters in INTERMEDIATE–POINTER. In this paragraph we want to STRING HIGH–VALUE in to the right of the rightmost character assigned to INTERMEDIATE–FIELD. We add 1 to the INTERMEDIATE–POINTER so that it refers to the character position to the right of the rightmost character; then we STRING in HIGH–VALUE.

The paragraph STRING–TEXT–WITH–REPLACEMENT first determines whether MAN or MEN is being replaced in the original text and SETs the REPLACEMENT–INDEX so that it refers to PERSON or PERSONS, respectively. If neither MAN nor MEN is being replaced, the REPLACEMENT–INDEX is SET to refer to nothing. The STRING statement then STRINGs the contents of both the INTERMEDIATE–FIELD and the suitable replacement INTO the RECEIVING–FIELD. The INTERMEDIATE–FIELD is indicated as being DELIMITED BY HIGH–VALUE, which is why we had to delimit the contents of the INTERMEDIATE–FIELD with HIGH–VALUE—so that this STRING statement would work. In your own COBOL system use as a delimiting character any character which could not be punched in a card.

The POINTER field is controlled automatically by the STRING statement so that each execution will pick up in the RECEIVING–FIELD where the previous one left off.

The STRING statement is considered to have OVERFLOWed if the fields being strung together do not fit INTO the receiving field or if the POINTER gets out of range in an illegal way. The OVERFLOW clause is optional, and, if an OVERFLOW occurs when no OVERFLOW clause has been specified, the program just goes on to the next statement. In this program, however, the paragraph LINE–TOO–LONG is executed should an OVERFLOW occur during the STRING operation.

If the edited text is not found to fit into 120 character positions, any characters that might have already been strung INTO positions 102 through 120 of the RECEIVING–FIELD are replaced by the words CHARACTERS MISSING preceded by a space. The two statements in LINE–TOO–LONG do it. Program 37 was run on the input data shown in Figure 12.18 and produced the output shown in Figure 12.19.

Figure 12.18 Input for Program 37.

```
MANY DISTINCTLY HUMAN ACTIVITIES EMANATED FROM ERAS AS DISTANT AS CAVEMAN DAYS.
MEN AND WOMEN ENGAGED IN MENDING, IN MANUFACTURING, IN MENTAL MANIPULATIONS,
USE MANKIND, HUMANS, OR HUMANKIND INSTEAD OF MAN OR MEN IN WORDS LIKE SALESMEN.
THE MAN-EATING TIGER MANAGED ANYWAY TO MANGLE THE WOMAN'S MANDIBLE.
MAN MEN WOMAN WOMEN MAN WOMAN MAN MEN WOMAN WOMEN MAN MEN WOMAN WOMEN MAN MEN
```

Figure 12.19 Output from Program 37

```
                    TEXT EDITOR

MANY DISTINCTLY HUMAN ACTIVITIES EMANATED FROM ERAS AS DISTANT AS CAVEMAN DAYS.
PERSONY DISTINCTLY HUPERSON ACTIVITIES EPERSONATED FROM ERAS AS DISTANT AS CAVEPERSON DAYS.

MEN AND WOMEN ENGAGED IN MENDING, IN MANUFACTURING, IN MENTAL MANIPULATIONS,
PERSONS AND WOPERSONS ENGAGED IN PERSONSDING, IN PERSONUFACTURING, IN PERSONSTAL PERSONIPULATIONS,

USE MANKIND, HUMANS, OR HUMANKIND INSTEAD OF MAN OR MEN IN WORDS LIKE SALESMEN.
USE PERSONKIND, HUPERSONS, OR HUPERSONKIND INSTEAD OF PERSON OR PERSONS IN WORDS LIKE SALESPERSONS.

THE MAN-EATING TIGER MANAGED ANYWAY TO MANGLE THE WOMAN'S MANDIBLE.
THE PERSON-EATING TIGER PERSONAGED ANYWAY TO PERSONGLE THE WOPERSON'S PERSONDIBLE.

MAN MEN WOMAN WOMEN MAN WOMAN MAN MEN WOMAN WOMEN MAN MEN WOMAN WOMEN MAN MEN
PERSON PERSONS WOPERSON WOPERSONS PERSON WOPERSON PERSON PERSONS WOPERSON WOPERSONS PERSON PERSONS WO CHARACTERS MISSING
```

Exercise 4. Write a program that reads in free-form name data, rearranges the fields, and prints the name with its first name first. The input name data begins in column 1 of each card and consists of a last name, a first name, an optional middle initial, and an optional tag like Jr. or CPA. The name is in the following format:

a. last name followed immediately by a comma;
b. first name followed by a space, if there is a middle initial; or first name followed by a space, if there is no middle initial and also no tag; or first name followed by a comma, if there is no middle initial but there is a tag;
c. middle initial, if any, followed by a period, if there is no tag; or followed by a period and a comma, if there is a tag;
d. tag, if any.

Examples of these possibilities are:

Input	Output
POPKIN, GARY	GARY POPKIN
POPKIN, GARY S.	GARY S. POPKIN
POPKIN, GARY, CDP	GARY POPKIN, CDP
POPKIN, GARY S., CDP	GARY S. POPKIN, CDP

In your own input data, make up names of different lengths to be sure your program works. Test your program with very short names and very long ones.

Summary

The SORT feature of COBOL permits programs to rearrange the order of records. Records may be SORTed directly from an input file, or else an INPUT PROCEDURE may be carried out before the records are SORTed. After the records are SORTed they may be placed directly onto an output file, or else an OUTPUT PROCEDURE may be carried out on the records in their new sequence. An SD entry must be used to describe the file that the SORT is to use for a work area. INPUT and OUTPUT PROCEDUREs must be defined as SECTIONs in the Procedure Division, and many COBOL systems further require that the last excutable instruction in an INPUT or OUTPUT PROCEDURE also be the last physical instruction in the PROCEDURE. The GO statement can be used to skip to the end of the PROCEDURE.

The RELEASE verb can be used only in an INPUT PROCEDURE. It is executed after all the necessary input processing has been carried out on an input record and the record is ready for SORTing. The RETURN verb is used in an OUTPUT PROCEDURE to fetch SORTed records for processing.

The SORT statement may name as many key fields as desired. All key fields must be described in an 01-level entry associated with the SD entry. Key fields may be SORTed in ASCENDING or DESCENDING order in any combination.

The STRING verb can be used to attach fields or parts of fields end to end in some larger receiving field. Whole sending fields can be strung together INTO a receiving field, in which case the sending fields are said to be DELIMITED BY SIZE. If parts of fields are to be strung together, the parts of the sending fields that will be involved in the operation may be DELIMITED BY any alphanumeric data. The POINTER option is used if more than one STRING statement is needed to construct an entire receiving field. The OVERFLOW option is used if there is the possibility that the completed string will be longer than the receiving field.

The UNSTRING verb can be used to select out certain parts of fields for processing. With the UNSTRING verb, a sending field may be scanned from left to right for one or more occurrences of one or more delimiters. The delimiters may be any alphanumeric data.

As delimiters are found, or the end of the sending fields reached, the characters that were examined in the search are unstrung INTO one or more receiving fields specified by the programmer. The programmer may also provide one or more fields to which the UNSTRING statement may assign the delimiters and may also provide one or more fields to which the UNSTRING statement may assign counts of the number of characters examined. A POINTER field may be used if more than one UNSTRING statement is needed in order to fully scan a field.

Fill-in Exercises

1. If records are to be processed before being SORTed, then the programmer must specify an ________ PROCEDURE.
2. If records are to be SORTed before being processed, then the programmer must specify an ________ PROCEDURE.
3. INPUT PROCEDUREs and OUTPUT PROCEDUREs must be defined as ________ in the Procedure Division.
4. The ________ entry defines the file that the SORT is to use as a work area.
5. In the SORT statement, the SORT keys must be listed in order from ________ to ________.
6. The ________ statement is used in place of a READ when already SORTed records are to be processed.
7. The ________ verb permits a program to attach fields or parts of fields together INTO a larger receiving field.
8. The ________ verb permits a program to select out pieces of a larger field for processing.
9. If a whole field is to be a sending field in a STRING operation, the field is said to be DELIMITED BY ________.
10. The required ________ clause names the receiving field in both the STRING and UNSTRING operations.

Review Exercises

1. Rewrite Program 8 to SORT the input data on account number before being processed. Have your program SORT the input file with the USING option. Use Report Writer to print the output.
2. Rewrite your solution for Exercise 2, Chapter 6, page 139. Arrange your input data in any order and have your program SORT it on customer number before processing. You may use Report Writer to produce the output or you may use the step-by-step report logic used in the original exercise.
3. Rewrite your solution for Review Exercise 3, Chapter 6, page 155. Arrange your input data in any order. Use an INPUT PROCEDURE to select for SORTing only those input records containing a credit amount. SORT the selected records in account number sequence.
4. Rewrite Program 31. Have your program SORT the transactions before they are used to update the master file. Include the transaction code as a KEY field so that, if there is more than one transaction with the same account number, a transaction with a new account

code 1 will fall first among them and a transaction with a delete code will fall last.

5. Write a program to read four-digit numbers with two decimal places and print the amount in words. Typical input and output is shown in Figure 12.RE5.
6. Modify Program 37 so that only if a blank follows MAN will it be replaced by PERSON, and only if a blank follows MEN will it be replaced by PERSONS. (Hint: make the delimiters 'MAN ' and 'MEN '.) The change improves the program somewhat but still leaves it far from perfect.

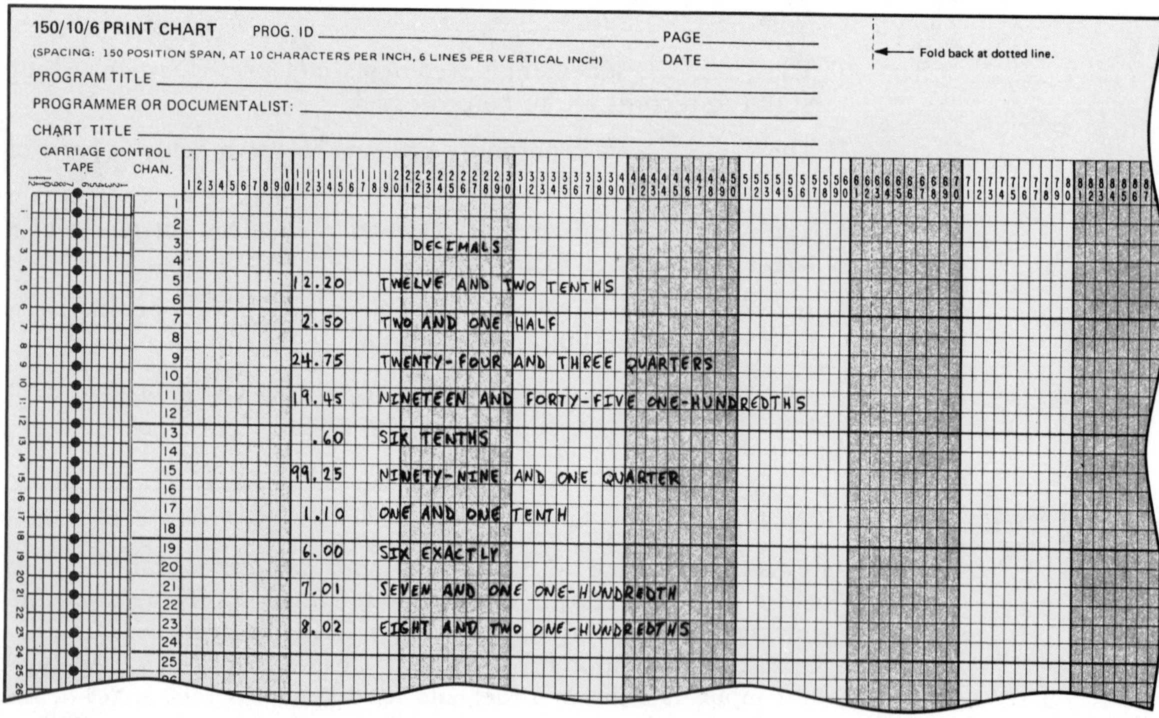

150/10/6 PRINT CHART PROG. ID ______ PAGE ______

(SPACING: 150 POSITION SPAN, AT 10 CHARACTERS PER INCH, 6 LINES PER VERTICAL INCH) DATE ______

Fold back at dotted line.

PROGRAM TITLE ______

PROGRAMMER OR DOCUMENTALIST: ______

CHART TITLE ______

CARRIAGE CONTROL TAPE CHAN.

1 2 3 4 5 6 7 8 9 10 11 12 13 14 15 16 17 18 19 20 21 22 23 24 25 26 27 28 29 30 31 32 33 34 35 36 37 38 39 40 41 42 43 44 45 46 47 48 49 50 51 52 53 54 55 56 57 58 59 60 61 62 63 64 65 66 67 68 69 70 71 72 73 74 75 76 77 78 79 80 81 82 83 84 85 86 87

1

2

3 DECIMALS

4

5 12.20 TWELVE AND TWO TENTHS

6

7 2.50 TWO AND ONE HALF

8

9 24.75 TWENTY-FOUR AND THREE QUARTERS

10

11 19.45 NINETEEN AND FORTY-FIVE ONE-HUNDREDTHS

12

13 .60 SIX TENTHS

14

15 99.25 NINETY-NINE AND ONE QUARTER

16

17 1.10 ONE AND ONE TENTH

18

19 6.00 SIX EXACTLY

20

21 7.01 SEVEN AND ONE ONE-HUNDREDTH

22

23 8.02 EIGHT AND TWO ONE-HUNDREDTHS

24

25

26

Figure 12.RE5 Output format for Review Exercise 5.

Appendix A

Debugging

Whenever you put a COBOL program into a computer, there is at least one other program in the computer also. This other program, called a **COBOL compiler,** accepts your COBOL program into storage, examines it for errors, and prepares it for execution. Each different computer make and model has its own COBOL compiler, which is why different COBOL systems behave slightly differently from one another. The COBOL compiler always produces its own output, aside from any output that your own program may produce. The output from the COBOL compiler tells the programmer about any errors that the compiler may have found and provides other information.

There are some errors that the compiler cannot find. Such errors may be detected only during the execution of the program. If an execution error is serious enough, the program may terminate its execution in the middle. In less serious cases, the program may execute to its end but produce erroneous output.

Different COBOL compilers produce different kinds of outputs. So even though no particular COBOL compiler has been mentioned in the body of the text, we must work with a particular compiler in this appendix so that we can look at its output. The compiler that was used to run the programs in the body of the text, and the one we will use in this appendix, is the widely used IBM OS Full American National Standard COBOL, Version 4, Release 1.5.

MISPLACED PERIODS AND MISSPELLED WORDS

We will first look at two of the most common errors made by beginning programming students. Figure A.1 shows an erroneous version of Program 9 and Figure A.2 shows the messages printed by the compiler. You can see at the very left of the program listing that the compiler has assigned its own sequence numbers to the COBOL statements. It refers to these sequence numbers in all messages.

Figure A.1 An erroneous version of Program 9, Part 1 of 2.

```
PP 5734-CB2 V4 RELEASE 1.5 10NOV77          IBM OS AMERICAN NATIONAL STANDARD COBOL

   1

00001    000010 IDENTIFICATION DIVISION.
00002    000020 PROGRAM-ID. PROG09.
00003    000030*AUTHOR. PHIL RUBENSTEIN.
00004    000040*
0000     000060*        THIS PROGRAM READS CUSTOMER ORDERS PUNCHED IN
00006    000070*        CARDS AND PRODUCES A DAILY ORDER REPORT.
00007    000075*
00008    000080**********************************************************
00009    000090
00010    000100 ENVIRONMENT DIVISION.
00011    000110 CONFIGURATION SECTION.
00012    000120 SPECIAL-NAMES.
00013    000130     C01 IS TO-NEW-PAGE.
00014    000140 INPUT-OUTPUT SECTION.
00015    000150 FILE-CONTROL.
00016    000160     SELECT ORDER-FILE-IN ASSIGN TO UT-S-SYSIN.
00017    000170     SELECT ORDER-REPORT  ASSIGN TO UT-S-SYSPRINT.
00018    000180
00019    000190**********************************************************
00020    000200
00021    000210 DATA DIVISION.
00022    000220 FILE SECTION.
00023    000230 FD  ORDER-FILE-IN.
00024    000240     LABEL RECORDS ARE OMITTED.
00025    000250
00026    000260 01  ORDER-RECORD-IN.
00027    000270     05 CUSTOMER-NUMBER-IN PIC X(7).
00028    000280     05 PART-NUMBER-IN     PIC X(8).
00029    000290     05 FILLER             PIC X(7).
00030    000300     05 QUANTITY-IN        PIC 9(3).
00031    000310     05 UNIT-PRICE-IN      PIC 9(4)V99.
00032    000320     05 HANDLING-IN        PIC 99V99.
00033    000330     05 FILLER             PIC X(45).
00034    000340
00035    000350 FD  ORDER-REPORT.
00036    000360     LABEL RECORDS ARE OMITTED
00037    000370
00038    000380 01  REPORT-LINE           PIC X(115).
00039    000390
00040    000400 WORKING-STORAGE SECTION.
00041    000410 01  MORE-INPUT            PIC X(3) VALUE 'YES'.
00042    000420
00043    000430 01  REPORT-TITLE.
00044    000440     05 FILLER             PIC X.
00045    000450     05 FILLER             PIC X(45) VALUE SPACES.
00046    000460     05 FILLER             PIC X(18) VALUE 'DAILY ORDER REPORT'.
00047    000470
00048    000480 01  COLUMN-HEADS-1.
00049    000490     05 FILLER             PIC X.
```

Figure A.1 Part 1 (continued).

```
  2

00050    000500       05 FILLER                  PIC X(10) VALUE SPACES.
00051    000510       05 FILLER                  PIC X(15) VALUE 'CUSTOMER'.
00052    000520       05 FILLER                  PIC X(12) VALUE 'PART'.
00053    000530       05 FILLER                  PIC X(13) VALUE 'QUANTITY'.
00054    000540       05 FILLER                  PIC X(13) VALUE 'UNIT'.
00055    000550       05 FILLER                  PIC X(18) VALUE 'MERCHANDISE'.
00056    000560       05 FILLER                  PIC X(9)  VALUE 'TAX'.
00057    000570       05 FILLER                  PIC X(16) VALUE 'HANDLING'.
00058    000580       05 FILLER                  PIC X(5)  VALUE 'TOTAL'.
00059    000590
00060    000600 01   COLUMN-HEADS-2.
00061    000610       05 FILLER                  PIC X.
00062    000620       05 FILLER                  PIC X(11) VALUE SPACES.
00063    000630       05 FILLER                  PIC X(13) VALUE 'NUMBER'.
00064    000640       05 FILLER                  PIC X(26) VALUE 'NUMBER'.
00065    000650       05 FILLER                  PIC X(15) VALUE 'PRICE'.
00066    000660       05 FILLER                  PIC X(6)  VALUE 'AMOUNT'.
00067    000670
00068    000680 01   BODY-LINE.
00069    000690       05 FILLER                        PIC X.
00070    000700       05 FILLER                        PIC X(11) VALUE SPACES.
00071    000710       05 CUSTOMER-NUMBER-OUT           PIC X(7).
00072    000720       05 FILLER                        PIC X(5) VALUE SPACES.
00073    000730       05 PART-NUMBER-OUT               PIC X(8).
00074    000740       05 FILLER                        PIC X(8) VALUE SPACES.
00075    000750       05 QUANTITY-OUT                  PIC ZZ9.
00076    000760       05 FILLER                        PIC X(6) VALUE SPACES.
00077    000770       05 UNIT-PRICE-OUT                PIC Z,ZZZ.99.
00078    000780       05 FILLER                        PIC X(7) VALUE SPACES.
00079    000790       05 MERCHANDISE-AMOUNT-OUT PIC ZZZ,ZZZ.99.
00080    000800       05 FILLER                        PIC X(6) VALUE SPACES.
00081    000810       05 TAX-OUT                       PIC Z,ZZZ.99.
00082    000820       05 FILLER                        PIC X(5) VALUE SPACES.
00083    000830       05 HANDLING-OUT                  PIC ZZ.99.
00084    000840       05 FILLER                        PIC X(5) VALUE SPACES.
00085    000850       05 ORDER-TOTAL-OUT               PIC Z,ZZZ,ZZZ.99
00086    000860
00087    000870 01   MERCHANDISE-AMOUNT-W          PIC 9(6)V99.
00088    000880 01   TAX-W                         PIC 9(4)V99.
00089    000890
00090    000900************************************************************
00091    000910
00092    000920 PROCEDURE DIVISION.
00093    000930       OPEN INPUT ORDER-FILE-IN, OUTPUT ORDER-REPORT.
00094    000940       WRITE REPORT-LINE FROM REPORT-TITLE AFTER ADVANCING
00095    000950           TO-NEW-PAGE.
00096    000960       WRITE REPORT-LINE FROM COLUMN-HEADS-1 AFTER ADVANCING 3.
00097    000970       WRITE REPORT-LINE FROM COLUMN-HEADS-2 AFTER ADVANCING 1.
00098    000980       MOVE SPACES TO REPORT-LINE.
00099    000990       WRITE REPORT-LINE AFTER ADVANCING 1.
00100    001000       READ ORDER-FILE-IN AT END MOVE 'NO' TO MORE-INPUT.
00101    001010       PERFORM MAIN-PROCESS UNTIL MORE-INPUT IS EQUAL TO 'NO'.
```

Figure A.1 An erroneous version of Program 9, Part 2 of 2.

```
    3

00102    001020          CLOSE ORDER-FILE-IN, ORDER-REPORT.
00103    001030          STOP RUN.
00104    001040
00105    001050      MAIN-PROCESS.
00106    001060          MULTIPLY QUANTITY-IN BY UNIT-PRICE-IN GIVING
00107    001070              MERCHANDISE-AMOUNT-W.
00108    001080          MULTIPLY MERCHANDISE-AMOUNT-W BY .07 GIVING TAX-W.
00109    001090          ADD MERCHANDISE-AMOUNT-W,TAX-W, HANDLING-IN GIVING
00110    001100              ORDER-TOTAL-OUT.
00111    001110          MOVE CUSTOMER-NUMBER-IN TO CUSTOMER-NUMBER-OUT.
00112    001120          MOVE PART-NUMBER-IN TO PART-NUMBER-OUT.
00113    001130          MOVE QUANTITY-IN TO QUANTITY-OUT.
00114    001140          MOVE UNIT-PRINCE-IN TO UNIT-PRICE-OUT.
00115    001150          MOVE MERCHANDISE-AMOUNT-W TO MERCHANDISE-AMOUNT-OUT.
00116    001160          MOVE TAX-W TO TAX-OUT.
00117    001170          MOVE HANDLING-IN TO HANDLING-OUT.
00118    001180          WRITE REPORT-LINE FROM BODY-LINE AFTER ADVANCING 1.
00119    001190          READ ORDER-FILE-IN AT END MOVE 'NO' TO MORE-INPUT.
```

Figure A.2 Error messages for Figure A.1.

```
    5

CARD   ERROR MESSAGE

16     IKF2133I-W    LABEL RECORDS CLAUSE MISSING. DD CARD OPTION WILL BE TAKEN.
24     IKF1004I-E    INVALID WORD LABEL . SKIPPING TO NEXT RECOGNIZABLE WORD.
17     IKF2133I-W    LABEL RECORDS CLAUSE MISSING. DD CARD OPTION WILL BE TAKEN.
36     IKF1004I-E    INVALID WORD LABEL . SKIPPING TO NEXT RECOGNIZABLE WORD.
87     IKF1043I-W    END OF SENTENCE SHOULD PRECEDE 01 . ASSUMED PRESENT.
109    IKF1120I-W    COMMA NOT FOLLOWED BY SPACE. ASSUMED.
114    IKF3001I-E    UNIT-PRINCE-IN NOT DEFINED. DISCARDED.
119    IKF4072I-W    EXIT FROM PERFORMED PROCEDURE ASSUMED BEFORE PROCEDURE-NAME .

       ERROR MESSAGE

       IKF6006I-E    SUPMAP SPECIFIED AND E-LEVEL DIAGNOSTIC HAS OCCURRED. PMAP CLIST LOAD DECK STATE FLOW SYMDMP IGNO
                     RED/
```

There are four kinds of error messages produced by this compiler. Each is designated by one of the letters *W, C, E,* or *D.* The meanings of the designations are:

- *W* Warning. The compiler takes its own corrective action.
- *C* Conditional. This may or may not be an error, depending on conditions at the time the program is executed.

E Error. A definite error has occurred requiring correction by the programmer.

D Disaster. An error of such magnitude has occurred that the compiler cannot even complete its examination of the program. The cause of the disaster must be remedied by the programmer and the program resubmitted to be examined again.

Each computer installation decides for itself the conditions under which the COBOL system will attempt to execute the program. In the installation where these programs were run, an error with severity higher than *W* would prevent execution. Some installations permit execution of programs with errors up through severity *C*. It is usually pointless to try to execute programs with *E*-level errors.

The first error message in Figure A.2, which refers to card 16, is typical of the misleading messages a compiler can produce. It says that the LABEL RECORDS clause is missing, but card 16 is a SELECT sentence and a SELECT sentence isn't supposed to have a LABEL RECORDS clause. The next message, referring to card 24, is in a sense even more confusing, for there in card 24 is the very LABEL RECORDS clause that is alleged to be missing. The real cause of the trouble is the period after ORDER–FILE–IN in card 23. Removing the period will eliminate both error messages. It often happens that the real cause of an error is in the card preceding the one indicated by the compiler. The next two messages, referring to cards 17 and 36, result from the same kind of error; see whether you can find it.

The error in card 87 is not severe, and the compiler takes corrective action. The message says that there is no end of sentence before the 01-level entry in card 87 and that there should be one. The compiler assumes that the 01 begins a new sentence, but the real cause of the trouble is the missing period in card 85. This is another example of the compiler sending us to the wrong card to find an error.

The message about card 109 refers to an error that really is in card 109. You can see that one of the commas in card 109 is not followed by a space. This is only a warning, and the compiler assumes that the space is there anyway.

In card 114 you can see that Phil has keyed in the programmer-supplied word UNIT–PRINCE–IN when he meant to key UNIT–PRICE–IN. The message from the compiler says that UNIT–PRINCE–IN is wrong and that the problem must be corrected by the programmer.

The last message, referring to card 119, results from an error in the compiler program itself. The compiler is objecting because we have a conditional statement, the READ with the conditional AT END clause, as the last statement in a PERFORMed paragraph. There is no restriction against any such construction in either the ANSI standard or the IBM specifications, and, of course, the construction works just as we would want it to. We ignore the warning.

BAD TROUBLE FROM MISPLACED PERIODS

In the program in Figure A.1 the compiler detected all the misplaced periods. In the erroneous version of Program 7 shown in Figure A.3 the compiler does not find any of them. The error messages from this run are shown in Figure A.4, and the most the compiler can find is the sequence error indicated with two asterisks in card 80. What is really wrong with this program is all the incorrect periods in the IF statement. Why didn't the compiler detect them?

Figure A.3 An erroneous version of Program 7.

```
PP 5734-CB2 V4 RELEASE 1.5 10NOV77         IBM OS AMERICAN NATIONAL STANDARD COBOL

    1

00001    000010 IDENTIFICATION DIVISION.
00002    000020 PROGRAM-ID. PROG07.
00003    000025*AUTHOR. PHIL RUBENSTEIN.
00004    000030*
00005    000040*     THIS PROGRAM READS DATA CARDS IN THE FOLLOWING FORMAT:
00006    000050*
00007    000060*     COLS. 1-8           PART NUMBER
00008    000070*     COLS. 9-28          PART DESCRIPTION
00009    000080*     COLS. 30-32         QUANTITY ON HAND
00010    000100*
00011    000110*     FOR EACH CARD, ONE LINE IS PRINTED IN THE FOLLOWING FORMAT:
00012    000120*
00013    000130*     PP 9-16             PART NUMBER
00014    000140*     PP 20-22            QUANTITY ON HAND
00015    000150*     PP 41-60            PART DESCRIPTION
00016    000155*
00017    000160*     IN ADDITION, THE WORD 'REORDER' IS TO BE PRINTED
00018    000170*     IN PRINT POSITIONS 65-71 NEXT TO ANY PART WHOSE
00019    000175*     QUANTITY ON HAND IS LESS THAN 400.
00020    000178*
00021    000180*     SKIP TO THE TOP OF A NEW PAGE BEFORE PRINTING THE FIRST
00022    000190*     LINE OF OUTPUT AND AFTER PRINTING THE LAST.
00023    000200*
00024    000210*****************************************************************
00025    000220
00026    000230 ENVIRONMENT DIVISION.
00027    000240 CONFIGURATION SECTION.
00028    000250 SPECIAL-NAMES.
00029    000260     C01 IS TO-NEW-PAGE.
00030    000270 INPUT-OUTPUT SECTION.
00031    000280 FILE-CONTROL.
00032    000290     SELECT INVENTORY-LIST      ASSIGN TO UT-S-SYSPRINT.
00033    000300     SELECT INVENTORY-FILE-IN   ASSIGN TO UT-S-SYSIN.
00034    000310
00035    000320*****************************************************************
```

Figure A.3 Continued.

```
00036    000330
00037    000340 DATA DIVISION.
00038    000350 FILE SECTION.
00039    000360 FD  INVENTORY-FILE-IN
00040    000370     LABEL RECORDS ARE OMITTED.
00041    000380
00042    000390 01  INVENTORY-RECORD-IN.
00043    000400     05 PART-NUMBER-IN          PIC X(8).
00044    000410     05 PART-DESCRIPTION-IN     PIC X(20).
00045    000420     05 QUANTITY-IN             PIC X(3).
00046    000430     05 FILLER                  PIC X(49).
00047    000440
00048    000450 FD  INVENTORY-LIST
00049    000460     LABEL RECORDS ARE OMITTED.
```

2

```
00050    000470
00051    000480 01  INVENTORY-LINE.
00052    000490     05 FILLER                  PIC X.
00053    000500     05 FILLER                  PIC X(8).
00054    000510     05 PART-NUMBER-OUT         PIC X(8).
00055    000520     05 FILLER                  PIC X(3).
00056    000530     05 QUANTITY-OUT            PIC X(3).
00057    000540     05 FILLER                  PIC X(18).
00058    000550     05 PART-DESCRIPTION-OUT    PIC X(20).
00059    000560     05 FILLER                  PIC X(4).
00060    000570     05 MESSAGE-SPACE           PIC X(7).
00061    000580
00062    000590 WORKING-STORAGE SECTION.
00063    000600 01  MORE-INPUT                 PIC X(3) VALUE 'YES'.
00064    000610
00065    000620*****************************************************************
00066    000630
00067    000640 PROCEDURE DIVISION.
00068    000650     OPEN INPUT INVENTORY-FILE-IN, OUTPUT INVENTORY-LIST.
00069    000660     MOVE SPACES TO INVENTORY-LINE.
00070    000670     WRITE INVENTORY-LINE AFTER ADVANCING TO-NEW-PAGE.
00071    000680     READ INVENTORY-FILE-IN AT END MOVE 'NO' TO MORE-INPUT.
00072    000690     PERFORM MAIN-PROCESS UNTIL MORE-INPUT IS EQUAL TO 'NO'.
00073    000700     MOVE SPACES TO INVENTORY-LINE.
00074    000710     WRITE INVENTORY-LINE AFTER ADVANCING TO-NEW-PAGE.
00075    000720     CLOSE INVENTORY-FILE-IN, INVENTORY-LIST.
00076    000730     STOP RUN.
00077    000740
00078    000750 MAIN-PROCESS.
00079    000802     IF QUANTITY-IN IS LESS THAN 400
00080  **000770         MOVE PART-NUMBER-IN TO PART-NUMBER-OUT.
00081    000780         MOVE PART-DESCRIPTION-IN TO PART-DESCRIPTION-OUT.
00082    000800         MOVE QUANTITY-IN TO QUANTITY-OUT.
00083    000804         MOVE 'REORDER' TO MESSAGE-SPACE.
00084    000810     WRITE INVENTORY-LINE AFTER ADVANCING 1.
00085    000820     READ INVENTORY-FILE-IN AT END MOVE 'NO' TO MORE-INPUT.
```

Figure A.4 Error messages for Figure A.3.

```
  4

CARD    ERROR MESSAGE

        IKF1100I-W      1 SEQUENCE ERROR IN SOURCE PROGRAM.
85      IKF4072I-W      EXIT FROM PERFORMED PROCEDURE ASSUMED BEFORE PROCEDURE-NAME .
```

According to the rules of COBOL, the construction shown in MAIN–PROCESS is correct. The compiler interpreted it as if it had been written:

```
IF QUANTITY–IN IS LESS THAN 400
    MOVE PART–NUMBER–IN TO PART–NUMBER–OUT.
MOVE PART–DESCRIPTION–IN TO PART–DESCRIPTION–OUT.
MOVE QUANTITY–IN TO QUANTITY–OUT.
MOVE 'REORDER' TO MESSAGE–SPACE.
```

The only MOVE that was executed conditionally was the first one. The other three MOVEs were executed for every input card whether the QUANTITY–IN was LESS THAN 400 or not. The compiler has no way of knowing what we really intended to do, and since the statements are all legal, it just processed them. The error is easily noticed when the report output appears, however, for every line contains the word REORDER regardless of the value of QUANTITY–IN.

INCORRECT HANDLING OF LEVEL NUMBERS

An incorrect version of Program 11 is shown in Figure A.5 and the error message for it shown in Figure A.6. This time the error message directs the programmer to the line preceding the real error. The message says that card 67 is a group item and has a PICTURE clause.

Figure A.5 An erroneous version of Program 11, Part 1 of 2.

```
PP 5734-CB2 V4 RELEASE 1.5 10NOV77         IBM OS AMERICAN NATIONAL STANDARD COBOL

    1

00001    000010 IDENTIFICATION DIVISION.
00002    000020 PROGRAM-ID. PROG11.
00003    000025*AUTHOR. PHIL RUBENSTEIN.
00004    000030*
00005    000040*         THIS PROGRAM READS A FILE OF EXAM GRADES AND
00006    000050*         COMPUTES EACH STUDENTS AVERAGE.
00007    000060*
00008    000070*****************************************************
```

Figure A.5 Part 1 (continued).

```
00009    000080
00010    000090 ENVIRONMENT DIVISION.
00011    000100 CONFIGURATION SECTION.
00012    000110 SPECIAL-NAMES.
00013    000120          C01 IS TO-NEW-PAGE.
00014    000130 INPUT-OUTPUT SECTION.
00015    000140 FILE-CONTROL.
00016    000150          SELECT EXAM-GRADE-FILE-IN     ASSIGN TO UT-S-SYSIN.
00017    000160          SELECT STUDENT-GRADE-REPORT  ASSIGN TO UT-S-SYSPRINT.
00018    000170
00019    000180*****************************************************
00020    000190
00021    000200 DATA DIVISION.
00022    000210 FILE SECTION.
00023    000220 FD   EXAM-GRADE-FILE-IN
00024    000230      LABEL RECORDS ARE OMITTED.
00025    000240
00026    000250 01   EXAM-GRADE-RECORD.
00027    000260      05 STUDENT-NUMBER-IN  PIC X(9).
00028    000270      05 FILLER             PIC X(8).
00029    000280      05 GRADE-1-IN         PIC 9(3).
00030    000290      05 GRADE-2-IN         PIC 9(3).
00031    000300      05 GRADE-3-IN         PIC 9(3).
00032    000310      05 GRADE-4-IN         PIC 9(3).
00033    000320      05 FILLER             PIC X(49).
00034    000330
00035    000340 FD   STUDENT-GRADE-REPORT
00036    000350      LABEL RECORDS ARE OMITTED.
00037    000360
00038    000370 01   REPORT-LINE           PIC X(79).
00039    000380
00040    000390 WORKING-STORAGE SECTION.
00041    000400 01   MORE-INPUT            PIC X(3) VALUE 'YES'.
00042    000410
00043    000420 01   REPORT-TITLE.
00044    000430      05 FILLER             PIC X.
00045    000440      05 FILLER             PIC X(35) VALUE SPACES.
00046    000450      05 FILLER             PIC X(20) VALUE 'STUDENT GRADE REPORT'.
00047    000460
00048    000470 01   COLUMN-HEADS-1.
00049    000480      05 FILLER             PIC X.
```

2

```
00050    000490      05 FILLER             PIC X(20) VALUE SPACES.
00051    000500      05 FILLER             PIC X(24) VALUE 'STUDENT'.
00052    000510      05 FILLER             PIC X(12) VALUE 'G R A D E S '.
00053    000580
00054    000590 01   COLUMN-HEADS-2.
00055    000600      05 FILLER             PIC X.
00056    000610      05 FILLER             PIC X(20) VALUE SPACES.
00057    000620      05 FILLER             PIC X(16) VALUE 'NUMBER'.
00058    000630      05 FILLER             PIC X(8)  VALUE 'EXAM 1'.
00059    000640      05 FILLER             PIC X(8)  VALUE 'EXAM 2'.
00060    000650      05 FILLER             PIC X(8)  VALUE 'EXAM 3'.
```

Figure A.5 An erroneous version of Program 11, Part 2 of 2.

```
00061   000653        05 FILLER                 PIC X(11) VALUE 'EXAM 4'.
00062   000656        05 FILLER                 PIC X(7)  VALUE 'AVERAGE'.
00063   000660
00064   000670 01   BODY-LINE.
00065   000680        05 FILLER                      PIC X.
00066   000690        05 FILLER                      PIC X(19) VALUE SPACES.
00067   000700        05 STUDENT-NUMBER-OUT          PIC X(9).
00068   000710        09 FILLER                      PIC X(9)  VALUE SPACES.
00069   000720        05 GRADE-1-OUT                 PIC ZZ9.
00070   000730        05 FILLER                      PIC X(5)  VALUE SPACES.
00071   000740        05 GRADE-2-OUT                 PIC ZZ9.
00072   000750        05 FILLER                      PIC X(5)  VALUE SPACES.
00073   000760        05 GRADE-3-OUT                 PIC ZZ9.
00074   000770        05 FILLER                      PIC X(5)  VALUE SPACES.
00075   000780        05 GRADE-4-OUT                 PIC ZZ9.
00076   000790        05 FILLER                      PIC X(7)  VALUE SPACES.
00077   000800        05 AVERAGE-GRADE               PIC ZZZ.9.
00078   000805
00079   000810 01   SUM-OF-GRADES                    PIC 9(3).
00080   000820 01   NUMBER-OF-EXAMS                  PIC 9 VALUE 4.
00081   000830
00082   000850*******************************************************
00083   001090
00084   001100 PROCEDURE DIVISION.
00085   001110        OPEN INPUT EXAM-GRADE-FILE-IN, OUTPUT STUDENT-GRADE-REPORT.
00086   001120        PERFORM REPORT-HEADINGS-ROUTINE.
00087   001130        READ EXAM-GRADE-FILE-IN AT END MOVE 'NO' TO MORE-INPUT.
00088   001140        PERFORM MAIN-PROCESS UNTIL MORE-INPUT IS EQUAL TO 'NO'.
00089   001160        CLOSE EXAM-GRADE-FILE-IN, STUDENT-GRADE-REPORT.
00090   001170        STOP RUN.
00091   001180
00092   001190 REPORT-HEADINGS-ROUTINE.
00093   001200        WRITE REPORT-LINE FROM REPORT-TITLE AFTER ADVANCING
00094   001210            TO-NEW-PAGE.
00095   001220        WRITE REPORT-LINE FROM COLUMN-HEADS-1 AFTER ADVANCING 3.
00096   001230        WRITE REPORT-LINE FROM COLUMN-HEADS-2 AFTER ADVANCING 1.
00097   001240        MOVE SPACES TO REPORT-LINE.
00098   001250        WRITE REPORT-LINE AFTER ADVANCING 1.
00099   001260
00100   001270 MAIN-PROCESS.
00101   001280        MOVE STUDENT-NUMBER-IN TO STUDENT-NUMBER-OUT.

    3

00102   001290        MOVE GRADE-1-IN TO GRADE-1-OUT.
00103   001300        MOVE GRADE-2-IN TO GRADE-2-OUT.
00104   001315        MOVE GRADE-3-IN TO GRADE-3-OUT.
00105   001325        MOVE GRADE-4-IN TO GRADE-4-OUT.
00106   001330        ADD GRADE-1-IN, GRADE-2-IN, GRADE-3-IN, GRADE-4-IN
00107   001335            GIVING SUM-OF-GRADES.
00108   001340        DIVIDE SUM-OF-GRADES BY NUMBER-OF-EXAMS
00109   001350            GIVING AVERAGE-GRADE ROUNDED.
00110   001360        WRITE REPORT-LINE FROM BODY-LINE AFTER ADVANCING 1.
00111   001370        READ EXAM-GRADE-FILE-IN AT END MOVE 'NO' TO MORE-INPUT.
```

Figure A.6 Error messages for Figure A.5.

```
 5

CARD    ERROR MESSAGE

67      IKF2034I-E      GROUP ITEM HAS PICTURE CLAUSE. CLAUSE DELETED.
111     IKF4072I-W      EXIT FROM PERFORMED PROCEDURE ASSUMED BEFORE PROCEDURE-NAME .

        ERROR MESSAGE

        IKF6006I-E      SUPMAP SPECIFIED AND E-LEVEL DIAGNOSTIC HAS OCCURRED. PMAP CLIST LOAD DECK STATE FLOW SYMDMP IGNO
                        RED/
```

Now it is true that a PICTURE clause may appear only in an elementary item and may not appear at the group level. But why does the compiler think that card 67 defines a group item? It is because of the erroneous level number in card 68. Changing the 09 to 05 would eliminate the error.

MISSPELLED RESERVED WORD

In the erroneous version of Program 12 shown in Figure A.7, a misspelled reserved word confuses the compiler so badly that it keeps finding errors where none exist. The error messages from this run are shown in Figure A.8.

Figure A.7 An erroneous version of Program 12, Part 1 of 3.

```
PP 5734-CB2 V4 RELEASE 1.5 10NOV77          IBM OS AMERICAN NATIONAL STANDARD COBOL

  1

0000     000040 IDENTIFICATION DIVISION.
00002    000050 PROGRAM-ID. PROG12.
00003    000060*AUTHOR. PHIL RUBENSTEIN.
00004    000070*
00005    000080*          THIS PROGRAM READS A FILE OF EXAM GRADES AND
00006    000090*          COMPUTES EACH STUDENTS AVERAGE AND THE CLASS AVERAGE.
00007    000100*
00008    000110*****************************************************************
00009    000120
00010    000130 ENVIRONMENT DIVISION.
00011    000140 CONFIGURATION SECTION.
00012    000150 SPECIAL-NAMES.
00013    000160      C01 IS TO-NEW-PAGE.
00014    000170 INPUT-OUTPUT SECTION.
00015    000180 FILE-CONTROL.
00016    000190      SELECT EXAM-GRADE-FILE-IN     ASSIGN TO UT-S-SYSIN.
00017    000200      SELECT STUDENT-GRADE-REPORT   ASSIGN TO UT-S-SYSPRINT.
00018    000210
00019    000220*****************************************************************
```

Figure A.7 An erroneous version of Program 12, Part 2 of 3.

```
00020    000230
00021    000240 DATA DIVISION.
00022    000250 FILE SECTION.
00023    000260 FD  EXAM-GRADE-FILE-IN
00024    000270     LABEL RECORDS ARE OMITTED.
00025    000280
00026    000290 01  EXAM-GRADE-RECORD.
00027    000300     05 STUDENT-NUMBER-IN  PIC X(9).
00028    000310     05 FILLER             PIC X(8).
00029    000320     05 GRADE-1-IN         PIC 9(3).
00030    000330     05 GRADE-2-IN         PIC 9(3).
00031    000340     05 GRADE-3-IN         PIC 9(3).
00032    000350     05 GRADE-4-IN         PIC 9(3).
00033    000360     05 FILLER             PIC X(51).
00034    000370
00035    000380 FD  STUDENT-GRADE-REPORT
00036    000390     LABEL RECORDS ARE OMITTED.
00037    000400
00038    000410 01  REPORT-LINE           PIC X(79).
00039    000420
00040    000430 WORKING-STORAGE SECTION.
00041    000440 01  MORE-INPUT            PIC X(3) VALUE 'YES'.
00042    000450
00043    000460 01  RUN-DATE.
00044    000470     05 RUN-YEAR           PIC 99.
00045    000480     05 RUN-MONTH          PIC 99.
00046    000490     05 RUN-DAY            PIC 99.
00047    000500
00048    000510 01  REPORT-TITLE.
00049    000520     05 FILLER             PIC X.


   2

00050    000530     05 FILLER             PIC X(35) VALUE SPACES.
00051    000540     05 FILLER             PIC X(39) VALUE 'CLASS AVERAGE REPORT'.
00052    000550     05 FILLER             PIC X(5)  VALUE 'DATE'.
00053    000560     05 RUN-MONTH          PIC Z9.
00054    000570     05 FILLER             PIC X     VALUE '/'.
00055    000580     05 RUN-DAY            PIC Z9.
00056    000590     05 FILLER             PIC X     VALUE '/'.
00057    000600     05 RUN-YEAR           PIC Z9.
00058    000610
00059    000620 01  COLUMN-HEADS-1.
00060    000630     05 FILLER             PIC X.
00061    000640     05 FILLER             PIC X(20) VALUE SPACES.
00062    000650     05 FILLER             PIC X(24) VALUE 'STUDENT'.
00063    000660     05 FILLER             PIC X(12) VALUE 'G R A D E S '.
00064    000670
00065    000680 01  COLUMN-HEADS-2.
00066    000690     05 FILLER             PIC X.
00067    000700     05 FILLER             PIC X(20) VALUE SPACES.
00068    000710     05 FILLER             PIC X(16) VALUE 'NUMBER'.
00069    000720     05 FILLER             PIC X(8)  VALUE 'EXAM 1'.
00070    000730     05 FILLER             PIC X(8)  VALUE 'EXAM 2'.
00071    000740     05 FILLER             PIC X(8)  VALUE 'EXAM 3'.
00072    000750     05 FILLER             PIC X(11) VALUE 'EXAM 4'.
00073    000760     05 FILLER             PIC X(7)  VALUE 'AVERAGE'.
```

Figure A.7 Part 2 (continued).

```
00074    000770
00075    000780 01   BODY-LINE.
00076    000790      05 FILLER                  PIC X.
00077    000800      05 FILLER                  PIC X(19) VALUE SPACES.
00078    000810      05 STUDENT-NUMBER-OUT      PIC X(9).
00079    000820      05 FILLER                  PIC X(9)  VALUE SPACES.
00080    000830      05 GRADE-1-OUT             PIC ZZ9.
00081    000840      05 FILLER                  PIC X(5)  VALUE SPACES.
00082    000850      05 GRADE-2-OUT             PIC ZZ9.
00083    000860      05 FILLER                  PIC X(5)  VALUE SPACES.
00084    000870      05 GRADE-3-OUT             PIC ZZ9.
00085    000880      05 FILLER                  PIC X(5)  VALUE SPACES.
00086    000890      05 GRADE-4-OUT             PIC ZZ9.
00087    000900      05 FILLER                  PIC X(7)  VALUE SPACES.
00088    000910      05 AVERAGE-GRADE           PIC ZZZ.9.
00089    000920
00090    000930 01   FINAL-AVERAGE-LINE.
00091    000940      05 FILLER                  PIC X.
00092    000950      05 FILLER                  PIC X(22) VALUE SPACES.
00093    000960      05 FILLER                  PIC X(15) VALUE 'AVERAGES'.
00094    000970      05 EXAM-1-AVERAGE-OUT      PIC ZZZ.9.
00095    000980      05 FILLER                  PIC X(3)  VALUE SPACES.
00096    000990      05 EXAM-2-AVERAGE-OUT      PIC ZZZ.9.
00097    001000      05 FILLER                  PIC X(3)  VALUE SPACES.
00098    001010      05 EXAM-3-AVERAGE-OUT      PIC ZZZ.9.
00099    001020      05 FILLER                  PIC X(3)  VALUE SPACES.
00100    001030      05 EXAM-4-AVERAGE-OUT      PIC ZZZ.9.
00101    001040      05 FILLER                  PIC X(5)  VALUE SPACES.
```

```
  3

00102    001050      05 CLASS-AVERAGE-OUT       PIC ZZZ.9.
00103    001060
00104    001070 01   NUMBER-OF-EXAMS            PIC 9     VALUE 4.
00105    001080 01    EXAM-1-SUM                PIC 9(5)  VALUE 0.
00106    001090 01   EXAM-2-SUM                 PIC 9(5)  VALUE 0.
00107    001100 01   EXAM-3-SUM                 PIC 9(5)  VALUE 0.
00108    001110 01   EXAM-4-SUM                 PIC 9(5)  VALUE 0.
00109    001120 01   NUMBER-OF-STUDENTS         PIC 9(3)  VALUE 0.
00110    001130
00111    001140***************************************************
00112    001150
00113    001160 PROCEDURE DIVISION.
00114    001170      OPEN INPUT EXAM-GRADE-FILE-IN, OUTPUT STUDENT-GRADE-REPORT.
00115    001180      PERFORM REPORT-HEADINGS-ROUTINE.
00116    001190      READ EXAM-GRADE-FILE-IN AT END MOVE 'NO' TO MORE-INPUT.
00117    001200      PERFORM MAIN-PROCESS UNTIL MORE-INPUT IS EQUAL TO 'NO'.
00118    001210      PERFORM FINAL-LINE-ROUTINE.
00119    001220      CLOSE EXAM-GRADE-FILE-IN, STUDENT-GRADE-REPORT.
00120    001230      STOP RUN.
0012     001240
00122    001250 REPORT-HEADINGS-ROUTINE.
00123    001260      ACCEPT RUN-DATE FROM DATE.
00124    001270      MOVE CORRESPONDING RUN-DATE TO REPORT-TITLE.
00125    001280      WRITE REPORT-LINE FROM REPORT-TITLE AFTER ADVANCING
00126    001290          TO-NEW-PAGE.
00127    001300      WRITE REPORT-LINE FROM COLUMN-HEADS-1 AFTER ADVANCING 3.
```

All the warning messages in this run are of the same type: They all warn that the receiving fields in arithmetic operations might be too small to hold the results. We can check that they are all large enough and then ignore the warnings.

The four *E*-level messages all stem from the same error. In statement 144 the word TO was misspelled. It was keyed in as tee–zero instead of tee–oh, so the compiler substituted something else for the word TO. Then the compiler expected all numeric operands while it looked for a GIVING or a TO to make a legal ADD statement.

The second error message relating to card 144 tells us that the compiler found an end of sentence before it found a TO or a GIVING. It discarded the end of sentence and dauntlessly went on to examine card 145, thinking that it was still examining parts of card 144. The third error message for card 144 tells us that the compiler found another word ADD while it was looking for TO or GIVING. Mercifully, the TO in card 145 ends the compiler's quest. Correcting the spelling of the word TO in card 144 eliminates all the error messages.

If there had been some error in card 145 also, the compiler may not have detected it in this run, because it was so confused by card 144. It would have detected any errors in card 145 after the error in card 144 was corrected and the program run again. The moral is: don't make a lot of errors close together.

Figure A.7 An erroneous version of Program 12, Part 3 of 3.

```
00128   001310        WRITE REPORT-LINE FROM COLUMN-HEADS-2 AFTER ADVANCING 1.
00129   001320        MOVE SPACES TO REPORT-LINE.
00130   001330        WRITE REPORT-LINE AFTER ADVANCING 1.
00131   001340
00132   001350 MAIN-PROCESS.
00133   001360        MOVE STUDENT-NUMBER-IN TO STUDENT-NUMBER-OUT.
00134   001370        MOVE GRADE-1-IN TO GRADE-1-OUT.
00135   001380        MOVE GRADE-2-IN TO GRADE-2-OUT.
00136   001390        MOVE GRADE-3-IN TO GRADE-3-OUT.
00137   001400        MOVE GRADE-4-IN TO GRADE-4-OUT.
00138   001410        COMPUTE AVERAGE-GRADE ROUNDED = (GRADE-1-IN +
00139   001420            GRADE-2-IN + GRADE-3-IN + GRADE-4-IN) /
00140   001430            NUMBER-OF-EXAMS.
00141   001440        ADD 1 TO NUMBER-OF-STUDENTS.
00142   001450        ADD GRADE-1-IN TO EXAM-1-SUM.
00143   001460        ADD GRADE-2-IN TO EXAM-2-SUM.
00144   001470        ADD GRADE-3-IN T0 EXAM-3-SUM.
00145   001480        ADD GRADE-4-IN TO EXAM-4-SUM.
00146   001490        WRITE REPORT-LINE FROM BODY-LINE AFTER ADVANCING 1.
00147   001500        READ EXAM-GRADE-FILE-IN AT END MOVE 'NO' TO MORE-INPUT.
00148   001510 FINAL-LINE-ROUTINE.
00149   001520        DIVIDE EXAM-1-SUM BY NUMBER-OF-STUDENTS GIVING
00150   001530            EXAM-1-AVERAGE-OUT ROUNDED.
00151   001540        DIVIDE EXAM-2-SUM BY NUMBER-OF-STUDENTS GIVING
00152   001550            EXAM-2-AVERAGE-OUT ROUNDED.
00153   001560        DIVIDE EXAM-3-SUM BY NUMBER-OF-STUDENTS GIVING
```

Figure A.7 Part 3 (continued).

```
     4

00154     001570             EXAM-3-AVERAGE-OUT ROUNDED.
00155     001580         DIVIDE EXAM-4-SUM BY NUMBER-OF-STUDENTS GIVING
00156     001590             EXAM-4-AVERAGE-OUT ROUNDED.
00157     001600         COMPUTE CLASS-AVERAGE-OUT ROUNDED =
00158     001610             (EXAM-1-SUM + EXAM-2-SUM + EXAM-3-SUM +
00159     001620             EXAM-4-SUM) / (NUMBER-OF-STUDENTS * NUMBER-OF-EXAMS).
00160     001630         WRITE REPORT-LINE FROM FINAL-AVERAGE-LINE
00161     001640             AFTER ADVANCING 3.
```

Figure A.8 Error messages for Figure A.7.

```
   6

CARD   ERROR MESSAGE

138    IKF5011I-W    AN INTERMEDIATE RESULT OR A SENDING FIELD MIGHT HAVE ITS HIGH ORDER DIGIT POSITION
                     TRUNCATED.
144    IKF3001I-E    TO NOT DEFINED. SUBSTITUTING TALLY .
144    IKF4056I-E    SYNTAX REQUIRES NUMERIC OPERAND . FOUND END-OF-SENT . DISCARDED.
144    IKF4056I-E    SYNTAX REQUIRES NUMERIC OPERAND . FOUND ADD . DISCARDED.
147    IKF4072I-W    EXIT FROM PERFORMED PROCEDURE ASSUMED BEFORE PROCEDURE-NAME .
149    IKF5011I-W    AN INTERMEDIATE RESULT OR A SENDING FIELD MIGHT HAVE ITS HIGH ORDER DIGIT POSITION
                     TRUNCATED.
151    IKF5011I-W    AN INTERMEDIATE RESULT OR A SENDING FIELD MIGHT HAVE ITS HIGH ORDER DIGIT POSITION
                     TRUNCATED.
153    IKF5011I-W    AN INTERMEDIATE RESULT OR A SENDING FIELD MIGHT HAVE ITS HIGH ORDER DIGIT POSITION
                     TRUNCATED.
155    IKF5011I-W    AN INTERMEDIATE RESULT OR A SENDING FIELD MIGHT HAVE ITS HIGH ORDER DIGIT POSITION
                     TRUNCATED.
157    IKF5011I-W    AN INTERMEDIATE RESULT OR A SENDING FIELD MIGHT HAVE ITS HIGH ORDER DIGIT POSITION
                     TRUNCATED.

       ERROR MESSAGE

       IKF6006I-E    SUPMAP SPECIFIED AND E-LEVEL DIAGNOSTIC HAS OCCURRED. PMAP CLIST LOAD DECK STATE FLOW SYMDMP IGNO
                     RED/
```

ERRONEOUS NESTED IF

Figure A.9 shows a version of Program 14 with one of its IF statements written incorrectly. The error messages from this run are shown in Figure A.10. The message relating to card 141 claims that the compiler found an ELSE without a matching IF and that the ELSE is discarded. Examination of card 141 shows no ELSE, but card 140 has an ELSE, presumably the ELSE in question. There seems to be a matching IF in card 138, but the erroneous period in card 139 leaves the ELSE disconnected from its IF. Removing the period eliminates the error.

Card 172 gets two error messages just because some stray numbers got into columns 7 and 8. First the compiler objects to the invalid character in column 7 and then takes the 0 in column 8 to be a paragraph name. (Rules for making up paragraph names are given in Chapter 2.) 0 is a legal paragraph name, and the compiler objects because there is no period after it and before the WRITE verb. The compiler assumes that there is such a period and goes on.

Figure A.9 An erroneous version of Program 14, Part 1 of 2.

```
PP 5734-CB2 V4 RELEASE 1.5 10NOV77          IBM OS AMERICAN NATIONAL STANDARD COBOL

   1

00001    000040 IDENTIFICATION DIVISION.
00002    000050 PROGRAM-ID. PROG14.
00003    000060*AUTHOR. PHIL RUBENSTEIN.
00004    000070*
00005    000080*     THIS PROGRAM PRODUCES A SALES  REPORT
00006    000090*     WITH THREE LEVELS OF TOTALS.
00007    000100*
00008    000110*****************************************************************
00009    000120
00010    000130 ENVIRONMENT DIVISION.
0001     000140 CONFIGURATION SECTION.
00012    000150 SPECIAL-NAMES.
00013    000160      C01 IS NEW-PAGE.
00014    000170 INPUT-OUTPUT SECTION.
00015    000180 FILE-CONTROL.
00016    000190      SELECT SALES-REPORT  ASSIGN TO UT-S-SYSPRINT.
00017    000200      SELECT SALES-FILE-IN ASSIGN TO UT-S-SYSIN.
00018    000210
00019    000220*****************************************************************
00020    000230
00021    000240 DATA DIVISION.
00022    000250 FILE SECTION.
00024    000260 FD  SALES-FILE-IN
00024    000270      LABEL RECORDS ARE OMITTED.
00025    000280
00026    000290 01   SALES-RECORD-IN.
00027    000300      05 STORE-NUMBER-IN         PIC 9(3).
00028    000310      05 SALESPERSON-NUMBER-IN   PIC 9(3).
00029    000320      05 CUSTOMER-NUMBER         PIC 9(6).
00030    000330      05 SALE-AMOUNT             PIC 9(4)V99.
00031    000340      05 FILLER                  PIC X(62).
00032    000350
00033    000360 FD  SALES-REPORT
00034    000370      LABEL RECORDS ARE OMITTED.
00035    000380
00036    000390 01   REPORT-LINE                PIC X(69).
00037    000400
00038    000410 WORKING-STORAGE SECTION.
00039    000420 01   MORE-INPUT                 PIC X(3) VALUE 'YES'.
00040    000430
00041    000440 01   PAGE-HEAD-1.
00042    000450      05 FILLER             PIC X.
00043    000460      05 FILLER             PIC X(35) VALUE SPACES.
00044    000470      05 FILLER             PIC X(12) VALUE 'SALES REPORT'.
00045    000480
00046    000490 01   PAGE-HEAD-2.
00047    000500      05 FILLER                  PIC X.
00048    000510      05 FILLER                  PIC X(19) VALUE SPACES.
00049    000520      05 FILLER                  PIC X(5)  VALUE 'DATE'.
```

Figure A.9 Part 1 (continued).

```
  2

00050    000530         05 RUN-MONTH                   PIC Z9.
00051    000540         05 FILLER                      PIC X      VALUE '/'.
00052    000550         05 RUN-DAY                     PIC 99.
00053    000560         05 FILLER                      PIC X      VALUE '/'.
00054    000570         05 RUN-YEAR                    PIC Z9.
00055    000580         05 FILLER                      PIC X(29) VALUE SPACES.
00056    000590         05 FILLER                      PIC X(5)  VALUE 'PAGE'.
00057    000600         05 PAGE-NUMBER-OUT             PIC Z9.
00058    000610
00059    000620 01      PAGE-HEAD-3.
00060    000630         05 FILLER                      PIC X.
00061    000640         05 FILLER                      PIC X(24) VALUE SPACES.
00062    000650         05 FILLER                      PIC X(9)  VALUE 'STORE'.
00063    000660         05 FILLER                      PIC X(10) VALUE 'SALES-'.
00064    000670         05 FILLER                      PIC X(13) VALUE 'CUSTOMER'.
00065    000680         05 FILLER                      PIC X(4)  VALUE 'SALE'.
00066    000690
00067    000700 01      PAGE-HEAD-4.
00068    000710         05 FILLER                      PIC X.
00069    000720         05 FILLER                      PIC X(25) VALUE SPACES.
00070    000730         05 FILLER                      PIC X(8)  VALUE 'NO.'.
00071    000740         05 FILLER                      PIC X(11) VALUE 'PERSON'.
00072    000750         05 FILLER                      PIC X(11) VALUE 'NUMBER'.
00073    000760         05 FILLER                      PIC X(6)  VALUE 'AMOUNT'.
00074    000770
00075    000780 01      DETAIL-LINE.
00076    000790         05 FILLER                      PIC X.
00077    000800         05 STORE-NUMBER-OUT            PIC B(25)9(3).
00078    000810         05 SALESPERSON-NUMBER-OUT        PIC B(6)9(3).
00079    000820         05 CUSTOMER-NUMBER-OUT         PIC B(7)9(6).
00080    000830         05 SALE-AMOUNT                 PIC B(4)Z,ZZZ.99.
00081    000840
00082    000850 01      SALESPERSON-TOTAL-LINE.
00083    000860         05 FILLER                      PIC X.
00084    000870         05 SALESPERSON-TOTAL-OUT PIC B(53)ZZ,ZZZ.99B.
00085    000880         05 FILLER                      PIC X     VALUE '*'.
00086    000890
00087    000900 01      STORE-TOTAL-LINE.
00088    000910         05 FILLER                      PIC X.
00089    000920         05 FILLER                      PIC X(26) VALUE SPACES.
00090    000930         05 FILLER           PIC X(20) VALUE 'TOTAL FOR STORE NO.'.
00091    000940         05 STORE-NUMBER-SAVE           PIC 9(3).
00092    000950         05 FILLER                      PIC X(2)  VALUE SPACES.
00093    000960         05 STORE-TOTAL-OUT             PIC $ZZZ,ZZZ.99BB.
00094    000970         05 FILLER                      PIC X(2)  VALUE '**'.
00095    000980
00096    000990 01      GRAND-TOTAL-LINE.
00097    001000         05 FILLER                      PIC X.
00098    001010         05 FILLER                      PIC X(34) VALUE SPACES.
00099    001020         05 FILLER                      PIC X(15) VALUE 'GRAND TOTAL'.
00100    001030         05 GRAND-TOTAL-OUT             PIC $Z,ZZZ,ZZZ.99B.
00101    001040         05 FILLER                      PIC X(3)  VALUE '***'.
```

Figure A.9 An erroneous version of Program 14, Part 2 of 2.

```
  3

00102    001050
00103    001060 01  PAGE-NUMBER-W                PIC 99 VALUE 0.
00104    001070 01  LINE-COUNT                   PIC 99 COMP.
00105    001080 01  PAGE-LIMIT                   PIC 99 VALUE 40 COMP.
00106    001090 01  NO-OF-LINES-IN-PAGE-HEADING PIC 9 VALUE 6 COMP.
00107    001100 01  LINE-SPACING                 PIC 9 COMP.
00108    001110 01  SALESPERSON-TOTAL-W          PIC 9(5)V99.
00109    001120 01  STORE-TOTAL-W                PIC 9(6)V99.
00110    001130 01  GRAND-TOTAL-W                PIC 9(7)V99 VALUE 0.
00111    001140 01  SALESPERSON-NUMBER-SAVE      PIC 9(3).
00112    001150
00113    001160 01  TODAYS-DATE.
00114    001170     05 RUN-YEAR                  PIC 99.
00115    001180     05 RUN-MONTH                 PIC 99.
00116    001190     05 RUN-DAY                   PIC 99.
00117    001200
00118    001210*****************************************************************
00119    001220
00120    001230 PROCEDURE DIVISION.
00121    001240     OPEN INPUT SALES-FILE-IN, OUTPUT SALES-REPORT.
00122    001250     READ SALES-FILE-IN   AT END MOVE 'NO' TO MORE-INPUT.
00123    001260     IF MORE-INPUT = 'YES'
00124    001270         PERFORM INITIALIZATION
00125    001280         PERFORM PRODUCE-REPORT-BODY UNTIL MORE-INPUT = 'NO'
00126    001290         PERFORM PRODUCE-FINAL-LINES.
00127    001300     CLOSE SALES-FILE-IN, SALES-REPORT.
00128    001310     STOP RUN.
00129    001320
00130    001330 INITIALIZATION.
00131    001340     ACCEPT TODAYS-DATE FROM DATE.
00132    001350     MOVE CORRESPONDING TODAYS-DATE TO PAGE-HEAD-2.
00133    001360     PERFORM SET-UP-FOR-NEW-STORE.
00134    001370     PERFORM SET-UP-FOR-NEW-SALESPERSON.
00135    001380     PERFORM PAGE-HEADING.
00136    001390
00137    001400 PRODUCE-REPORT-BODY.
00138    001410     IF STORE-NUMBER-IN NOT = STORE-NUMBER-SAVE
00139    001420         PERFORM PRODUCE-STORE-TOTAL-LINE.
00140    001430     ELSE
00141    001440         IF SALESPERSON-NUMBER-IN
00142    001450             NOT = SALESPERSON-NUMBER-SAVE
00143    001460             PERFORM PRODUCE-SALESPERSON-TOTAL-LINE.
00144    001470     PERFORM PRODUCE-A-DETAIL-LINE.
00145    001480     READ SALES-FILE-IN AT END MOVE 'NO' TO MORE-INPUT.
00146    001490     PERFORM PRODUCE-A-DETAIL-LINE.
00147    001500
00148    001510 SET-UP-FOR-NEW-STORE.
00149    001520     MOVE STORE-NUMBER-IN TO STORE-NUMBER-SAVE,
00150    001530         STORE-NUMBER-OUT.
00151    001540     MOVE 0 TO STORE-TOTAL-W.
00152    001550     MOVE 3 TO LINE-SPACING.
00153    001560 SET-UP-FOR-NEW-SALESPERSON.
```

Figure A.9 Part 2 (continued).

```
  4

00154   001570       MOVE SALESPERSON-NUMBER-IN TO SALESPERSON-NUMBER-OUT,
00155   001580           SALESPERSON-NUMBER-SAVE.
00156   001590       MOVE 0 TO SALESPERSON-TOTAL-W.
00157   001600       MOVE 2 TO LINE-SPACING.
00158   001610
00159   001620   PRODUCE-STORE-TOTAL-LINE.
00160   001630       PERFORM PRODUCE-SALESPERSON-TOTAL-LINE.
00161   001640       PERFORM PRINT-A-STORE-TOTAL-LINE.
00162   001650       PERFORM SET-UP-FOR-NEW-STORE.
00163   001660
00164   001670   PRODUCE-SALESPERSON-TOTAL-LINE.
00165   001680       PERFORM PRINT-A-SLSPSN-TOTAL-LINE.
00166   001690       PERFORM SET-UP-FOR-NEW-SALESPERSON.
00167   001700
00168   001710   PRODUCE-FINAL-LINES.
00169   001720       PERFORM PRINT-A-SLSPSN-TOTAL-LINE.
00170   001730       PERFORM PRINT-A-STORE-TOTAL-LINE.
00171   001740       MOVE GRAND-TOTAL-W TO GRAND-TOTAL-OUT.
00172   00175080       WRITE REPORT-LINE FROM GRAND-TOTAL-LINE AFTER 2.
00173   001760
00174   001770   PRINT-A-STORE-TOTAL-LINE.
00175   001780       MOVE STORE-TOTAL-W TO STORE-TOTAL-OUT.
00176   001790       WRITE REPORT-LINE FROM STORE-TOTAL-LINE AFTER 2.
00177   001800       ADD STORE-TOTAL-W TO GRAND-TOTAL-W.
00178   001810       ADD 2 TO LINE-COUNT.
00179   001820
00180   001830   PRINT-A-SLSPSN-TOTAL-LINE.
00181   001840       MOVE SALESPERSON-TOTAL-W TO SALESPERSON-TOTAL-OUT.
00182   001850       WRITE REPORT-LINE FROM SALESPERSON-TOTAL-LINE AFTER 2.
00183   001860       ADD SALESPERSON-TOTAL-W TO STORE-TOTAL-W.
00184   001870       ADD 2 TO LINE-COUNT.
00185   001880
00186   001890   PRODUCE-A-DETAIL-LINE.
00187   001900       MOVE CORR SALES-RECORD-IN TO DETAIL-LINE.
00188   001910       ADD SALE-AMOUNT IN SALES-RECORD-IN TO
00189   001920           SALESPERSON-TOTAL-W.
00190   001930       IF LINE-COUNT GREATER THAN PAGE-LIMIT
00191   001940           PERFORM PAGE-HEADING.
00192   001950       WRITE REPORT-LINE FROM DETAIL-LINE AFTER LINE-SPACING.
00193   001960       ADD LINE-SPACING TO LINE-COUNT.
00194   001970       MOVE 1 TO LINE-SPACING.
00195   001980       MOVE SPACES TO DETAIL-LINE.
00196   001990
00197   002000   PAGE-HEADING.
00198   002010       ADD 1 TO PAGE-NUMBER-W.
00199   002020       MOVE PAGE-NUMBER-W TO PAGE-NUMBER-OUT.
00200   002030       WRITE REPORT-LINE FROM PAGE-HEAD-1 AFTER NEW-PAGE.
00201   002040       WRITE REPORT-LINE FROM PAGE-HEAD-2 AFTER 1.
00202   002050       WRITE REPORT-LINE FROM PAGE-HEAD-3 AFTER 3.
00203   002060       WRITE REPORT-LINE FROM PAGE-HEAD-4 AFTER 1.
00204   002070       MOVE 2 TO LINE-SPACING.
00205   002080       MOVE NO-OF-LINES-IN-PAGE-HEADING TO LINE-COUNT.
00206   002090       MOVE STORE-NUMBER-SAVE TO STORE-NUMBER-OUT.
00207   002100       MOVE SALESPERSON-NUMBER-SAVE TO
00208   002110           SALESPERSON-NUMBER-OUT.
```

Figure A.10 Error messages for Figure A.9.

```
  7

CARD   ERROR MESSAGE

141    IKF4017I-E     ELSE UNMATCHED BY CONDITION IS DISCARDED.
172    IKF1150I-W     ILLEGAL CHARACTER IN COLUMN 7, BLANK ASSUMED.
172    IKF1043I-W     END OF SENTENCE SHOULD PRECEDE WRITE . ASSUMED PRESENT.

       ERROR MESSAGE

       IKF6006I-E     SUPMAP SPECIFIED AND E-LEVEL DIAGNOSTIC HAS OCCURRED. PMAP CLIST LOAD DECK STATE FLOW SYMDMP IGNO
                      RED/
```

Exercise 1. Figure A.E1.1 shows an erroneous version of Program 15. Figure A.E1.2 shows the error messages. Find and correct all errors.

Figure A.E1.1 Erroneous program for Exercise 1, Part 1 of 3.

```
PP 5734-CB2 V4 RELEASE 1.5 10NOV77        IBM OS AMERICAN NATIONAL STANDARD COBOL

   1

00001    000010 IDENTIFICATION DIVISION.
00002    000020 PROGRAM-ID. PROG15.
00003    000030*
00004    000040*    THIS PROGRAM COMPUTES SALESPERSON COMMISSIONS FOR JUNIOR
00005    000050*    SALESPERSONS, ASSOCIATE SALESPERSONS, AND SENIOR SALESPERSONS
00006    000060*
00007    000070*****************************************************************
00008    000080
00009    000090 ENVIRONMENT DIVISION.
00010    000100 CONFIGURATION SECTION.
00011    000110 SPECIAL-NAMES.
00012    000120     C01 IS NEW-PAGE.
00013    000130 INPUT-OUTPUT SECTION.
00014    000140 FILE-CONTROL.
00015    000150     SELECT COMMISSION-REPORT ASSIGN TO UT-S-SYSPRINT.
00016    000160     SELECT SALES-FILE-IN      ASSIGN TO UT-S-SYSIN.
00017    000170
00018    000180*****************************************************************
00019    000190
00020    000200' DATA DIVISION.
00021    000210 FILE SECTION.
00022    000220 FD  SALES-FILE-IN
00023    000230     LABEL RECORDS ARE OMITTED.
```

Figure A.E1.1 Part 1 (continued).

```
00024   000240
00025   000250 01  SALES-RECORD-IN.
00026   000260     05 SALESPERSON-NUMBER-IN     PIC 9(5).
00027   000270     05 CLASS-IN                  PIC X.
00028   000280        88 SALESPERSON-IS-JUNIOR     VALUES 'A' THRU 'F'.
00029   000290        88 SALESPERSON-IS-ASSOCIATE VALUE 'G'.
00030   000300        88 SALESPERSON-IS-SENIOR     VALUE 'H'.
00031   000310        88 CLASS-CODE-IS-MISSING     VALUE SPACE.
00032   000320     05 SALE-1-IN                 PIC 9(6)V99
00033   000330     05 SALE-2-IN                 PIC 9(6)V99.
00034   000340     05 SALE-3-IN                 PIC 9(6)V99.
00035   000350     05 FILLER                    PIC X(50).
00036   000360
00037   000370 FD  COMMISSION-REPORT.
00038   000380     LABEL RECORDS ARE OMITTED.
00039   000390
00040   000400 01  REPORT-LINE                  PIC X(89).
00041   000410
00042   000420 WORKING-STORAGE SECTION.
00043   000430 01  MORE-INPUT                   PIC X(3) VALUE 'YES'.
00044   000440
00045   000450 01  PAGE-HEAD-1.
00046   000460     05 FILLER                    PIC X.
00047   000470     05 FILLER                    PIC X(17) VALUE SPACES.
00048   000480     05 FILLER             PIC X(12) VALUE 'COMMISSION REGISTER'.
00049   000490
00050   000500 01  PAGE-HEAD-2.
00051   000510     05 FILLER                    PIC X.
00052   000520     05 FILLER                    PIC X(43) VALUE SPACES.
00053   000530     05 FILLER                    PIC X(5)  VALUE 'DATE'.
00054   000540     05 RUN-MONTH                 PIC Z9.
00055   000550     05 FILLER                    PIC X     VALUE '/'.
00056   000560     05 RUN-DAY                   PIC 99.
00057   000570     05 FILLER                    PIC X     VALUE '/'.
00058   000580     05 RUN-YEAR                  PIC Z9.
00059   000590
00060   000600 01  PAGE-HEAD-3.
00061   000610     05 FILLER                    PIC X.
00062   000620     05 FILLER                    PIC X(10) VALUE SPACES.
00063   000630     05 FILLER                    PIC X(12) VALUE 'SALES-'.
00064   000640     05 FILLER                    PIC X(11) VALUE 'CLASS'.
00065   000650     05 FILLER                    PIC X(10) VALUE 'SALE'.
00066   000660     05 FILLER                    PIC X(10) VALUE 'COMMISSION'.
00067   000670
00068   000680 01  PAGE-HEAD-4.
00069   000690     05 FILLER                    PIC X.
00070   000700     05 FILLER                    PIC X(10) VALUE SPACES.
00071   000710     05 FILLER                    PIC X(22) VALUE 'PERSON'.
00072   000720     05 FILLER                    PIC X(6)  VALUE 'AMOUNT'.
00073   000730
00074   000740 01  PAGE-HEAD-5.
00075   000750     05 FILLER                    PIC X.
00076   000760     05 FILLER                    PIC X(6)  VALUE 'NUMBER'.
00077   000770
00078   000780 01  DETAIL-LINE.
00079   000790     05 FILLER                    PIC X.
00080   000800     05 SALESPERSON-NUMBER-OUT    PIC B(10)X(5).
```

Figure A.E1.1 Erroneous program for Exercise 1, Part 2 of 3.

```
00081   000810         05 CLASS-TITLE-OUT              PIC B(5)X(8)B.
00082   000820         05 SALE-AMOUNT-OUT              PIC ZZZ,ZZZ.99B.
00083   000830         05 COMMISSION-OUT               PIC ZZ,ZZZ.99.
00084   000840
00085   000850 01   ERROR-LINE.
00086   000860         05 FILLER                       PIC X.
00087   000870         05 SALESPERSON-NUMBER-ERR       PIC B(10)X(5)B(5).
00088   000880         05 FILLER                       PIC X(8)   VALUE 'ERROR -'.
00089   000890         05 ERROR-MESSAGE                PIC X(15).
0009    000900         05 FILLER PIC X(29) VALUE 'CLASS SHOULD BE A, B, C, D, E'.
00091   000910         05 FILLER PIC X(13) VALUE ', F, G, OR H.'.
00092   000920
00093   000930 01   ERROR-MESSAGES.
00094   000940         05 CLASS-MISSING-MESSAGE        PIC X(15) VALUE 'CLASS MISSING'.
00095   000950         05 INVALID-CLASS-MESSAGE.
00096   000960            10 FILLER                    PIC X(9) VALUE 'CLASS IS'
00097   000970            10 INVALID-CLASS             PIC X.
00098   000980            10 FILLER                    PIC X     VALUE '.'.
00099   000990
00100   001000 01   AVERAGE-SALE-AMOUNT               PIC S9(4)V99.
00101   001010 01   NUMBER-OF-SALES                   PIC S9      VALUE 3.
00102   001020 01   SALE-QUOTAS.
00103   001030         05 JUNIOR-SALE-QUOTA            PIC S9      VALUE 0
00104   001040         05 ASSOCIATE-SALE-QUOTA         PIC S9(5) VALUE 10000.
00105   001050         05 SENIOR-SALE-QUOTA            PIC S9(5) VALUE 50000.
00106   001060 01   SALE-QUOTA                        PIC S9(5).
00107   001070
00108   001080 01   COMMISSION-RATES.
00109   001090         05 LOW-RATE                     PIC SV99.
00110   001100         05 HIGH-RATE                    PIC SV99.
00111   001110         05 COMMISSION-RATE-1            PIC SV99   VALUE .10.
00112   001120         05 COMMISSION-RATE-2            PIC SV99   VALUE .10.
00113   001130         05 COMMISSION-RATE-3            PIC SV99   VALUE .05.
00114   001140         05 COMMISSION-RATE-4            PIC SV99   VALUE .20.
00115   001150         05 COMMISSION-RATE-5            PIC SV99   VALUE .20.
00116   001160         05 COMMISSION-RATE-6            PIC SV99   VALUE .30.
00117   001170         05 COMMISSION-RATE-7            PIC SV99   VALUE .30.
00118   001180         05 COMMISSION-RATE-8            PIC SV99   VALUE .40.
00119   001190
00120   001200*****************************************************************
00121   001210
00122   001220 PROCEDURE DIVISION.
00123   001230         OPEN INPUT SALES-FILE-IN, OUTPUT COMMISSION-REPORT.
00124   001240         READ SALES-FILE-IN AT END MOVE 'NO' TO MORE-INPUT.
00125   001250         IF MORE-INPUT = 'YES'
00126   001260             PERFORM INITIALIZATION
00127   001270             PERFORM MAIN-PROCESS UNTIL MORE-INPUT = 'NO'.
00128   001280         CLOSE SALES-FILE-IN, COMMISSION-REPORT.
00129   001290         STOP RUN.
00130   001300
00131   001310 INITIALIZATION.
00132   001320         ACCEPT TODAYS-DATE FROM DATE.
00133   001330         MOVE CORR TODAYS-DATE TO PAGE-HEAD-2.
00134   001340         WRITE REPORT-LINE FROM PAGE-HEAD-1 AFTER NEW-PAGE.
00135   001350         WRITE REPORT-LINE FROM PAGE-HEAD-2 AFTER 1.
00136   001360         WRTIE REPORT-LINE FROM PAGE-HEAD-3 AFTER 3.
00137   001370         WRITE REPORT-LINE FROM PAGE-HEAD-4 AFTER 1.
00138   001380         WRITE REPORT-LINE FROM PAGE-HEAD-5 AFTER 1.
```

Figure A.E1.1 Part 2 (continued).

```
00139    001390
00140    001400 MAIN-PROCESS.
00141    001410      IF CLASS-CODE-IS-MISSING
00142    001420           PERFORM CLASS-MISSING-ROUTINE
00143    001430      ELSE
00144    001440           IF SALESPERSON-IS-JUNIOR
00145    001450                PERFORM JUNIOR-SALESPERSON-COMMISSION
00146    001460           ELSE
00147    001470                COMPUTE AVERAGE-SALE-AMOUNT = (SALE-1-IN +
00148    001480                     SALE-2-IN + SALE-3-IN) / NUMBER-OF-SALES
00149    001490                IF SALESPERSON-IS-ASSOCIATE AND AVERAGE-SALE-AMOUNT
00150    001500                     LESS THAN ASSOCIATE-SALE-QUOTA
00151    001510                     PERFORM LEVEL-2-COMMISSION
00152    001520                ELSE
00153    001530                     IF SALESPERSON-IS-ASSOCIATE
00154    001540                          PERFORM LEVEL-3-COMMISSION
00155    001550                     ELSE
00156    001560                          IF SALESPERSON-IS-SENIOR AND AVERAGE-SALE-AMOUNT
00157    001570                               LESS THAN SENIOR-SALE-QUOTA
00158    001580                               PERFORM LEVEL-4-COMMISSION
00159    001590                          ELSE
00160    001600                               IF SALESPERSON-IS-SENIOR
00161    001610                                    PERFORM LEVEL-5-COMMISSION
00162    001620                               ELSE
00163    001630                                    PERFORM INVALID-CLASS-ROUTINE.
00164    001640      READ SALES-FILE-IN AT END MOVE 'NO' TO MORE-INPUT.
00165    001650
00166    001660 JUNIOR-SALESPERSON-COMMISSION.
00167    001670      MOVE 'JUNIOR' TO CLASS-TITLE-OUT.
00168    001680      MOVE JUNIOR-SALE-QUOTA TO SALE-QUOTA.
00169    001690      MOVE COMMISSION-RATE-1 TO LOW-RATE.
00170    001700      MOVE COMMISSION-RATE-2 TO HIGH-RATE.
00171    001710      PERFORM WRITE-A-SALESPERSON-GROUP.
00172    001720
00173    001730 LEVEL-2-COMMISSION.
00174    001740      MOVE 'ASSOCIATE' TO CLASS-TITLE-OUT.
00175    001750      MOVE ASSOCIATE-SALE-QUOTA TO SALE-QUOTA.
00176    001760      MOVE COMMISSION-RATE-3 TO LOW-RATE.
00177    001770      MOVE COMMISSION-RATE-4 TO HIGH-RATE.
00178    001780      PERFORM WRITE-A-SALESPERSON-GROUP.
00179    001790
00180    001800 LEVEL-3-COMMISSION.
00181    001810      MOVE 'ASSOCIATE' TO CLASS-TITLE-OUT.
00182    001820      MOVE ASSOCIATE-SALE-QUOTA TO SALE-QUOTA.
00183    001830      MOVE COMMISSION-RATE-5 TO LOW-RATE.
00184    001840      MOVE COMMISSION-RATE-6 TO HIGH-RATE.
00185    001850      PERFORM WRITE-A-SALESPERSON-GROUP.
00186    001860
00187    001870 LEVEL-4-COMMISSION.
00188    001880      MOVE 'SENIOR' TO CLASS-TITLE-OUT.
00189    001890      MOVE SENIOR-SALE-QUOTA TO SALE-QUOTA.
00190    001900      MOVE COMMISSION-RATE-5 TO LOW-RATE.
00191    001910      MOVE COMMISSION-RATE-6 TO HIGH-RATE.
00192    001920      PERFORM WRITE-A-SALESPERSON-GROUP.
00193    001930
00194    001940 LEVEL-5-COMMISSION.
00195    001950      MOVE 'SENIOR' TO CLASS-TITLE-OUT.
00196    001960      MOVE SENIOR-SALE-QUOTA TO SALE-QUOTA.
```

Figure A.E1.1 Erroneous program for Exercise 1, Part 3 of 3.

```
00197   001970        MOVE COMMISSION-RATE-7 TO LOW-RATE.
00198   001980        MOVE COMMISSION-RATE-8 TO HIGH-RATE.
00199   001990        PERFORM WRITE-A-SALESPERSON-GROUP.
00200   002000
00201   002010    WRITE-A-SALESPERSON-GROUP.
00202   002020        IF SALE-1-IN LESS THAN SALE-QUOTA
00203   002030            MULTIPLY SALE-1-IN BY LOW-RATE GIVING COMMISSION-OUT
00204   002040        ELSE
00205   002050            MULTIPLY SALE-1-IN BY HIGH-RATE GIVING COMMISSION-OUT.
00206   002060        MOVE SALESPERSON-NUMBER-IN TO SALESPERSON-NUMBER-OUT.
00207   002070        MOVE SALE-1-IN TO SALE-AMOUNT-OUT.
00208   002080        WRITE REPORT-LINE FROM DETAIL-LINE AFTER 2.
00209   002090        MOVE SPACES TO DETAIL-LINE.
00210   002100        IF SALE-2-IN LESS THAN SALE-QUOTA
00211   002110            MULTIPLY SALE-2-IN BY LOW-RATE GIVING COMMISSION-OUT
00212   002120        ELSE
00213   002130            MULTIPLY SALE-2-IN BY HIGH-RATE GIVING COMMISSION-OUT.
00214   002140        MOVE SALE-2-IN TO SALE-AMOUNT-OUT.
00215   002150        WRITE REPORT-LINE FROM DETAIL-LINE AFTER 1.
00216   002160        IF SALE-3-IN LESS THAN SALE-QUOTA
00217   002170            MULTIPLY SALE-3-IN BY LOW-RATE GIVING COMMISSION-OUT
00218   002180        ELSE
00219   002190            MULTIPLY SALE-3-IN BY HIGH-RATE GIVING COMMISSION-OUT.
00220   002200        MOVE SALE-3-IN TO SALE-AMOUNT-OUT.
00221   002210        WRITE REPORT-LINE FROM DETAIL-LINE AFTER 1.
00222   002220
00223   002230    CLASS-MISSING-ROUTINE.
00224   002240        MOVE CLASS-MISSING-MESSAGE TO ERROR-MESSAGE.
00225   002250        PERFROM WRITE-ERROR-LINE.
00226   002260
00227   002270    INVALID-CLASS-ROUTINE.
00228   002280        MOVE CLASS-IN TO INVLAID-CLASS.
00229   002290        MOVE INVALID-CLASS-MESSAGE TO ERROR-MESSAGE.
00230   002300        PERFORM WRITE-ERROR-LINE.
00231   002310
00232   002320    WRITE-ERROR-LINE.
00233   002330        MOVE SALESPERSON-NUMBER-IN TO SALESPERSON-NUMBER-ERR.
00234   002340        WRITE REPORT-LINE FROM ERROR-LINE AFTER 2.
```

Figure A.E1.2 Error messages for Exercise 1.

```
15    IKF2133I-W    LABEL RECORDS CLAUSE MISSING. DD CARD OPTION WILL BE TAKEN.
38    IKF1004I-E    INVALID WORD LABEL . SKIPPING TO NEXT RECOGNIZABLE WORD.
48    IKF2126I-C    VALUE CLAUSE LITERAL TOO LONG, TRUNCATED TO PICTURE SIZE.
97    IKF1043I-W    END OF SENTENCE SHOULD PRECEDE 10 . ASSUMED PRESENT.
101   IKF2190I-W    PICTURE CLAUSE IS SIGNED, VALUE CLAUSE UNSIGNED. ASSUMED POSITIVE.
104   IKF1043I-W    END OF SENTENCE SHOULD PRECEDE 05 . ASSUMED PRESENT.
103   IKF2190I-W    PICTURE CLAUSE IS SIGNED, VALUE CLAUSE UNSIGNED. ASSUMED POSITIVE.
104   IKF2190I-W    PICTURE CLAUSE IS SIGNED, VALUE CLAUSE UNSIGNED. ASSUMED POSITIVE.
105   IKF2190I-W    PICTURE CLAUSE IS SIGNED, VALUE CLAUSE UNSIGNED. ASSUMED POSITIVE.
111   IKF2190I-W    PICTURE CLAUSE IS SIGNED, VALUE CLAUSE UNSIGNED. ASSUMED POSITIVE.
112   IKF2190I-W    PICTURE CLAUSE IS SIGNED, VALUE CLAUSE UNSIGNED. ASSUMED POSITIVE.
113   IKF2190I-W    PICTURE CLAUSE IS SIGNED, VALUE CLAUSE UNSIGNED. ASSUMED POSITIVE.
114   IKF2190I-W    PICTURE CLAUSE IS SIGNED, VALUE CLAUSE UNSIGNED. ASSUMED POSITIVE.
115   IKF2190I-W    PICTURE CLAUSE IS SIGNED, VALUE CLAUSE UNSIGNED. ASSUMED POSITIVE.
116   IKF2190I-W    PICTURE CLAUSE IS SIGNED, VALUE CLAUSE UNSIGNED. ASSUMED POSITIVE.
117   IKF2190I-W    PICTURE CLAUSE IS SIGNED, VALUE CLAUSE UNSIGNED. ASSUMED POSITIVE.
118   IKF2190I-W    PICTURE CLAUSE IS SIGNED, VALUE CLAUSE UNSIGNED. ASSUMED POSITIVE.
132   IKF3001I-E    TODAYS-DATE NOT DEFINED. DISCARDED.
133   IKF3001I-E    TODAYS-DATE NOT DEFINED. DISCARDED.
136   IKF1086I-W    WRTIE SHOULD BEGIN A-MARGIN.
136   IKF1043I-W    END OF SENTENCE SHOULD PRECEDE REPORT-LINE . ASSUMED PRESENT.
136   IKF4003I-E    EXPECTING NEW STATEMENT. FOUND DNM=2-199 . DELETING TILL NEXT VERB OR
                    PROCEDURE-NAME.
156   IKF3001I-E    AVERAGE-SALE-AMO NOT DEFINED. TEST DISCARDED.
164   IKF4072I-W    EXIT FROM PERFORMED PROCEDURE ASSUMED BEFORE PROCEDURE-NAME .
203   IKF5011I-W    AN INTERMEDIATE RESULT OR A SENDING FIELD MIGHT HAVE ITS HIGH ORDER DIGIT POSITION
                    TRUNCATED.
205   IKF5011I-W    AN INTERMEDIATE RESULT OR A SENDING FIELD MIGHT HAVE ITS HIGH ORDER DIGIT POSITION
                    TRUNCATED.
211   IKF5011I-W    AN INTERMEDIATE RESULT OR A SENDING FIELD MIGHT HAVE ITS HIGH ORDER DIGIT POSITION
                    TRUNCATED.
213   IKF5011I-W    AN INTERMEDIATE RESULT OR A SENDING FIELD MIGHT HAVE ITS HIGH ORDER DIGIT POSITION
                    TRUNCATED.
217   IKF5011I-W    AN INTERMEDIATE RESULT OR A SENDING FIELD MIGHT HAVE ITS HIGH ORDER DIGIT POSITION
                    TRUNCATED.
219   IKF5011I-W    AN INTERMEDIATE RESULT OR A SENDING FIELD MIGHT HAVE ITS HIGH ORDER DIGIT POSITION
                    TRUNCATED.
225   IKF1086I-W    PERFROM SHOULD BEGIN A-MARGIN.
225   IKF1043I-W    END OF SENTENCE SHOULD PRECEDE WRITE-ERROR-LINE . ASSUMED PRESENT.
225   IKF4003I-E    EXPECTING NEW STATEMENT. FOUND PROCEDURE-NAME REFERENCE . DELETING TILL NEXT VERB
                    OR PROCEDURE-NAME.
228   IKF3001I-E    INVLAID-CLASS NOT DEFINED. DISCARDED.

      ERROR MESSAGE

      IKF6006I-E    SUPMAP SPECIFIED AND E-LEVEL DIAGNOSTIC HAS OCCURRED. PMAP CLIST LOAD DECK STATE FLOW SYMDMP IGNO
                    RED/
```

EXECUTION ERRORS

There are some kinds of errors that the compiler cannot detect. Some of these errors may become very obvious during program execution. We will discuss here some of the errors that cause program execution to terminate in the middle. This abnormal ending of program execution is called an **ABEND.** The word is sometimes incorrectly used as a verb in: "Be careful or your program will ABEND;" better to say "Be careful or you will get an ABEND."

INPUT/OUTPUT ERRORS

Abnormal terminations occur when files cannot be successfully OPENed and read from. An unsuccessful input/output operation is signaled by a message such as the ABEND message in Figure A.11. The ABEND code S001 indicates that this program terminated on an input/output error. Further details about the error can be found in the diagnostic-aid message shown in Figure A.12. The program that caused the generation of all these messages is shown in Figure A.13.

The message in Figure A.12 tells us that verb number 1 in card 87 is the operation that caused the abnormal termination. Looking at card 87 in Figure A.13, we see that the verb in question is a READ. The most common cause of an unsuccessful READ is a wrong-length-record condition, which occurs when the record description associated with the input file specifies a record length that does not agree with the actual length of the records on the file. Since this is a card file, the records are 80 characters long. The description of the input record EXAM–GRADE–RECORD, though, is only 78 characters long. Changing the last FILLER to X(51) eliminates the problem.

Another kind of input/output error occurs when one or more files cannot be OPENed. An S001 message will signal an input/output error, and the diagnostic-aid message will indicate that an OPEN verb failed. When that happens, check your SELECT sentences, and check particularly that the ASSIGN clauses are exactly right.

Figure A.11 ABEND message showing termination code S001.

```
130826 M2 R= IEF403I GSPNY191 STARTED
131009 M2 R= IEF450I GSPNY191.GO      .          ABEND S001          TIME=13.10.00
131014 M2 R= IEF404I GSPNY191 ENDED
```

Figure A.12 Diagnostic-aid message showing the statement number (card number) causing the termination.

```
PROGRAM          PROG11

LAST PSW BEFORE ABEND = FFA50037505FD6C8        SYSTEM COMPLETION CODE = 001

LAST CARD NUMBER/VERB NUMBER EXECUTED -- CARD NUMBER 000087/VERB NUMBER 01.

                                                END OF COBOL DIAGNOSTIC AIDS
```

Figure A.13 Program causing the generation of the messages shown in Figures A.11 and A.12, Part 1 of 2.

```
PP 5734-CB2 V4 RELEASE 1.5 10NOV77          IBM OS AMERICAN NATIONAL STANDARD COBOL

   1

00001    000010 IDENTIFICATION DIVISION.
00002    000020 PROGRAM-ID. PROG11.
00003    000025*AUTHOR. PHIL RUBENSTEIN.
00004    000030*
00005    000040*       THIS PROGRAM READS A FILE OF EXAM GRADES AND
00006    000050*       COMPUTES EACH STUDENTS AVERAGE.
00007    000060*
00008    000070*****************************************************
00009    000080
00010    000090 ENVIRONMENT DIVISION.
00011    000100 CONFIGURATION SECTION.
00012    000110 SPECIAL-NAMES.
00013    000120     C01 IS TO-NEW-PAGE.
00014    000130 INPUT-OUTPUT SECTION.
00015    000140 FILE-CONTROL.
00016    000150     SELECT EXAM-GRADE-FILE-IN     ASSIGN TO UT-S-SYSIN.
00017    000160     SELECT STUDENT-GRADE-REPORT  ASSIGN TO UT-S-SYSPRINT.
00018    000170
00019    000180*****************************************************
00020    000190
00021    000200 DATA DIVISION.
00022    000210 FILE SECTION.
00023    000220 FD  EXAM-GRADE-FILE-IN
00024    000230     LABEL RECORDS ARE OMITTED.
00025    000240
00026    000250 01  EXAM-GRADE-RECORD.
00027    000260     05 STUDENT-NUMBER-IN  PIC X(9).
00028    000270     05 FILLER             PIC X(8).
00029    000280     05 GRADE-1-IN         PIC 9(3).
00030    000290     05 GRADE-2-IN         PIC 9(3).
00031    000300     05 GRADE-3-IN         PIC 9(3).
00032    000310     05 GRADE-4-IN         PIC 9(3).
00033    000320     05 FILLER             PIC X(49).
00034    000330
00035    000340 FD  STUDENT-GRADE-REPORT
00036    000350     LABEL RECORDS ARE OMITTED.
00037    000360
00038    000370 01  REPORT-LINE           PIC X(79).
00039    000380
00040    000390 WORKING-STORAGE SECTION.
00041    000400 01  MORE-INPUT            PIC X(3) VALUE 'YES'.
00042    000410
00043    000420 01  REPORT-TITLE.
00044    000430     05 FILLER             PIC X.
00045    000440     05 FILLER             PIC X(35) VALUE SPACES.
00046    000450     05 FILLER             PIC X(20) VALUE 'STUDENT GRADE REPORT'.
00047    000460
00048    000470 01  COLUMN-HEADS-1.
00049    000480     05 FILLER             PIC X.
```

Figure A.13 Program causing the generation of the messages shown in Figures A.11 and A.12, Part 2 of 2.

```
  2

00050    000490          05 FILLER                  PIC X(20) VALUE SPACES.
00051    000500          05 FILLER                  PIC X(24) VALUE 'STUDENT'.
00052    000510          05 FILLER                  PIC X(12) VALUE 'G R A D E S '.
00053    000580
00054    000590 01       COLUMN-HEADS-2.
00055    000600          05 FILLER                  PIC X.
00056    000610          05 FILLER                  PIC X(20) VALUE SPACES.
00057    000620          05 FILLER                  PIC X(16) VALUE 'NUMBER'.
00058    000630          05 FILLER                  PIC X(8)  VALUE 'EXAM 1'.
00059    000640          05 FILLER                  PIC X(8)  VALUE 'EXAM 2'.
00060    000650          05 FILLER                  PIC X(8)  VALUE 'EXAM 3'.
00061    000653          05 FILLER                  PIC X(11) VALUE 'EXAM 4'.
00062    000656          05 FILLER                  PIC X(7)  VALUE 'AVERAGE'.
00063    000660
00064    000670 01       BODY-LINE.
00065    000680          05 FILLER                       PIC X.
00066    000690          05 FILLER                       PIC X(19) VALUE SPACES.
00067    000700          05 STUDENT-NUMBER-OUT           PIC X(9).
00068    000710          05 FILLER                       PIC X(9)  VALUE SPACES.
00069    000720          05 GRADE-1-OUT                  PIC ZZ9.
00070    000730          05 FILLER                       PIC X(5)  VALUE SPACES.
00071    000740          05 GRADE-2-OUT                  PIC ZZ9.
00072    000750          05 FILLER                       PIC X(5)  VALUE SPACES.
00073    000760          05 GRADE-3-OUT                  PIC ZZ9.
00074    000770          05 FILLER                       PIC X(5)  VALUE SPACES.
00075    000780          05 GRADE-4-OUT                  PIC ZZ9.
00076    000790          05 FILLER                       PIC X(7)  VALUE SPACES.
00077    000800          05 AVERAGE-GRADE                PIC ZZZ.9.
00078    000805
00079    000810 01       SUM-OF-GRADES                   PIC 9(3).
00080    000820 01       NUMBER-OF-EXAMS                 PIC 9 VALUE 4.
00081    000830
00082    000850*****************************************************
00083    001090
00084    001100 PROCEDURE DIVISION.
00085    001110          OPEN INPUT EXAM-GRADE-FILE-IN, OUTPUT STUDENT-GRADE-REPORT.
00086    001120          PERFORM REPORT-HEADINGS-ROUTINE.
00087    001130          READ EXAM-GRADE-FILE-IN AT END MOVE 'NO' TO MORE-INPUT.
00088    001140          PERFORM MAIN-PROCESS UNTIL MORE-INPUT IS EQUAL TO 'NO'.
00089    001160          CLOSE EXAM-GRADE-FILE-IN, STUDENT-GRADE-REPORT.
00090    001170          STOP RUN.
00091    001180
00092    001190 REPORT-HEADINGS-ROUTINE.
00093    001200          WRITE REPORT-LINE FROM REPORT-TITLE AFTER ADVANCING
00094    001210              TO-NEW-PAGE.
00095    001220          WRITE REPORT-LINE FROM COLUMN-HEADS-1 AFTER ADVANCING 3.
00096    001230          WRITE REPORT-LINE FROM COLUMN-HEADS-2 AFTER ADVANCING 1.
00097    001240          MOVE SPACES TO REPORT-LINE.
00098    001250          WRITE REPORT-LINE AFTER ADVANCING 1.
00099    001260
00100    001270 MAIN-PROCESS.
00101    001280          MOVE STUDENT-NUMBER-IN TO STUDENT-NUMBER-OUT.
```

Figure A.13 Part 2 (continued).

```
  3

00102     001290          MOVE GRADE-1-IN TO GRADE-1-OUT.
00103     001300          MOVE GRADE-2-IN TO GRADE-2-OUT.
00104     001315          MOVE GRADE-3-IN TO GRADE-3-OUT.
00105     001325          MOVE GRADE-4-IN TO GRADE-4-OUT.
00106     001330          ADD GRADE-1-IN, GRADE-2-IN, GRADE-3-IN, GRADE-4-IN
00107     001335              GIVING SUM-OF-GRADES.
00108     001340          DIVIDE SUM-OF-GRADES BY NUMBER-OF-EXAMS
00109     001350              GIVING AVERAGE-GRADE ROUNDED.
00110     001360          WRITE REPORT-LINE FROM BODY-LINE AFTER ADVANCING 1.
00111     001370          READ EXAM-GRADE-FILE-IN AT END MOVE 'NO' TO MORE-INPUT.
```

DATA EXCEPTIONS

Another most common cause of abnormal termination occurs during arithmetic operations. An ABEND message with code S0C7 indicates that an arithmetic operation or a numeric comparison has failed. The ABEND code is S–zero–C–7; if you pronounce it S–oh–C–7 be sure you know that it's a number, not a letter. Whatever you do, don't pronounce it sock–7. When an S0C7 failure occurs, use the diagnostic-aid message to find the very verb that caused the trouble. Then proceed as follows:

If an arithmetic statement causing the abnormal termination uses the GIVING option, then one or more of the sending fields contains nonnumeric data. Even though a field may be defined as purely numeric, there are still several ways for it to get illegal nonnumeric data assigned to it. For one, an input data item that is supposed to be numeric might erroneously not be; a programming error might have caused a nonnumeric value to be assigned to the field or perhaps no value at all to be assigned. Fields with no value assigned to them will often behave as though they contain nonnumeric data.

If the arithmetic causing the termination does not use the GIVING option or if the termination is caused by a numeric comparison, then any or all of the fields involved in the operation might be at fault by containing nonnumeric data. In particular, remember that the receiving field in an arithmetic operation without the GIVING option must have a numeric value assigned before the operation begins, even if that value is 0. If the receiving field is defined in working storage, be sure that it has either a numeric value assigned to it by the program before arithmetic is attempted or that it has a VALUE clause.

LOGIC ERRORS

In spite of all the error detection capabilities we have discussed, you will still find that there are times when programs run to completion without a hitch, but the final printed output is just wrong. You may find that even though the compiler gave you no error messages, your arithmetic results may be wrong, record selection procedures may be wrong, column headings may not be where you want them, vertical spacing of the paper may be wrong, some output fields may not print or may print the incorrect information, and so on. These kinds of difficulties are attributable to **logic errors.** Usually, with a little thought, you can locate the precise portion of the program that contains the error. For example, if some addition does not seem to be working properly, a good place to start looking for the error is in the ADD instruction that is supposed to be doing the addition. If some output field does not print properly, look at the statement whose job it is to assign the value of that output field.

If all else fails, you may use the debugging aids that are a part of the compiler. These consist of four statements **READY TRACE, RESET TRACE, EXHIBIT,** and **ON.**

After a READY TRACE statement is executed, the compiler traces the execution of your program and prints the names of the paragraphs and sections in the Procedure Division in the order in which they are executed. Execution of a RESET TRACE turns off the trace feature and allows the program to execute normally.

The EXHIBIT statement prints out the values of fields. The EXHIBIT statement can be made to print every time it is executed, or it can be made to print only when the values assigned to fields change.

The EXHIBIT statement can be a considerable aid in debugging if you use it to show the values of the input data fields for each record read. You will then be able to see which program paths ought to be executed, and you will also be able to notice whether any of the input data fields themselves are in error. Much unexpected program behavior can be attributed to incorrect input data.

The ON statement counts the number of times that particular program paths are executed and takes specified actions on specified occurrences of path executions.

Appendix B

American National Standard list of COBOL reserved words

This list of reserved words is the 1974 ANSI standard. Your own COBOL system may make additions to or deletions from this list. Check your COBOL manual.

ACCEPT
ACCESS
ADD
ADVANCING
AFTER
ALL
ALPHABETIC
ALSO
ALTER
ALTERNATE
AND
ARE
AREA
AREAS
ASCENDING
ASSIGN
AT
AUTHOR

BEFORE
BLANK
BLOCK
BOTTOM
BY

CALL
CANCEL
CD
CF
CH
CHARACTER
CHARACTERS
CLOCK-UNITS
CLOSE
COBOL
CODE
CODE-SET
COLLATING
COLUMN
COMMA
COMMUNICATION
COMP
COMPUTATIONAL
COMPUTE
CONFIGURATION
CONTAINS
CONTROL
CONTROLS
COPY
CORR
CORRESPONDING
COUNT
CURRENCY

DATA
DATE
DATE-COMPILED
DATE-WRITTEN
DAY
DE
DEBUG-CONTENTS
DEBUG-ITEM
DEBUG-LINE
DEBUG-NAME
DEBUG-SUB-1
DEBUG-SUB-2
DEBUG-SUB-3
DEBUGGING
DECIMAL-POINT
DECLARATIVES
DELETE
DELIMITED
DELIMITER
DEPENDING
DESCENDING
DESTINATION
DETAIL
DISABLE
DISPLAY
DIVIDE

DIVISION
DOWN
DUPLICATES
DYNAMIC

EGI
ELSE
EMI
ENABLE
END
END–OF–PAGE
ENTER
ENVIRONMENT
EOP
EQUAL
ERROR
ESI
EVERY
EXCEPTION
EXIT
EXTEND

FD
FILE
FILE–CONTROL
FILLER
FINAL
FIRST
FOOTING
FOR
FROM

GENERATE
GIVING
GO
GREATER
GROUP

HEADING
HIGH–VALUE
HIGH–VALUES

I–O
I–O–CONTROL
IDENTIFICATION
IF

IN
INDEX
INDEXED
INDICATE
INITIAL
INITIATE
INPUT
INPUT–OUTPUT
INSPECT
INSTALLATION
INTO
INVALID
IS

JUST
JUSTIFIED

KEY

LABEL
LAST
LEADING
LEFT
LENGTH
LESS
LIMIT
LIMITS
LINAGE
LINAGE–COUNTER
LINE
LINE–COUNTER
LINES
LINKAGE
LOCK
LOW–VALUE
LOW–VALUES

MEMORY
MERGE
MESSAGE
MODE
MODULES
MOVE
MULTIPLE
MULTIPLY

NATIVE
NEGATIVE
NEXT
NO
NOT
NUMBER
NUMERIC

OBJECT–COMPUTER
OCCURS
OF
OFF
OMITTED
ON
OPEN
OPTIONAL
OR
ORGANIZATION
OUTPUT
OVERFLOW

PAGE
PAGE–COUNTER
PERFORM
PF
PH
PIC
PICTURE
PLUS
POINTER
POSITION
POSITIVE
PRINTING
PROCEDURE
PROCEDURES
PROCEED
PROGRAM
PROGRAM–ID

QUEUE
QUOTE
QUOTES

RANDOM
RD
READ

RECEIVE
RECORD
RECORDS
REDEFINES
REEL
REFERENCES
RELATIVE
RELEASE
REMAINDER
REMOVAL
RENAMES
REPLACING
REPORT
REPORTING
REPORTS
RERUN
RESERVE
RESET
RETURN
REVERSED
REWIND
REWRITE
RF
RH
RIGHT
ROUNDED
RUN

SAME
SD
SEARCH
SECTION
SECURITY
SEGMENT
SEGMENT–LIMIT
SELECT
SEND
SENTENCE
SEPARATE
SEQUENCE
SEQUENTIAL
SET
SIGN
SIZE
SORT
SORT–MERGE
SOURCE
SOURCE–COMPUTER
SPACE
SPACES
SPECIAL–NAMES
STANDARD
STANDARD–1
START
STATUS
STOP
STRING
SUB–QUEUE–1
SUB–QUEUE–2
SUB–QUEUE–3
SUBTRACT
SUM
SUPPRESS
SYMBOLIC
SYNC
SYNCHRONIZED

TABLE
TALLYING
TAPE
TERMINAL
TERMINATE
TEXT
THAN
THROUGH
THRU
TIME
TIMES
TO
TOP
TRAILING
TYPE

UNIT
UNSTRING
UNTIL
UP
UPON
USAGE
USE
USING

VALUE
VALUES
VARYING

WHEN
WITH
WORDS
WORKING–STORAGE
WRITE

ZERO
ZEROES
ZEROS

+
–
*
/
**
>
<
=

Appendix C

Verb Formats

American National Standard formats of COBOL verbs used in this book:

```
                        ⎧DATE⎫
ACCEPT identifier FROM  ⎨DAY ⎬
                        ⎩TIME⎭

    ⎧identifier-1⎫ ⎡, identifier-2⎤
ADD ⎨            ⎬ ⎢               ⎥ ... TO identifier-m [ROUNDED]
    ⎩literal-1   ⎭ ⎣, literal-2    ⎦
   [, identifier-n [ROUNDED]] ... [; ON SIZE ERROR imperative-statement]

    ⎧identifier-1⎫   ⎧identifier-2⎫ ⎡, identifier-3⎤
ADD ⎨            ⎬ , ⎨            ⎬ ⎢               ⎥ ...
    ⎩literal-1   ⎭   ⎩literal-2   ⎭ ⎣, literal-3    ⎦
   GIVING identifier-m [ROUNDED] [, identifier-n [ROUNDED]] ...
   [; ON SIZE ERROR imperative-statement]

    ⎧CORRESPONDING⎫
ADD ⎨             ⎬ identifier-1 TO identifier-2 [ROUNDED]
    ⎩CORR         ⎭
   [; ON SIZE ERROR imperative-statement]

CLOSE file-name-1 [, file-name-2] ...

COMPUTE identifier-1 [ROUNDED] [, identifier-2 [ROUNDED]] ...
   = arithmetic-expression [; ON SIZE ERROR imperative-statement]

DELETE file-name RECORD [; INVALID KEY imperative-statement]

       ⎧identifier-1⎫
DIVIDE ⎨            ⎬ INTO identifier-2 [ROUNDED]
       ⎩literal-1   ⎭
   [, identifier-3 [ROUNDED]] ... [; ON SIZE ERROR imperative-statement]

       ⎧identifier-1⎫      ⎧identifier-2⎫
DIVIDE ⎨            ⎬ INTO ⎨            ⎬ GIVING identifier-3 [ROUNDED]
       ⎩literal-1   ⎭      ⎩literal-2   ⎭
   [, identifier-4 [ROUNDED]] ... [; ON SIZE ERROR imperative-statement]
```

DIVIDE { identifier–1 | literal–1 } BY { identifier–2 | literal–2 } GIVING identifier–3 [ROUNDED]
[, identifier–4 [ROUNDED]] . . . [; ON SIZE ERROR imperative–statement]

DIVIDE { identifier–1 | literal–1 } INTO { identifier–2 | literal–2 } GIVING identifier–3 [ROUNDED]
REMAINDER identifier–4 [; ON SIZE ERROR imperative–statement]

DIVIDE { identifier–1 | literal–1 } BY { identifier–2 | literal–2 } GIVING identifier–3 [ROUNDED]
REMAINDER identifier–4 [; ON SIZE ERROR imperative–statement]

GENERATE { data–name | report–name }

GO TO [procedure–name–1]

IF condition; { statement–1 | NEXT SENTENCE } { ; ELSE statement–2 | ; ELSE NEXT SENTENCE }

INITIATE report–name–1 [, report–name–2] . . .

MOVE { identifier–1 | literal } TO identifier–2 [, identifier–3] . . .

MOVE { CORRESPONDING | CORR } identifier–1 TO identifier–2

MULTIPLY { identifier–1 | literal–1 } BY identifier–2 [ROUNDED]
[, identifier–3 [ROUNDED]] . . . [; ON SIZE ERROR imperative–statement]

MULTIPLY { identifier–1 | literal–1 } BY { identifier–2 | literal–2 } GIVING identifier–3 [ROUNDED]
[, identifier–4 [ROUNDED]] . . . [; ON SIZE ERROR imperative–statement]

OPEN { INPUT file–name–1 [, file–name–2] . . . | OUTPUT file–name–3 [, file–name–4] . . . | I–O file–name–5 [, file–name–6] . . . } . . .

PERFORM procedure–name–1 [{ THROUGH | THRU } procedure–name–2]

PERFORM procedure–name–1 [{ THROUGH | THRU } procedure–name–2] { identifier–1 | integer–1 } TIMES

PERFORM procedure–name–1 [{ THROUGH | THRU } procedure–name–2] UNTIL condition–1

PERFORM procedure–name–1 [{ THROUGH | THRU } procedure–name–2]

$$\underline{\text{VARYING}} \left\{\begin{array}{l}\text{identifier–2}\\ \text{index–name–1}\end{array}\right\} \underline{\text{FROM}} \left\{\begin{array}{l}\text{identifier–3}\\ \text{index–name–2}\\ \text{literal–1}\end{array}\right\}$$

$$\underline{\text{BY}} \left\{\begin{array}{l}\text{identifier–4}\\ \text{literal–3}\end{array}\right\} \underline{\text{UNTIL}} \text{ condition–1}$$

$$\left[\underline{\text{AFTER}} \left\{\begin{array}{l}\text{identifier–5}\\ \text{index–name–3}\end{array}\right\} \underline{\text{FROM}} \left\{\begin{array}{l}\text{identifier–6}\\ \text{index–name–4}\\ \text{literal–3}\end{array}\right\}\right.$$

$$\underline{\text{BY}} \left\{\begin{array}{l}\text{identifier–7}\\ \text{literal–4}\end{array}\right\} \underline{\text{UNTIL}} \text{ condition–2}$$

$$\left[\underline{\text{AFTER}} \left\{\begin{array}{l}\text{identifier–8}\\ \text{index–name–5}\end{array}\right\} \underline{\text{FROM}} \left\{\begin{array}{l}\text{identifier–9}\\ \text{index–name–6}\\ \text{literal–5}\end{array}\right\}\right.$$

$$\left.\underline{\text{BY}} \left\{\begin{array}{l}\text{identifier–10}\\ \text{literal–6}\end{array}\right\} \underline{\text{UNTIL}} \text{ condition–3}\right]\Bigg]$$

$$\underline{\text{READ}} \text{ file–name RECORD [}\underline{\text{INTO}}\text{ identifier] [; AT }\underline{\text{END}}\text{ imperative–statement]}$$

$$\underline{\text{READ}} \text{ file–name RECORD [}\underline{\text{INTO}}\text{ identifier] [; }\underline{\text{INVALID}}\text{ KEY imperative–statement]}$$

$$\underline{\text{RELEASE}} \text{ record–name [}\underline{\text{FROM}}\text{ identifier]}$$

$$\underline{\text{RETURN}} \text{ file–name RECORD [}\underline{\text{INTO}}\text{ identifier] ; AT }\underline{\text{END}}\text{ imperative–statement}$$

$$\underline{\text{REWRITE}} \text{ record–name [}\underline{\text{FROM}}\text{ identifier]}$$

$$\underline{\text{SEARCH}} \text{ identifier–1} \left[\underline{\text{VARYING}} \left\{\begin{array}{l}\text{identifier–2}\\ \text{index–name–1}\end{array}\right\}\right] \text{ [; AT }\underline{\text{END}}\text{ imperative–statement–1]}$$

$$\text{; }\underline{\text{WHEN}} \text{ condition–1} \left\{\begin{array}{l}\text{imperative–statement–2}\\ \underline{\text{NEXT}}\ \underline{\text{SENTENCE}}\end{array}\right\}$$

$$\left[\text{; }\underline{\text{WHEN}} \text{ condition–2} \left\{\begin{array}{l}\text{imperative–statement–3}\\ \underline{\text{NEXT}}\ \underline{\text{SENTENCE}}\end{array}\right\}\right] \ldots$$

$$\underline{\text{SEARCH}}\ \underline{\text{ALL}} \text{ identifier–1 [; AT }\underline{\text{END}}\text{ imperative–statement–1]}$$

$$\text{; }\underline{\text{WHEN}} \left\{\begin{array}{l}\text{data–name–1} \left\{\begin{array}{l}\text{IS }\underline{\text{EQUAL}}\text{ TO}\\ \text{IS }\underline{=}\end{array}\right\} \left\{\begin{array}{l}\text{identifier–3}\\ \text{literal–1}\\ \text{arithmetic–expression–1}\end{array}\right\}\\ \text{condition–name–1}\end{array}\right\}$$

$$\left[\underline{\text{AND}} \left\{\begin{array}{l}\text{data–name–2} \left\{\begin{array}{l}\text{IS }\underline{\text{EQUAL}}\text{ TO}\\ \text{IS }\underline{=}\end{array}\right\} \left\{\begin{array}{l}\text{identifier–4}\\ \text{literal–2}\\ \text{arithmetic–expression–2}\end{array}\right\}\\ \text{condition–name–2}\end{array}\right\}\right] \ldots$$

$$\left\{\begin{array}{l}\text{imperative–statement–2}\\ \underline{\text{NEXT}}\ \underline{\text{SENTENCE}}\end{array}\right\}$$

```
                                                          { identifier-3 }
SET { identifier-1 [, identifier-2 ] . . .      }    TO   { index-name-3 }
    { index-name-1 [, index-name-2 ] . . .      }         { integer-1    }

                                       { UP BY   }  { identifier-4 }
SET index-name-4 [, index-name-5] . . .{ DOWN BY }  { integer-2    }

SORT file-name-1 ON { ASCENDING  } KEY data-name-1 [, data-name-2] . . .
                    { DESCENDING }

          [ ON { ASCENDING  } KEY data-name-3 [, data-name-4] . . . ] . . .
          [    { DESCENDING }                                       ]

   { INPUT PROCEDURE IS section-name-1 [ { THROUGH } section-name-2 ] }
   {                                   [ { THRU    }                ] }
   { USING file-name-2 [, file-name-3] . . .                          }

   { OUTPUT PROCEDURE IS section-name-3 [ { THROUGH } section-name-4 ] }
   {                                    [ { THRU    }                ] }
   { GIVING file-name-4                                                }

STOP { RUN     }
     { literal }

                                                                 { identifier-3 }
STRING { identifier-1 } [ , identifier-2 ] . . . DELIMITED BY    { literal-3    }
       { literal-1    } [ , literal-2    ]                       { SIZE         }

                                                           { identifier-6 }
   [ , { identifier-4 } [ , identifier-5 ] . . . DELIMITED BY { literal-6 }   ] . . .
   [   { literal-4    } [ , literal-5    ]                 { SIZE         }   ]
   INTO identifier-7 [WITH POINTER identifier-8]
   [; ON OVERFLOW imperative-statement]

SUBTRACT { identifier-1 } [ , identifier-2 ] . . . FROM identifier-m [ROUNDED]
         { literal-1    } [ , literal-2    ]
   [, identifier-n [ROUNDED]] . . . [; ON SIZE ERROR imperative-statement]

SUBTRACT { identifier-1 } [ , identifier-2 ] . . . FROM { identifier-m }
         { literal-1    } [ , literal-2    ]            { literal-m    }
   GIVING identifier-n [ROUNDED] [, identifier-o [ROUNDED]] . . .
   [; ON SIZE ERROR imperative-statement ]

SUBTRACT { CORRESPONDING } identifier-1 FROM identifier-2 [ROUNDED]
         { CORR          }
   [; ON SIZE ERROR imperative-statement]

TERMINATE report-name-1 [, report-name-2] . . .
```

```
UNSTRING identifier-1

  [ DELIMITED BY [ALL] {identifier-2} [, OR [ALL] {identifier-3}] ... ]
                       {literal-1   }              {literal-2   }

  INTO identifier-4 [, DELIMITER IN identifier-5] [, COUNT IN identifier-6]
  [, identifier-7 [, DELIMITER IN identifier-8] [, COUNT IN identifier-9]] ...
  [WITH POINTER identifier-10] [TALLYING IN identifier-11]
  [; ON OVERFLOW imperative-statement]

USE BEFORE REPORTING identifier.

WRITE record-name [FROM identifier-1]
  [ {BEFORE} ADVANCING { {identifier-2} [LINE ] } ]
  [ {AFTER } ADVANCING { {integer     } [LINES] } ]
  [                    { mnemonic-name          } ]

WRITE record-name [FROM identifier] [; INVALID KEY imperative-statement]
```

Glossary

American National Standard Glossary of COBOL terms used in this book:

Abbreviated Combined Relation Condition. The combined condition that results from the explicit omission of a common subject or a common subject and common relational operator in a consecutive sequence of relation conditions.

Access Mode. The manner in which records are to be operated upon within a file.

Actual Decimal Point. The physical representation, using either of the decimal point characters period (.) or comma (,), of the decimal point position in a data item.

Alphabetic Character. A character that belongs to the following set of letters: A, B, C, D, E, F, G, H, I, J, K, L, M, N, O, P, Q, R, S, T, U, V, W, X, Y, Z, and the space.

Alphanumeric Character. Any character in the computer's character set.

Arithmetic Expression. An arithmetic expression can be an identifier or a numeric elementary item, a numeric literal, such identifiers and literals separated by arithmetic operators, two arithmetic expressions separated by an arithmetic operator, or an arithmetic expression enclosed in parentheses.

Arithmetic Operator. A single character, or a fixed two-character combination, that belongs to the following set:

Character	*Meaning*
+	addition
−	subtraction
*	multiplication
/	division
**	exponentiation

Ascending Key. A key upon the values of which data is ordered starting with the lowest value of key up to the highest value of key in accordance with the rules for comparing data items.

Assumed Decimal Point. A decimal point position which does not involve the existence of an actual character in a data item. The assumed decimal point has logical meaning but no physical representation.

At End Condition. A condition caused:

1. During the execution of a READ statement for a sequentially accessed file.
2. During the execution of a RETURN statement, when no next logical record exists for the associated sort or merge file.
3. During the execution of a SEARCH statement, when the search operation terminates without satisfying the condition specified in any of the associated WHEN phrases.

Block. A physical unit of data that is normally composed of one or more logical records. For mass storage files, a block may contain a portion of a logical record. The size of a block has no direct relationship to the size of the file within which the block is contained or to the size of the logical record(s) that are either continued within the block or that overlap the block. The term is synonymous with physical record.

Body Group. Generic name for a report group of TYPE DETAIL, CONTROL HEADING or CONTROL FOOTING.

Character. The basic indivisible unit of the language.

Character Position. A character position is the amount of physical storage required to store a single standard data format character described as usage in DISPLAY. Further characteristics of the physical storage are defined by the implementor.

Character-String. A sequence of contiguous characters which form a COBOL word, a literal, a PICTURE character-string, or a comment-entry.

Class Condition. The proposition, for which a truth value can be determined, that the content of an item is wholly alphabetic or is wholly numeric.

Clause. A clause is an ordered set of consecutive COBOL character-strings whose purpose is to specify an attribute of an entry.

COBOL Character Set. The complete COBOL character set consists of the 51 characters listed below:

Character	*Meaning*
0,1,...,9	digit
A,B,...,Z	letter
	space (blank)
+	plus sign
–	minus sign (hyphen)
*	asterisk
/	stroke (virgule, slash)
=	equal sign
$	currency sign
,	comma (decimal point)
;	semicolon
.	period (decimal point)
"	quotation mark
(	left parenthesis
)	right parenthesis
>	greater than symbol
<	less than symbol

COBOL Word. (See Word)

Collating Sequence. The sequence in which the characters that are acceptable in a computer are ordered for purposes of sorting, merging, and comparing.

Column. A character position within a print line. The columns are numbered from 1, by 1, starting at the leftmost character position of the print line and extending to the rightmost position of the print line.

Combined Condition. A condition that is the result of connecting two or more conditions with the 'AND' or the 'OR' logical operator.

Comment-Entry. An entry in the Identification Division that may be any combination of characters from the computer character set.

Comment Line. A source of program line represented by an asterisk in the indicator area of the line and any characters from the computer's character set in area A and area B of that line. The comment line serves only for documentation in a program. A special form of comment line represented by a stroke (/) in the indicator area of the line and any characters from the computer's character set in area A and area B of that line causes page ejection prior to printing the comment.

Complex Condition. A condition in which one or more logical operators act upon one or more conditions. (See Negated Simple Condition, Combined Condition, Negated Combined Condition.)

Condition. A status of a program at execution time for which a truth value can be determined. Where the term 'condition' (condition-1, condition-2, . . .) appears in these language specifications in or in reference to 'condition' (condition-1, condition-2, . . .) of a general format, it is a conditional expression consisting of either a simple condition optionally parenthesized, or a combined condition consisting of the syntactically correct combination of simple conditions, logical operators, and parentheses, for which a truth value can be determined.

Condition-Name. A user-defined word assigned to a specific value, set of values, or range of values, within the complete set of values that a conditional variable may possess; or the user-defined word assigned to a status of an implementor-defined switch or device.

Condition-Name Condition. The proposition, for which a truth value can be determined, that the value of a conditional variable is a member of the set of values attributed to a condition-name associated with the conditional variable.

Conditional Expression. A simple condition or a complex condition specified in an IF, PERFORM, or SEARCH statement. (See Simple Condition and Complex Condition.)

Conditional Statement. A conditional statement specifies that the truth value of a condition is to be determined and that the subsequent action of the object program is dependent on this truth value.

Conditional Variable. A data item one or more values of which has a condition-name assigned to it.

Configuration Section. A section of the Environment Division that describes overall specifications of source and object computers.

Connective. A reserved word that is used to:

1. Associate a data-name, paragraph-name, condition-name, or text-name with its qualifier.
2. Link two or more operands written in a series.
3. Form conditions (logical connectives). (See Logical Operator.)

Contiguous Items. Items that are described by consecutive entries in the Data Division, and that bear a definite hierarchic relationship to each other.

Control Break. A change in the value of a data item that is referenced in the CONTROL clause. More generally, a change in the value of a data item that is used to control the hierarchical structure of a report.

Control Break Level. The relative position within a control hierarchy at which the most major control break occurred.

Control Data Item. A data item, a change in whose contents may produce a control break.

Control Data-Name. A data-name that appears in a CONTROL clause and refers to a control data item.

Control Footing. A report group that is presented at the end of the control group of which it is a member.

Control Group. A set of body groups that is presented for a given value of a control data item or of FINAL. Each control group may begin with a CONTROL HEADING, end with a CONTROL FOOTING, and contain DETAIL report groups.

Control Heading. A report group that is presented at the beginning of the control group of which it is a member.

Control Hierarchy. A designated sequence of report subdivisions defined by the positional order of FINAL and the data-names within a CONTROL clause.

Counter. A data item used for storing numbers or number representations in a manner that permits these numbers to be increased or decreased by the value of another number, or to be changed or reset to zero or to an arbitrary positive or negative value.

Current Record. The record which is available in the record area associated with the file.

Current Record Pointer. A conceptual entity that is used in the selection of the next record.

Data Clause. A clause that appears in a data description entry in the Data Division and provides information describing a particular attribute of a data item.

Data Description Entry. An entry in the Data Division that is composed of a level-number followed by a data-name, if required, and then followed by a set of data clauses, as required.

Data Item. A character or a set of contiguous characters (excluding in either case literals) defined as a unit of data by the COBOL program.

Data-Name. A user-defined word that names a data item described in a data description entry in the Data Division. When used in the general formats, 'data-name' represents a word which can neither be subscripted, indexed, nor qualified unless specifically permitted by the rules for that format.

Delimiter. A character or a sequence of contiguous characters that identify the end of a string of characters and separates that string of characters from the following string of characters. A delimiter is not part of the string of characters that it delimits.

Descending Key. A key upon the values of which data is ordered starting with the highest value of key down to the lowest value of key, in accordance with the rules for comparing data items.

Digit Position. A digit position is the amount of physical storage required to store a single digit. This amount may vary depending on the usage of the data item describing the digit position. Further characteristics of the physical storage are defined by the implementor.

Division. A set of zero, one, or more sections or paragraphs, called the division body, that are formed and combined in accordance with a specific set of rules. There are four (4) divisions in a COBOL program: Identification, Environment, Data, and Procedure.

Division Header. A combination of words followed by a period and a space that indicates the beginning of a division. The division headers are:

IDENTIFICATION DIVISION.
ENVIRONMENT DIVISION.
DATA DIVISION.
PROCEDURE DIVISION [USING data-name-1 [data-name-2] . . .].

Editing Character. A single character or a fixed two-character combination belonging to the following set:

Character	*Meaning*
B	space
0	zero
+	plus
–	minus
CR	credit

DB	debit
Z	zero suppress
*	check protect
$	currency sign
,	comma (decimal point)
.	period (decimal point)
/	stroke (virgule, slash)

Elementary Item. A data item that is described as not being further logically subdivided.

End of Procedure Division. The physical position in a COBOL source program after which no further procedures appear.

Entry. Any descriptive set of consecutive clauses terminated by a period and written in the Identification Division, Environment Division, or Data Division of a COBOL source program.

Environment Clause. A clause that appears as part of an Environment Division entry.

Figurative Constant. A compiler generated value referenced through the use of certain reserved words.

File. A collection of records.

File clause. A clause that appears as part of either of the following Data Division entries:
File description (FD)
Sort-merge file description (SD)

FILE-CONTROL. The name of an Environment Division paragraph in which the data files for a given source program are declared.

File Description Entry. An entry in the File Section of the Data Division that is composed of the level indicator FD, followed by a file-name, and then followed by a set of file clauses as required.

File-Name. A user-defined word that names a file described in a file description entry or a sort-merge file description entry within the File Section of the Data Division.

File Organization. The permanent logical file structure established at the time that a file is created.

File Section. The section of the Data Division that contains file description entries and sort-merge file description entries together with their associated record descriptions.

Format. A specific arrangement of a set of data.

Group Item. A named contiguous set of elementary or group items.

High Order End. The leftmost character of a string of characters.

I-O Mode. The state of a file after execution of an OPEN statement, with the I-O phrase specified, for that file and before the execution of a CLOSE statement for that file.

Identifier. A data-name, followed as required, by the syntactically correct combination of qualifiers, subscripts, and indices necessary to make unique reference to a data item.

Imperative Statement. A statement that begins with an imperative verb and specifies an unconditional action to be taken. An imperative statement may consist of a sequence of imperative statements.

Index. A computer storage position or register, the contents of which represent the identification of a particular element in a table.

Index-Name. A user-defined word that names an index associated with a specific table.

Indexed Data-Name. An identifier that is composed of a data-name, followed by one or more index-names enclosed in parentheses.

Indexed File. A file with indexed organization.

Indexed Organization. The permanent logical file structure in which each record is identified by the value of one or more keys within that record.

Input File. A file that is opened in the input mode.

Input Mode. The state of a file after execution of an OPEN statement, with the INPUT phrase specified, for that file and before the execution of a CLOSE statement for that file.

Input-Output File. A file that is opened in the I-O mode.

Input-Output Section. The section of the Environment Division that names the files and the external media required by an object program and which provides information required for transmission and handling of data during execution of the object program.

Input Procedure. A set of statements that is executed each time a record is released to the sort file.

Integer. A numeric literal or a numeric data item that does not include any character positions to the right of the assumed decimal point. Where the term 'integer' appears in general formats, integer must not be a numeric data item, and must not be signed, nor zero unless explicitly allowed by the rules of that format.

Invalid Key Condition. A condition, at object time, caused when a specific value of the key associated with an indexed or relative file is determined to be invalid.

Key. A data item which identifies the location of a record, or a set of data items which serve to identify the ordering of data.

Key of Reference. The key currently being used to access records within an indexed file.

Key Word. A reserved word whose presence is required when the format in which the word appears is used in a source program.

Level Indicator. Two alphabetic characters that identify a specific type of file or a position in hierarchy.

Level-Number. A user-defined word which indicates the position of a data item in the hierarchical structure of a logical record or which indicates special properties of a data description entry. A level-number is expressed as a one or two digit number. Level-numbers in the range 1 through 49 indicate the position of a data item in the hierarchical structure of a logical record. Level-numbers in the range 1 through 9 may be written either as a single digit or as a zero followed by a significant digit. Level-number 88 identifies special properties of a data description entry.

Line. (See Report Line)

Line Number. An integer that denotes the vertical position of a report line on a page.

Literal. A character-string whose value is implied by the ordered set of characters comprising the string.

Logical Operator. One of the reserved words AND, OR, or NOT. In the formation of a condition, both or either of AND and OR can be used as logical connectives. NOT can be used for logical negation.

Logical Record. The most inclusive data item. The level-number for a record is 01. (See Report Writer Logical Record)

Low Order End. The rightmost character of a string of characters.

Mass Storage. A storage medium on which data may be organized and maintained in both a sequential and nonsequential manner.

Mass Storage File. A collection of records that is assigned to a mass storage medium.

Mnemonic-Name. A user-defined word that is associated in the Environment Division with a specified implementor-name.

Negated Combined Condition. The 'NOT' logical operator immediately followed by a parenthesized combined condition.

Negated Simple Condition. The 'NOT' logical operator immediately followed by a simple condition.

Next Executable Sentence. The next sentence to which control will be transferred after execution of the current statement is complete.

Next Executable Statement. The next statement to which control will be transferred after execution of the current statement is complete.

Next Record. The record which logically follows the current record of a file.

Noncontiguous Items. Elementary data items, in the Working-Storage Section, which bear no hierarchic relationship to other data items.

Nonnumeric Item. A data item whose description permits its contents to be composed of any combination of characters taken from the computer's character set. Certain categories of nonnumeric items may be formed from more restricted character sets.

Nonnumeric Literal. A character-string bounded by quotation marks. The string of characters may include any character in the computer's character set. To represent a single quotation mark character within a nonnumeric literal, two contiguous quotation marks must be used.

Numeric Character. A character that belongs to the following set of digits: 0, 1, 2, 3, 4, 5, 6, 7, 8, 9.

Numeric Item. A data item whose description restricts its contents to a value represented by characters chosen from the digits '0' through '9'; if signed, the item may also contain a '+', '−', or other representation of an operational sign.

Numeric Literal. A literal composed of one or more numeric characters that also may contain either a decimal point, or an algebraic sign, or both. The decimal point must not be the rightmost character. The algebraic sign, if present, must be the leftmost character.

Object of Entry. A set of operands and reserved words, within a Data Division entry, that immediately follows the subject of the entry.

Object Program. A set or group of executable machine language instructions and other material designed to interact with data to provide problem solutions. In this context, an object program is generally the machine language result of the operation of a COBOL compiler on a source program. Where there is no danger of ambiguity, the word 'program' alone may be used in place of the phrase 'object program'.

Object Time. The time at which an object program is executed.

Open Mode. The state of a file after execution of an OPEN statement for that file and before the execution of a CLOSE statement for that file. The particular open mode is specified in the OPEN statement as either INPUT, OUTPUT, or I-O.

Operand. Whereas the general definition of operand is 'that component which is operated upon', for the purposes of this publication, any lowercase word (or words) that appears in a statement or entry format may be considered to be an operand and, as such, is an implied reference to the data indicated by the operand.

Operational Sign. An algebraic sign, associated with a numeric data item or a numeric literal, to indicate whether its value is positive or negative.

Optional Word. A reserved word that is included in a specific format only to improve the readability of the language and whose presence is optional to the user when the format in which the word appears is used in a source program.

Output File. A file that is opened in the output mode.

Output Mode. The state of a file after execution of an OPEN statement, with the OUTPUT phrase specified, for that file and before the execution of a CLOSE statement for that file.

Output Procedure. A set of statements to which control is given during execution of a SORT statement after the sort function is completed.

Page. A vertical division of a report representing a physical separation of report data, the separation being based on internal reporting requirements and/or external characteristics of the reporting medium.

Page Body. That part of the logical page in which lines can be written and/or spaced.

Page Footing. A report group that is presented at the end of a report page as determined by the Report Writer Control System.

Page Heading. A report group that is presented at the beginning of a report page and determined by the Report Writer Control System.

Paragraph. In the Procedure Division, a paragraph-name followed by a period and a space and by zero, one, or more sentences. In the Identification and Environment Divisions, a paragraph header followed by zero, one, or more entries.

Paragraph Header. A reserved word, followed by a period and a space that indicates the beginning of a paragraph in the Identification and Environment Divisions. The permissible paragraph headers are:

In the Identification Division:
PROGRAM–ID.
AUTHOR.
INSTALLATION.
DATE–WRITTEN.
DATE–COMPILED.
SECURITY.

In the Environment Division:
SOURCE–COMPUTER.
OBJECT–COMPUTER.
SPECIAL–NAMES.
FILE–CONTROL.
I–O–CONTROL.

Paragraph-Name. A user-defined word that identifies and begins a paragraph in the Procedure Division.

Phrase. A phrase is an ordered set of one or more consecutive COBOL character-strings that form a portion of a COBOL procedural statement or of a COBOL clause.

Physical Record. (See Block)

Prime Record Key. A key whose contents uniquely identify a record within an indexed file.

Printable Group. A report group that contains at least one print line.

Printable Item. A data item, the extent and contents of which are specified by an elementary report entry. This elementary report entry contains a COLUMN NUMBER clause, a PICTURE clause, and a SOURCE, SUM or VALUE clause.

Procedure. A paragraph or group of logically successive paragraphs, or a section or group of logically successive sections, within the Procedure Division.

Procedure-Name. A user-defined word which is used to name a paragraph or section in the Procedure Division. It consists of a paragraph-name (which may be qualified), or a section-name.

Program-Name. A user-defined word that identifies a COBOL source program.

Punctuation Character. A character that belongs to the following set:

Character	*Meaning*
,	comma
;	semicolon
.	period
"	quotation mark
(	left parenthesis
)	right parenthesis
	space
=	equal sign

Qualified Data-Name. An identifier that is composed of a data-name followed by one or more sets of either of the connectives OF and IN followed by a data-name qualifier.

Qualifier.

1. A data-name which is used in a reference together with another data name at a lower level in the same hierarchy.
2. A section-name which is used in a reference together with a paragraph-name specified in that section.

Random Access. An access mode in which the program-specified value of a key data item identifies the logical record that is obtained from, deleted from, or placed into a relative or indexed file.

Record. (See Logical Record)

Record Area. A storage area allocated for the purpose of processing the record described in a record description entry in the File Section.

Record Description. (See Record Description Entry)

Record Description Entry. The total set of data description entries associated with a particular record.

Record Key. A key whose contents identify a record within an indexed file.

Record-Name. A user-defined word that names a record described in a record description entry in the Data Division.

Reference Format. A format that provides a standard method for describing COBOL source programs.

Relation. (See Relational Operator)

Relation Character. A character that belongs to the following set:

Character	*Meaning*
>	greater than
<	less than
=	equal to

Relation Condition. The proposition, for which a truth value can be determined, that the value of an arithmetic expression or data item has a specific relationship to the value of another arithmetic expression or data item. (See Relational Operator)

Relational Operator. A reserved word, a relation character, a group of consecutive reserved words, or a group of consecutive reserved words and relation characters used in the construction of a relation condition. The permissible operators and their meaning are:

Relational Operator	*Meaning*
IS [NOT] GREATER THAN IS [NOT] >	Greater than or not greater than
IS [NOT] LESS THAN IS [NOT] <	Less than or not less than
IS [NOT] EQUAL TO IS [NOT] =	Equal to or not equal to

Relative File. A file with relative organization.

Relative Organization. The permanent logical file structure in which each record is uniquely identified by an integer value greater than zero, which specifies the record's logical ordinal position in the file.

Report Clause. A clause, in the Report Section of the Data Division, that appears in a report description entry or a report group description entry.

Report Description Entry. An entry in the Report Section of the Data Division that is composed of the level indicator RD, followed by a report name, followed by a set of report clauses as required.

Report File. An output file whose file description entry contains a REPORT clause. The contents of a report file consist of records that are written under control of the Report Writer Control System.

Report Footing. A report group that is presented only at the end of a report.

Report Group. In the Report Section of the Data Division, an 01 level-number entry and its subordinate entries.

Report Group Description Entry. An entry in the Report Section of the Data Division that is composed of the level-number 01, the optional data-name, a TYPE clause, and an optional set of report clauses.

Report Heading. A report group that is presented only at the beginning of a report.

Report Line. A division of a page representing one row of horizontal character positions. Each character position of a report line is aligned vertically beneath the corresponding character position of the report line above it. Report lines are numbered from 1, by 1, starting at the top of the page.

Report-Name. A user-defined word that names a report described in a report description entry within the Report Section of the Data Division.

Report Section. The section of the Data Division that contains one or more report description entries and their associated report group description entries.

Report Writer Control System (RWCS). An object time control system, provided by the implementor, that accomplishes the construction of reports.

Report Writer Logical Record. A record that consists of the Report Writer print line and associated control information necessary for its selection and vertical positioning.

Reserved Word. A COBOL word specified in the list of words which may be used in COBOL source programs, but which must not appear in the programs as user-defined words.

Run Unit. A set of one or more object programs which function, at object time, as a unit to provide problem solutions.

RWCS. (See Report Writer Control System)

Section. A set of zero, one, or more paragraphs or entries, called a section body, the first of which is preceded by a section header. Each section consists of the section header and the related section body.

Section Header. A combination of words followed by a period and a space that indicates the beginning of a section in the Environment, Data, and Procedure Division.

In the Environment and Data Divisions, a section header is composed of reserved words followed by a period and a space. The permissible section headers are:

In the Environment Division:
CONFIGURATION SECTION.
INPUT–OUTPUT SECTION.

In the Data Division:
FILE SECTION.
WORKING–STORAGE SECTION.
REPORT SECTION.

In the Procedure Division, a section header is composed of a section-name, followed by the reserved word SECTION, followed by a period and a space.

Section-Name. A user-defined word which names a section in the Procedure Division.

Sentence. A sequence of one or more statements, the last of which is terminated by a period followed by a space.

Separator. A punctuation character used to delimit character-strings.

Sequential Access. An access mode in which logical records are obtained from or placed into a file in a consecutive predecessor-to-successor logical record sequence determined by the order of records in the file.

Sequential File. A file with sequential organization.

Sequential Organization. The permanent logical file structure in which a record is identified by a predecessor-successor relationship established when the record is placed into the file.

Sign Condition. The proposition, for which a truth value can be determined, that the algebraic value of a data item or an arithmetic expression is either less than, greater than, or equal to zero.

Simple Condition. Any single condition chosen from the set:
relation condition
class condition
condition-name condition
sign condition
(simple-condition)

Sort File. A collection of records to be sorted by a SORT statement. The sort file is created and can be used by the sort function only.

Sort-Merge File Description Entry. An entry in the File Section of the Data Division that is composed of the level indicator SD, followed by a file-name, and then followed by a set of file clauses as required.

Source Item. An identifier designated by a SOURCE clause that provides the value of a printable item.

Source Program. Although it is recognized that a source program may be represented by other forms and symbols, in this document it always refers to a syntactically correct set of COBOL statements beginning with an Identification Division and ending with the end of the Procedure Division. In contexts where there is no danger of ambiguity, the word 'program' alone may be used in place of the phrase 'source program'.

Special Character. A character that belongs to the following set:

Character	*Meaning*
+	plus sign
–	minus sign
*	asterisk
/	stroke (virgule, slash)
=	equal sign
$	currency sign
,	comma (decimal point)
;	semicolon
.	period (decimal point)
"	quotation mark
(	left parenthesis
)	right parenthesis
>	greater than symbol
<	less than symbol

Special-Character Word. A reserved word which is an arithmetic operator or a relation character.

SPECIAL-NAMES. The name of an Environment Division paragraph in which implementor-names are related to user specified mnemonic-names.

Special Registers. Compiler generated storage areas whose primary use is to store information produced in conjunction with the use of specific COBOL features.

Standard Data Format. The concept used in describing the characteristics of data in a COBOL Data Division under which the characteristics or properties of the data are expressed in a form oriented to the appearance of the data on a printed page of infinite length and breadth, rather than a form oriented to the manner in which the data is stored internally in the computer, or on a particular external medium.

Statement. A syntactically valid combination of words and symbols written in the Procedure Division beginning with a verb.

Subject of Entry. An operand or reserved word that appears immediately following the level indicator or the level-number in a Data Division entry.

Subscript. An integer whose value identifies a particular element in a table.

Subscripted Data-Name. An identifier that is composed of a data-name followed by one or more subscripts enclosed in parentheses.

Sum Counter. A signed numeric data item established by a SUM clause in the Report Section of the Data Division. The sum counter is used by the Report Writer Control System to contain the result of designated summing operations that take place during production of a report.

Table. A set of logically consecutive items of data that are defined in the Data Division by means of the OCCURS clause.

Table Element. A data item that belongs to the set of repeated items comprising a table.

Truth Value. The representation of the result of the evaluation of a condition in terms of one of two values

true
false

Unary Operator. A plus (+) or a minus (−) sign, which precedes a variable or a left parenthesis in an arithmetic expression and which has the effect of multiplying the expression of +1 or −1, respectively.

Unit. A module of mass storage, the dimensions of which are determined by each implementor.

User-Defined Word. A COBOL word that must be supplied by the user to satisfy the format of a clause or statement.

Variable. A data item whose value may be changed by execution of the object program. A variable used in an arithmetic expression must be a numeric elementary item.

Verb. A word that expresses an action to be taken by a COBOL compiler or object program.

Word. A character-string of not more than 30 characters which forms a user-defined word, a system-name, or a reserved word.

Working-Storage Section. The section of the Data Division that describes working storage data items, composed either of noncontiguous items or of working storage records or of both.

Answers to Fill-in Exercises

Chapter 1

1. COmmon Business Oriented Language
2. program
3. programmer
4. output
5. IDENTIFICATION, ENVIRONMENT, DATA, PROCEDURE
6. IDENTIFICATION
7. DATA
8. PROCEDURE
9. OPEN, CLOSE
10. 01

Chapter 2

1. B
2. record
3. 02, 49
4. FILLER
5. nonnumeric, numeric
6. Working Storage
7. VALUE
8. priming
9. AFTER ADVANCING
10. SPECIAL–NAMES

Chapter 3

1. alphanumeric, numeric
2. 18
3. 9
4. IS EQUAL TO, IS NOT EQUAL TO, IS LESS THAN, IS NOT LESS THAN, IS GREATER THAN, IS NOT GREATER THAN
5. flowchart, control chart
6. Z
7. NEXT SENTENCE
8. ELSE NEXT SENTENCE
9. dot, V
10. reserved, required, optional

Chapter 4

1. TO, GIVING
2. blanks
3. numeric
4. numeric, edited numeric
5. hierarchy
6. charts, diagrams
7. English
8. working storage
9. INTO
10. FROM

Chapter 5

1. COMPUTE
2. exponentiation, multiplication, division, addition, subtraction
3. unary operators
4. ACCEPT
5. CORRESPONDING
6. data name qualification
7. SIGN
8. left
9. +
10. negative

Chapter 6

1. field, break
2. total
3. crossing
4. detail line
5. COMPUTATIONAL
6. COMPUTATIONAL, DISPLAY
7. figurative constants
8. group item
9. elementary item
10. ADVANCING, LINES

Chapter 7

1. first
2. indenting
3. period
4. nested
5. AND, OR
6. ANDs, ORs
7. clear
8. subject
9. name
10. VALUE

Chapter 8

1. alphanumeric
2. redefined
3. NUMERIC, ALPHABETIC
4. reasonableness
5. SIZE ERROR
6. left
7. period, ELSE
8. conditional
9. before
10. No

Chapter 9

1. OCCURS
2. INDEXED BY
3. ASCENDING KEY, DESCENDING KEY
4. TAX–AMOUNT (TAX–SUBSCRIPT)
5. integer
6. VARYING
7. SET, VARYING, SEARCH, SEARCH
8. three
9. index name
10. WHEN

Chapter 10

1. REPORT IS
2. RD
3. REPORT HEADING, PAGE HEADING, CONTROL HEADING, DETAIL, CONTROL FOOTING, PAGE FOOTING, REPORT FOOTING
4. COLUMN
5. LINE, NEXT GROUP
6. INITIATE
7. GENERATE
8. TERMINATE
9. PAGE
10. CONTROL

Chapter 11

1. magnetic tape, magnetic disk, data cell
2. sequential
3. magnetic media are reusable, input and output operations are faster, magnetic storage is more compact
4. sequential, indexed, relative, direct
5. sequentially
6. RECORD KEY
7. BLOCK CONTAINS
8. READ, WRITE
9. REWRITE
10. LOW–VALUE

Chapter 12

1. INPUT
2. OUTPUT
3. SECTIONs
4. SD
5. major, minor
6. RETURN
7. STRING
8. UNSTRING
9. SIZE
10. INTO

Index